D1233838

CMOS Analog Design Using All-Region MOSFET Modeling

Covering the essentials of analog circuit design, this book takes a unique design approach based on a MOSFET model valid for all operating regions, rather than the standard square-law model. Opening chapters focus on device modeling, integrated circuit technology, and layout, whilst later chapters go on to cover noise and mismatch, and analysis and design of the basic building blocks of analog circuits, such as current mirrors, voltage references, voltage amplifiers, and operational amplifiers. An introduction to continuous-time filters is also provided, as are the basic principles of sampled-data circuits, especially switched-capacitor circuits. The final chapter then reviews MOSFET models and describes techniques to extract design parameters. With numerous design examples and exercises also included, this is ideal for students taking analog CMOS design courses and also for circuit designers who need to shorten the design cycle.

MÁRCIO CHEREM SCHNEIDER is a Professor in the Electrical Engineering Department at the Federal University of Santa Catarina, Brazil, where he has worked since 1976. He has also spent a year at the Swiss Federal Institute of Technology (EPFL) and has worked as a Visiting Associate Professor in the Department of Electrical and Computer Engineering at Texas A&M University. His current research interests mainly focus on MOSFET modeling and transistor-level design, in particular of analog and RF circuits.

CARLOS GALUP-MONTORO is currently a Visiting Scholar in the Electrical Engineering Department at the University of California, Berkeley, and a Professor in the Electrical Engineering Department at the Federal University of Santa Catarina, Brazil, where he has worked since 1990. His main research interests are in field-effect-transistor modeling and transistor-level design.

CMOS Analog Design Using All-Region MOSFET Modeling

MÁRCIO CHEREM SCHNEIDER
AND
CARLOS GALUP-MONTORO

Federal University of Santa Catarina, Brazil

CAMBRIDGE
UNIVERSITY PRESS

CAMBRIDGE UNIVERSITY PRESS
Cambridge, New York, Melbourne, Madrid, Cape Town, Singapore,
São Paulo, Delhi, Dubai, Tokyo

Cambridge University Press
The Edinburgh Building, Cambridge CB2 8RU, UK

Published in the United States of America by Cambridge University Press, New York

www.cambridge.org
Information on this title: www.cambridge.org/9780521110365

© Cambridge University Press 2010

This publication is in copyright. Subject to statutory exception
and to the provisions of relevant collective licensing agreements,
no reproduction of any part may take place without the written
permission of Cambridge University Press.

First published 2010

Printed in the United Kingdom at the University Press, Cambridge

A catalogue record for this publication is available from the British Library

ISBN 978 0 521 11036 5 Hardback

Additional resources for this publication at www.cambridge.org/9780521110365

Cambridge University Press has no responsibility for the persistence or
accuracy of URLs for external or third-party Internet websites referred to
in this publication, and does not guarantee that any content on such
websites is, or will remain, accurate or appropriate.

To our wives Rita and Marlene

Contents

Preface

Analog integrated circuits in bipolar technology, beginning with operational amplifiers and advancing to data conversion and communication circuits, were developed in the 1960s and matured during the 1970s. During this period, the metal–oxide–semiconductor (MOS) technology evolved for digital circuits because of its better efficiency in terms of silicon-area use and power consumption compared with bipolar digital technologies. To reduce the system cost and power consumption, chips including digital and analog circuits appeared in MOS technology in the late 1970s. The first analog circuits in MOS technology were for audio-frequency applications. With the scaling of the MOS technology, driven by the need for large-scale integration levels, enhanced performance and reduced cost, even radio-frequency (RF) applications in MOS technology have become possible. Compared with digital design, analog design requires much more careful device modeling, and for this reason analog designers were at the origin of many MOS modeling enhancements.

The strong similarities between the basic operating principles of many bipolar and MOS analog building blocks and circuits have led some textbook authors to combine their presentation. On the other hand, there are profound differences between bipolar and MOS circuits in terms of the electrical performance and design approaches, and for this reason other texts focus only on MOS analog circuits. In this textbook we take this area of specialization a step further, focusing on analog MOS circuits at transistor level, using an accurate but simple MOS transistor model for design in order to reduce the distance between hand design and simulation results. In place of the common approach of furnishing separate analytical formulas for the strong- and weak-inversion operation regions of a building block, we provide simple formulas that are valid in all operation regions, including moderate inversion. This unified design approach is particularly suitable for analog design in advanced complementary-metal–oxide–semiconductor (CMOS) technologies. In effect, for deep-submicron MOS technologies good design tradeoffs are often obtained with transistors operating in weak and moderate inversion. It should be observed that the conventional approach based on the asymptotic models of strong and weak inversion does not allow meaningful exploration of the design space.

The book starts with a short comparison between bipolar and MOS analog circuits. The main differences between bipolar and MOS transistors are emphasized, since superficial similarities between them often lead to erroneous results. The drawbacks of some classical MOS field-effect-transistor (FET) models, particularly those related to the choice of the source terminal as the reference, are explained. Chapter 2 presents an accurate model for the MOS transistor. Large- and small-signal models for low and high frequency, which are valid in all the operating regions, are presented. The important concept of inversion level is developed and explicit expressions for all large- and small-signal parameters of transistors in terms of the inversion levels are provided. The main small-geometry effects are summarized. An overview of CMOS technology for designers and the basic properties of passive devices in CMOS technology are the subjects of Chapter 3. The models for integrated resistors and capacitors are developed with the necessary depth for analog design. Some good practices for designing MOS transistor layouts are summarized. Chapter 4 gives a unified modeling for mismatch and noise. With the shrinking of the MOSFET dimensions and reduction in the supply voltage of advanced technologies, the consideration of matching and noise has become even more important for analog design. Thus, we have included a detailed presentation of mismatch and noise in Chapter 4 so that they can be considered in the subsequent study of the basic circuits and building blocks.

Chapter 5 starts with the simple current mirror, one of the basic building blocks of analog circuits. The main cascode configurations and some advanced mirror topologies are then presented. We make a complete large- and small-signal analysis and include errors due to finite output resistance, mismatch, and noise. Chapter 6 deals with current sources and voltage references. Self-biased current sources and voltage references are described, emphasizing bandgap references. The whole chapter is dedicated to the basic bias building blocks, because bias and dc behavior are of the utmost importance in relation to analog circuits. In Chapter 7 the basic gain stages are described. Common-source, common-gate, source-follower, cascode, and differential amplifiers are thoroughly analyzed. The use of an all-region one-equation MOSFET model allows the complete exploration of the design space, and the choice of the best operating region (weak, moderate, or strong inversion) for each transistor involved. The important topic of CMOS design scaling and reuse is summarized at the end of the chapter. Chapter 8 deals with the design of operational amplifiers. The main topologies used in CMOS technology are presented, including single- and two-stage operational amplifiers. Fully differential amplifiers, including the folded-cascode type, and common-mode feedback circuits are described.

The following two chapters of the book introduce the basic circuit techniques for frequency-selective filters and some building blocks for data converters. In Chapter 9 the MOSFET-C filter technique derived from active *RC* filters is presented, followed by the basics of operational transconductance amplifier-capacitor (OTA-C) filters, including on-chip tuning circuits. Digitally-programmable filters using MOSFET-only current dividers (MOCDs) are also discussed.

In Chapter 10, following the analysis of analog MOS switches and sample-and-hold circuits, sampled-data techniques are introduced. Switched-capacitor building blocks for integrated filters and converters are described. The important topic of switched-capacitor filters fully compatible with digital MOS technology is covered. Finally, some complementary modeling topics considered important for circuit design are summarized in the appendices.

Chapter 11 provides an overview of compact MOSFET models, which play a significant role in the analysis and design of integrated circuits. This chapter also describes some procedures employed to extract fundamental design parameters associated with the MOSFET model used in this textbook.

This book is intended for an in-depth first course in analog CMOS design, for senior-undergraduate and first-year graduate students, as well as for self-study in the case of practicing engineers. The required background for the students is one or two introductory courses in electronics and in semiconductor devices.

Since analog-circuit design requires knowledge in the areas of device modeling, integrated-circuit technology, and layout, in addition to signals and circuits, the study of Chapters 1–4 is essential for any use of the book. A course focused on transistor-level design could be restricted to Chapters 1–8. A 15-week semester is sufficient to cover the whole book.

We are very grateful to our former PhD students Professors Ana Isabela Araújo Cunha, Oscar da Costa Gouveia Filho, Alfredo Arnaud, and Hamilton Klimach, who made invaluable contributions to the research on MOSFET modeling in our group and to the CNPq and CAPES, Brazilian agencies for scientific development, for their support of the research in our laboratory.

As in the past few years, we continue to have the collaboration of Dr. Siobhan Wiese for the revision of our texts. We are lucky to have a native English speaker with a scientific background to count on, as this is not always a rewarding task. We are also very grateful to João Romão for the skillful preparation of the figures. Last but not least, the hard work of Gustavo Leão Moreira for the simulations in Chapter 11 is gratefully acknowledged.

1 Introduction to analog CMOS design

This chapter begins by explaining briefly why there is still a need for analog design and introduces its main tradeoffs. The need for accurate component modeling follows. Then, the essentials of p–n junctions and bipolar and field-effect transistors (FETs) for circuit design are recalled. The main differences between bipolar-transistor and FET operations are emphasized and drawbacks of some popular FET models for circuit design are commented on. Finally, two single-stage amplifiers, one in bipolar and another in MOS technology, are designed in order to make the differences between these technologies clear.

1.1. Analog design

1.1.1. The need for analog design

Analog circuits such as audio and radio amplifiers have been in use since the early days of electronics. Analog systems carry the signals in the form of physical variables such as voltages, currents, or charges, which are continuous functions of time. The manipulation of these variables must often be carried out with high accuracy.

On the other hand, in digital systems the link of the variables with the physical world is indirect, since each signal is represented by a sequence of numbers. Clearly, the types of electrical performance that must be achieved by analog and digital electronic circuits are quite different, and for this reason they are generally studied in separate university courses.

Nowadays, analog circuits continue to be used for direct signal processing in some very-high-frequency or specialized applications, but their main use is in interfacing computers to the analog world. The development of the very-large-scale-integration (VLSI) technology has led to computers being pervasive in telecommunications, consumer electronics, biomedicine, robotics, the automotive industry, etc. As a consequence, the analog circuits needed around them are also pervasive.

Interfacing computers or digital signal processors to the analog world requires various analog functions, among them amplification, filtering, sampling, (de)multiplexing, and analog-to-digital (A/D) and digital-to-analog (D/A) conversions. Since analog circuits are needed together with digital ones in almost any complex chip and the technology for VLSI is the complementary metal–oxide–semiconductor (CMOS), most of the current analog circuits are CMOS circuits. For this reason, this book is focused on CMOS analog circuits and the treatment of bipolar transistors is essentially limited to this introductory chapter and to a few sections that cover the use of bipolar transistors compatible with CMOS technologies at appropriate points in the book.

The spread of analog circuits continues to increase with the evolution of technology. The current extensive research efforts directed toward sensors and actuators [1] for applications in numerous industrial products will lead to a demand for analog circuits in all subsequent products.

1.1.2.   Tradeoffs in analog design

Analog circuits present a large variety of circuit functions, performance objectives, and specification methodologies [2]. The basic elements of modern analog circuits are MOS transistors, which are highly non-linear devices.

Even a simple cell as an operational amplifier (op amp) has many different but interrelated specifications including noise, distortion, power consumption, gain, phase margin, common-mode range, offset, temperature stability, supply sensitivity, and decoupling from other circuitry.

Numerous circuit topologies are often considered in the design phase. The broad choice of geometries and operating conditions for each device increases enormously the parametric complexity of this apparently simple design. The final designed circuit must satisfy the specifications considering the possible variations in the fabrication process, operating temperature, and power-supply voltage. As a consequence of the variety of performance objectives and of the design complexity, there are numerous op-amp designs available [3], each of them suitable for a specific situation.

The high degree of accuracy required and the difficulties involved in precisely modeling the device characteristics and interferences make the analog design problem even more complicated. Each design involves many complex, multi-variable interactions, and no widely applicable systematic design procedure is available [4].

Following the design, the actual performance of an analog circuit is dependent on the detailed electrical parameters of each of the many devices constituting the integrated circuit.

From the considerations summarized in this section, it is clear that analog-circuit-design expertise is not easy to achieve. To cope with the complexity of the analog design area, this book focuses on CMOS analog circuits and emphasizes the basic analog building blocks and their application to operational amplifier design.

1.1.3.   The importance of component modeling

Additional difficulties in designing analog circuits come from inadequate device modeling [5] and poor knowledge of the technological parameters relevant to analog design, such as those related to excess noise and to device mismatch. Many of the device models used for hand design and for circuit simulation are inadequate for analog design and, sometimes, can cause a design to fail. This is particularly true for certain models of the MOSFET.

Analog circuits rely on details of the device characteristics to a much greater extent than digital circuits. In digital circuits operating with a relatively large supply voltage (say above 3 V), transistors operate essentially as switches, and an approximate model is sufficient for the transition between the on and off states. For low-voltage digital circuits (say below 1 V), transistors must be considered as dimmers rather than switches; thus, an accurate model of the transistor for all the operating regions becomes necessary. Design of analog circuits requires, in general, very careful device modeling in all phases of the design procedure. An initial design with simple models is the first step in the design procedure. In this initial design, all transistor currents and sizes must be determined in order to satisfy the specifications. Transistor sizing and current levels can be easily derived from simple expressions, as we will show at the end of this chapter in Section 1.3.2. A systematic presentation of the design models is the subject of Chapter 2.

At the end of the design process, complicated and accurate MOSFET models are employed for design verification using circuit simulators such as SPICE or ELDO. MOSFET models for circuit simulation and the basic parameter extraction for design are the subjects of Chapter 11.

1.2.   Bipolar and metal–oxide–semiconductor field-effect transistors

Transistors are semiconductor devices that constitute the basic building blocks of modern electronic circuits and systems. Transistors are essentially of two types: bipolar junction transistors (BJTs) and field-effect transistors (FETs). Analog bipolar circuits have been under development since the invention of the transistor. The first electronic products in bipolar technology, hearing aids and AM radios, appeared in the mid 1950s. The modern metal–oxide–semiconductor field-effect transistor (MOSFET) appeared in 1960 and MOS digital integrated circuits (ICs), memories, and microprocessors became available in the early 1970s. The first analog MOS circuits, converters, and switched capacitor filters were launched in the mid and late 1970s. Since MOS analog circuits began to be developed when analog bipolar technology was already mature, many concepts and techniques from analog bipolar technology were transferred to analog MOS circuits. Bipolar and MOS electronic circuits have many commonalities because transistors, whatever their type, have some basic common characteristics. One terminal, called the emitter or source, furnishes the carriers that are collected by a second terminal, called the collector or drain. The amount of carriers able to cross from the first to the second terminal is controlled by a third electrode, called the base in BJTs and the gate in FETs. The simplest model of either a BJT or an FET represents either transistor as a controlled current source. Although the similarities between transistors of different types are useful to facilitate understanding of the basic principle of operation of many circuits, for design purposes some superficial similarities can induce mistakes. Thus, for design, we must have a clear understanding and an accurate modeling of the specific behaviors of BJTs and FETs.

1.2.1.   p–n Junctions

Let us begin with the p–n junctions [6]–[8], which are essential to the operation of bipolar and field-effect transistors. Electrons and holes with opposite electric charges are free to

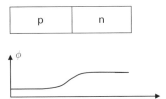

Fig. 1.1   The internal electrostatic potential of a p–n junction.

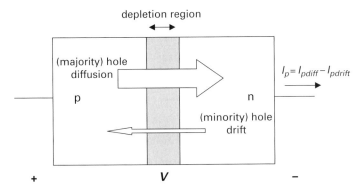

Fig. 1.2   Hole current in a forward-biased p–n junction.

move in semiconductors. Since free particles concentrate in the region where their energy is minimum, the electrostatic potential in the p-type region with plenty of holes, with positive charge, must be lower than that in the n-type region, where the electron concentration is high and holes are scarce. Similarly, electrons concentrate in the n-type region, where the potential is high and consequently their energy is low, because of their negative charge. The internal electrostatic potential in a p–n junction is shown in Figure 1.1.

In thermal equilibrium there is no net flow of carriers through the sample. Just a few holes on the p-side have enough energy to overcome the potential barrier and reach the n-side, where they recombine. Owing to these holes, a current I_{pdiff} flows but it is balanced by the current I_{pdrift} originating from the holes coming down the potential hill from the n-side. In equilibrium $I_{pdrift} = I_{pdiff} = I_{pS}$. A similar reasoning holds for the electrons. Now suppose that a positive bias V is applied to the p-region with respect to the n-region. The height of the barrier decreases by an amount V when the ohmic drops at the contacts and in the p- and n-regions are negligible. The potential-barrier region extends across the depletion region shown in Figure 1.2, where the electron (hole) carrier density is (well) below its value in the neutral n-region (p-region). The hole current injected from the p-region into the n-region is proportional to the number of holes with enough energy to overcome the potential barrier, and the reduction of the barrier by an amount V produces an exponential increase (given by the Boltzmann factor) in the number of holes with enough energy to cross the barrier. The hole current over the barrier will now be increased by the Boltzmann factor, i.e. $I_{pdiff} = I_{pS} \exp(V/\phi_t)$, where

$\phi_t \cong 26\,\text{mV}$ at $300\,\text{K}$ is the thermal voltage. On the other hand, the hole current from the n-region will not have changed after the application of voltage V because it is dependent on the generation rate of hole–electron pairs which, in turn, is dependent on the local properties of the semiconductor near the junction.

As shown in Figure 1.2, the net hole current through the junction is

$$I_p = I_{pdiff} - I_{pdrift} = I_{pS}(e^{V/\phi_t} - 1). \tag{1.2.1}$$

The electron current is added to the hole current, giving the total as

$$I = I_p + I_n = I_S(e^{V/\phi_t} - 1), \tag{1.2.2}$$

where $I_S = I_{pS} + I_{nS}$ is the reverse saturation current, which is dependent both on the diode parameters and on the temperature.

The junction is a good rectifier of ac voltages higher than a few times ϕ_t. When the junction is forward-biased ($V > 0$), there is no limitation to the exponential increase in the current modeled by (1.2.2). A realistic model of the high-current region, however, must include other phenomena such as high-injection effects, series resistance, and self-heating. For a reverse-biased ($V < 0$) junction, the current is limited to I_S. A realistic model of the reverse-current region must include excess currents due both to recombination mechanisms and to high-field effects, which were not included in the simple model given above. Despite its limitations with regard to very high and very low currents, the idealized model of (1.2.2) is usually valid for a current range of up to six or seven orders of magnitude in junction diodes implemented with vertical n–p–n transistors, which is the subject of the next subsection.

1.2.2. Bipolar junction transistors

We will focus this subsection mainly on the planar n–p–n bipolar transistor [9] shown schematically in Figure 1.3, which is the most used component of the bipolar technology. The n–p–n intrinsic structure, where the main transistor action takes place, is shown by the dashed rectangle. To build the transistor dc model, we start with its operation in the so-called forward active mode, where the base–emitter junction is forward-biased and the collector–base junction is reverse-biased, as shown in Figure 1.4. Since the doping of the emitter region is much higher than that of the base region, the forward current of the base–emitter junction is constituted mostly by electrons injected from the emitter into the base. The hole current through the emitter–base junction is just a very small fraction of

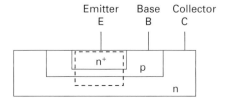

Fig. 1.3   Simplified structure of the ordinary vertical n–p–n transistor.

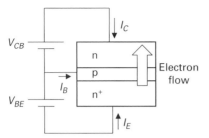

Fig. 1.4   An n–p–n bipolar junction transistor biased in the forward active region, i.e. the base–emitter junction is forward-biased and the base–collector junction is reverse-biased.

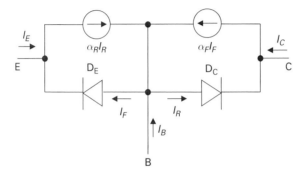

Fig. 1.5   The Ebers–Moll equivalent circuit of an n–p–n transistor.

the electron current. Most of the electrons coming from the emitter are attracted by the high potential of the collector region and diffuse through the base without recombining, since the base region is very thin. In the forward active mode, the potential barrier across the collector–base junction impedes the holes from being injected into the collector.

The collector current is a fraction of the emitter current since some of the electrons injected by the emitter into the base recombine without reaching the collector and, mainly, because a (small) part of the emitter current is composed of holes coming from the base. Thus,

$$I_C = -\alpha_F I_E, \tag{1.2.3}$$

where α_F ($\alpha_F < 1$) is the common-base current gain in the forward active mode.

The Ebers–Moll model [10] of the BJT represents the terminal currents of a bipolar transistor as a superposition of the effects of the two junctions, for all bias conditions, in terms of easily measurable transistor parameters. The equivalent circuit of the Ebers–Moll model of an n–p–n transistor is shown in Figure 1.5. I_F, the current through diode D_E, is the emitter current for $V_{BC}=0$, while I_R, the current through D_C, is the collector current for $V_{BE}=0$. The forward (I_F) and reverse (I_R) currents are described by

$$I_F = I_{ES}(e^{V_{BE}/\phi_t} - 1), \tag{1.2.4}$$

$$I_R = I_{CS}(e^{V_{BC}/\phi_t} - 1), \tag{1.2.5}$$

which are similar to (1.2.2) for the p–n junction.

The gains of the two current-controlled current sources in Figure 1.5 are α_F and α_R, the latter being the common-base current gain in the inverted mode, that is, with the collector operating as emitter and the emitter as collector. The terminal currents are

$$I_C = \alpha_F I_F - I_R, \tag{1.2.6}$$

$$I_E = \alpha_R I_R - I_F, \tag{1.2.7}$$

$$I_B = -(I_C + I_E) = (1 - \alpha_F)I_F + (1 - \alpha_R)I_R. \tag{1.2.8}$$

The Ebers–Moll model has only three independent parameters since the four parameters in the equations above are related by the reciprocity relation [9], [10]

$$\alpha_F I_{ES} = \alpha_R I_{CS} = I_S, \tag{1.2.9}$$

which holds as long as the minority-carrier density in the base is small compared with the thermal-equilibrium majority-carrier density. When the base–collector junction is reverse-biased, the combination of (1.2.4), (1.2.6), and (1.2.9) yields the familiar expression for the collector current in the forward active mode, i.e.

$$I_C \cong \alpha_F I_F = \alpha_F I_{ES}(e^{V_{BE}/\phi_t} - 1) \cong I_S e^{V_{BE}/\phi_t}. \tag{1.2.10}$$

Bipolar transistors are, in general, asymmetric devices, i.e. the collector and emitter regions are not interchangeable because they are optimized to operate in the forward active mode, typically having $\alpha_F > \alpha_R$. For the high-performance vertical BJT structure, the collector doping is much lower than the emitter doping, and typical values of the common-base current gains are $\alpha_F = 0.99$ and $\alpha_R = 0.65$. For the parasitic horizontal structure with the collector region surrounding the emitter region, common values of the common-base current gains are $\alpha_F = 0.98$ and $\alpha_R = 0.75$.

For an ideal device with common-base current gains $\alpha_F = \alpha_R = 1$, the base current is zero. Using (1.2.4) through (1.2.9) yields

$$I_C = -I_E = I_F - I_R = I_S(e^{V_{BE}/\phi_t} - e^{V_{BC}/\phi_t}). \tag{1.2.11}$$

The equation above clearly represents a (non-linear) symmetric device, i.e. one in which the output terminals are equivalent (or interchangeable). It is worth observing that the Ebers–Moll model also represents properly the not so common symmetric bipolar transistor. It is remarkable that the (more than) fifty-year-old Ebers–Moll model, although with some complements introduced by the Gummel–Poon model [11], continues to be the basic framework for BJT modeling and bipolar circuit design.

1.2.3.   MOS field-effect transistors

The MOS structure, consisting of a metal–oxide (SiO_2)–semiconductor (Si) sandwich, is shown in Figure 1.6. In the common strong-inversion model of the field effect it is assumed that the MOS structure operates as a linear capacitor. For the case of the p-type substrate shown in Figure 1.6, the electron charge is assumed to be zero for gate voltages below the

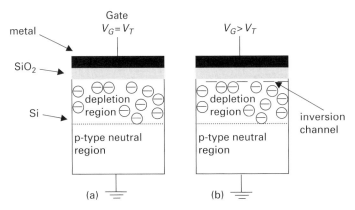

Fig. 1.6   The strong-inversion model of the field effect. (a) The MOS structure at threshold. The charge of the minority carriers (electrons) is negligible. The stored charge in the semiconductor is assumed to be due only to the depletion of majority carriers (holes). (b) The MOS structure above threshold. The depletion charge is assumed to be the same as at threshold while all the semiconductor charge variation is due to an electron (inversion) channel at the semiconductor surface.

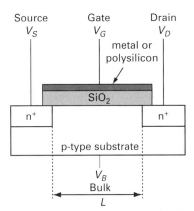

Fig. 1.7   An idealized enhancement-mode nMOS transistor. The width W is in the direction perpendicular to the page.

so-called threshold voltage and to increase linearly with the gate voltage for above-threshold operation. The basic flaw in the strong-inversion model is that the electron density follows Boltzmann's (exponential) law and, consequently, the electron charge cannot be zero for any finite applied voltage. Thus, the electron charge cannot vanish for subthreshold operation; consequently, the conventional definition of threshold cannot be applied in practice. Most of the numerous methods [12] developed to determine the threshold voltage are certainly due to the lack of a physically correct definition of threshold. Even so, since the strong-inversion model of the MOS transistor was for some time very popular, and still is, to a lesser extent, it should be summarized here and its shortcomings commented on.

A typical planar nMOS transistor structure is shown in Figure 1.7. The (parasitic) n–p–n bipolar transistor associated with this structure is inactive during the normal operation of

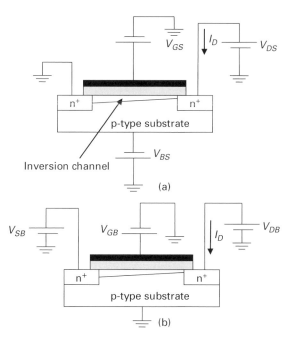

Fig. 1.8   Bias voltages applied to an nMOS transistor. (a) The common choice of the source as the reference terminal and (b) the choice of the substrate as the reference terminal. Since source and drain terminals are equivalent in the conventional MOS transistor, the choice of the substrate (or the gate) as the reference terminal preserves the symmetry between source and drain.

the MOSFET because the source–substrate and drain–substrate junctions are reverse-biased during the normal operation of the MOSFET. Conduction between source and drain occurs if an inversion (electron) channel connects them.

In the commonly used strong-inversion model [8], the applied voltages are referred to the source node, as Figure 1.8(a) shows. The inversion-channel charge density Q'_I above threshold is approximated as a linear function of the local potential to give

$$Q'_I = -C'_{ox}(V_{GS} - V - V_T), \tag{1.2.12}$$

where C'_{ox} is the oxide capacitance per unit area and V is the channel-to-source voltage, which varies from zero at the source to V_{DS} at the drain. To derive a first-approximation three-terminal model of the MOSFET, the threshold voltage V_T is assumed to be constant along the channel. As will be seen later on in this subsection, the four-terminal model is derived by including the effect of the fourth terminal (bulk) on the threshold voltage.

From expression (1.2.12) for the inversion charge density and through the integration of the equation for the electron current density along the channel assuming that diffusion is negligible, the drain current of the three-terminal device given by (1.2.14) is obtained.

For gate voltages below V_T the drain current is assumed to be zero, i.e.

$$I_D = 0 \quad \text{for } V_{GS} < V_T, \tag{1.2.13}$$

whereas

$$I_D = \beta \left(V_{GS} - V_T - \frac{V_{DS}}{2} \right) V_{DS}$$

for $V_{GS} > V_T$, $V_{DS} < V_{DSsat} = V_{GS} - V_T$.   (1.2.14)

It is important to note that the strong-inversion expression (1.2.14) holds only for the inequalities indicated. Note that, in this simplified model of the drain current in the three-terminal device, the transistor is represented by two parameters, namely the threshold voltage V_T and the transconductance parameter β given by

$$\beta = \mu C'_{ox} \frac{W}{L},$$ (1.2.15)

where μ is the carrier mobility, W the channel width, and L the channel length.

For drain-to-source voltages above $V_{DSsat} = V_{GS} - V_T$, the drain current is assumed to saturate, i.e.

$$I_D = I_{Dsat} = \frac{\beta}{2} (V_{GS} - V_T)^2 \quad \text{for } V_{GS} > V_T, \ V_{DS} > V_{GS} - V_T.$$ (1.2.16)

In effect, the plot of (1.2.14) in the I_D–V_{DS} plane is a parabola with a maximum at $V_{DS} = V_{DSsat}$. For $V_{DS} > V_{DSsat}$ (1.2.14) has no physical meaning and, in a first-order approximation, the transistor is assumed to operate as an ideal current source, the value of the current being given by (1.2.16).

In the commonly used four-terminal model of the MOSFET, the threshold voltage is a function of the source-to-substrate bias V_{SB} given [8] by

$$V_T = V_{T0} + \gamma \left(\sqrt{2\phi_F + V_{SB}} - \sqrt{2\phi_F} \right),$$ (1.2.17)

where ϕ_F is the Fermi potential, γ is the body-effect factor given by

$$\gamma = \frac{\sqrt{2q\varepsilon_s N_A}}{C'_{ox}},$$ (1.2.18)

where $q = 1.6 \times 10^{-19}$ C is the electronic charge, ε_s is the permittivity of silicon, and N_A is the substrate doping, which is assumed to be uniform. The effect of the source–substrate bias is usually referred to as the body effect.

Expression (1.2.17) represents the effect of the depletion charge density at the source on the threshold voltage. In fact, for $V_{DS} \neq 0$, the depletion charge is non-uniform along the channel and, thus, the threshold voltage varies along the channel. Therefore, there is some incon-sistency in using (1.2.17) in association with Equation (1.2.14) for the drain current since the latter has been deduced using the hypothesis that the depletion charge, and consequently V_T, is independent of the channel-to-source voltage V along the channel. As a consequence, the commonly employed model of the MOSFET which applies the body effect at the source to the whole channel does not preserve the physical interchangeability of source and drain.

To obtain a symmetric MOSFET model, the variation of the threshold voltage must be considered along the whole channel. To emphasize the symmetry between source and drain in the model expressions, we will refer all the voltages to the grounded substrate [13]–[15] as indicated in Figure 1.8(b). We rewrite the inversion charge density given by (1.2.12) as

$$Q'_I = -C'_{ox}(V_{GB} - V_{SB} - (V_{CB} - V_{SB}) - V_T)$$
$$= -C'_{ox}(V_{GB} - V_{TB}),$$
(1.2.19)

where [16]

$$V_{TB} = V_T + V_{CB}$$
(1.2.20)

is the gate-to-substrate threshold and V_{CB} is the channel-to-substrate potential ($V_{SB} \leq V_{CB} \leq V_{DB}$).

The variation of the depletion charge density Q'_B along the channel produces a variation in the threshold voltage given by

$$V_{TB} = V_{T0} - \frac{Q'_B(V_{CB}) - Q'_B(0)}{C'_{ox}} + V_{CB},$$
(1.2.21)

where V_{T0} is the equilibrium ($V_{CB} = 0$) threshold voltage.

The simplest approach to account for the variation of the depletion charge along the channel is to assume a linear increase in the magnitude of the depletion charge with increasing channel–substrate voltage (i.e. assuming a constant depletion capacitance) [14]. Thus,

$$V_{TB} = V_{T0} + nV_{CB},$$
(1.2.22)

where

$$n = 1 + \frac{C'_b}{C'_{ox}}$$
(1.2.23)

and

$$C'_b = -\frac{dQ'_B}{dV_{CB}}.$$
(1.2.24)

Here n is the so-called slope factor, which is considered constant along the channel, and is usually in the range 1.2–1.5 for submicron technologies. C'_b is the depletion capacitance, which is assumed to be constant along the channel.

The use of (1.2.22) together with (1.2.19) to integrate the drift component of the current density along the channel results in the symmetric model given as

$$I_D = \beta \left[V_{GB} - V_{T0} - \frac{n}{2}(V_{SB} + V_{DB}) \right](V_{DB} - V_{SB}).$$
(1.2.25)

In the above equation all the potentials are referred to the substrate, as shown in Figure 1.8(b).

An alternative form of (1.2.25), which splits the drain current into a forward and a reverse term as in the Ebers–Moll BJT model, is [15]

$$I_D = I_F - I_R, \tag{1.2.26}$$

with

$$I_{F(R)} = \frac{\beta}{2n}\left(V_{GB} - nV_{SB(DB)} - V_{T0}\right)^2. \tag{1.2.27}$$

When

$$V_{DB} = \frac{V_{GB} - V_{T0}}{n} \tag{1.2.28}$$

the reverse current $I_R=0$. In this case, the (forward) saturation current is given by

$$I_{Dsat} = I_F = \frac{\beta}{2n}(V_{GB} - V_{T0} - nV_{SB})^2. \tag{1.2.29}$$

On comparing (1.2.16) and (1.2.29) in the case of $V_{SB}=0$, we observe that the saturation current given by (1.2.16) gives a value of the drain current n times higher than that of (1.2.29). In effect, since (1.2.16) neglects the variation of the depletion charge density along the channel, it overestimates the value of the drain current.

The strong-inversion model is clearly inadequate to model MOSFETs operating at low currents. In the subthreshold or weak-inversion region the inversion (channel) charge density is much lower than the depletion charge density; thus, it is appropriate to model the MOS structure as shown in Fig. 1.9. Assuming that the carrier charge is concentrated within an infinitely thin layer at the semiconductor surface (the charge-sheet approximation [17]), the inversion charge density is given in terms of the surface potential ϕ_s by the Boltzmann exponential $\exp(\phi_s/\phi_t)$. The potential at the semiconductor surface ϕ_s is given by the capacitive division of the gate voltage V_{GB}. Neglecting the variation of the depletion capacitance with the applied gate voltage, the surface potential varies linearly with the gate voltage, i.e. $d\phi_s/dV_{GB}=1/n$, where n is the slope factor presented previously in the context of the symmetric strong-inversion MOSFET model.

In weak inversion the surface potential is dependent, in a first-order calculation, on the gate-to-substrate voltage only and is thus constant along the channel. Therefore, the drain

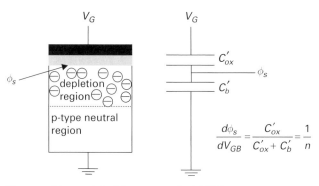

Fig. 1.9   The capacitive model of the two-terminal MOS structure.

current in subthreshold operation is a pure diffusion current. The weak-inversion model of the MOSFET current [15], [16]

$$I_D = I_F - I_R = I_0 \frac{W}{L} \left(e^{(V_{GB} - V_{T0} - nV_{SB})/n\phi_t} - e^{(V_{GB} - V_{T0} - nV_{DB})/n\phi_t} \right) \qquad (1.2.30)$$

is similar to expression (1.2.11) for the collector current of a symmetric BJT. For $V_{DS} > 4\phi_t$ ($\sim100\,\mathrm{mV}$ at room temperature), $I_F \gg I_R$ and the transistor is in forward saturation, i.e.

$$I_{Dsat} \cong I_F = I_0 \frac{W}{L} e^{(V_{GB} - V_{T0} - nV_{SB})/(n\phi_t)}, \qquad (1.2.31)$$

where I_0 is the specific or "square" ($W = L$) saturation current for an FET with the source terminal at threshold ($V_{GB} = V_{T0} + nV_{SB}$).

There are physical reasons for the similarities between the BJT current and the MOSFET subthreshold current. Both result from the diffusion of minority carriers, the densities of which are governed by Boltzmann's law. The main difference between (1.2.10) and (1.2.31) is that for the BJT current the control variable is simply the base–emitter voltage V_{BE}, but for the MOSFET the gate and source voltages have different weights due to the body effect. For a bipolar transistor, the effect of increasing the base voltage by a certain amount is strictly equivalent to decreasing the emitter voltage by the same amount. For a MOSFET, however, the applied voltage V_{GB} is attenuated by the capacitive voltage division and becomes V_{GB}/n at the semiconductor interface, but the source voltage has no attenuation. One of the problems associated with using the voltage V_{GS} as the main control variable of the MOSFET state is to mistakenly assume it to be equivalent to the V_{BE} control voltage of the BJT.

An important difference between BJTs and MOSFETs is that for the former we have one transconductance, whereas for the latter we must consider two transconductances since they are four-terminal devices. The transconductance of the BJT in the forward active region is calculated, taking the derivative of (1.2.10), as

$$g_m = \frac{dI_C}{dV_{BE}} = \frac{I_C}{\phi_t}. \qquad (1.2.32)$$

For a saturated MOSFET in weak inversion, the partial derivatives of (1.2.31) yield

$$g_m = g_{mg} = \frac{\partial I_F}{\partial V_{GB}} = \frac{I_F}{n\phi_t}, \qquad (1.2.33)$$

$$g_{ms} = -\frac{\partial I_F}{\partial V_{SB}} = \frac{I_F}{\phi_t}. \qquad (1.2.34)$$

Thus, the two transconductances[1] are related by

$$g_{mg} = \frac{g_{ms}}{n}. \qquad (1.2.35)$$

---

[1] We will use both symbols g_m and g_{mg} for the derivative of the drain current with respect to the gate voltage.

The body effect reduces the gate transconductance with respect to the source transconductance. In fact, in weak inversion the variation in the channel voltage contributes to variation in the inversion charge only, whereas the variation in the gate voltage contributes to variations both in the inversion charge and in the depletion charge. The equation above is valid also in strong-inversion saturation, as can be verified by taking the derivatives of (1.2.29) and recalling the definition of β given in (1.2.15):

$$g_{ms} = ng_{mg} = \frac{W}{L}\mu C'_{ox}(V_{GB} - V_{T0} - nV_{SB}). \tag{1.2.36}$$

Observing that

$$C'_{ox}(V_{GB} - V_{T0} - nV_{SB}) = -Q'_{IS}, \tag{1.2.37}$$

we can rewrite the transconductances g_{mg} and g_{ms} in terms of the inversion charge density at the source Q'_{IS} as

$$g_{ms} = ng_{mg} = -\frac{W}{L}\mu Q'_{IS}. \tag{1.2.38}$$

Even though (1.2.38) has been deduced from the strong-inversion model, it is quite general, being valid from weak to strong inversion.

1.2.4.   Important differences between BJTs and MOSFETs

Since analog circuit design depends strongly on the representation of the transistors, it is useful to be aware of the differences between BJT and MOSFET models. Since bipolar analog design is a much more mature field than analog MOS design, MOSFET modeling has been strongly influenced by bipolar modeling. This influence sometimes has negative outcomes and the designer must be aware of them. The choice of the source as the reference terminal following the use of the emitter as the reference is maybe the best example. Although the choice of the emitter as the reference seems to be the natural choice for the common BJT, because emitter and collector are not interchangeable, this is not the case for the MOSFET. The common four-terminal MOSFET model that uses the threshold voltage at the source gives an incorrect model of the body effect and does not preserve the intrinsic source–drain symmetry of the transistor. The following list summarizes some important differences between BJTs and MOSFETs.

(A) BJTs are three-terminal devices and MOSFETs are four-terminal devices

Since the MOSFET is a four-terminal device (gate, source, drain, and bulk), in general, three control voltages, e.g. V_{GB}, V_{SB}, and V_{DB}, rather than two (V_{BE} and V_{CE}) as in the case of the three-terminal BJT, are needed to set the state of the device.

For the MOSFET 16 coefficients in the linear (small-signal) model are defined to relate the four-terminal (transport and capacitive) currents to the four applied voltages. However, only nine independent small-signal coefficients are necessary to model the relationship between three independent voltages and currents. In comparison, the linear model of the BJT, which is represented by a two-port network, has only four independent coefficients.

(B) Differences in the internal symmetries of the most commonly used BJTs and MOSFETs

In the BJT the emitter and collector are, in general, not interchangeable, but the source and drain terminals in a common MOSFET are. For a BJT operating in the active mode, V_{BE} is the natural control voltage of the current. For MOS devices operating in saturation, we propose the use of V_{GB} and V_{SB} as the control voltages instead of V_{GS} and V_{BS}, owing to the transistor symmetry and the body effect.

(C) The BJT exponential current law versus the MOS current law

The current in a BJT is essentially modeled by an exponential dependence on the base–emitter voltage. The inclusion of high-injection effects modifies the ideality coefficient (the BJT analog to the MOSFET slope factor) of the Boltzmann exponential and the effect of recombination on the current requires the inclusion of exponentials with different ideality factors, but the exponential behavior is still preserved. In the case of a MOSFET, the drain current follows an exponential law in weak inversion, but in strong inversion it is a quadratic or linear function of the gate-to-bulk voltage for high or low V_{DS}, respectively. In between weak and strong inversion, for about two orders of magnitude of variation in the drain current, neither the strong-inversion nor the weak-inversion model is accurate. The differences between weak and strong inversion are not due to second-order effects but to fundamental physics; whereas in weak inversion the current flows mainly by diffusion, in strong inversion drift is the predominant transport mechanism.

(D) The geometric degrees of freedom for MOSFETs in analog design

Since the common BJT has a vertical structure, the base width is fixed and is not a design parameter. The emitter area is the only geometric design parameter of the common BJT, but, in general, most bipolar transistors are designed with a fixed size. Some exceptions, however, may be found in an IC design. An example is a high-current transistor of the output stage of an amplifier, which is commonly made by connecting several standard transistors in parallel [8]. On the other hand, designers can play with both the width W and the length L in MOSFET design. Summarizing, for analog design, the collector current I_C (and sometimes the emitter area) is the only design parameter of BJTs, but three degrees of freedom are available for MOSFETs: I_D, W, and L.

(E) Quality of BJT and MOSFET models

The more than fifty-year-old Ebers–Moll model is an elegant and simple model valid in all the operating regions of the BJT. The Ebers–Moll model, flavored with some results of the Gummel–Poon model, still provides the framework for complex (and accurate) computer-implemented models of the BJT for circuit simulation. On the other hand, the most commonly employed MOSFET models are valid only in specific regions of operation. Owing to this regional approach, some fundamental parameters such as the gate transconductance and the gate-to-source capacitance, just to name a couple, can be discontinuous. Although enormous progress in relation to complex computer-implemented models has been made in recent years, the common MOSFET models based on the regional approach are still used in textbooks on analog circuit design. Thus, there is a large discrepancy between hand-design and simulation tools for analog MOS circuits, which this book aims to reduce.

1.3.   Analog bipolar and MOS integrated circuits

In integrated circuits, the smaller the area of a given component, the more functionality can be placed on one die and, therefore, the lower the cost per function performed. Thus, since the cost of a device is proportional to the area it occupies on the die, the use of large capacitors and resistors must be avoided. MOS circuit designers must avoid, to a much greater extent than bipolar designers, the use of resistors since, in many cases, the low values of dc currents would require very large resistors, in the range of MΩ or above. Thus, MOS analog circuits are essentially transistor-only circuits,[2] with linear capacitors and resistors having mainly an auxiliary role. There are many other differences between BJT and MOS analog circuits, which include the high input resistance and elevated excess low-frequency noise of the MOSFET, and the specificities of the temperature dependence and matching of V_{BE} voltages of BJTs. Noise and mismatch characteristics will be developed in the following chapters, but, to understand some basic differences between analog MOS and BJT circuits, a very simple design example such as that at the end of this chapter is sufficient.

1.3.1.   Analysis and design of integrated circuits

Circuit synthesis is the main concern of designers, but circuit theory is strongly focused on analysis. Circuit analysis is a mature field and simulation tools are widespread, but analog design tools are still in their infancy. To know why, it is useful to recall the definitions of analysis and synthesis.

    Circuit analysis is the determination of the performance of a circuit given its topology, the technology and dimensions of the transistors, the values of the passive elements, and the bias voltages and currents. Powerful circuit analysis programs are intensively used by designers, SPICE developed by the University of California at Berkeley being the most widespread. A computer simulation of a circuit is a verification tool that, when properly interpreted, allows us to gain insight into the operation of the circuit. Modern transistor models for circuit simulation are very accurate, but, particularly in the case of the MOSFET, they are very complex. However, in all cases, a circuit can be simulated only after a design has been completed.

    Circuit synthesis is the inverse of analysis. The design involves such degrees of freedom as circuit topology, fabrication technology, transistor geometries, and values for the passive elements to fulfill predetermined specifications. In contrast with analysis, formal methods are almost non-existent for transistor-level design. The design depends

---

[2] In CMOS RF circuits, resistors (and other passive elements) are widely employed.

strongly on the designer's wisdom; in contrast to the uniqueness of analysis results, design solutions are as abundant as designers.

Transistor models for design must be sufficiently simple, but still accurate and containing few parameters, in order to be useful. Models close to the device physics have a greater chance of being both simple and accurate. The example in the next section highlights the desirable properties of device models for design and some basic differences between MOSFET and BJT analog design [18].

1.3.2. Design of common-emitter and common-source amplifiers

The intrinsic MOS gain stage shown in Figure 1.10 consists of a single common-source transistor M_1 loaded by an ideal direct-current source I_B. C_L represents the capacitance of the output node. The MOS transistor is assumed to be saturated. Similar comments apply to the common-emitter amplifier. The bulk terminal of the MOSFET is not represented, in order to emphasize the analogy with the BJT. For simplicity we assume that $V_{BS}=0$, but it is worth noting that in this way we are reducing by one the number of degrees of freedom for the design of the common-source amplifier.

The equivalent small-signal circuits of the intrinsic gain stages are represented in Figure 1.11. The MOSFET and BJT capacitive effects are not included in the small-signal models since they are not relevant to the purpose of comparing the first-order behaviors

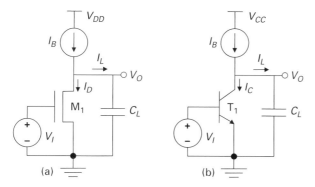

Fig. 1.10   Intrinsic gain stages: (a) common-source and (b) common-emitter amplifiers with ideal current source bias.

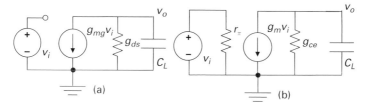

Fig. 1.11   Small-signal equivalent circuits of the (a) common-source and (b) common-emitter amplifiers. r_π models the input resistance of the BJT.

of the amplifiers. The conductance g_{ds} (g_{ce}) represents the output conductance of the real MOSFET (BJT) in saturation (the forward active region).

In the simple models of (1.2.10) for the BJT or (1.2.29) for the MOSFET, the transistors are represented as ideal current sources and thus $g_{ds} = g_{ce} = 0$. Note that, even though Equations (1.2.10) and (1.2.29) are accurate for determining the dc currents, they do not allow us to calculate the transistor output conductance. The use of more elaborate transistor models, which include the so-called Early effect, allows g_{ds} and/or g_{ce} to be calculated.

Since at this time we do not have an appropriate model of the output conductance, let us concentrate on the high-frequency gain, which is controlled by both the load capacitance C_L and the transconductance g_m (or g_{mg}), for the comparison between the two amplifiers.

The asymptotic small-signal voltage gain of both intrinsic gain stages is plotted in Figure 1.12. The low-frequency voltage gain, which is dependent on the transistor transconductance and the output conductance, and the transition or unity-gain angular frequency defined as

$$|A_v(\omega_u)| = 1 \tag{1.3.1}$$

are the main parameters of the intrinsic gain stages.

For both small-signal circuits in Figure 1.11, we can write at high frequency

$$v_o = A_v(\omega)v_i = -\frac{g_m}{j\omega C_L} v_i. \tag{1.3.2}$$

We can use the simple model above to illustrate a typical design problem. Usually designers must synthesize a circuit satisfying some predefined specifications. Let us now consider the following basic design problem: determine the transistor current and sizes to obtain a given gain–bandwidth product GB for a given capacitive load C_L.

To solve this problem we must start from GB and determine the transconductance from (1.3.2) as

$$g_m = \omega_u C_L = 2\pi \cdot GB \cdot C_L. \tag{1.3.3}$$

The bipolar transistor design is straightforward, since its transconductance given by (1.2.32) is proportional to the collector current. Thus,

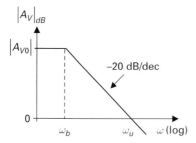

Fig. 1.12   The asymptotic frequency response of the magnitude of the voltage gain of the intrinsic amplifiers.

$$I_C = g_m \phi_t = 2\pi \cdot GB \cdot C_L \cdot \phi_t \tag{1.3.4}$$

and the result is independent of the transistor dimensions. In our example $GB = 10\,\text{MHz}$, $C_L = 10\,\text{pF}$, and $\phi_t = 26\,\text{mV}$; thus $I_C \cong 16.3\,\mu\text{A}$.

The MOS design problem is more involved. The MOSFET gate transconductance given in (1.2.36) and repeated below for convenience is, for $V_{SB} = 0$,

$$g_m = g_{mg} = \frac{\beta}{n}(V_{GB} - V_{T0}). \tag{1.3.5}$$

The geometric ratio W/L obtained from (1.3.5) is

$$\frac{W}{L} = \frac{n g_{mg}}{\mu C'_{ox}(V_{GB} - V_{T0})}. \tag{1.3.6}$$

From (1.2.29) and (1.3.5) we obtain the expression for the strong-inversion drain current I_F as

$$I_F = \frac{g_{mg}}{2}(V_{GB} - V_{T0}). \tag{1.3.7}$$

Finally, combining (1.3.6) and (1.3.7) yields the relationship between the current I_F and the aspect ratio W/L,

$$I_F = \frac{n g_{mg}^2}{2\mu C'_{ox} W/L}. \tag{1.3.8}$$

To illustrate the design calculations, let us consider that the design specifications are $GB = 10\,\text{MHz}$ and $C_L = 10\,\text{pF}$, and assume that the technological parameters are $\mu C'_{ox} = 80 \times 10^{-6}\,\text{A/V}^2$ and $n = 1.35$. The gate transconductance obtained from (1.3.3) is $g_{mg} = 628\,\mu\text{A/V}$. The values of the drain current corresponding to various W/L factors are shown in Table 1.1. The synthesis based on Equation (1.3.8), which is valid in strong inversion only, is compared with that obtained using an all-region MOSFET model [19] (valid in weak, moderate, and strong inversion). Table 1.1 makes clear the higher complexity of the MOS analog design compared with bipolar design. For BJT design we have a fixed value for the collector current defined by the transconductance, but for MOSFET design we have an infinite number of solutions. There is a tradeoff between

Table 1.1 Synthesis of an MOS intrinsic gain stage

W/L	$I_F\ (\mu\text{A})^a$	$I_D\ (\mu\text{A})^b$
∞	0	22
500	6.6	28.6
100	33.2	55.2
50	66.4	88.4
10	332	354

[a] Strong-inversion model.
[b] Accurate all-region MOSFET model [19].

power consumption and area, for transistors with constant L. We can reduce the drain current and thus the power consumption by increasing the W/L ratio and, consequently, the transistor area. It is important to note that for synthesis with the strong-inversion model we can reduce the drain current to an arbitrarily small value, which clearly has no physical meaning. The accurate model of the MOSFET gives a minimum value for the current at around $22\,\mu A$. It should be noted that this value is higher than the collector current of $16.3\,\mu A$ for the bipolar transistor in our example.

In this particular example, it is easy to guess the relationship between the strong-inversion and all-region models by comparing the values of the current given in the two columns in Table 1.1. The real current is $22\,\mu A$ higher than the current calculated using the strong-inversion model. Thus,

$$I_D = I_F + I_{WI}, \tag{1.3.9}$$

where $I_{WI} = 22\,\mu A$ is the minimum current (in weak inversion) which gives the desired value of the transconductance. In effect, using (1.2.33) we have

$$I_{WI} = n g_{mg} \phi_t = g_{ms} \phi_t, \tag{1.3.10}$$

which gives, for this example,

$$I_{WI} = 1.35 \times 628 \times 26 \times 10^{-3} = 22 \ \mu A.$$

Consequently, we can write, using (1.3.8) and (1.3.10) in (1.3.9),

$$I_D = g_{ms} \phi_t \left(1 + \frac{g_{ms}}{2\mu n C'_{ox} \phi_t W/L} \right), \tag{1.3.11}$$

which has all the qualities required for an appropriate design model: it is very simple, valid in all operating regions, and has only a few parameters, all of them associated with the device physics.

To explore the MOSFET design space it is useful to rewrite (1.3.11) as

$$I_D = I_{WI} \left[1 + \frac{(W/L)_{th}}{W/L} \right], \tag{1.3.12}$$

where $(W/L)_{th}$ is defined by

$$g_{ms} = (W/L)_{th} \mu \left(2 n C'_{ox} \phi_t \right) = -(W/L)_{th} \mu Q'_{ISth}. \tag{1.3.13}$$

The above equation, obtained from (1.2.38), gives the desired transconductance for an inversion charge density of $-2n C'_{ox} \phi_t$ at the source. We can use this specific (small) value of the inversion charge density to define a physics-based threshold condition. The current consumption for a design with the MOSFET operating at this threshold would be twice the minimum (weak-inversion) current. In our example the current at threshold is $44\,\mu A$ and $(W/L)_{th} = 151$. Thus, the current per aspect ratio $I_D/(W/L)$ at threshold is $0.3\,\mu A$.

Figure 1.13 summarizes the importance of (1.3.12). Let us assume for the time being that the channel length is fixed. If we reduce W/L in order to decrease the area, we pay for

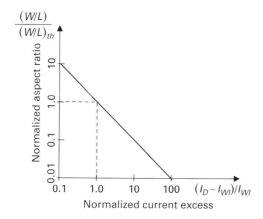

Fig. 1.13   The relationship between the normalized aspect ratio and the normalized current excess in a MOSFET design

this reduction with an increase in current consumption. Equivalently, if we try to save power by reducing the excess current, more silicon area is spent. Although it has been introduced as an empirical fitting expression, we will see in the next chapter that (1.3.12) is one of the many important results of the development of a physics-based design model.

Problems

1.1 (a) By applying Ohm's law to an n-channel element of length dy, derive the differential relationship

$$dV = -\frac{I_D\,dy}{W\mu Q_I'(V)},\tag{P1.1.1}$$

where dV is the potential drop in the channel element, I_D the drain current, W the channel width, μ the carrier mobility, and $Q_I'(V)$ the inversion charge density.

(b) Derive the general expression for the long-channel drain current

$$I_D = -\frac{W}{L}\mu \int_{V_{SB}}^{V_{DB}} Q_I'(V)dV.\tag{P1.1.2}$$

1.2 (a) Derive the strong-inversion symmetric model of the drain current of Equation (1.2.25).

(b) Show that (1.2.25) can be written as

$$I_D = I_F - I_R\tag{P1.2.1}$$

with

$$I_{F(R)} = \mu C_{ox}'\frac{W}{2nL}\left(V_{GB} - V_{T_0} - nV_{SB(DB)}\right)^2\tag{P1.2.2}$$

or

$$I_{F(R)} = \mu \frac{W}{L} \frac{Q'^2_{IS(D)}}{2nC'_{ox}},$$ (P1.2.3)

where $Q'_{IS(D)}$ is the inversion charge density at the source (drain).

1.3 Design nMOS transistors in 0.35-μm technology operating at low V_{DS} with equivalent resistances of 10 Ω, 1 kΩ, and 100 kΩ. Consider $V_{SB}=0$ and $V_{GS} - V_T = 1$V. Minimize the area of the designed transistors considering 0.5 μm ≤ L and 2.5 μm ≤ W. Assume that $C'_{ox} = 5 \times 10^{-7}$ F/cm² and $\mu C'_{ox} = 300$ μA/V². Estimate the number of minimum-dimension transistors ($L_{min} = 0.5$ μm, $W_{min} = 2.5$ μm) that would occupy the same area as the three resistors.

1.4 The source, drain, gate, and bulk tranconductances are defined as

$$g_{ms} = -\frac{\partial I_D}{\partial V_S}, \qquad g_{md} = \frac{\partial I_D}{\partial V_D}, \qquad g_{mg} = \frac{\partial I_D}{\partial V_G}, \qquad g_{mb} = \frac{\partial I_D}{\partial V_B}. \qquad \text{(P1.4.1)}$$

Prove that the relationship among the four transconductances is

$$g_{ms} = g_{mg} + g_{mb} + g_{md} \qquad \text{(P1.4.2)}$$

and calculate the four transconductances for a saturated transistor in weak and in strong inversion. Recalculate the small-signal transconductances for $V_{DS}=0$.

1.5 Using the strong-inversion symmetric model of (1.2.25) and/or the weak-inversion model of (1.2.30), show that the composite transistor of Figure P1.5 is dc-equivalent to a transistor with an aspect ratio of

$$(W/L)_{eq} = \frac{(W/L)_S (W/L)_D}{(W/L)_S + (W/L)_D}. \qquad \text{(P1.5.1)}$$

For $W_S = W_D = W_{eq}$, determine the equivalent length of the composite transistor.

1.6 Show that the (saturation) drain current per aspect ratio at the charge-based threshold of $Q'_{ISth} = -2nC'_{ox}\phi_t$ is given by

$$I_{Fth}/(W/L)_{th} = 4\mu C'_{ox} n\phi_t^2. \qquad \text{(P1.6.1)}$$

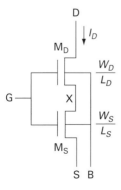

Fig. P1.5   A composite MOSFET constituted by the series association of two MOSFETs with common gate and substrate terminals.

Assuming that $\phi_t = 26$ mV, $n = 1.2$, $\varepsilon_{ox} = 0.34$ pF/cm and $\mu = 400$ cm^2/V per s, find the MOS technology for which

$$I_{Fth}/(W/L)_{th} = 1 \,\mu A, \tag{P1.6.2}$$

using the empirical relationship between oxide thickness and channel length given below:

$$t_{ox} = L_{min}/50. \tag{P1.6.3}$$

1.7 For the devices of the design example in Section 1.3.2 (the BJT and the MOSFETs with aspect ratios and bias currents reported in Table 1.1) calculate
(a) V_{BE} and V_{GS} assuming that $I_S = 10^{-16}$ A and $V_{T0} = 0.8$ V, and
(b) the voltage gain assuming that $1/g_{ce} = V_{ABJT}/I_C$ and $1/g_{ds} = V_{AMOS}/I_D$ with $V_{ABJT} = 20$ V and $V_{AMOS} = 10$ V.
   Compare and discuss the results.

1.8 To include the effect of the parasitic drain–bulk junction capacitance in the design of the common-source amplifier of Section 1.3.2, assume a simple model of the drain–bulk junction capacitance such that it is independent of the voltage and its value per unit width is C'_J. Assuming that the transition or unity-gain frequency of the transistor f_T can be approximated by

$$f_T \cong g_{mg} \bigg/ \left(2\pi \frac{1}{2} C'_{ox} WL \right), \tag{P1.8.1}$$

show that the width W of the transistor for the specified GB is given by

$$W = \left(\frac{2C_L}{LC'_{ox}} \frac{GB}{f_T} \right) \bigg/ \left(1 - \frac{2C'_J}{LC'_{ox}} \frac{GB}{f_T} \right). \tag{P1.8.2}$$

Assuming that $f_T = 4GB$, $C_L = 1$ pF, $C'_{ox} = 2$ fF/μm^2, $C'_J = 1$ fF/μm, and $L = 1 \,\mu$m, calculate the error in the determination of W when the parasitic drain capacitance is neglected.

1.9 Show that for a MOSFET operating in strong inversion

$$I_F/(\phi_t g_{ms}) = \sqrt{I_F \bigg/ \left(2\mu C'_{ox} n \frac{W}{L} \phi_t^2 \right)} = \frac{1}{2} \sqrt{i_f}, \tag{P1.9.1}$$

where the normalized forward current i_f is given by

$$i_f = \frac{I_F}{\mu C'_{ox} n (W/L)(\phi_t^2/2)}. \tag{P1.9.2}$$

On the other hand, in weak inversion the exponential variation of the current with the source voltage yields

$$I_F/(\phi_t g_{ms}) = 1. \tag{P1.9.3}$$

Since in weak inversion $i_f \to 0$, an asymptotic all-region model of the current over transconductance ratio is simply given by the addition of (P1.9.1) and (P1.9.3). Thus,

$$I_F/(\phi_t g_{ms}) \cong 1 + \sqrt{i_f}/2. \tag{P1.9.4}$$

Compare the asymptotic model of (P1.9.4) with the physics-based model of (P1.9.5),

$$I_F/(\phi_t g_{ms}) = \left(1 + \sqrt{1 + i_f}\right)/2, \tag{P1.9.5}$$

which will be derived in Chapter 2. Plot the two normalized curves on the same graph and determine the highest relative error of (P1.9.4) with respect to (P1.9.5) and the value of the normalized forward current at which it occurs.

1.10 Derive an interpolation all-region model from Equation (P1.9.4). (Hint: use $g_{ms} = -dI_F/dV_S$ to integrate (P1.9.4) between the pinch-off voltage and a generic voltage to obtain $V_P - V_S = F(i_f, i_p)$. Use transistor symmetry to derive $V_P - V_D = F$ (i_r, i_p), where i_p and i_r are the normalized reference (pinch-off) and reverse currents, respectively. Owing to the MOSFET symmetry, i_r is also defined by (P1.9.2) on replacing I_F with I_R.)

References

[1] Berkeley Sensor and Actuator Center, http://www-bsac.eecs.berkeley.edu/.
[2] H. Zumbahlen, editor, *Linear Circuit Design Handbook*, Amsterdam: Elsevier/Newnes, 2008.
[3] W. M. C. Sansen, *Analog Design Essentials*, Dordrecht: Springer, 2006.
[4] B. Gilbert, "Design for manufacture," in *Trade-Offs in Analog Circuit Design: The Designer's Companion*, ed. C. Toumazou, G. Moschytz, and B. Gilbert, Boston, MA: Kluwer, 2002, pp. 7–74.
[5] Y. Tsividis, *Mixed Analog–Digital VLSI Devices and Technology: An Introduction*, New York: McGraw-Hill, 1996.
[6] M. Born, *Atomic Physics*, 8th edn., New York: Dover Publications, 1989.
[7] R. S. Muller and T. I. Kamins with M. Chan, *Device Electronics for Integrated Circuits*, 3rd edn., New York: John Wiley & Sons, 2003.
[8] P. R. Gray, P. J. Hurst, S. H. Lewis, and R. G. Meyer, *Analysis and Design of Analog Integrated Circuits*, 4th edn., New York: John Wiley & Sons, 2001.
[9] W. Shockley, M. Sparks, and G. K. Teal, "p–n Junction transistors," *Physical Review*, vol. **83**, no. 1, pp. 151–162, July 1, 1951.
[10] J. J. Ebers and J. L. Moll, "Large-signal behavior of junction transistors," *Proceedings of the IRE*, vol. **42**, no. 12, pp. 1761–1772, Dec. 1954.
[11] H. K. Gummel and H. C. Poon, "An integral charge control model of bipolar transistors," *The Bell System Technical Journal*, vol. **49**, no. 3, pp. 827–852, May–June 1970.
[12] A. Ortiz-Conde, F. J. García Sánchez, J. J. Liou *et al.*, "A review of recent MOSFET threshold voltage extraction methods," *Microelectronics Reliability*, vol. **42**, pp. 583–596, 2002.
[13] D. Frohman-Bentchkowsky and L. Vadasz, "Computer-aided design and characterization of digital MOS integrated circuits," *IEEE Journal of Solid-State Circuits*, vol. **4**, no. 2, pp. 57–64, Apr. 1969.
[14] H. Wallinga and K. Bult, "Design and analysis of CMOS analog signal processing circuits by means of a graphical MOST model," *IEEE Journal of Solid-State Circuits*, vol. **24**, no. 3, pp. 672–680, June 1989.

[15] C. Enz, F. Krummenacher, and E. A. Vittoz, "An analytical MOS transistor model valid in all regions of operation and dedicated to low-voltage and low-current applications," *Journal of Analog Integrated Circuits and Signal Processing*, vol. **8**, no. 7, pp. 83–114, July 1995.

[16] Y. Tsividis, *Operation and Modeling of the MOS Transistor*, 2nd edn., Boston, MA: McGraw-Hill, 1999.

[17] J. R. Brews, "A charge-sheet model of the MOSFET," *Solid-State Electronics*, vol. **22**, no. 12, pp. 991–997, Dec. 1979.

[18] P. Jespers, "MOSFET modeling for low-power design," *X Congress of the Brazilian Microelectronics Society*, August 1995, pp. 63–77.

[19] C. Galup-Montoro and M. C. Schneider, *MOSFET Modeling for Circuit Analysis and Design*, Hackensack, NJ: World Scientific, 2007.

2 Advanced MOS transistor modeling

Analog designers require MOSFET models that are accurate for (numerical) circuit simulation, but they also need models that are simple enough to allow design (at a symbolic level) that captures the essential non-linearity of the transistors. Clearly, the analysis and design models must be consistent in order to allow a smooth transition between design and analysis. Ever since the origins of MOS circuit design [1], until recently, the most popular compact models for design have been based on the threshold-voltage (V_T) formulation. This class of models relies on approximate solutions that are accurate only in the limit cases of strong- and weak-inversion operation, and no accurate model is available for the transition (moderate-inversion) region between them. V_T models are no longer acceptable for analog circuits in advanced, low-voltage technologies, where the moderate-inversion region is increasingly important. For this reason, compact models strongly based on the MOS transistor theory [2], which are accurate in all the operating regions of the transistor, will be summarized in this chapter. The fully consistent and simple charge-based MOSFET model will be reviewed in Section 2.1. The model presented here preserves the symmetry of the conventional rectangular-geometry transistor, an important property on which several circuits are based. A design-oriented current-based model is summarized in Section 2.2. The dynamic models, including expressions for nine linearly independent capacitive coefficients of the four-terminal transistor and a non-quasi-static model are presented in Section 2.3, and the main short-channel effects are rapidly reviewed in Section 2.4.

2.1 Fundamentals of the MOSFET model

2.1.1 Electrons and holes in semiconductors

In equilibrium, electrons and holes in an ideal crystalline semiconductor behave as ideal gases when the doping concentration is not too high (typically of the order of 10^{18} cm^{-3} or less), as is the case for the bulk material of MOS devices [3]. Consequently, electrons and holes follow the Boltzmann distribution law and their concentrations (number per unit volume) are proportional to

$$e^{-\text{Energy}/(kT)},\tag{2.1.1}$$

where $k = 1.38 \times 10^{-23}$ J/K is the Boltzmann constant and T is the absolute temperature in degrees Kelvin (K).

It follows that the electron and hole concentrations in equilibrium, designated here by the symbols n and p, respectively, are related to the electrostatic potential ϕ by

$$\frac{p(\phi_1)}{p(\phi_2)} = e^{-\frac{q(\phi_1-\phi_2)}{kT}}, \tag{2.1.2}$$

$$\frac{n(\phi_1)}{n(\phi_2)} = e^{\frac{q(\phi_1-\phi_2)}{kT}}, \tag{2.1.3}$$

where $q = 1.6 \times 10^{-19}$ C is the electronic charge. Because of their negative charge, electrons are attracted to regions of higher electric potential. The opposite is true for holes. From Equations (2.1.2) and (2.1.3) it follows that, in thermal equilibrium, the product pn is constant.

Calling the concentration of electrons (and holes) in an intrinsic semiconductor n_i, the so-called mass-action law is written as

$$np = n_i^2. \tag{2.1.4}$$

The interest in the equilibrium condition not only derives from the fact that it serves as a reference state, but also arises because semiconductor devices can often be considered to be operating in a quasi-equilibrium regime.

For MOS compact models, the case of a homogeneous substrate (constant doping concentration) is the most important. Calling n_0 and p_0 the equilibrium electron and hole concentrations, respectively, deep in the bulk of a homogeneous semiconductor, where charge neutrality holds, and choosing the potential reference $\phi = 0$ in the neutral bulk, it follows that

$$p = p_0 e^{-\frac{q\phi}{kT}} = p_0 e^{-u}, \tag{2.1.5}$$

$$n = n_0 e^{\frac{q\phi}{kT}} = n_0 e^{u}, \tag{2.1.6}$$

where $u = \phi/\phi_t$ is the normalized electrostatic potential and $\phi_t = kT/q$ is the thermal voltage, which is approximately 25.9 mV at $T = 300$ K.

The charge density inside the semiconductor results from an imbalance between positive and negative charges. Four types of charges, namely electrons, holes, ionized acceptors (negatively charged) and ionized donors (positively charged), must be considered (Appendix A2.1). Thus, the volume charge density ρ is given by

$$\rho = q(p - n + N_D - N_A), \tag{2.1.7}$$

where N_D and N_A are the ionized donor and acceptor densities, respectively.

Deep in the bulk of a uniformly doped semiconductor, where charge neutrality holds, the carrier concentrations at equilibrium are obtained from charge neutrality,

$$p_0 - n_0 + N_D - N_A = 0, \tag{2.1.8}$$

and the mass-action law $n_0 p_0 = n_i^2$.

Example 2.1

Calculate the built-in potential at room temperature of (a) a silicon p–n junction with doping densities of $N_A = 10^{17}$ atoms/cm^3 and $N_D = 10^{18}$ atoms/cm^3, and (b) a silicon n^+–p junction with $N_A = 10^{17}$ atoms/cm^3. (c) Comment on the effect of a positive (negative) bias applied to the junction.

Answer

(a) In equilibrium, if we choose the potential origin $\phi = 0$ in the region where the semiconductor is intrinsic, i.e., where $p_0 = n_0 = n_i$, we have, from (2.1.2) and (2.1.3),

$$p_0 = n_i e^{-\phi/\phi_t} \qquad \text{and} \qquad n_0 = n_i e^{\phi/\phi_t}.$$

Far from the junction, in the n- and p-regions, we have

$$n_0 \cong N_D = n_i e^{\phi_{\text{n-region}}/\phi_t} \qquad \text{and} \qquad p_0 \cong N_A = n_i e^{-\phi_{\text{p-region}}/\phi_t}.$$

Consequently, the potentials of the neutral (far from the metallurgical junction) n- and p-regions are given by

$$\phi_{\text{n-region}} = \phi_t \ln(N_D/n_i) \qquad \text{and} \qquad \phi_{\text{p-region}} = -\phi_t \ln(N_A/n_i).$$

As expected, the potential in the p-region is lower than that in the n-region because the holes (electrons) are abundant in the region where their energy is at the minimum value. The built-in potential is given by

$$\phi_{bi} = \phi_{\text{n-region}} - \phi_{\text{p-region}} = \phi_t \ln\left(\frac{N_D}{n_i}\right) - \left(-\phi_t \ln\left(\frac{N_A}{n_i}\right)\right) = \phi_t \ln\left(\frac{N_D N_A}{n_i^2}\right).$$

In this example, $\phi_{bi} \cong 26 \ln(10^{15}) \cong 900\,\text{mV}$.

(b) The Boltzmann distribution law is adequate for doping concentrations below $10^{18}\,\text{cm}^{-3}$. For a highly doped n^+-material, with a doping concentration above $10^{19}\,\text{cm}^{-3}$, the potential $\phi_{\text{n-region}}$ is not dependent on the logarithm of the doping N_D but remains approximately constant at a value of 0.56 V for silicon at room temperature. This value corresponds to half of the bandgap voltage of silicon. Thus,

$$\phi_{bi} = 0.56\,\text{V} + \phi_{\text{p-region}} = 0.56\,\text{V} + \phi_t \ln\left(\frac{N_A}{n_i}\right) \cong 979\,\text{mV}.$$

(c) A positive (negative) bias of the junction (i.e. making the p-region positive (negative) with respect to the n-region) decreases (increases) the internal barrier height of the junction.

2.1.2   The two-terminal MOS structure

The MOS structure, which consists of a metal–oxide (SiO_2)–semiconductor (Si) sandwich, made the fabrication of the first practical surface field-effect transistor possible

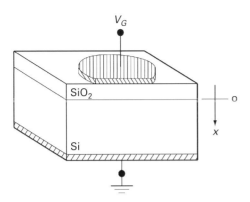

Fig. 2.1   The metal–oxide–semiconductor capacitor.

in 1960, and it remains the core structure of very-large-scale-integration (VLSI) circuits. The MOS structure (Figure 2.1) functions as a two-terminal capacitor in which a conducting (metal) plate is separated from the semiconductor substrate by a thin (oxide) insulator. The ideal MOS structure has the following main properties: (1) the oxide is a perfect insulator with no charge inside it or at its interfaces; (2) the semiconductor is uniformly doped; (3) the semiconductor is sufficiently thick that a field-free region (the "bulk") lies far from the interface; and (4) the potential contact between the metal and semiconductor is zero. The analysis of the ideal MOS structure is of great use because all the idealized properties, except property (4), are approached in many real MOS structures. However, the inclusion of the potential contact value in the model requires a simple voltage shift of the device characteristics, as will be shown in Section 2.1.2.1.

If we substitute for the semiconductor a second metal plate, the relationship between the stored gate charge Q_G and the applied gate-to-substrate voltage V_G is simply

$$V_G = \frac{Q_G}{C_{ox}}, \tag{2.1.9}$$

where the oxide capacitance is given in terms of the capacitor area A, oxide thickness t_{ox}, and permittivity of oxide ε_{ox} by the conventional expression for a parallel-plate capacitor

$$C_{ox} = \frac{A\varepsilon_{ox}}{t_{ox}}. \tag{2.1.10}$$

In MOS theory, charge and capacitance per unit area are the preferred variables. On writing

$$Q'_G = \frac{Q_G}{A}, \qquad C'_{ox} = \frac{C_{ox}}{A} = \frac{\varepsilon_{ox}}{t_{ox}}, \tag{2.1.11}$$

it follows that

$$V_G = \frac{Q'_G}{C'_{ox}}. \tag{2.1.12}$$

From now on, unless stated otherwise, the superscript prime will be used to denote a quantity per unit area.

For the ideal MOS capacitor, taking ϕ_s as the potential at the surface of the semi-conductor, it follows that

$$V_G - \phi_s = \frac{Q'_G}{C'_{ox}} \qquad (2.1.13)$$

since $V_G - \phi_s$ is the voltage drop across the oxide capacitor. The fundamental expression that relates the applied voltage V_G to the surface potential ϕ_s and the semiconductor space charge Q'_C $(= -Q'_G)$ is

$$V_G = \phi_s - \frac{Q'_C}{C'_{ox}}. \qquad (2.1.14)$$

Example 2.2

(a) Calculate the oxide capacitance per unit area C'_{ox} for $t_{ox} = 5$ and 20 nm assuming $\varepsilon_{ox} = 3.9\varepsilon_0$, where $\varepsilon_0 = 8.85 \times 10^{-14}$ F/cm is the permittivity of free space. (b) Determine the area of a 1-pF metal–oxide–metal capacitor for the two oxide thicknesses given in (a).

Answer

(a) $C'_{ox} = 690\,\text{nF/cm}^2 = 6.9\,\text{fF/µm}^2$ for $t_{ox} = 5\,\text{nm}$ and $C'_{ox} = 172\,\text{nF/cm}^2 = 1.7\,\text{fF/µm}^2$ for $t_{ox} = 20\,\text{nm}$. The capacitors have areas of 145 and 580 µm² for oxide thicknesses of 5 and 20 nm, respectively.

2.1.2.1   The flat-band voltage

In equilibrium (with the two terminals shorted/open) the contact potential between the gate and the semiconductor substrate of the MOS induces charges in the gate and the semiconductor for $V_{GB} = 0$. The existence of charges inside the insulator and at the semiconductor–insulator interface also induces a semiconductor charge at zero bias. For high-quality Si–SiO$_2$ interfaces this effect can be almost negligible, but sometimes, for stressed devices, for example, it must be considered.

The effect of the contact potential and oxide charges can be counterbalanced by applying a gate–bulk voltage called the flat-band voltage V_{FB}. Thus, for $V_{GB} = V_{FB}$, the charge induced is zero and the potential is constant inside the semiconductor (this is called the flat-band condition). The inclusion of the flat-band voltage in Equation (2.1.14) yields

$$V_G - V_{FB} = \phi_s - \frac{Q'_C}{C'_{ox}}. \qquad (2.1.15)$$

Since the effects of both the contact potential and (fixed) oxide charges are counter-balanced by a constant voltage (V_{FB}), their inclusion as effects on the MOS characteristic is simply represented by a voltage shift in the equation for the idealized MOS device, as (2.1.15) shows.

Example 2.3

(a) Determine the expressions for the flat-band voltages of an n$^+$ polysilicon gate on p-type silicon and a p$^+$ polysilicon gate on n-type silicon. (b) Calculate the flat-band voltage for an n$^+$ polysilicon gate on a p-type silicon structure with $N_A = 10^{17}$ atoms/cm^3.

Answer

(a) In modern technologies the effect of the charges trapped in the oxide and at the semiconductor–oxide interface can be neglected in a first approximation for the calculation of the flat-band voltage. In equilibrium, by analogy with an n$^+$–p junction, the potential of the n$^+$-region is positive with respect to that of the p-region and is given by the expression developed in Example 2.1. The flat-band condition with no charges induced in the semiconductor is obtained by counteracting the built-in potential, that is, by applying a negative potential to the n$^+$-gate with respect to the p-type semiconductor of value

$$V_{FB\_n^+p} = -\phi_{bi\_n^+p} = -0.56\,\text{V} - \phi_t \ln\left(\frac{N_A}{n_i}\right).$$

For the p$^+$–n structure

$$V_{FB\_p^+n} = -\phi_{bi\_p^+n} = 0.56\,\text{V} + \phi_t \ln\left(\frac{N_D}{n_i}\right).$$

Note the symmetry of the expressions above.

(b) For $N_A = 10^{17}$ atoms/cm^3, it follows that

$$V_{FB} = -0.56\,\text{V} - \phi_t \ln(10^7) = -980\,\text{mV}.$$

2.1.3 Accumulation, depletion, and inversion (for p-type substrates)

Using Equations (2.1.5) through (2.1.8), the expression for the volume charge density inside the semiconductor "plate" of the MOS capacitor in terms of the normalized potential u becomes

$$\rho = q(p_0 e^{-u} - n_0 e^u + n_0 - p_0). \tag{2.1.16}$$

For $V_G < V_{FB}$ the potential inside the semiconductor is negative ($u < 0$) and the charge induced in the semiconductor is positive because the positive hole term in (2.1.16) is above its flat-band value (p_0) and the negative electron term is below its flat-band value (n_0). For a p-type substrate $p_0 \gg n_0$; consequently, the positive charge in the semiconductor is essentially due to the contribution from the holes; the p-type substrate is in the so-called accumulation (of majority carriers) region.

For $V_G > V_{FB}$ the potential inside the semiconductor is positive ($u > 0$) and the charge induced in the semiconductor is negative because the positive hole term in (2.1.16) is below its flat-band value and the negative electron term is above its flat-band value. The relative contribution of electrons and holes depends on the value of u.

The local concentration of holes prevails over the concentration of electrons when

$$p_0 e^{-u} > n_0 e^u \qquad (2.1.17)$$

or, using (2.1.4),

$$\phi < \frac{\phi_t}{2} \ln\left(\frac{p_0}{n_0}\right) = \frac{\phi_t}{2} \ln\left(\frac{p_0^2}{n_i^2}\right) = \phi_t \ln\left(\frac{p_0}{n_i}\right) = \phi_F, \qquad (2.1.18)$$

where ϕ_F is called the Fermi potential (of the uniformly doped p-type substrate). For $\phi > \phi_F$ the concentration of minority carriers (n) becomes higher than that of majority carriers (p); thus, there is an inversion of the type of prevailing carriers and the semi-conductor is said to be in inversion.

2.1.4   The small-signal equivalent circuit of the two-terminal MOS (for p-type substrates)

If Q'_G is the gate charge per unit area, then, from the charge-conservation principle, the total charge variation in the MOS capacitor is zero, i.e. $\Delta Q'_G + \Delta Q'_C = 0$. Thus, the small-signal capacitance per unit area is given by

$$C'_{gb} = \frac{dQ'_G}{dV_G} = -\frac{dQ'_C}{dV_G}. \qquad (2.1.19)$$

Substitution of (2.1.15) into (2.1.19) yields

$$C'_{gb} = -\frac{dQ'_C}{d\phi_s - \dfrac{dQ'_C}{C'_{ox}}} = \frac{1}{-\dfrac{d\phi_s}{dQ'_C} + \dfrac{1}{C'_{ox}}} \qquad (2.1.20)$$

or

$$\frac{1}{C'_{gb}} = \frac{1}{C'_c} + \frac{1}{C'_{ox}}, \qquad (2.1.21)$$

where $C'_c = -dQ'_C/d\phi_s$ is the semiconductor (space-charge) capacitance per unit area. Thus, the capacitance of the MOS structure is equivalent to the series combination of the oxide and the semiconductor capacitances, as shown in Figure 2.2(a). To calculate the semiconductor capacitance, we need the value of the total charge density, which in the (p-type) semiconductor is given by

$$Q'_C = \int_0^\infty \rho \, dx = q\left[\int_0^\infty (p - p_0)dx + \int_0^\infty (n_0 - n)dx\right] = Q'_B + Q'_I, \qquad (2.1.22)$$

where the index B (bulk) is used for the majority-carrier charge (holes, for p-type substrate) and the index I (inversion) for the minority carrier, electrons in this case. The lower ($x = 0$) and

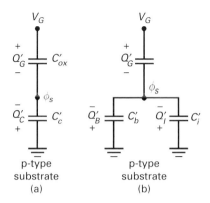

Fig. 2.2   Equivalent circuits for the MOS capacitor. (a) The two-capacitance model. (b) Splitting of the
semiconductor capacitance into electron C'$_i$ and hole C'$_b$ storage capacitances.

upper $(x \rightarrow \infty)$ limits of integration refer to positions at the semiconductor interface with the
oxide and deep in the semiconductor bulk, respectively.

Changing the integration variable from length x to potential ϕ gives

$$Q'_C = Q'_B + Q'_I = q \int_0^{\phi_s} \frac{p - p_0}{F} \, d\phi + \int_0^{\phi_s} \frac{n_0 - n}{F} \, d\phi, \qquad (2.1.23)$$

where $F = -d\phi/dx$ is the electric field. Note that ϕ_s (the surface potential) is the potential
at the semiconductor interface $(x = 0)$, whereas the potential deep in the bulk $(x \rightarrow \infty)$ is
taken arbitrarily as zero. The contributions of the hole and electron capacitances to the
total semiconductor capacitance are

$$C'_b = -\frac{dQ'_B}{d\phi_s} = \frac{q}{F_s}(p_0 - p_s), \qquad C'_i = -\frac{dQ'_I}{d\phi_s} = \frac{q}{F_s}(n_s - n_0), \qquad (2.1.24)$$

respectively, where F_s, p_s, and n_s are the electric field and the hole and electron
concentrations at the semiconductor interface, respectively. The equivalent circuit for
the ideal MOS capacitor, with the semiconductor capacitance decomposed into the bulk
and inversion capacitances, is shown in Figure 2.2(b).

The expression for the bulk charge in accumulation $(\phi_s < 0)$ and depletion $(0 < \phi_s < \phi_t)$
is, as shown in Appendix A2.1, given by

$$-Q'_B = \varepsilon_s F_s = \mathrm{sgn}(\phi_s) \sqrt{2q\varepsilon_s N_A} \sqrt{\phi_s + \phi_t(e^{-\phi_s/\phi_t} - 1)}, \qquad (2.1.25)$$

where ϕ_s and F_s are the electrostatic potential and electric field, respectively, at the
semiconductor interface $(x = 0)$. Now, using the result of (2.1.25) in the definition of
(2.1.24) for the bulk capacitance, we find

$$C'_b = \frac{\sqrt{2q\varepsilon_s N_A}(1 - e^{-u_s})}{2 \, \mathrm{sgn}(u_s) \sqrt{\phi_s + \phi_t(e^{-u_s} - 1)}}. \qquad (2.1.26)$$

In depletion and inversion $(u_s = \phi_s/\phi_t > 3)$ we can neglect the exponential terms and
obtain the classical expression for the (bulk) depletion capacitance,

$$C'_b \cong \frac{\sqrt{2q\varepsilon_s N_A}}{2\sqrt{\phi_s - \phi_t}}. \tag{2.1.27}$$

Equation (2.1.27) is used even in strong inversion for compact modeling. In strong inversion the depletion capacitance decreases more rapidly with the surface potential than if it were calculated using (2.1.27) because of the electron contribution to the surface field, which is usually neglected in compact modeling.

The expression for the (weak) inversion capacitance can be obtained by combining (2.1.24) and (2.1.25). In effect, in weak inversion the electron charge is negligible compared with the depletion charge.

Neglecting in (2.1.24) both the bulk minority-carrier terms (n_0) and, similarly to the case of the depletion capacitance, the exponential term in the denominator, which is relevant only in accumulation, it follows that

$$C'_i = \frac{\sqrt{2q\varepsilon_s N_A} e^{u_s - 2u_F}}{2\sqrt{\phi_s - \phi_t}}, \tag{2.1.28}$$

where $u_F = \phi_F/\phi_t$. Neglecting the variation of the denominator in (2.1.28), the inversion capacitance (and, consequently, the inversion charge density) becomes an exponential function of the surface potential. Thus, we can write the inversion capacitance in terms of the inversion charge simply as

$$C'_i = -\frac{dQ'_I}{\phi_t \, du_s} = -\frac{Q'_I}{\phi_t}. \tag{2.1.29}$$

In strong inversion, it can be shown [2] that

$$C'_i \cong -\frac{Q'_I}{2\phi_t}, \tag{2.1.30}$$

but, since in strong inversion $C'_i \gg C'_{ox}$, the series association in Figure 2.2 is approximately equal to C'_{ox} and the exact value of C'_i is not relevant for computing the inversion charge density. Additionally, approximation (2.1.29) is also needed in order to obtain the expression for the current using the charge-sheet approximation, as will be shown in Section 2.1.8.

2.1.5   The three-terminal MOS structure and the unified charge-control model (UCCM)

The MOS transistor needs contact regions, which must be of a type opposite to that of the substrate, to access the inversion channel. Thus, the modeling of the three-terminal structure or gate-controlled diode of Figure 2.3 is an intermediate stage in the development of transistor models. Since the n^+ (source) region is in electrical contact with the inversion region but junction-isolated from the p bulk, the hole concentration is still given by (2.1.5), repeated below, but for the electrons their concentration is now controlled by $\phi - V_C$ instead of ϕ. Thus,

$$p = p_0 e^{-\frac{q\phi}{kT}} = p_0 e^{-u}; \qquad n = n_0 e^{\frac{q(\phi - V_C)}{kT}} = n_0 e^{u - u_C}.$$

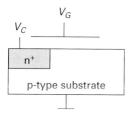

Fig. 2.3   A three-terminal MOS structure.

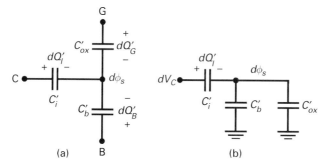

(a)   B   (b)

Fig. 2.4   (a) The capacitive model of the three-terminal MOS structure. (b) The small-signal equivalent
circuit for both V_G and V_B constant.

The equations above can be viewed as a quasi-equilibrium model of the three-terminal
MOS structure.

Holes are in equilibrium in the bulk and, consequently, follow the Boltzmann distribution
law as in the case of the two-terminal structure. The electrons of the inversion channel are in
equilibrium with the electrons of the (source) diffusion and, consequently, follow
Boltzmann statistics with their energy given by $-q(\phi - V_C)$. The product pn is now given by

$$pn = n_i^2 e^{-u_C} = n_i^2 e^{-V_C/\phi_t}. \qquad (2.1.31)$$

Thus, the electrons are no longer in equilibrium with the holes, due to the (usually
reverse) bias of the (source–bulk) junction $V_C = u_C \phi_t$.

As shown in Figure 2.4, the inversion charge is, for small-signal operation, controlled
by $d(\phi_s - V_C)$, while the majority-carrier charge is controlled by $d\phi_s$.

Figure 2.4 represents the capacitive model of the three-terminal MOS device of
Figure 2.3. The small-signal equivalent of Figure 2.4(b) is essential for determining
the dependence of the inversion charge density on the channel voltage V_C in an MOS
transistor. The inversion charge is stored in the series association of the inversion
capacitance with the parallel connection of the oxide and depletion capacitances. Thus,

$$\frac{dQ_I'}{dV_C} = \frac{(C_b' + C_{ox}')C_i'}{C_i' + C_b' + C_{ox}'}. \qquad (2.1.32)$$

To develop an explicit expression for the inversion charge in terms of the applied
voltages, (2.1.32) is rewritten below for convenience as

$$dQ'_I\left(\frac{1}{C'_{ox}+C'_b}+\frac{1}{C'_i}\right)=dV_C.$$ (2.1.33)

To derive a simple model for the relationship between charges and applied voltages, we make the following assumptions.

(i) The depletion capacitance per unit area is constant along the channel and is calculated assuming the inversion charge to be negligible in the potential-balance equation. C'_b is calculated using (2.1.27) for $\phi_s=\phi_{sa}$, with the value of the surface potential ϕ_{sa} determined in terms of the gate-to-substrate potential V_G neglecting the inversion charge in (2.1.15), the potential-balance equation. At this point, we calculate the values of both ϕ_{sa} and C'_b for the depletion/inversion regime only; thus, we assume that the exponential term in (2.1.26), which is important in accumulation, is negligible for depletion/inversion. Summarizing, the bulk capacitance per unit area (excluding accumulation) is given by

$$C'_b\cong\frac{\sqrt{2q\varepsilon_s N_A}}{2\sqrt{\phi_{sa}-\phi_t}}=\frac{\gamma C'_{ox}/2}{\sqrt{V_G-V_{FB}-\phi_t+\gamma^2/4}-\gamma/2}=(n-1)C'_{ox},$$ (2.1.34)

where $\gamma=\sqrt{2q\varepsilon_s N_A}/C'_{ox}$ is the body-effect coefficient, which is dependent on technology through substrate doping and the oxide capacitance. Typical values of γ and n, the slope factor, lie in the ranges 0.1–1 $V^{1/2}$ and 1.1–1.5, respectively.

(ii) The inversion capacitance is approximated by $C'_i=-Q'_I/\phi_t$ for the inversion regime, including strong inversion.

Using the hypotheses (i) and (ii) above allows us to obtain, from (2.1.33), the following relationship:

$$dQ'_I\left(\frac{1}{nC'_{ox}}-\frac{\phi_t}{Q'_I}\right)=dV_C,$$ (2.1.35)

where

$$n=1+\frac{C'_b}{C'_{ox}}.$$ (2.1.36)

$n=n(V_G)$,[1] which can be calculated from (2.1.34), is called the slope factor because, as will be seen in Section 2.2.3, the logarithmic slope of the current versus normalized gate voltage curve in weak inversion is $1/n$.

Integrating (2.1.35) from an arbitrary channel potential V_C to a reference potential V_P yields the unified charge-control model (UCCM) as

$$V_P-V_C=\phi_t\left[\frac{Q'_{IP}-Q'_I}{nC'_{ox}\phi_t}+\ln\left(\frac{Q'_I}{Q'_{IP}}\right)\right],$$ (2.1.37)

where Q'_{IP} is the value of Q'_I for $V=V_P$.

---

[1] We use the symbol n to represent both the electron concentration and the slope factor, following the customary notation.

Example 2.4

For $t_{ox} = 5$ and $20\,\text{nm}$ determine the minimum doping N_A for which the slope factor $n < 1.25$ for $\phi_{sa} = 2\phi_F$. Comment on the results.

Answer

For $\phi_{sa} = 2\phi_F$ we calculate the slope factor n by combining (2.1.36) and (2.1.34), yielding

$$n = 1 + \frac{C'_b}{C'_{ox}} \cong 1 + \frac{\sqrt{2q\varepsilon_s N_A}}{2C'_{ox}\sqrt{\phi_{sa}}} = 1 + \frac{\sqrt{2q\varepsilon_s N_A}}{2C'_{ox}\sqrt{2\phi_F}}.$$

Thus, for $n = 1.25$ the substrate doping is given by

$$N_A = \frac{(0.25)^2 \cdot 2\phi_F \cdot 4C'^2_{ox}}{2q\varepsilon_s},$$

where ϕ_F is a weak (logarithmic) function of N_A. Using $2\phi_F = 0.8\,\text{V}$ for the first calculation we find, after two iterations, that $N_A > 4.9 \times 10^{15}$ atoms/cm^3 for $t_{ox} = 5\,\text{nm}$ and $N_A > 2.3 \times 10^{14}$ atoms/cm^3 for $t_{ox} = 20\,\text{nm}$. The ratio of the former doping density to the latter is approximately 21. Neglecting the variation of the Fermi level with the doping density, this ratio is $16 = (20\,\text{nm}/5\,\text{nm})^2$. Thus, to maintain the slope factor constant from one (*old*) to another (*new*) technology node with a thinner oxide, it is necessary to increase the substrate doping by a factor of the order of $(t_{ox\_old}/t_{ox\_new})^2$.

Example 2.5

Consider the UCCM given by (2.1.37). "Regional" models for weak inversion (WI) or strong inversion (SI) can be derived from the UCCM by dropping either the linear term or the logarithmic term in the charge density, respectively. (a) Determine the expression for the charge density in terms of both the gate voltage and the channel voltage for (a1) the SI approximation and (a2) the WI approximation. (b) Calculate the value of the inversion charge density, normalized with respect to $Q'_{IP} = -nC'_{ox}\phi_t$, for which the value of the voltage $V_P - V_C$ calculated using the SI approximation differs from that calculated using the UCCM by 10%. (c) Determine the value of the inversion charge density, normalized with respect to Q'_{IP}, for which the value of the voltage $V_P - V_C$ calculated using the WI approximation differs from that calculated using the UCCM by 10%. (d) Comment on the results.

Answer

(a1) For $V_P - V_C \gg \phi_t$, (2.1.37) reduces to the SI approximation $-Q'_I \cong nC'_{ox}(V_P - V_C)$. On normalizing the inversion charge density with respect to $Q'_{IP} = -nC'_{ox}\phi_t$ we obtain $q'_I \cong (V_P - V_C)/\phi_t$.

(a2) For $V_P - V_C \ll -\phi_t$, (2.1.37) reduces to the WI approximation

$$V_P - V_C = \phi_t \left[\ln \left(\frac{Q'_I}{Q'_{IP}} \right) - 1 \right]$$

or, equivalently,

$$q'_I = e^{\frac{V_P - V_C + \phi_t}{\phi_t}}.$$

(b) The SI approximation has an error of less than 10% for $q'_I > 20$.

(c) The WI approximation has an error of less than 10% for $q'_I < 0.22$.

(d) Let us define moderate inversion (MI) as the region in which the SI and WI approximations give errors greater than 10% for the control voltage $V_P - V_C$. Considering (b) and (c), the inversion charge-density variation from the lower to the upper limit of the MI region is approximately two orders of magnitude (20/0.22), and, for this reason, the MI region cannot be approximated by either the WI or the SI equation in an accurate MOSFET model.

2.1.6   The pinch-off voltage

The channel charge density corresponding to the effective channel capacitance times the thermal voltage, or thermal charge [4], will be used as a reference to define pinch-off and as the normalization charge

$$Q'_{IP} = -(C'_{ox} + C'_b)\phi_t = -nC'_{ox}\phi_t. \tag{2.1.38}$$

The name pinch-off is retained herein for historical reasons and means the channel potential corresponding to a small (but well-defined) amount of carriers in the channel. The channel-to-substrate voltage (V_C) for which the channel charge density equals Q'_{IP} is called the pinch-off voltage V_P. Since we know the explicit expression for the inversion charge in weak inversion, we will use it to determine the pinch-off voltage. From (2.1.28) we write the following expression for the weak-inversion capacitance due to the inclusion of the channel contact

$$C'_i = \frac{\sqrt{2q\varepsilon_s N_A} e^{u_{sa} - 2u_F - u_C}}{2\sqrt{\phi_{sa} - \phi_t}}. \tag{2.1.39}$$

As explained previously, the non-zero channel potential V_C means that the electron charge density is controlled by $\phi - V_C$ instead of only ϕ. This leads us to include the (normalized) channel potential in the exponential term in (2.1.39). We have also replaced ϕ_s (u_s) from (2.1.28) with ϕ_{sa} (u_{sa}), since in weak inversion the surface potential is dependent only on V_G, not on V_C. Now, by equating (2.1.29) with (2.1.39), we find that

$$-Q'_I = \frac{\sqrt{2q\varepsilon_s N_A}}{2\sqrt{\phi_s - \phi_t}} \phi_t e^{(\phi_{sa} - 2\phi_F)/\phi_t} e^{-V_C/\phi_t}$$

$$= C'_{ox}(n-1)\phi_t e^{(\phi_{sa} - 2\phi_F)/\phi_t} e^{-V_C/\phi_t}. \tag{2.1.40}$$

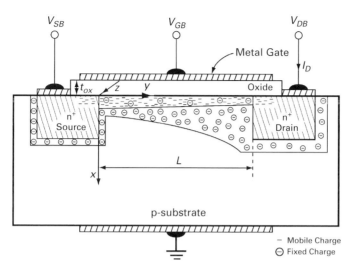

Fig. 2.5   A cross section of a MOSFET (adapted from [1]).

In weak inversion, due to the prevalence of the logarithmic term, we can rewrite (2.1.37) as

$$Q_I' = Q_{IP}' e^{(V_P - V_C + \phi_t)/\phi_t} = -nC_{ox}' \phi_t e^{(V_P - V_C + \phi_t)/\phi_t}. \tag{2.1.41}$$

Since we must have the same value of the inversion charge using (2.1.40) or (2.1.41) it follows that V_P must be given by

$$V_P = \phi_{sa} - 2\phi_F - \phi_t \left[1 + \ln\left(\frac{n}{n-1}\right)\right]. \tag{2.1.42}$$

2.1.7   The Pao–Sah exact *I–V* model

As shown in Figure 2.5, the MOSFET is inherently a two-dimensional structure. The (gate) input voltage is applied in the *x*-direction perpendicular to the semiconductor surface in order to modulate the inversion charge which flows near the surface in the *y*-direction when a voltage is applied between source and drain.

The first step in creating a compact model [5] is the decomposition of the two-dimensional problem into two one-dimensional problems. For a long-channel device, the *gradual channel approximation* is valid, i.e. the longitudinal (*y*-direction) component of the electric field can be assumed to be much smaller than the transversal (*x*-direction) component. Thus, the three-capacitor equivalent circuit of Figure 2.4 is valid for each cross section of the channel.

The other assumptions for the calculation of the current I_D that flows from drain to source are as follows.

(i) The hole current (for the n-channel) is negligible. This assumption is quite acceptable since under normal operation conditions the source–substrate and drain–substrate junctions are reverse- or zero-biased.

(ii) The current is laminar, that is, it flows in the *y*-direction only.

The electron current density considering drift and diffusion is given by [3]

$$J_n = qn\mu_n\left(-\frac{d\phi}{dy}\right) + qD_n\frac{dn}{dy},$$ (2.1.43)

where μ_n is the mobility and D_n the diffusion coefficient for electrons.

The derivative of the electron concentration can be calculated in terms of the potential derivatives using

$$n = n_0 e^{\frac{q(\phi-V_C)}{kT}} = n_0 e^{u-u_C}.$$ (2.1.44)

The channel potential V_C along the channel is such that

$$V_S \leq V_C \leq V_D,$$ (2.1.45)

where $V_{S(D)}$ is the source (drain) voltage. From (2.1.44) it follows that

$$\frac{dn}{dy} = \frac{n}{\phi_t}\left(\frac{d\phi}{dy} - \frac{dV_C}{dy}\right).$$ (2.1.46)

Using the Einstein relationship ($D_n = \mu_n\phi_t$) [3] between the electron mobility μ_n and the diffusion coefficient D_n, and substituting (2.1.46) into (2.1.43), yields

$$J_n = -qn\mu_n\frac{d\phi}{dy} + qn\mu_n\left(\frac{d\phi}{dy} - \frac{dV_C}{dy}\right) = -qn\mu_n\frac{dV_C}{dy}.$$ (2.1.47)

The drain current is obtained by integrating the total current density over the cross-sectional area of the channel,

$$I_D = -\int_0^W\int_0^{x_i} J_n\,dx\,dz = -W\int_0^{x_i} J_n\,dx,$$ (2.1.48)

where W is the transistor width and x_i is taken as the x-coordinate at which the electron concentration equals the intrinsic concentration n_i (the bottom end of the inversion channel).

On substituting (2.1.47) into (2.1.48) and assuming the mobility μ_n to be independent of bias and position we obtain

$$I_D = qW\int_0^{x_i} n\mu_n\frac{dV_C}{dy}\,dx = -W\mu_n Q_I'\frac{dV_C}{dy}$$ (2.1.49)

because V_C (and dV_C/dy) is independent of x and, by definition,

$$Q_I' = -q\int_0^{x_i} n\,dx.$$ (2.1.50)

Since the current is constant along the channel, the integration of (2.1.49) along the channel, from source to drain, yields

$$I_D = -\frac{\mu_n W}{L}\int_{V_S}^{V_D} Q_I'\,dV_C,$$ (2.1.51)

where L is the channel length. Expression (2.1.51) is very general and includes both drift and diffusion mechanisms, thus giving an exact model of the long-channel MOSFET.

Because there is no general analytically integrable expression for the inversion charge density in terms of the channel potential, (2.1.51) cannot be directly used to obtain compact models valid in all operating regions. A supplementary approximation, assuming that the channel can be modeled by a charge sheet, is used in most of the compact models.

2.1.8 A charge-sheet formula for the current

From Figure 2.4, it is clear that

$$dQ'_I = C'_i(dV_C - d\phi_s). \tag{2.1.52}$$

On substituting (2.1.29) into (2.1.52) we have

$$dV_C = d\phi_s - \phi_t \frac{dQ'_I}{Q'_I}, \tag{2.1.53}$$

which, substituted into expression (2.1.49), gives the charge-sheet expression for the current [6] as

$$I_D = I_{drift} + I_{diff} = -\mu_n W Q'_I \frac{d\phi_s}{dy} + \mu_n W \phi_t \frac{dQ'_I}{dy}. \tag{2.1.54}$$

The expression above, which gives explicitly the drift and diffusion terms for carriers confined in a sheet, has been accepted widely and is often used as the starting point from which to derive compact models. However, it is not equivalent to the general Pao–Sah current expression, since a supplementary simplification, namely the proportionality between inversion capacitance and inversion charge, is needed in order to derive it from the Pao–Sah formula.

2.1.9 A charge-control compact model

From the capacitive equivalent circuit of the three-terminal MOS in Figure 2.4(b) it follows that, for constant V_G,

$$dQ'_I = (C'_{ox} + C'_b)d\phi_s = nC'_{ox} d\phi_s. \tag{2.1.55}$$

C'_b is the depletion capacitance calculated assuming the inversion charge to be negligible (see (2.1.26)) and n is the slope (or linearization) factor, which depends only on the gate voltage. The linear relation between inversion charge density and surface potential along the channel is at the base of the current compact model because the substitution of (2.1.55) into the charge-sheet current expression, (2.1.54), allows the current to be written as a function of the inversion charge density as

$$I_D = -\frac{\mu_n W}{nC'_{ox}}(Q'_I - \phi_t nC'_{ox})\frac{dQ'_I}{dy}. \tag{2.1.56}$$

From (2.1.56), one can readily conclude that the diffusion and drift components are equal when the local inversion charge density is equal to the pinch-off charge $-nC'_{ox}\phi_t$. In other words, we have chosen the pinch-off voltage as the channel voltage for which the drift and diffusion components of the current are equal. Above threshold (pinch-off), drift is the prevailing conduction mechanism, whereas diffusion dominates below threshold [4].

Finally, integrating (2.1.56) along the channel [4] yields

$$I_D = \frac{\mu_n W}{L}\left[\frac{Q_{IS}'^2 - Q_{ID}'^2}{2nC'_{ox}} - \phi_t(Q'_{IS} - Q'_{ID})\right]. \tag{2.1.57}$$

In (2.1.57) the quadratic term corresponds to the drift current and the linear term to the diffusion current. The slope factor n depends on ϕ_{sa} according to

$$n = 1 + \gamma\left/\left(2\sqrt{\phi_{sa} - \phi_t}\right)\right.; \tag{2.1.58}$$

ϕ_{sa}, which is the surface potential calculated from (2.1.15) assuming the inversion charge to be negligible, is given by

$$\sqrt{\phi_{sa} - \phi_t} = \sqrt{V_G - V_{FB} - \phi_t + \gamma^2/4} - \gamma/2. \tag{2.1.59}$$

The inversion charge densities Q'_{IS} and Q'_{ID} at the source and drain are calculated in terms of the applied voltages using the UCCM given in (2.1.37).

Equation (2.1.57) can be written as

$$I_D = I_F - I_R, \tag{2.1.60}$$

where

$$I_{F(R)} = \frac{W}{L}\mu_n\left(\frac{Q_{IS(D)}'^2}{2nC'_{ox}} - \phi_t Q'_{IS(D)}\right) \tag{2.1.61}$$

is called the forward (reverse) component of the drain current. Equations (2.1.60) and (2.1.61) emphasize the symmetry of the rectangular-geometry MOSFET. As V_{DB} becomes large (forward saturation), Q'_{ID} approaches zero and so does I_R; consequently, the drain current approaches a constant value called the forward saturation current I_F. Similarly, when V_{SB} becomes large (reverse saturation), Q'_{IS} approaches zero and so does I_F; thus, the drain current approaches a constant value equal to the reverse saturation current $-I_R$. The decomposition of the drain current into forward and reverse currents [7], [8] was inspired by the Ebers–Moll equations of the bipolar transistor [9].

2.1.10   Threshold voltage

Since V_T-based models are useful for gaining an insight into the operation of the transistor and for back-of-an-envelope calculations, it is interesting to link the previous general calculation of the inversion charge density to the classical strong- and weak-inversion approximations.

We start by giving a rigorous (and practical) definition for the threshold voltage, which is clearly impossible when the transistor model is restricted to the strong-inversion approximation. In effect, in the strong-inversion approximation the threshold corresponds to zero current, a state that does not occur in the real world.

We define the equilibrium threshold voltage V_{T0}, measured for $V_C=0$, as the gate voltage for which the channel charge density equals Q'_{IP} or, in other words, as the gate voltage for which the pinch-off voltage $V_P=0$. Using (2.1.42) we find that

$$\phi_{sa}|_{V_P=0, V_G=V_{T0}} = 2\phi_F + \phi_t \left[1 + \ln\left(\frac{n}{n-1}\right)\right]. \tag{2.1.62}$$

Now, using (2.1.59) with $V_G = V_{T0}$, we find that

$$V_{T0} = V_{FB} + 2\phi_F + \phi_t \left[1 + \ln\left(\frac{n}{n-1}\right)\right] + \gamma\sqrt{2\phi_F + \phi_t \ln\left(\frac{n}{n-1}\right)}$$

$$\cong V_{FB} + 2\phi_F + \gamma\sqrt{2\phi_F}. \tag{2.1.63}$$

The approximation in (2.1.63) is the classical definition of the threshold voltage.

For hand design, it is useful to have a simplified expression for the pinch-off voltage. The linear approximation of V_P around $V_G = V_{T0}$ can be written [7] as

$$V_P \cong \frac{V_G - V_{T0}}{n} \tag{2.1.64}$$

because, using (2.1.42), the slope dV_P/dV_G is given by

$$\frac{dV_P}{dV_G} \cong \frac{d\phi_{sa}}{dV_G} = \frac{C'_{ox}}{C'_b + C'_{ox}} = \frac{1}{n} \tag{2.1.65}$$

with n given by (2.1.58). For hand calculations of the drain current, n can be assumed constant for several decades of current.

Figure 2.6 shows a plot of the pinch-off voltage against the gate voltage. The pinch-off voltage varies almost linearly with the gate voltage in inversion. Substituting the first-order approximation of V_P of (2.1.64) into (2.1.37) gives [9]

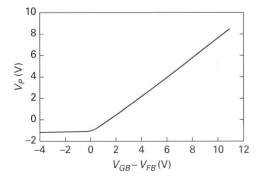

Fig. 2.6 The pinch-off voltage as a function of the gate-to-substrate voltage V_{GB}.

$$Q'_{IP} - Q'_I + nC'_{ox}\phi_t \ln\left(\frac{Q'_I}{Q'_{IP}}\right) = C'_{ox}(V_G - V_{T0} - nV_C). \tag{2.1.66}$$

Equation (2.1.66) is useful for hand analysis, but the exponential dependence of Q'_I on V_P in weak inversion precludes (2.1.66) or other expressions based on approximations of the pinch-off voltage from being used for accurate modeling.

Example 2.6

(a) Estimate V_{T0} for an n-channel transistor with an n^+ polysilicon gate, $N_A = 10^{17}$ atoms/cm^3 and $t_{ox} = 5$ nm. (b) Determine the effect of temperature on V_{T0}.

Answer

(a) The flat-band voltage, calculated in Example 2.3, is -0.98 V. Using (2.1.18) with $p_0 = N_A$, $n_i = 10^{10}$ cm^{-3}, and $\phi_t = 26$ mV, it follows that $\phi_F = 0.419$ V. C'_{ox}, calculated in Example 2.2, is 690 nF/cm^2. Therefore, $\gamma = \sqrt{2q\varepsilon_s N_A}/C'_{ox} = 0.264\sqrt{V}$.

Using the approximate expression for the threshold voltage in (2.1.63), we obtain

$$V_{T0} \cong -0.98 + 0.838 + 0.264\sqrt{0.838} = 0.1 \text{ V.}$$

For this low value of the threshold voltage, the off-current (for $V_{GS} = 0$) is too high for digital circuits. The standard solution to control the magnitude of the threshold voltage without an exaggerated increase in the substrate doping (and of the slope factor) has been the use of a non-uniform high-low channel doping. A shallow, p-type implantation for an n-channel is carried out to increase the magnitude of the depletion charge (and of the threshold voltage), but without significantly increasing the depletion capacitance (and the slope factor n).

(b) Using (2.1.18), we can rewrite the flat-band voltage calculated in Example 2.3 as $V_{FB} = \text{constant} - \phi_F$ and the threshold voltage, using the approximated expression in (2.1.63), as $V_{T0} = \text{constant} + \phi_F + \gamma\sqrt{2\phi_F}$, where the constant is a factor independent of temperature and $\phi_F = \phi_t \ln(N_A/n_i)$. The intrinsic concentration n_i can be expressed approximately [3] as

$$n_i^2 = DT^3 \exp\left(-\frac{E_G}{kT}\right),$$

where D is independent of temperature and E_G is the silicon bandgap extrapolated to 0 K, which is around 1.2 eV. Neglecting the variation of the pre-exponential factor in the commercial temperature range, we will consider that $n_i^2 = C^2 \exp[-E_G/(kT)]$, with C being independent of temperature. Using the expression above for n_i^2, we rewrite the Fermi potential as $\phi_F = \phi_t \ln(N_A/C) + E_G/(2q)$ and, thus, its temperature coefficient is

$$\frac{d\phi_F}{dT} = -\frac{1}{T}\left(\frac{E_G}{2q} - \phi_F\right).$$

Finally,

$$\frac{dV_{T0}}{dT} = \frac{dV_{T0}}{d\phi_F}\frac{d\phi_F}{dT} = -\frac{1}{T}\left(1 + \frac{\gamma}{\sqrt{2\phi_F}}\right)\left(\frac{E_G}{2q} - \phi_F\right).$$

In this example, $dV_{T0}/dT = -0.7$ mV/°C. For the sake of simplicity, the dependence of the silicon bandgap E_G on temperature was assumed to contribute negligibly to the threshold-voltage temperature coefficient.

2.2 A design-oriented MOSFET model

In this section we present an MOS transistor model suitable for integrated-circuit design [10]. The small-signal parameters of the MOSFET are described by single-piece functions of the forward (I_F) and reverse (I_R) components of the drain current, and of the slope factor n. For hand analysis of a MOSFET operating from weak to strong inversion, n can be considered constant and the state of the transistor is modeled in a first approximation by two variables only, namely I_F and I_R. For a transistor in forward (reverse) saturation the transistor state is modeled by I_F (I_R) only. The strong- and weak-inversion regions are analyzed as asymptotic cases of the single-piece function that describes the overall device behavior. Sometimes we refer to this design-oriented model as the current-based model.

2.2.1 Forward and reverse components of the drain current

For the sake of convenience, we rewrite here expressions (2.1.60) and (2.1.61) for the drain current as

$$I_D = I_F - I_R = I(V_G, V_S) - I(V_G, V_D) \qquad (2.2.1)$$

with

$$I_{F(R)} = \mu_n C'_{ox} n \frac{W}{L}\frac{\phi_t^2}{2}\left[\left(\frac{Q'_{IS(D)}}{nC'_{ox}\phi_t}\right)^2 - 2\frac{Q'_{IS(D)}}{nC'_{ox}\phi_t}\right]. \qquad (2.2.2)$$

Note that, for a long-channel device, the forward (reverse) current is dependent on both the gate voltage and the source (drain) voltage, being independent of the drain (source) voltage. Equations (2.2.1) and (2.2.2) emphasize the source–drain symmetry of the MOSFET.

Now let us explain how to determine the forward and reverse components of the drain current from the transistor output characteristics, for the example shown in Figure 2.7 for a long-channel MOSFET. There is a region, usually called the saturation region, where the drain current is almost independent of V_D. This means that, in saturation, $I(V_G, V_D) \ll I(V_G, V_S)$. Therefore, $I(V_G, V_S)$ can be interpreted as the drain current in forward saturation. Similarly, in reverse saturation, I_D is independent of the source voltage.

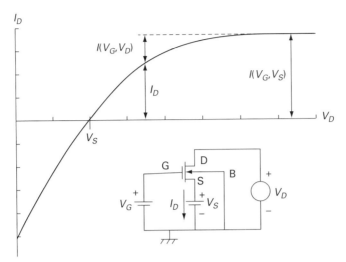

Fig. 2.7   Output characteristics of a long-channel NMOS transistor for constant V_S and V_G. Voltages are referenced to the substrate.

Expression (2.2.2) can be rewritten in the form

$$q'_{IS(D)} = -\frac{Q'_{IS(D)}}{nC'_{ox}\phi_t} = \sqrt{1 + i_{f(r)}} - 1, \tag{2.2.3}$$

where $q'_{IS(D)}$ is the normalized inversion charge density at the source (drain) and

$$i_{f(r)} = \frac{I_{F(R)}}{I_S} = \frac{I(V_G, V_{S(D)})}{I_S} \tag{2.2.4}$$

is the forward (reverse) normalized current or inversion coefficient, at the source (drain), where

$$I_S = \mu_n C'_{ox} n \frac{\phi_t^2}{2} \frac{W}{L} \tag{2.2.5}$$

is the normalization current, which is four times smaller than its homonym presented in [7]. The factor $I_{SH} = \mu_n C'_{ox} n \phi_t^2/2$, herein denominated the sheet normalization current, is a technological parameter that is slightly dependent on V_G, through μ_n and n. As a rule of thumb, values of i_f greater than 100 characterize strong inversion, and the transistor operates in weak inversion up to $i_f = 1$. Intermediate values of i_f, from 1 to 100, indicate moderate inversion.

Figure 2.8 shows the normalization current of a long-channel MOS transistor versus the gate voltage. The variation of the normalization current around its average value is about ±30% for a gate voltage ranging from 0.6 to 5 V.

Using the normalized form of the UCCM,

$$V_P - V_{S(D)} = \phi_t\left(q'_{IS(D)} - 1 + \ln q'_{IS(D)}\right), \tag{2.2.6}$$

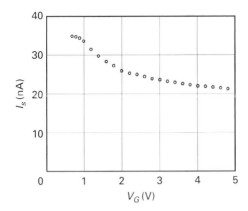

Fig. 2.8 The normalization current of an NMOS transistor ($t_{ox}=280$ Å, $W=L=25$ µm) versus the gate voltage.

and (2.2.3), we find the following relationship between normalized currents and voltages:

$$V_P - V_{S(D)} = \phi_t\left[\sqrt{1+i_{f(r)}} - 2 + \ln\left(\sqrt{1+i_{f(r)}} - 1\right)\right]. \qquad (2.2.7)$$

Expression (2.2.7) is a universal relationship for long-channel MOSFETs, which is valid for any technology, gate voltage, dimensions, and temperature. Owing to its similarity to the UCCM equation and to the change of variable from charge to current, we will simply call it the unified current-control model, or the UICM for short.

Example 2.7

(a) Calculate the charge density and the sheet resistance ($1/(\mu n C'_{ox}\phi_t)$) at pinch-off for an n-channel MOSFET in a 0.35-µm technology for which $t_{ox}=7$ nm, $n=1.2$, and $\mu_n=400$ cm^2/V per s at 300 K. Assuming that $W_{min}=2L_{min}$, determine the minimum nominal dimensions of a 1-MΩ resistor operating at zero dc current implemented with an n-channel MOSFET at (a1) pinch-off and (a2) $i_f=99$. (b) Determine the minimum nominal dimensions of a current source of 1 µA implemented with a saturated n-channel MOSFET operating at (b1) $i_f=1$ and (b2) $i_f=100$.

Answer

(a) The pinch-off charge is $nC'_{ox}\phi_t=1.2\times(493$ nF/cm$^2)\times0.026$ V $=15$ nC/cm^2. The sheet resistance $1/(\mu n C'_{ox}\phi_t)=162$ kΩ at pinch-off.

    (a1) The equivalent resistance of a MOSFET in the linear region ($V_{DS}=0$) is $(L/W)/(\mu n C'_{ox}\phi_t)$. A 1-M$\Omega$ resistor requires $L/W=6.17$. Since $W=W_{min}=0.7$ µm, it follows that $L=4.3$ µm.

    (a2) For $i_f=99$, $q'_{IS}=\sqrt{1+i_f}-1=\sqrt{100}-1=9$. The equivalent resistance of the MOSFET is $(L/W)/(\mu q'_I C'_{ox}\phi_t)=(L/W)\cdot18$ kΩ. Thus, $L/W=55.5$, with $W=0.7$ µm and $L=38.8$ µm.

(b) The forward saturation current of a MOSFET is $I_F = (W/L)I_{SH}i_f$. In this example, the sheet normalization current is $I_{SH} = \mu_n C'_{ox} n\phi_t^2/2 = 400 \times 493 \times 1.2 \times 10^{-9} \times 0.026^2/2 = 80$ nA. (b1) At $i_f = 1$, a current of 1 µA is obtained with $W/L = 12.5$. The minimum area for the current source is obtained with $L = 0.35$ µm and $W = 4.4$ µm. (b2) At $i_f = 100$, a current of 1 µA is obtained with $W/L = 0.125$. The minimum area for the current source is obtained with $W = 0.7$ µm and $L = 5.6$ µm.

Example 2.8

Write the UICM for the p-channel device.

Answer

Conventionally, for a p-channel device, the current flows from source to drain. Therefore, the drain current is written as

$$I_D = I_F - I_R,$$

where $I_{F(R)}$ is the forward (reverse) current, which flows from source (drain) to drain (source). For the p-channel device, the inversion charge densities at source and drain (or normalized forward and reverse currents) increase for increasing values of the channel voltage and decreasing values of the gate voltage. Thus, the UICM is written as

$$V_{S(D)} - V_P = \phi_t \left[\sqrt{1 + i_{f(r)}} - 2 + \ln \left(\sqrt{1 + i_{f(r)}} - 1 \right) \right],$$

where all the potentials are referred to the substrate of the p-channel transistor, which is usually biased at the highest voltage applied to the circuit. Thus, V_S and V_D are negative (or zero) during the normal operation of a p-channel MOSFET. As expected, strong-inversion operation in p-MOSFETs occurs for $V_S - V_P \cong V_S - (V_G - V_{T0})/n \gg \phi_t$. It should be noted that, for an enhancement-mode p-MOSFET, the threshold voltage V_{T0} is negative. For $V_{SB} = 0$, the source is in the threshold condition when the gate-to-substrate voltage is equal to V_{T0}. Strong inversion in p-channel transistors occurs for a sufficiently negative value of the voltage V_G.

2.2.2   Universal dc characteristics

Figure 2.9 shows the common-gate characteristics of a MOSFET, which are plots of the drain current in saturation versus V_S at constant V_G. The semi-log plot exhibits a low-current region characterized by a straight line associated with the exponential dependence of the current on the source voltage. For higher currents the strong-inversion current follows approximately a parabolic behavior. The curve knee represents the moderate-inversion region.

Figure 2.10 shows the common-source characteristics of the MOSFET employed to obtain the data in Figure 2.9. As in Figure 2.9, one identifies the weak-inversion region, characterized by the exponential dependence of the drain current on the gate voltage, and

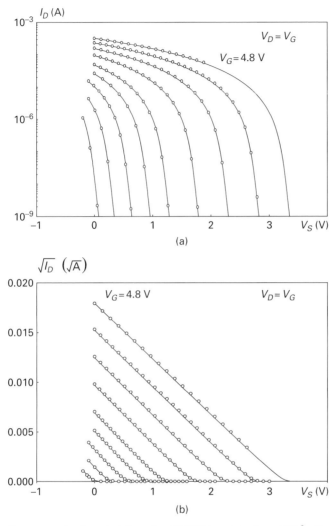

Fig. 2.9 Common-gate characteristics of an NMOS transistor ($t_{ox} = 280$ Å, $W = L = 25$ μm) in saturation ($V_G = 0.8$, 1.2, 1.6, 2.0, 2.4, 3.0, 3.6, 4.2, and 4.8 V): (—) simulated and (○) measured data. Plots of (a) log I_D versus source voltage and (b) the square root of I_D versus source voltage.

the curve knee, which is associated with moderate inversion. The application of (2.2.6) to the source and drain terminals allows us to eliminate V_P and write a universal relationship for the MOSFET output characteristics as

$$\frac{V_{DS}}{\phi_t} = q'_{IS} - q'_{ID} + \ln\left(\frac{q'_{IS}}{q'_{ID}}\right) \tag{2.2.8}$$

or

$$\frac{V_{DS}}{\phi_t} = \sqrt{1 + i_f} - \sqrt{1 + i_r} + \ln\left(\frac{\sqrt{1 + i_f} - 1}{\sqrt{1 + i_r} - 1}\right). \tag{2.2.9}$$

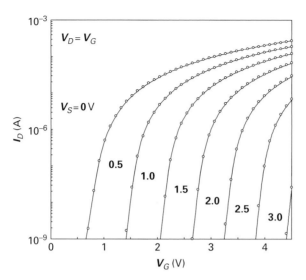

Fig. 2.10   Common-source characteristics of an NMOS transistor ($t_{ox}=280$ Å, $W=L=25$ μm) in saturation ($V_S=0$, 0.5, 1.0, 1.5, 2.0, 2.5, and 3.0 V): (—) simulated and (○) measured data.

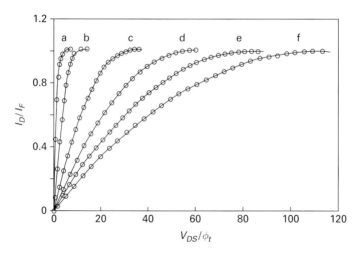

Fig. 2.11   Normalized output characteristics of an NMOS transistor ($t_{ox}=280$ Å, $W=L=25$ μm). I_F has been measured for $V_D=V_G$ and $V_S=0$: (○) measured data and (—) calculated from [10]. (a) $i_f=4.5\times10^{-2}$ ($V_G=0.7$ V). (b) $i_f=65$ ($V_G=1.2$ V). (c) $i_f=9.5\times10^2$ ($V_G=2.0$ V). (d) $i_f=3.1\times10^3$ ($V_G=2.8$ V). (e) $i_f=6.8\times10^3$ ($V_G=3.6$ V). (f) $i_f=1.2\times10^4$ ($V_G=4.4$ V).

Expression (2.2.9) demonstrates that the normalized output characteristics of a long-channel MOSFET are independent of technology and transistor dimensions, corroborating again the universality and consistency of the transistor model that has been derived. In Figure 2.11 we compare the measured output characteristics, for several gate voltages, and the curves obtained by using (2.2.9). Once again, the measured results agree quite well with the theoretical model.

Now, we define V_{DSsat}, which is the value of V_{DS} at which the ratio $q'_{ID}/q'_{IS} = \xi$, where ξ is an arbitrary number much smaller than unity. As we shall see, this definition of saturation is extremely practical for long-channel MOSFETs, since the ratio of the inversion charge density at the drain to that at the source coincides with the ratio of the transistor conductance at an arbitrary V_{DS} to that determined for $V_{DS}=0$. Note that $1-\xi$ represents the saturation level of the MOSFET. If $\xi \to 1$, the transistor is operating in the linear region, or close to $V_{DS}=0$. If $\xi \to 0$, the current tends to saturate or, in other words, the saturation level tends to be maximum. From (2.2.8) and $q'_{ID}/q'_{IS} = \xi$, we have

$$V_{DSsat} = \phi_t \left[\ln\left(\frac{1}{\xi}\right) + (1-\xi)\left(\sqrt{1+i_f} - 1\right) \right]. \qquad (2.2.10)$$

The definition in (2.2.10) is extremely useful for circuit design since it gives the boundary between the triode and saturation regions in terms of the inversion level. In Figure 2.12 we present the theoretical drain-to-source saturation voltage for two values of ξ. Note that, in weak inversion, V_{DSsat} (of the order of $4.5\phi_t$ for $\xi=0.01$) is independent of the inversion level, whereas in strong inversion it is proportional to the square root of the inversion level. Our definition of saturation is arbitrary but gives designers a very good first-order approximation of the minimum V_{DS} required to keep the MOSFET in the "constant-current region."

2.2.3 MOSFET operation in weak and strong inversion

The different modes of operation of the MOS transistor can be defined according to the difference between the values of the source or drain voltage and the pinch-off voltage or, equivalently, according to the ratio of the source or drain inversion charge density to the pinch-off inversion charge density.

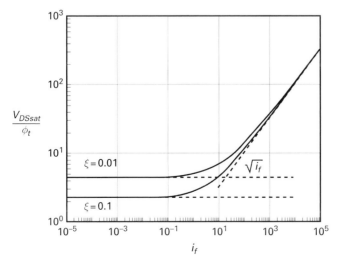

Fig. 2.12 Drain-to-source saturation voltage versus inversion coefficient, calculated for $\xi=0.1$ and $\xi=0.01$.

Roughly speaking, weak inversion (WI) is a condition for which $-Q'_I < nC'_{ox}\phi_t$, whereas, for strong inversion (SI), $-Q'_I \gg nC'_{ox}\phi_t$. The UCCM of (2.1.37) reduces to

$$\phi_t \ln\left(\frac{Q'_I}{Q'_{IP}}\right) = V_P - V_C + \phi_t \tag{2.2.11}$$

or

$$Q'_I = Q'_{IP} e^{(V_P - V_C + \phi_t)/\phi_t} \tag{2.2.12}$$

for WI, with $V_P - V_C < 0$.

On the other hand, for SI, $V_P - V_C \gg \phi_t$, and the UCCM becomes

$$-Q'_I = nC'_{ox}(V_P - V_C). \tag{2.2.13}$$

Expressions (2.2.12) and (2.2.13) are the well-known exponential and linear charge regimes for WI and SI, respectively.

The charge-based current expression (2.1.57) can also be simplified. In WI, the quadratic terms in the charges can be neglected and (2.1.57) reduces to

$$I_D = -\frac{\mu_n W}{L}\phi_t\left(Q'_{IS} - Q'_{ID}\right). \tag{2.2.14}$$

Combining (2.2.12) and (2.2.14) yields [7]

$$I_D = I_0\left(e^{(V_P - V_S)/\phi_t} - e^{(V_P - V_D)/\phi_t}\right)$$
$$= I_0 e^{(V_P - V_S)/\phi_t}\left(1 - e^{-V_{DS}/\phi_t}\right), \tag{2.2.15}$$

where

$$I_0 = \mu_n \frac{W}{L} nC'_{ox}\phi_t^2 e^1. \tag{2.2.16}$$

The drain saturation voltage is the drain voltage at which the reverse current becomes arbitrarily smaller than the forward current. In WI, the current for constant V_G and V_S saturates for $V_{DS} > 4\phi_t$ at a value equal to

$$I_D = I_0 e^{(V_P - V_S)/\phi_t} \cong I_0 e^{(V_G - V_{T0})/n\phi_t} e^{-V_S/\phi_t}. \tag{2.2.17}$$

The WI characteristics of a long-channel MOSFET are summarized in Figure 2.13 [11]. The rate of change of the drain current is one decade per $2.3n\phi_t$ of gate-voltage variation and one decade per $-2.3\phi_t$ (around -58 mV at 20 °C) of source-voltage variation. The output characteristics in WI saturate for a drain-to-source voltage around $4\phi_t$, or around 100 mV at 20 °C.

In SI, the linear terms in the charges in (2.1.57) can be dropped; in this case, the drain current becomes

$$I_D = \frac{\mu_n W}{L}\left(\frac{Q'^2_{IS} - Q'^2_{ID}}{2nC'_{ox}}\right). \tag{2.2.18}$$

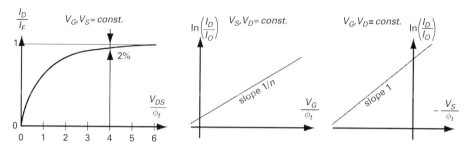

Fig. 2.13   Forward characteristics in weak inversion (adapted from [11]).

Combining (2.2.13) and (2.2.18) yields

$$I_D = \mu_n C'_{ox} n \frac{W}{2L} \left[(V_P - V_S)^2 - (V_P - V_D)^2 \right]. \tag{2.2.19}$$

Equation (2.2.19) shows once again the source–drain symmetry of the transistor, in contrast to classical SI textbook formulas that do not preserve the structural and, consequently, electrical symmetry of the device. The reader should be advised against using expression (2.2.19) when the terms in parentheses for an n (p)-channel transistor are less (greater) than zero. If, for example, $V_P - V_D < 0$ for an NMOS transistor and the source is strongly inverted, one calculates the current using (2.2.19) with $V_P - V_D = 0$. In strong inversion, the value of the current tends to saturate when the second term in (2.2.19) becomes much smaller than the first term, i.e., $V_D \geq V_P$ or, equivalently, $V_D \geq (V_G - V_{T0})/n$. In this case, the drain current becomes

$$I_D \cong \mu_n C'_{ox} n \frac{W}{2L} (V_P - V_S)^2 \cong \mu_n C'_{ox} \frac{W}{2nL} (V_G - V_{T0} - n V_S)^2. \tag{2.2.20}$$

For $n = 1$ (negligible body effect), (2.2.20) is converted into the classical textbook expression for the saturated MOSFET, i.e.

$$I_D = \mu_n C'_{ox} \frac{W}{2L} (V_G - V_{T0} - V_S)^2 = \mu_n C'_{ox} \frac{W}{2L} (V_{GS} - V_{T0})^2. \tag{2.2.21}$$

2.2.4   Small-signal transconductances

At low frequencies, the variation of the drain current due to small variations in the gate, source, and drain voltages is

$$\Delta I_D = g_{mg} \Delta V_G - g_{ms} \Delta V_S + g_{md} \Delta V_D + g_{mb} \Delta V_B, \tag{2.2.22}$$

where

$$g_{mg} = \frac{\partial I_D}{\partial V_G}, \qquad g_{ms} = -\frac{\partial I_D}{\partial V_S}, \qquad g_{md} = \frac{\partial I_D}{\partial V_D}, \qquad g_{mb} = \frac{\partial I_D}{\partial V_B} \tag{2.2.23}$$

are the gate, source, drain, and bulk transconductances, respectively.

When the variation of the gate, source, drain, and bulk voltages is the same, $\Delta I_D = 0$. Therefore, we can conclude that

$$g_{mg} + g_{md} + g_{mb} = g_{ms}. \tag{2.2.24}$$

Thus, three transconductances are enough to characterize the low-frequency small-signal behavior of the MOSFET.

Applying the definitions of source and drain transconductances in (2.2.23) to the Pao–Sah drain-current formula (2.1.51) results in

$$g_{ms(d)} = -\mu_n \frac{W}{L} Q'_{IS(D)}. \tag{2.2.25}$$

Using (2.2.25) and expression (2.2.3), which gives the relationship between inversion charge and inversion level, the source and drain transconductances can be expressed as

$$g_{ms(d)} = \frac{2I_S}{\phi_t}\left(\sqrt{1 + i_{f(r)}} - 1\right). \tag{2.2.26}$$

The above expression for the transconductance is very useful for circuit design because it is very compact, is valid for any inversion level, and uses easily measurable parameters. Moreover, (2.2.26) is a universal relationship for MOSFETs. The only technology-dependent parameter in (2.2.26) is the normalization current, which depends on the transistor aspect ratio as well. It can be shown (see Problem 2.4) that

$$g_{mg} = \frac{g_{ms} - g_{md}}{n}. \tag{2.2.27}$$

Equation (2.2.27) gives the conventional (gate) transconductance in terms of the source and drain transconductances. For a long-channel MOSFET in saturation $i_r \ll i_f$; consequently, $g_{mg} \cong g_{ms}/n$.

Figure 2.14 compares measured and simulated values of both source and gate transconductances of a long-channel transistor in saturation, thus demonstrating the satisfactory accuracy of the proposed model. It can be noted that all the above calculated transconductances approximate to their well-known asymptotic values in weak and strong inversion [7]. For instance, deep in weak inversion, that is, for $i_f \ll 1$, $\sqrt{1 + i_f}$ can be approximated by $1 + i_f/2$. Therefore, g_{ms} tends toward its expected value, I_F/ϕ_t. On the other hand, in very strong inversion, g_{ms} is proportional to $\sqrt{I_F}$ since i_f is much greater than unity.

An important design parameter in analog circuits is the transconductance-to-current ratio (g_m/I_D), a measure of the efficiency of translation of current (power) into transconductance (speed); thus, speed per unit power consumed. g_m/I_D gives an indication of the inversion level, is strongly related to the performance of a circuit, and provides a tool for calculating transistor dimensions [12]. In the following we will demonstrate that this design parameter can be expressed in terms of a normalized saturation current.

The substitution of I_S for I_F/i_f in (2.2.26) allows one to write the ratio of the source (drain) transconductance to the forward (reverse) saturation current as

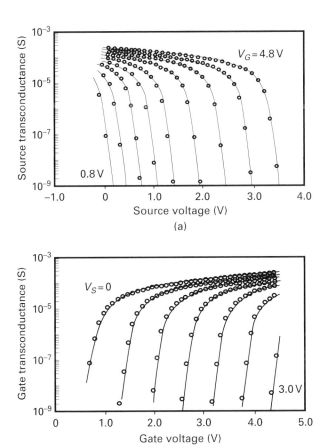

Fig. 2.14 (a) Source transconductance ($V_G = 0.8$, 1.2, 1.6, 2.0, 2.4, 3.0, 3.6, 4.2, and 4.8 V) and (b) gate transconductance ($V_S = 0$, 0.5, 1.0, 1.5, 2.0, 2.5, and 3.0 V) of an NMOS transistor with $t_{ox} = 280$ Å and $W = L = 25$ μm: (–) simulated curves and (○) measured curves.

$$\frac{g_{ms(d)}\phi_t}{I_{F(R)}} = \frac{2}{\sqrt{1 + i_{f(r)}} + 1}. \tag{2.2.28}$$

Equation (2.2.28) is a universal expression for MOS transistors, as is the transconductance-to-current ratio for bipolar transistors. Expression (2.2.28) allows designers to compute the available transconductance-to-current ratio in terms of the inversion level i_f.

The universality of expression (2.2.28) is confirmed in Figures 2.15 and 2.16, where measured and simulated current-to-transconductance ratios are plotted for various technologies and channel lengths. All the measurements were taken in saturation, where $I_D \cong I_F$, and are in close agreement with the model we have developed.

Using lower-case symbols to represent dynamic quantities, we rewrite (2.2.22) as

$$i_d = g_{mg}v_g - g_{ms}v_s + g_{md}v_d + g_{mb}v_b. \tag{2.2.29}$$

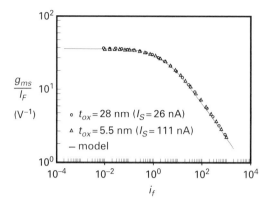

Fig. 2.15   Plots of g_{ms}/I_F of NMOS transistors for various technologies ($W = L = 25$ µm for $t_{ox} = 280$ Å, $W = 25$ µm and $L = 20$ µm for $t_{ox} = 55$ Å; $V_{GB} = 2$ V).

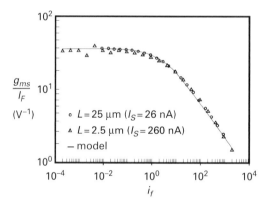

Fig. 2.16   Plots of g_{ms}/I_F of NMOS transistors with various channel lengths ($W = 25$ µm, $t_{ox} = 280$ Å, and $V_{GB} = 2$ V).

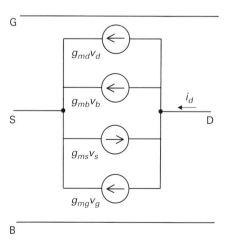

Fig. 2.17   The low-frequency small-signal model of the MOSFET.

The low-frequency small-signal model of the MOSFET that corresponds to expression (2.2.29) is represented in Figure 2.17 (see Appendix A2.4 for an alternative low-frequency model). The long-channel transconductances g_{ms} and g_{md} are given by (2.2.28), while g_{mg} is determined from (2.2.27). The bulk transconductance is calculated from (2.2.24) and (2.2.27), yielding

$$g_{mb} = (n - 1)g_{mg}. \tag{2.2.30}$$

2.3 Dynamic MOSFET models

The charge stored in an MOS transistor cannot be modeled by the charge stored in a set of two-terminal reciprocal capacitors (of maximum number equal to six) connected between the four terminals of the device. For a non-reciprocal device such as the MOSFET, the two-terminal capacitor model is inappropriate. This is because for a four-terminal device nine independent coefficients are necessary in order to model the relationships among three independent voltages and currents, and only six two-terminal reciprocal capacitances can be defined.

To obtain consistent solutions for the transient currents, the current entering each terminal of the transistor is split into a transport component (I_T) and a capacitive-charging term, which contributes to the total charge associated with each terminal. Thus, for the drain current, for example, we write

$$I_D(t) = I_T(t) + \frac{dQ_D}{dt}. \tag{2.3.1}$$

To calculate the stored charges, we assume that the charge stored in the transistor is dependent only on the instantaneous terminal voltages and not on their past variation. Clearly this approximation fails for rapid changes in the terminal voltages because the charge distribution in the channel cannot follow the terminal voltages instantaneously owing to the distributed nature of the channel.

2.3.1 Stored charges

The definitions of gate and bulk stored charges are unambiguous, but the definition of source and drain stored charges used to be somewhat controversial, because they are not necessarily the charges that physically enter into the (source and drain) terminals. The definitions below [13], for source and drain stored charges,

$$Q_S = W \int_0^L \left(1 - \frac{y}{L}\right) Q_I' \, dy \tag{2.3.2}$$

and

$$Q_D = W \int_0^L \frac{y}{L} Q_I' \, dy, \tag{2.3.3}$$

respectively, are now widely accepted because they give the same values for the charging currents as those resulting from the integration of the continuity equation.

As expected,

$$I_D(t) - I_S(t) = \frac{dQ_D}{dt} + \frac{dQ_S}{dt} = \frac{dQ_I}{dt}, \tag{2.3.4}$$

where $I_S(t)$ is the source current and

$$Q_I = W \int_0^L Q_I' \, dy \tag{2.3.5}$$

is the total inversion charge stored in the channel. The difference between the drain and source currents is exactly the amount required to change the inversion charge in the channel. The symmetry of the expressions for the drain- and source-associated charges is self-evident.

Concerning the transient gate and bulk charges, the (leakage) transport current is usually neglected, and in this case the transient currents reduce to the charging currents

$$I_G(t) = \frac{dQ_G}{dt} \tag{2.3.6}$$

and

$$I_B(t) = \frac{dQ_B}{dt}, \tag{2.3.7}$$

where

$$Q_G = W \int_0^L Q_G' \, dy,$$
$$Q_B = W \int_0^L Q_B' \, dy. \tag{2.3.8}$$

The stored charges given by expressions (2.3.2), (2.3.3), and (2.3.8) are easily calculated by changing the integration variable from length y to inversion charge density Q_I'. On substituting the basic approximation (2.1.55) into expression (2.1.54) for the charge-sheet drain current, we have

$$dy = -\frac{\mu_n W}{nC_{ox}' I_D}(Q_I' - nC_{ox}'\phi_t)dQ_I', \tag{2.3.9}$$

which can readily be used to calculate the integrals for the total charges. It is convenient to define a new variable

$$Q_{It}' = Q_I' - nC_{ox}'\phi_t. \tag{2.3.10}$$

Since n is constant along the channel (it is dependent only on V_{GB}),

$$dQ_{It}' = dQ_I' \tag{2.3.11}$$

and (2.3.9) may be rewritten as

$$dy = -\frac{\mu_n W}{nC'_{ox}I_D}Q'_{It}\,dQ'_{It}.$$

(2.3.12)

Substituting (2.3.10) and (2.3.12) into (2.3.5) yields

$$Q_I = -\frac{\mu_n W^2}{I_D nC'_{ox}}\left[\int_{Q'_F}^{Q'_R}(Q'_{It} + nC'_{ox}\phi_t)Q'_{It}\,dQ'_{It}\right],$$

(2.3.13)

where

$$Q'_{F(R)} = Q'_{IS(D)} - nC'_{ox}\phi_t.$$

(2.3.14)

The integration of (2.3.13) and the substitution of I_D by its value calculated in terms of Q'_F and Q'_R yields

$$Q_I = WL\left(\frac{2}{3}\frac{Q'^2_F + Q'_F Q'_R + Q'^2_R}{Q'_F + Q'_R} + nC'_{ox}\phi_t\right),$$

(2.3.15)

which is similar to the conventional expression for the charge in strong inversion, except for the rightmost term. We can rewrite (2.3.15) as an explicit function of the inversion charge densities at source and drain as

$$Q_I = WL\frac{(2/3)(Q'^2_{IS} + Q'_{IS}Q'_{ID} + Q'^2_{ID}) - nC'_{ox}\phi_t(Q'_{IS} + Q'_{ID})}{Q'_{IS} + Q'_{ID} - 2nC'_{ox}\phi_t}.$$

(2.3.16)

The two expressions above give the total channel charge in all inversion regions. Let us now interpret two very simple cases of transistor operation. Deep in strong inversion, the terms in (2.3.16) containing the thermal charge $nC'_{ox}\phi_t$ are negligible and in saturation $Q'_{ID} \cong 0$. Consequently, $Q_I = (2/3)WLQ'_{IS}$, which is the classical result for the charge in SI saturation. Deep in weak inversion, the quadratic terms are negligible because $|Q'_I| \ll nC'_{ox}\phi_t$. Thus, (2.3.16) reduces to

$$Q_I = WL\frac{Q'_{IS} + Q'_{ID}}{2}.$$

(2.3.17)

Things are very simple in weak inversion: the total charge is proportional to the average charge density in the channel. In fact, for a diffusion-only current, the gradient of the charge in the channel is constant.

The derivation of the source- and drain-associated inversion charges is shown in Appendix A2.2, and the calculation of the bulk and gate charges is left as an exercise.

2.3.2 Capacitive coefficients

Since we express all charges Q_j ($j = G, S, D, B$) in terms of the instantaneous values of the terminal voltages, by applying the chain rule of differentiation we obtain the expression for the charging currents as

Table 2.1 A complete set of nine capacitive coefficients for the intrinsic MOSFET

$$C_{gs} = \frac{2}{3} C_{ox} \frac{1 + 2\alpha}{(1+\alpha)^2} \frac{q'_{IS}}{1 + q'_{IS}} \qquad C_{gd} = \frac{2}{3} C_{ox} \frac{\alpha^2 + 2\alpha}{(1+\alpha)^2} \frac{q'_{ID}}{1 + q'_{ID}}$$

$$C_{bs(d)} = (n-1)C_{gs(d)} \qquad C_{gb} = C_{bg} = \frac{n-1}{n}(C_{ox} - C_{gs} - C_{gd})$$

$$C_{sd} = -\frac{4}{15} nC_{ox} \frac{\alpha + 3\alpha^2 + \alpha^3}{(1+\alpha)^3} \frac{q'_{ID}}{1 + q'_{ID}}$$

$$C_{ds} = -\frac{4}{15} nC_{ox} \frac{1 + 3\alpha + \alpha^2}{(1+\alpha)^3} \frac{q'_{IS}}{1 + q'_{IS}}$$

$$C_{dg} - C_{gd} = C_m = (C_{sd} - C_{ds})/n$$

$$\frac{dQ_j}{dt} = \frac{\partial Q_j}{\partial V_G} \frac{dV_G}{dt} + \frac{\partial Q_j}{\partial V_S} \frac{dV_S}{dt} + \frac{\partial Q_j}{\partial V_D} \frac{dV_D}{dt} + \frac{\partial Q_j}{\partial V_B} \frac{dV_B}{dt}. \tag{2.3.18}$$

The four-by-four matrix of the MOSFET intrinsic capacitances for quasi-static operation is defined by

$$C_{jk} = -\left.\frac{\partial Q_j}{\partial V_k}\right|_0, \quad j \neq k, \tag{2.3.19}$$

$$C_{jj} = \left.\frac{\partial Q_j}{\partial V_j}\right|_0, \tag{2.3.20}$$

where Q_j can be any one of the charges Q_S, Q_D, Q_B, and Q_G, and V_j and V_k can be any of the voltages V_S, V_D, V_B, and V_G. The notation "0" indicates that the derivatives are calculated at the bias point. The term C_{jk} can be interpreted as a mutual capacitance. C_{jk} determines the current transferred out of node j because of a voltage change on node k, all the other node voltages remaining constant.

Because the MOSFET is an active device, the capacitances C_{jk} are non-reciprocal, that is, in general, $C_{jk} \neq C_{kj}$ for $j \neq k$. Only 9 of the 16 capacitive coefficients are linearly independent, due to charge conservation and the fact that only three voltage differences from four terminal voltages can be chosen independently.

From the expressions for the MOSFET charges, we can derive explicit formulas for the capacitive coefficients in (2.3.19) and (2.3.20). The use of the normalized inversion charges $q'_{IS(D)} = Q'_{IS(D)}/(-nC'_{ox}\phi_t)$ and of the linearity coefficient[2] $\alpha = Q'_R/Q'_F$ gives us the simple expressions for the capacitive coefficients in Table 2.1 [2].

---

[2] The value of α indicates the degree of linearity of the profile of the inversion charge density along the channel. Note that $\alpha \cong 1$ for weak inversion or when $V_{DS} \to 0$, whereas in strong-inversion saturation $\alpha \to 0$ [14].

The symmetry between the source and drain terminals is apparent in the formulas of Table 2.1. Interchanging the source and drain terminals corresponds to exchanging α for $1/\alpha$ and $q'_{IS(D)}$ for $q'_{ID(S)}$. For $V_{DS} = 0$, $q'_{IS} = q'_{ID}$ and $\alpha = 1$; consequently $C_{jk} = C_{kj}$. However, the capacitive coefficients are non-reciprocal in the general case. The asymptotic cases of strong and weak inversion are readily obtained from the formulas of Table 2.1. In SI, $q'_{IS(D)} \gg 1$; consequently $q'_{IS(D)}/\left(1 + q'_{IS(D)}\right) \simeq 1$, and we re-encounter the conventional expressions for the capacitive coefficients in SI. In WI $\alpha \cong 1$, $1 + q'_{IS(D)} \cong 1$, and we obtain, for example, $C_{gs(d)} = C_{ox}q'_{IS(D)}/2$.

The small-signal schematic diagram of Figure 2.18, which comprises five capacitances, three transcapacitances, and three transconductances, preserves the inherent symmetry of the MOSFET.

The shapes of the curves of the capacitances as functions of the gate-to-bulk voltage in Figure 2.19 can be understood as follows. For low-values of V_G, the only relevant capacitance is C_{gb}, because the channel charge is very small and, consequently, capacitances C_{gs} and C_{gd} are negligible. At the transition between weak and strong inversion C_{gs} increases sharply because the channel charge increases rapidly with the gate voltage and becomes dominant over the depletion charge. Capacitance C_{gd} remains negligible

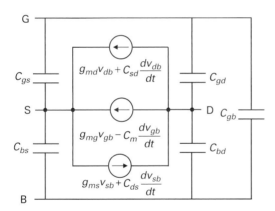

Fig. 2.18   The simplified small-signal MOSFET model.

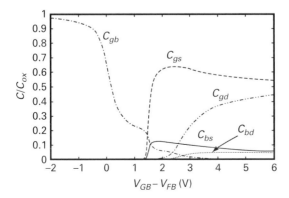

Fig. 2.19   Capacitances C_{gs}, C_{gd}, C_{gb}, C_{bs}, and C_{bd} for $V_{DS} = 1$ V.

since the transistor operates in saturation and the drain voltage has no significant effect on the channel charge. Finally, for high values of the gate voltage the transistor enters the linear region and the two capacitances C_{gs} and C_{gd} approach asymptotically the same value of half of the total gate capacitance. In strong inversion C_{gb} is negligible because the channel charge screens the bulk from the variation in the gate voltage.

2.3.3   Capacitances of the extrinsic transistor

The charge storage in the extrinsic parts (Figure 2.20) of the MOS transistor can be modeled with up to six capacitances, one between each pair of terminals (Figure 2.21). The unavoidable overlap between the gate and the source and drain diffusions generates the overlap capacitances. In parallel with the overlap capacitances, the outer fringing and top capacitances must be included. Fringing capacitances are particularly important for small-dimension devices. The substrate–source and substrate–drain junctions must also be modeled by the non-linear diode capacitances. The very small drain-to-source proximity capacitance can usually be neglected and the extrinsic gate-to-bulk capacitance can be incorporated into the gate wiring capacitance. A more complete model for the extrinsic part should include parasitic resistances as well.

2.3.4   A non-quasi-static small-signal model

A complete non-quasi-static (NQS) model suitable for very-high-frequency operation can be derived by solving the continuity equation coupled with the transport equation. Solving the two equations using an iterative procedure, we obtain the values of the

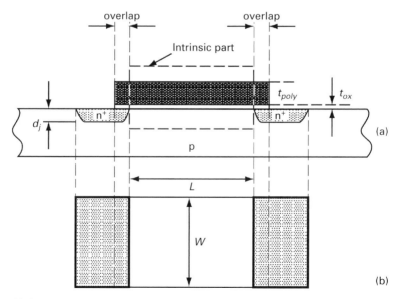

Fig. 2.20   (a) Cross section and (b) top view of the idealized MOS transistor showing the intrinsic and extrinsic parts (adapted from [15]).

Table 2.2 Time constants of the NQS MOSFET model

$$\tau_1 = \frac{\tau}{1 + q'_{IS}} \frac{4}{15} \frac{1 + 3\alpha + \alpha^2}{(1+\alpha)^3}$$

$$\tau_2 = \frac{\tau}{1 + q'_{IS}} \frac{1}{15} \frac{2 + 8\alpha + 5\alpha^2}{(1+\alpha)^2(1+2\alpha)} \qquad \tau_3 = \frac{\tau}{1 + q'_{IS}} \frac{1}{15} \frac{5 + 8\alpha + 2\alpha^2}{(1+\alpha)^2(2+\alpha)}$$

$$\tau = \frac{L^2}{\mu\phi_t} \qquad q'_{IS(D)} = -\frac{Q'_{IS(D)}}{nC'_{ox}\phi_t} \qquad \alpha = \frac{1 + q'_{ID}}{1 + q'_{IS}}$$

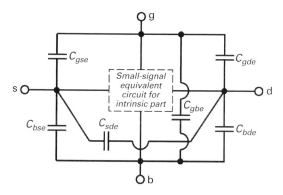

Fig. 2.21   Extrinsic transistor capacitances (adapted from [15]).

NQS time constants [2] given in Table 2.2. The first-order small-signal NQS model in Figure 2.22 is a simplified model that includes only the main time constants.

From the expressions in Tables 2.1 and 2.2, it can be readily verified that

$$C_{sd(ds)} = -g_{md(s)}\tau_1; \qquad C_m = C_{dg} - C_{gd} = C_{gs} - C_{sg} = g_{mg}\tau_1. \qquad (2.3.21)$$

Some comments are in order for the first-order NQS small-signal model of the MOS transistor in Figure 2.22. (i) Under low-frequency operation ($\omega\tau_1 \ll 1$) the circuit in Figure 2.22 reduces to the conventional three-transconductance, five-capacitance model shown in Figure 2.23. (ii) For frequencies such that $(\omega\tau_1)^2 \ll 1$, the voltage-controlled current sources are represented not only by transconductances but also by additional terms corresponding to transcapacitances, as in Figure 2.18; also, the capacitive components are equivalent to series connections of a capacitor and a resistor. To give an example, let us calculate the admittance $y_{gs} = -i_g/v_s$. Keeping all the voltages constant except the source voltage, we calculate the admittance as

$$y_{gs} = -\frac{j\omega C_{gs}}{1 + j\omega(\tau_1 - \tau_2)} \rightarrow -\frac{1}{y_{gs}} = \frac{1}{j\omega C_{gs}} + \frac{\tau_1 - \tau_2}{C_{gs}}. \qquad (2.3.22)$$

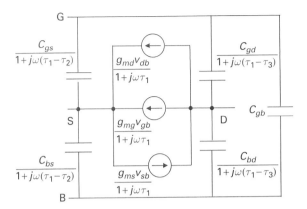

Fig. 2.22   The simplified high-frequency MOSFET model.

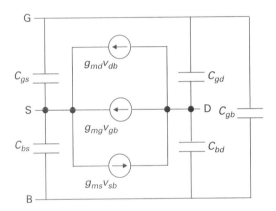

Fig. 2.23   The quasi-static small-signal MOSFET model.

Therefore, the component seen between gate and source is equivalent to a series association of a capacitor C_{gs} and a resistor with a value equal to $(\tau_1 - \tau_2)/C_{gs}$. Of course, the series resistance is not relevant for low and moderate frequencies, but it can play an important role for high-frequency operation. (iii) For frequencies such that $\omega\tau_1$ is of the order of 1 or higher, the model in Figure 2.22 is no longer valid. For the purpose of simulation for fast signal variation, the designer can consider the transistor as consisting of N equal series-connected channel segments [2], with the length of each segment being smaller than a value that ensures the inequality $\omega\tau_1 \ll 1$. With the exception of the extreme segments, each segment consists only of the intrinsic parts.

Example 2.9

Find an expression for the resistance in series with capacitance C_{gs} of the NQS small-signal MOSFET model in terms of the drain current for strong-inversion saturation.

Answer

From Table 2.1 we have

$$C_{gs} = \frac{2}{3} C_{ox} \frac{1 + 2\alpha}{(1+\alpha)^2} \frac{q'_{IS}}{1 + q'_{IS}} \quad \text{with } \alpha = \frac{q'_{ID} + 1}{q'_{IS} + 1}.$$

In SI saturation $q'_{IS} \gg 1$ and $q'_{ID} \to 0$; thus, $\alpha \to 0$ and

$$C_{gs} \cong \frac{2}{3} C_{ox} = \frac{2}{3} C'_{ox} WL.$$

From the expressions for τ_1 and τ_2 in Table 2.2,

$$\tau_1 = \frac{\tau}{1 + q'_{IS}} \frac{4}{15} \frac{1 + 3\alpha + \alpha^2}{(1+\alpha)^3} \cong \frac{\tau}{q'_{IS}} \frac{4}{15} = \frac{L^2/(\mu\phi_t)}{\sqrt{I_D/I_S}} \frac{4}{15} = \frac{4}{15} \sqrt{\frac{WL^3 nC'_{ox}}{2\mu I_D}},$$

$$\tau_2 \cong \frac{\tau}{1 + q'_{IS}} \frac{2}{15} = \frac{\tau_1}{2}.$$

The equivalent resistance in series with C_{gs} is

$$R_{gs} = \frac{\tau_1 - \tau_2}{C_{gs}} = \frac{\tau_1}{2C_{gs}} = \frac{4}{2 \cdot 15} \sqrt{\frac{WL^3 nC'_{ox}}{2\mu I_D}} \frac{1}{(2/3)WLC'_{ox}}$$

$$= \frac{1}{5} \sqrt{\frac{Ln}{2\mu WC'_{ox} I_D}} = \frac{1}{5} \frac{n}{g_{ms}} = \frac{1}{5g_{mg}}.$$

2.3.5 A quasi-static small-signal model

When the operating frequency is such that $\omega\tau_1 \ll 1$, it follows that $\omega\tau_{2(3)} \ll 1$ and the NQS model of Figure 2.22 reduces to the classic five-capacitor model of Figure 2.23. An alternative derivation of the schematic diagram of Figure 2.23 can be obtained from the complete capacitive model of Figure 2.18, considering sinusoidal steady-state operation. In effect, for $\omega\tau_1 \ll 1$ it follows from (2.3.21) that the component of the drain current with magnitude equal to $\omega C_m v_{GB}$ is much smaller than $g_m v_{gb}$ and likewise $-\omega C_{sd} v_{db}$ and $-\omega C_{ds} v_{sb}$ are much smaller than $g_{md} v_{db}$ and $g_{ms} v_{sb}$, respectively. The quasi-static model of Figure 2.23 will be used extensively in this book to calculate the frequency responses of circuits.

Example 2.10

(a) Determine I_D, V_{DSsat}, and the (quasi-static) small-signal parameters of a saturated n-channel MOSFET operating at (a1) $i_f = 3$ and (a2) $i_f = 99$ with $V_{SB} = 0$. The transistor dimensions are $W = 10$ μm and $L = 1$ μm and it is built on the basis of a 0.35-μm technology in which $t_{ox} = 7$ nm, $n = 1.2$, and $\mu_n = 400$ cm^2/V per s at 300 K. (b) Determine an upper frequency limit for the validity of the models the parameters of which were calculated in (a).

Answer

(a) The technological parameters are the same as in Example 2.7; thus, $C'_{ox} = 493$ nF/cm^2 and $I_{SH} = \mu_n C'_{ox} n \phi_t^2 / 2 = 80$ nA. Of the three transconductances and five capacitances of the model in Figure 2.23, only g_m, C_{gs}, and C_{gb} must be considered. In effect, because $V_{SB} = 0$ the currents associated with g_{ms} and C_{bs} are zero. Additionally, $g_{md} = 0$, $C_{gd} = 0$ and $C_{bd} = 0$ since the transistor is saturated.[3]

The saturation drain current is given by $I_F = (W/L) I_{SH} i_f$, whereas V_{DSsat}, which is given by (2.2.10), is $V_{DSsat} \cong \phi_t (3 + \sqrt{1 + i_f})$ when we adopt $1/\xi \cong 55$. The transconductance, which is obtained by combining (2.2.26) and (2.2.27), is $g_{mg} = [2 I_S / (n \phi_t)](\sqrt{1 + i_f} - 1)$ (note that $g_{md} \to 0$) and the small-signal capacitances are obtained by taking the expressions of Table 2.1:

$$C_{gs} = \frac{2}{3} C_{ox} \frac{1 + 2\alpha}{(1 + \alpha)^2} \frac{q'_{IS}}{1 + q'_{IS}} ; \qquad C_{gb} = \frac{n - 1}{n} (C_{ox} - C_{gs} - C_{gd}),$$

where

$$C_{ox} = W L C'_{ox} = 49 \text{ fF},$$

$$\alpha = \frac{q'_{ID} + 1}{q'_{IS} + 1} = \frac{1}{q'_{IS} + 1},$$

$$q'_{IS} = -\frac{Q'_{IS}}{n C'_{ox} \phi_t} = \sqrt{1 + i_f} - 1.$$

(a1) From the above formulas we have, for $i_f = 3$,

$$I_D = I_F = (W/L) I_{SH} i_f = 10 \times 80 \times 10^{-9} \times 3 = 2.4 \, \mu\text{A},$$

$$V_{DSsat} \cong \phi_t (3 + \sqrt{1 + i_f}) = 26 \times 10^{-3} \times (3 + 2) = 130 \, \text{mV},$$

$$q'_{IS} = \sqrt{1 + 3} - 1 = 1,$$

$$\alpha = \frac{1}{1 + 1} = 0.5,$$

$$g_{mg} = \frac{2 \times 10 \times 80 \times 10^{-9}}{1.2 \times 0.026} = 51 \, \mu\text{A/V},$$

$$C_{gs} = \frac{2}{3} 49 \times 10^{-15} \frac{1 + 1}{(1 + 0.5)^2} \frac{1}{1 + 1} = 14.5 \, \text{fF},$$

$$C_{gb} = \frac{0.2}{1.2} (49 - 14.5 - 0) \times 10^{-15} = 5.75 \, \text{fF}.$$

---

[3] Even though the transistor channel is not long, we have assumed that g_{md}, C_{gd}, and C_{bd}, which are controlled by the drain voltage, are equal to zero in saturation.

(a2) For $i_f = 99$ we have

$$I_D = 79\,\mu\text{A},$$

$$V_{DSsat} = 338\,\text{mV},$$

$$q'_{IS} = \sqrt{1 + 99} - 1 = 9,$$

$$\alpha = \frac{1}{1 + 9} = 0.1,$$

$$g_{mg} = \frac{2 \times 10 \times 80 \times 10^{-9}}{1.2 \times 0.026} \times 9 = 461\,\mu\text{A/V},$$

$$C_{gs} = \frac{2}{3} 49 \times 10^{-15} \frac{1 + 0.2}{(1 + 0.2)^2} \frac{9}{10} = 24.5\,\text{fF},$$

$$C_{gb} = \frac{0.2}{1.2} (49 - 24.5 - 0) \times 10^{-15} = 4.1\,\text{fF}.$$

On comparing (a1) and (a2), we observe that an increase in the current by a factor of 33 increases the transconductance only by a factor of 9, due to the reduced amplifying efficiency of the transistor operating at the upper limit of moderate inversion.

(b) The NQS time constant is

$$\tau_1 = \frac{\tau}{1 + q'_{IS}} \frac{4}{15} \frac{1 + 3\alpha + \alpha^2}{(1 + \alpha)^3}; \quad \tau = \frac{L^2}{\mu\phi_t} = \frac{10^{-8}}{400 \times 0.026} = 961\,\text{ps}.$$

For $i_f = 3$, $\tau_1 = 104$ ps and $f_{max} = [1/(2\pi\tau_1)]/10 = 153$ MHz. For $i_f = 99$, $\tau_1 = 25$ ps and $f_{max} = [1/(2\pi\tau_1)]/10 = 636$ MHz.

2.3.6   The intrinsic transition frequency

An important figure of merit for a MOSFET is the unity-gain or intrinsic cut-off frequency, defined as the frequency at which the short-circuit current gain in the common-source configuration (Figure 2.24) drops to 1. Using the quasi-static model of Figure 2.23, we obtain the three-capacitor model of Figure 2.24. In effect, since the only non-constant voltage is V_G, only capacitances connected to the gate or current sources controlled by V_G may be considered. For a saturated transistor, the intrinsic $C_{gd} = 0$. Small-signal

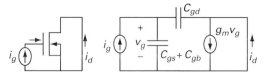

Fig. 2.24   A conceptual circuit for the definition of the intrinsic transition frequency and small-signal equivalent circuit.

analysis of the equivalent circuit in Figure 2.24 gives the intrinsic cut-off frequency of a MOSFET as

$$f_T = \frac{g_{mg}}{2\pi\left(C_{gs} + C_{gb}\right)} = \frac{g_{ms}}{2\pi n\left(C_{gs} + C_{gb}\right)}. \tag{2.3.23}$$

For the sake of simplicity we assume for the time being that the sum of the gate-to-source and gate-to-bulk capacitances is independent of bias. This approximation is quite acceptable since n is a slowly varying function of bias and C_{gs} is a negligible fraction of C_{ox} in weak inversion and $(2/3)C_{ox}$ in strong inversion. Thus, using $C_{gs} + C_{gb} \simeq C_{ox}/2 = (1/2)WLC'_{ox}$, without significant loss of accuracy, f_T can be approximated by

$$f_T \cong \frac{\mu\phi_t}{2\pi L^2} 2\left(\sqrt{1 + i_f} - 1\right) \tag{2.3.24}$$

for any inversion level.

Example 2.11

Determine the inversion level for which the transition frequency of a minimum (nominal)-length NMOS transistor in the 0.35-μm technology is 10 GHz at room temperature.

Answer

Assuming that $n = 1.2$ and $\mu_n = 400 \text{ cm}^2/\text{V}$ per s at 300 K, it follows from (2.3.24) that $i_f = [1 + \pi L^2 f_T/(\mu_n \phi_t)]^2 - 1 = 21$. Thus, operation in moderate inversion can be considered for designing, for example, an amplifier at 1 GHz.

2.4   Short-channel effects in MOSFETs

In the previous sections we considered the MOSFET to be long and wide. We also assumed the mobility to be independent of the electric field. Here we include the effects of both the mobility variation with the transverse field and velocity saturation on the drain current as well as other physical effects such as channel-length modulation (CLM) and drain-induced barrier lowering (DIBL) [2].

2.4.1   Effective mobility

The mobility is determined by several scattering mechanisms through which the carriers exchange momentum (and kinetic energy) with the semiconductor. In MOS transistors, the carriers flow near the interface of the semiconductor with the oxide, and extra scattering mechanisms at the interface lower the mobility of the carriers of the inversion layer (surface mobility) to values of the order of half of the bulk mobility.

The mobility is dependent on the electric-field component perpendicular to the current flow (transversal field), which can concentrate the carriers near the surface and subject them to additional scattering. As for the transport in the semiconductor bulk, the velocity of surface carriers saturates for high electric fields in the direction of the current flow (longitudinal field).

The usual procedure employed to include mobility variation in compact modeling is to substitute for the constant mobility in the drain-current equation an effective mobility, which is dependent on the applied voltages. A classical approximation assumes that the mobility results from the combination of bulk mobility and surface mobility, which is assumed to be inversely proportional to the transverse field.

For the sake of simplicity, we can use the average value of the transverse field along the channel to define the effective mobility in the channel as

$$\mu_{eff} = \frac{\mu_0}{1 - \alpha_\theta \left(\dfrac{Q'_{BS} + \eta Q'_{IS}}{2\varepsilon_s} + \dfrac{Q'_{BD} + \eta Q'_{ID}}{2\varepsilon_s} \right)}, \tag{2.4.1}$$

where μ_0, the low-field mobility, and α_θ, the scattering constant, are considered as fitting parameters. The terms in the denominator of (2.4.1) account for the "average" bulk and inversion charges along the channel. The value of η would be 1/2 for a uniform electron concentration across the thickness of the inversion layer. Usually η is also considered an empirical parameter.

Another simplification [2] considers the effective transversal field to be constant along the channel and equal to its value at pinch-off. Thus

$$\mu_{eff} = \frac{\mu_0}{1 - \alpha_\theta Q'_{Ba}/\varepsilon_s}. \tag{2.4.2}$$

In this case, the effective mobility is dependent only on the depletion charge Q'_{Ba} calculated for negligible inversion charge density $(Q'_I = 0)$, which is a function of the gate-to-substrate voltage only.

2.4.2 Velocity saturation

In Section 2.1.7 we assumed that the drift velocity of carriers v_d is proportional to the parallel electric field F_y, which is true for low electric fields only. For high fields, the velocity tends to saturate, as Figure 2.25 shows. This phenomenon, referred to as velocity saturation, is one of the most important short-channel effects in MOSFETs. In the following, we analyze the consequences of velocity saturation on the transistor charges and drain current. To this end, we rewrite the Pao–Sah expression for the drain current

$$I_D = -\mu W Q'_I(y) \frac{dV_C}{dy} \tag{2.4.3}$$

with the mobility μ given by

$$\mu = \frac{\mu_s}{1 + F_y/F_C}. \tag{2.4.4}$$

In (2.4.4) F_C is the critical field and $\mu_s = v_{sat}/F_C$ is the low-field mobility.

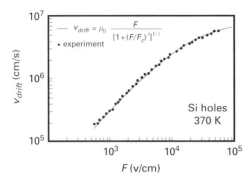

Fig. 2.25   Hole drift velocity versus electric field in silicon (adapted from [16]).

On writing the longitudinal field in terms of the surface potential and using the linear relationship between surface potential and inversion charge of (2.1.55), expression (2.4.4) becomes

$$\mu = \frac{\mu_s}{1 + [1/(nC'_{ox}F_C)]dQ'_I/dy}. \tag{2.4.5}$$

On the other hand, using (2.1.35), we rewrite dV_C/dy as

$$\frac{dV_C}{dy} = \left(\frac{1}{nC'_{ox}} - \frac{\phi_t}{Q'_I}\right)\frac{dQ'_I}{dy}. \tag{2.4.6}$$

Finally, combining (2.4.3), (2.4.5), and (2.4.6), and integrating the resulting equation along the transistor channel, from source to drain, results in

$$I_D = -\frac{\mu_s W}{nC'_{ox}L}\frac{1}{1 + (Q'_{ID} - Q'_{IS})/(LF_CnC'_{ox})}\left[\frac{(Q'_{ID} + Q'_{IS})}{2} + Q'_{IP}\right](Q'_{ID} - Q'_{IS}). \tag{2.4.7}$$

Equation (2.4.7) is a compact expression for the drain current in terms of the inversion charge densities at the source and drain ends of the channel. It includes the effects of both the transverse, via μ_s, and parallel electric fields. One can readily notice that, for low longitudinal fields, (2.4.7) reduces to expression (2.1.61) for the forward and reverse components of the drain current derived for long-channel devices.

On normalizing the charges with respect to the pinch-off charge $Q'_{IP} = -nC'_{ox}\phi_t$ and the current with respect to the normalization current $I_S = (W/L)\mu_s nC'_{ox}\phi_t^2/2$, (2.4.7) can be rewritten as

$$i_D = \frac{q'_{IS} + q'_{ID} + 2}{1 + \zeta(q'_{IS} - q'_{ID})}(q'_{IS} - q'_{ID}), \tag{2.4.8}$$

where the dimensionless short-channel parameter ζ ($\zeta \to 0$ for $L \to \infty$) is defined as

$$\zeta = \phi_t/(LF_C) = \phi_t\mu_s/(Lv_{sat}). \tag{2.4.9}$$

ζ can be viewed as the ratio of the diffusion-related velocity $\mu_s(\phi_t/L)$ at the source of a saturated transistor to the saturation velocity v_{sat}.

When electrons at the drain end of the channel reach the saturation velocity, the drain current is expressed as

$$I_D = -W v_{sat} Q'_{IDsat}.$$ (2.4.10)

In (2.4.10) Q'_{IDsat} is the inversion charge density at the drain end of the channel. Normalizing (2.4.10), in the same way as carried out for the current, yields

$$i_{Dsat} = \frac{I_{Dsat}}{I_S} = \frac{2}{\zeta} q'_{IDsat}.$$ (2.4.11)

The saturation condition, relating the source charge to the drain charge in saturation, is obtained by imposing the equality of the general expression for the drain current with the saturated current. Thus,

$$i_{Dsat} = \frac{q'_{IS} + q'_{IDsat} + 2}{1 + \zeta(q'_{IS} - q'_{IDsat})} (q'_{IS} - q'_{IDsat}) = \frac{2}{\zeta} q'_{IDsat},$$ (2.4.12)

which is a simple quadratic equation relating the source (q'_{IS}) and drain saturation (q'_{IDsat}) charge densities. It is preferable to express q'_{IS} in terms of q'_{IDsat} because the resultant algebraic expression is simpler and more suitable for numerical calculations. Thus,

$$q'_{IS} = \sqrt{1 + \frac{2}{\zeta} q'_{IDsat}} - 1 + q'_{IDsat}.$$ (2.4.13)

Using (2.4.11) in (2.4.13) yields

$$q'_{IS} = \sqrt{1 + i_{Dsat}} - 1 + \frac{\zeta}{2} i_{Dsat}.$$ (2.4.14)

We note that, for $\zeta \to 0$, the value of q'_{IS} given by expression (2.4.14) is the same as that given by (2.2.3), which was derived for obtaining the relationship between the inversion charge density at the source and the forward (saturation) current of the long-channel transistor. Equation (2.4.14) is a simple physics-based generalization of the long-channel expression, which is very useful for design.

Finally, the effect of velocity saturation on the transconductances, stored charges, and intrinsic capacitances is shown in Appendix A2.3.

2.4.3   Channel-length modulation

The gradual channel approximation is valid for $q'_{ID} > q'_{IDsat}$. For $q'_{ID} \to q'_{IDsat}$ the transistor is operating in the so-called saturation region. The usual approach employed to find an analytical formulation for the saturation region divides the channel into two sections. In one of them, closer to the source, the gradual channel approximation is valid, whereas in the second one, closer to the drain, two-dimensional effects must be accounted for. Using this formulation, the current can be calculated from the expression derived under the gradual channel approximation but considering the effective channel length of the device to be reduced by the length ΔL of the drain section. In addition, the

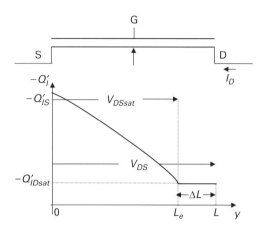

Fig. 2.26   An illustration of the inversion charge density along the channel for a transistor operating in saturation; ΔL represents the channel-length shortening.

voltage drop in the drain section must be accounted for in order to calculate the drain-to-source saturation voltage V_{DSsat}. The dependence of the effective channel length on the drain-to-source voltage, which is illustrated in Figure 2.26, is referred to as the channel-length modulation (CLM).

The channel-length shortening ΔL is, according to a popular model [17], given by

$$\Delta L = L_C \ln\left(1 + \frac{V_{DS} - V_{DSsat}}{V_p}\right) \qquad (2.4.15)$$

for $V_{DS} \geq V_{DSsat}$ and by $\Delta L = 0$ for $V_{DS} < V_{DSsat}$. L_C and V_p are considered as fitting parameters in compact models.

In order to determine the influence of the CLM on the current, we rewrite (2.4.8) as

$$I_D = I_S \frac{1}{1 - \Delta L/L} \frac{q'_{IS} + q'_{ID} + 2}{1 + \zeta(q'_{IS} - q'_{ID})} (q'_{IS} - q'_{ID}). \qquad (2.4.16)$$

We note that ΔL in (2.4.16) represents the CLM, which can be given either as in (2.4.15) or by some other approximation.

2.4.4   Drain-induced barrier lowering

For short-channel transistors, source and drain play a role similar to the gate in addition to their specific functions. For a shorter channel length, the two-dimensional nature of the electric-field pattern changes the surface potential profile. Field lines emanating from source and drain terminate at channel charges in addition to the field lines emanating from the gate. An increase in the drain voltage produces an increase in the surface potential in the channel and, consequently, a reduction in the potential barrier seen by the electrons at the source. For this reason, this effect is called drain-induced barrier lowering (DIBL).

The inclusion of the DIBL effect in MOSFET models is generally through the threshold voltage. An analytical model of the dependence of the threshold voltage on technological parameters and bias is shown in [18] as

$$V_T \cong V_{T,lc} - \frac{6t_{ox}}{d_1}[2(\phi_{bi} - V_{BS}) + V_{DS}]\exp\left(-\frac{\pi L}{4d_1}\right), \qquad (2.4.17)$$

where $V_{T,lc}$ is the long-channel threshold voltage and ϕ_{bi} is the built-in potential of the source–substrate junction. Equation (2.4.17) predicts the variation in V_T with L, t_{ox}, and bias. d_1 represents the depletion depth of a long-channel device for a band bending equal to $2\phi_F$. The dependence of V_T on substrate concentration is through $d_1 \propto (N_A)^{-1/2}$. Note that the short-channel effect on the threshold voltage is proportional to the oxide thickness, a result that is consistent with various models.

In order to emphasize the source–drain symmetry of the MOSFET, we rewrite expression (2.4.17) as

$$V_T \cong V_{T,lc} - \sigma[(\phi_{bi} + V_{SB}) + (\phi_{bi} + V_{DB})], \qquad (2.4.18)$$

where σ is the magnitude of the DIBL factor.

2.4.5 Output conductance in saturation

In the design of CMOS analog circuits, the Early voltage (or the output conductance) of a transistor in saturation is a fundamental parameter since it affects, for example, the accuracy of current mirrors and the gain of voltage amplifiers. In the circuit-design-oriented approach, the simplest model of the output conductance assumes it to be proportional to the drain current and inversely proportional to the Early voltage V_A as given below:

$$g_{md} = \frac{dI_D}{dV_D} = I_D/V_A. \qquad (2.4.19)$$

V_A is a constant parameter in first-order models such as SPICE1. However, a constant Early voltage is inadequate to model the output conductance for the simulation of analog circuits. An improved model of the Early voltage considers it to be proportional to the channel length and independent of both current level and drain voltage, i.e.

$$V_A = V_E L. \qquad (2.4.20)$$

In this case, the output conductance is simply $g_{md} = I_D/(V_E L)$, with V_E being a technology parameter. Even though this model of the output conductance is not accurate, it provides a simple expression to quickly evaluate how the transistor output conductance is affected by the drain current and the channel length. Approximation (2.4.20) will be extensively used throughout this book.

A physics-based model of the output conductance in saturation must include velocity-saturation, CLM, and DIBL effects. Weak avalanche effects and the substrate current-induced body effect [2], which can be significant, especially when the electric field close to the drain is high, will not be considered here for the determination of the output conductance.

The MOSFET output conductance in saturation can be taken as the limit of the value of g_{md} in (A2.3.5a) of Table A2.3.3 when the inversion charge density at the drain end of the channel becomes equal to $-I_D/(Wv_{sat})$, as given by (2.4.10). Additionally, including CLM, we arrive at the more complete expression of the output conductance in saturation given in (A2.3.8) of Appendix A2.3. A very simple interpretation of (A2.3.8) for low to moderate inversion levels is given in the example that follows.

Example 2.12

Using expression (A2.3.8), find a simplified result for the ratio of the MOSFET output conductance to the drain current for the cases of low and moderate inversion levels. Assume that $\zeta I_D/I_S < 1$.

Answer

For low-to-moderate inversion levels, the output conductance-to-current ratio is given by

$$\frac{g_{md}}{I_D} = \frac{2\sigma}{n\phi_t} \frac{1}{\sqrt{1 + I_D/I_S} + 1} + \frac{1}{L - \Delta L} \frac{d\Delta L}{dV_D}.$$

For the case of weak inversion, the previous expression becomes

$$\frac{g_{md}}{I_D} = \frac{\sigma}{n\phi_t} + \frac{1}{L - \Delta L} \frac{d\Delta L}{dV_D}.$$

Since in weak inversion the inversion charge is much lower than the depletion charge, the last term in the previous equation, which corresponds to CLM, is independent of the inversion charge and, consequently, of the current. Therefore, the output conductance-to-current ratio is independent of the current level for operation in weak inversion, a conclusion corroborated by the experimental results shown in Figure 2.27.

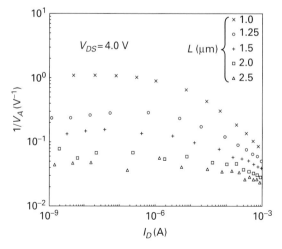

Fig. 2.27   Output conductance-to-current ratio versus drain current for MOSFETs in a 0.75-μm CMOS technology.

As shown in Example 2.12, g_{md}/I_D can be written as the sum of the terms corresponding to each of the physical effects contributing to the variation in the current in saturation, i.e.,

$$\frac{g_{md}}{I_D} = \frac{1}{V_A} = \frac{1}{V_{ADIBL}} + \frac{1}{V_{ACLM}}. \tag{2.4.21}$$

When the current level is such that $\zeta I_D/I_S < 1$, the two components of the Early voltage can be approximated by

$$\frac{V_{ADIBL}}{\phi_t} \cong \frac{n}{2\sigma}\left(\sqrt{1 + \frac{I_D}{I_S}} + 1\right), \tag{2.4.22}$$

$$\frac{V_{ACLM}}{\phi_t} \cong \frac{1}{\phi_t}\left(\frac{1}{L_e}\frac{d\Delta L}{dV_D}\right)^{-1} = \frac{V_p}{\phi_t}\frac{L_e}{L_C}\left(1 + \frac{V_{DS} - V_{DSsat}}{V_p}\right). \tag{2.4.23}$$

Expression (2.4.23) is derived by assuming that the carrier velocity in the saturated part of the channel is equal to the saturation velocity and that carriers flow within the depth of the junction [2], [17]. Note that formula (2.4.23) is approximately valid for the case of high inversion levels. For the case of low and moderate inversion levels, different results are obtained, e.g. as in [19].

In weak inversion $I_D/I_S < 1$; therefore, as previously explained, the Early voltage V_A is independent of the current level. Typically, the DIBL component of the Early voltage can be neglected for high inversion levels, CLM being the dominant factor in the output conductance.

In order to verify the consistency of the output conductance model, MOSFETs from a 0.75-μm technology with various channel lengths were subjected to measurements under several bias conditions. Figure 2.27 shows the variation in the Early voltage, for several channel lengths, in terms of the normalized drain current. One can conclude that the Early voltage increases with increasing channel length, is almost independent of the current level in weak inversion, and increases in moderate and strong inversion.

2.4.6 Gate tunneling currents

For oxide thicknesses below 4 nm, high current leakages through the oxide can occur due to the quantum-mechanical tunneling of electrons. The gate leakage current can not only negatively affect the device performance but also significantly increase the standby power consumption of a chip. For these reasons, a compact model for the gate current is mandatory for advanced technologies.

For an n-MOS transistor operating in inversion, the intrinsic gate current is due to electrons tunneling from the inversion layer. The gate current density satisfies the continuity equation, which is given by

$$WJ_G(y) = \frac{\partial I}{\partial y}, \tag{2.4.24}$$

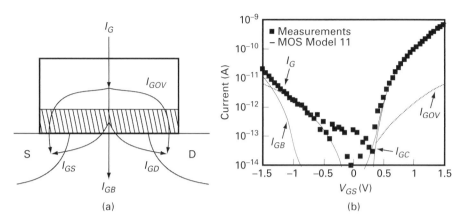

Fig. 2.28   (a) The various gate current components in a MOSFET. (b) Measured and modeled gate current as a function of gate bias V_{GS} for $V_{DS} = V_{SB} = 0$ V. $I_{GC} = I_{GS} + I_{GD}$ is the channel current (adapted from [20]).

where I is the channel current. The Ward–Dutton method, which was presented in Section 2.3.1 to determine the source and drain components of the charging currents, is also based on the current-continuity equation. Consequently, we can also apply the Ward–Dutton method to the splitting of the gate current between source and drain [2]. Thus,

$$I_{GS} = W \int_0^L \left(1 - \frac{y}{L} \right) J_G \, dy, \tag{2.4.25}$$

$$I_{GD} = W \int_0^L \frac{y}{L} J_G \, dy. \tag{2.4.26}$$

The total intrinsic gate current is given by

$$I_G = I_{GS} + I_{GD} = W \int_0^L J_G \, dy. \tag{2.4.27}$$

In accumulation, the intrinsic gate current consists of electrons tunneling from gate to bulk. Because no channel current is flowing, the calculation of the gate-to-bulk current is straightforward, being given simply by

$$I_{GB} = W \int_0^L J_{GB} \, dy. \tag{2.4.28}$$

Overlap currents can be included by considering these regions as specific MOS structures with their specific flat-band voltage and body factor, see for example [20]. On including all the above components, the gate-current model of MM11 gives an accurate description of I_G, as shown in Figure 2.28.

2.4.7   Bulk current

In addition to the bulk current occasionally resulting from tunneling through the gate oxide, other mechanisms can create bulk currents. Clearly, the drain–bulk and source–bulk diodes

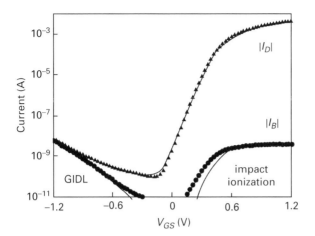

Fig. 2.29   Measured (symbols) and simulated (lines) drain and bulk currents as a function of V_{GS} at high drain bias ($V_{DS} = 1.2$ V) (adapted from [20]).

contribute with their reverse currents, but here we will restrict our analysis to the intrinsic transistor. Two mechanisms, band-to-band tunneling (BBT) under the drain-to-gate overlap region, which is called gate-induced drain leakage (GIDL), and impact ionization, can contribute to the bulk current.

For advanced CMOS technologies, because the supply voltage is of the same order as, or even lower than, the bandgap of silicon, the carriers are no longer able to attain enough energy to create electron–hole pairs. Consequently, impact ionization is no longer important for advanced silicon technologies.

The bulk current for an n-MOS transistor simulated using MM11 [20] is displayed in Figure 2.29. The GIDL is dominant for negative values of the gate-to-source voltage, while impact ionization dominates for the conducting transistor.

Appendices

A2.1   Semiconductor charges

Through a process called doping, foreign atoms (impurities) of two types (donors and acceptors) are introduced into semiconductors in order to modify the carrier densities [2]. Each donor atom can donate one electron to the crystal lattice while each acceptor atom can accept one electron from the valence band, creating a hole. Donor materials come from group V of the periodic table; the most commonly used materials for n-type silicon are phosphorus and arsenic. Acceptors are elements from group III of the periodic table; the most common acceptor material is boron. Because the dopant atoms are originally neutral, donors become positively ionized when donating one electron, whereas acceptors become negatively ionized when accepting one electron. Thus, in the general case, four types of charge are present inside a semiconductor: the fixed positive charge of ionized donors, the fixed negative charge of ionized acceptors, the mobile positive charge of holes, and the mobile negative charge of electrons. We consider here the case in which all donors and acceptors are ionized, i.e.

$$N_D = N_D^+ \quad \text{and} \quad N_A = N_A^-, \tag{A2.1.1}$$

where (N_D^+) N_D and (N_A^-) N_A are the (ionized) donor and acceptor atoms per unit volume. On this basis, the net positive charge density is given by

$$\rho = q(N_D - N_A + p - n). \tag{A2.1.2}$$

Deep in the bulk of a uniformly doped semiconductor, where charge neutrality holds and $\phi = 0$, the carrier concentrations p_0 and n_0 at equilibrium are obtained from charge neutrality,

$$p_0 - n_0 + N_D - N_A = 0, \tag{A2.1.3}$$

and from the law of mass action $n_0 p_0 = n_i^2$. On substituting (A2.1.3), (2.1.5), and (2.1.6) into (A2.1.2) and using the law of mass action, the volumetric charge density becomes written as

$$\rho = q[(p - p_0) - (n - n_0)] = q p_0 \left[(e^{-u} - 1) - \frac{n_i^2}{p_0^2}(e^{u} - 1) \right]. \tag{A2.1.4}$$

The semiconductor charge can be calculated from Poisson's equation. For the time being, we assume that the charge-sheet approximation (CSA) [6] can be used to calculate the bulk charge density Q_B'. In the CSA, the inversion layer of electrons (in a p-type substrate) is assumed to be a sheet of negligible thickness at the semiconductor–oxide interface. Therefore, the voltage drop across the inversion layer can be neglected. Then, dropping the electron term in (A2.1.4), we write Poisson's equation as

$$\frac{d^2\phi}{dx^2} = -\frac{\rho}{\varepsilon_s} \cong -\frac{q p_0}{\varepsilon_s}(e^{-u} - 1), \tag{A2.1.5}$$

which has to satisfy the following boundary conditions:

$$\phi = 0, \quad \frac{d\phi}{dx} = 0 \quad \text{for } x = \infty. \tag{A2.1.6}$$

Deep in the interior of the material, both the potential and the electric field decrease to zero. Using now the definition of the one-dimensional electric field, it directly follows that $dx = -d\phi/F$. The differential equation (A2.1.5) can thus be written as

$$\frac{1}{2\phi_t}\frac{dF^2}{du} = -\frac{q p_0}{\varepsilon_s}(e^{-u} - 1). \tag{A2.1.7}$$

After solving the integral of (A2.1.7) and considering the boundary conditions in (A2.1.6), we find that

$$F^2 = \frac{2q\phi_t p_0}{\varepsilon_s}(e^{-u} + u - 1) \cong \frac{2q N_A \phi_t}{\varepsilon_s}(e^{-u} + u - 1). \tag{A2.1.8}$$

Since in a p-type substrate $p_0 \gg n_0$, we have employed the approximation $p_0 \cong N_A$ in (A2.1.8), the expression which gives the electric field in terms of the normalized local potential. For a positive field going into the semiconductor, the potential decreases in the inward direction, and, because the origin of the potential is at infinity, the potential is positive inside the semiconductor. Consequently, to calculate the electric field, the appropriate sign of the square root corresponds to $F > 0$ when $u > 0$ and $F < 0$ when $u < 0$. Thus, from (A2.1.8), we have

$$F = \text{sgn}(u)\sqrt{\frac{2qN_A\phi_t}{\varepsilon_s}(e^{-u} + u - 1)}. \qquad (A2.1.9)$$

To calculate the depletion charge in the semiconductor, we enclose the semiconductor in a rectangular Gaussian box from $x = 0^+$ (very close to the interface but not enclosing the electron charge) down to deep in the substrate ($x \to \infty$). Noting that at $x = 0^+$ the normalized potential and the electric field are represented as u_s and F_s, respectively, that the electric field and the potential are equal to zero deep in the substrate, and that the components of the electric field in the y- and z-directions are zero, it follows that Gauss' law for the rectangular box is written as

$$\text{charge enclosed/unit area} = Q'_B = -\varepsilon_s F_s$$
$$= -\text{sgn}(u_s)\sqrt{2q\varepsilon_s N_A \phi_t(e^{-u_s} + u_s - 1)}. \qquad (A2.1.10)$$

In order to calculate approximately the extension of the depletion layer into the substrate for the depletion and inversion regimes of operation, we use the depletion approximation, i.e. we neglect both types of carriers across the depletion width x_d. In this case, Poisson's equation is solved by dropping the terms corresponding to electrons and holes but keeping the impurity term of the volume charge density. This is equivalent to retaining the linear term in parentheses in (A2.1.10) and dropping the exponential and constant terms. In this case, (A2.1.10) becomes

$$Q'_B = -\varepsilon_s F_s = -\sqrt{2q\varepsilon_s N_A \phi_s}. \qquad (A2.1.11)$$

Since the depletion charge associated with a width x_d of a uniformly doped semiconductor is

$$Q'_B = -qN_A x_d, \qquad (A2.1.12)$$

equating (A2.1.11) and (A2.1.12) gives

$$x_d = \sqrt{[2\varepsilon_s/(qN_A)]\phi_s}. \qquad (A2.1.13)$$

A2.2 Drain- and source-associated inversion charges

The drain- and source-associated charges [2], [15] Q_D and Q_S are

$$Q_D = W\int_0^L \frac{y}{L}Q'_I\,dy, \qquad Q_S = W\int_0^L \left(1 - \frac{y}{L}\right)Q'_I\,dy. \qquad (A2.2.1)$$

The relationship between the differential channel length and the inversion charge density, given by (2.3.12), is repeated below:

$$dy = -\frac{\mu_n W}{n C'_{ox} I_D} Q'_{It} \, dQ'_{It}, \qquad (A2.2.2)$$

where Q'_{It}, which is a shifted version of the inversion charge density, is given by

$$Q'_{It} = Q'_I - n C'_{ox} \phi_t. \qquad (A2.2.3)$$

In order to determine the drain and source charges, the coordinate y must be expressed in terms of the inversion charge density Q'_I. The integration of (A2.2.2) from the source ($y=0$) to an arbitrary point y of the channel leads to

$$y = \frac{\mu_n W}{2 n C'_{ox} I_D} \left(Q'^2_F - Q'^2_{It} \right). \qquad (A2.2.4)$$

The substitution of (A2.2.2), (A2.2.3), and (A2.2.4) into the expression for Q_D in (A2.2.1) yields

$$Q_D = -\frac{\mu_n^2 W^3}{2 L \left(n C'_{ox} I_D \right)^2} \left[\int_{Q'_F}^{Q'_R} \left(Q'^2_F - Q'^2_{It} \right) \left(Q'_{It} + n C'_{ox} \phi_t \right) Q'_{It} \, dQ'_{It} \right]. \qquad (A.2.2.5)$$

After calculating the integral, we can use

$$I_D = \frac{\mu_n W}{C'_{ox} L} \frac{Q'^2_F - Q'^2_R}{2n}, \qquad (A2.2.6)$$

which is derived from expressions (2.1.60) and (2.1.61), for the current I_D in the denominator of (A.2.2.5). Next, we remove the common factor $\left(Q'_R - Q'_F \right)^2$ from both numerator and denominator to finally obtain

$$Q_D = WL \left[\frac{6 Q'^3_R + 12 Q'_F Q'^2_R + 8 Q'^2_F Q'_R + 4 Q'^3_F}{15 \left(Q'_F + Q'_R \right)^2} + \frac{n}{2} C'_{ox} \phi_t \right]. \qquad (A2.2.7)$$

Q_S may be determined in a similar way, or through $Q_S = Q_I - Q_D$, or simply by taking advantage of the source–drain symmetry, as follows:

$$Q_S = WL \left[\frac{6 Q'^3_F + 12 Q'_R Q'^2_F + 8 Q'^2_R Q'_F + 4 Q'^3_R}{15 \left(Q'_F + Q'_R \right)^2} + \frac{n}{2} C'_{ox} \phi_t \right]. \qquad (A2.2.8)$$

On introducing the channel linearity coefficient [14] α,

$$\alpha = \frac{Q'_R}{Q'_F} = \frac{Q'_{ID} - n C'_{ox} \phi_t}{Q'_{IS} - n C'_{ox} \phi_t} = \frac{q'_{ID} + 1}{q'_{IS} + 1} \qquad (A2.2.9)$$

the total, the source-, and the drain-associated inversion charges can be written as in Table A2.2.1.

Table A2.2.1 Total inversion, source-, and drain-associated inversion charges as functions of the inversion charge density at source and the channel linearity coefficient α

$$Q_I = WL\left[\frac{2}{3}\frac{1+\alpha+\alpha^2}{1+\alpha}(Q'_{IS} - nC'_{ox}\phi_t) + nC'_{ox}\phi_t\right] \quad \text{(A2.2.10)}$$

$$Q_S = WL\left[\frac{6 + 12\alpha + 8\alpha^2 + 4\alpha^3}{15(1+\alpha)^2}(Q'_{IS} - nC'_{ox}\phi_t) + \frac{n}{2}C'_{ox}\phi_t\right] \quad \text{(A2.2.11)}$$

$$Q_D = WL\left[\frac{4 + 8\alpha + 12\alpha^2 + 6\alpha^3}{15(1+\alpha)^2}(Q'_{IS} - nC'_{ox}\phi_t) + \frac{n}{2}C'_{ox}\phi_t\right] \quad \text{(A2.2.12)}$$

Table A2.3.1 The UCCM, pinch-off voltage, threshold voltage, and slope factor

$$\frac{V_P - V_{S(D)B}}{\phi_t} = \frac{-Q'_{IS(D)}}{nC'_{ox}\phi_t} - 1 + \ln\left(\frac{-Q'_{IS(D)}}{nC'_{ox}\phi_t}\right) \quad \text{(A2.3.1)}$$

$$V_P \cong \left(\sqrt{V_{GB} - V_{FB} - \phi_t + \gamma^2/4 + \sigma(V_{SB} + V_{DB})} - \gamma/2\right)^2 - 2\phi_F \quad \text{(A2.3.2a)}$$

$$V_P \simeq \frac{V_{GB} - V_T}{n} = \frac{V_{GB} - [V_{T0} - \sigma(V_{SB} + V_{DB})]}{n} \quad \text{(A2.3.2b)}$$

$$V_T = V_{T0} - \sigma(V_{SB} + V_{DB}); \quad V_{T0} = V_{FB} + \phi_t + 2\phi_F + \gamma\sqrt{2\phi_F} \quad \text{(A2.3.2c)}$$

$$n = 1 + \frac{\gamma}{2\sqrt{V_P + 2\phi_F}} \quad \text{(A2.3.2d)}$$

A2.3 Summary of n-channel MOSFET equations: UCCM, current, charges, transconductances, and capacitances including short-channel effects

See Tables A2.3.1–A2.3.5.

A2.4 An alternative low-frequency small-signal model of the MOSFET in saturation

The low-frequency small-signal drain current in the MOS transistor is

$$i_d = -g_{ms}v_s + g_{md}v_d + g_{mg}v_g + g_{mb}v_b, \quad \text{(A2.4.1)}$$

which can alternatively be rewritten [21] as

$$i_d = -(g_{ms} - g_{md})v_s + g_{md}(v_d - v_s) + g_{mg}v_g + g_{mb}v_b. \quad \text{(A2.4.2)}$$

Table A2.3.2 The drain current, virtual charges, and channel linearity factor

$$\frac{I_D}{I_S} = \frac{q'_{IS} + q'_{ID} + 2}{1 + \zeta(q'_{IS} - q'_{ID})}\left(q'_{IS} - q'_{ID}\right) \tag{A2.3.3a}$$

$$\frac{I_D}{I_S} = q'^2_{VS}(1 - \alpha^2) \tag{A2.3.3b}$$

$$q'_{VS(D)} = \frac{Q'_{VS(D)}}{-nC'_{ox}\phi_t} \qquad Q'_{VS(D)} = Q'_{IS(D)} - nC'_{ox}\phi_t + \frac{I_D}{Wv_{sat}} \tag{A2.3.4a}$$

$$\alpha = \frac{Q'_{VD}}{Q'_{VS}} \tag{A2.3.4b}$$

Table A2.3.3 Transconductances

$$\frac{g_{md}}{2I_S/\phi_t} = \frac{q'_{ID}\left(1 - \dfrac{\zeta I_D/(2I_S)}{q'_{ID} + 1}\right) + \dfrac{\sigma}{n}(q'_{IS} - q'_{ID})\left(1 - \dfrac{\zeta I_D/(2I_S)}{(q'_{IS} + 1)(q'_{ID} + 1)}\right)}{1 + \zeta(q'_{IS} - q'_{ID})} \tag{A2.3.5a}^a$$

$$\frac{g_{ms}}{2I_S/\phi_t} = \frac{q'_{IS}\left(1 - \dfrac{\zeta I_D/(2I_S)}{q'_{IS} + 1}\right) + \dfrac{\sigma}{n}(q'_{ID} - q'_{IS})\left(1 - \dfrac{\zeta I_D/(2I_S)}{(q'_{IS} + 1)(q'_{ID} + 1)}\right)}{1 + \zeta(q'_{IS} - q'_{ID})} \tag{A2.3.5b}^a$$

$$\frac{g_m}{2I_S/\phi_t} \cong \frac{g_{ms} - g_{md}}{2nI_S/\phi_t}$$

$$= \frac{(q'_{IS} - q'_{ID})\left[1 - \dfrac{2\sigma}{n}\left(1 - \dfrac{[\zeta I_D/(2I_S)][1 - n/(2\sigma)]}{(q'_{IS} + 1)(q'_{ID} + 1)}\right)\right]}{1 + \zeta(q'_{IS} - q'_{ID})} \tag{A2.3.6}$$

$$\cong \frac{(q'_{IS} - q'_{ID})\left[1 - \dfrac{2\sigma}{n} - \dfrac{\zeta I_D/(2I_S)}{(q'_{IS} + 1)(q'_{ID} + 1)}\right]}{1 + \zeta(q'_{IS} - q'_{ID})}$$

$$g_{mb} = (n - 1)g_m \tag{A2.3.7}$$

$$g_{md} = \frac{2I_S}{\phi_t}\frac{\dfrac{\sigma}{n}\left(\sqrt{1 + \dfrac{I_D}{I_S}} - 1\right)\left(1 - \dfrac{\zeta I_D/(2I_S)}{\left(\sqrt{1 + \dfrac{I_D}{I_S}} + \dfrac{\zeta I_D}{2I_S}\right)\left(1 + \dfrac{\zeta I_D}{2I_S}\right)}\right)}{1 + \zeta\left(\sqrt{1 + \dfrac{I_D}{I_S}} - 1\right)} + \frac{I_D}{L - \Delta L}\frac{d\Delta L}{dV_D} \tag{A2.3.8}^b$$

^{a} For the triode region (CLM not included).
^{b} In saturation.

Table A2.3.4 Transistor charges

$$Q_I = W(L - \Delta L) \left[\frac{2}{3} \frac{Q'^2_{VS} + Q'_{VS}Q'_{VD} + Q'^2_{VD}}{Q'_{VS} + Q'_{VD}} + nC'_{ox}\phi_t \right] - \frac{LI_D}{v_{sat}} \tag{A2.3.9}$$

$$Q_D = \frac{W(L - \Delta L)^2}{L} \left(\frac{2}{15} \frac{3Q'^3_{VD} + 6Q'^2_{VD}Q'_{VS} + 4Q'_{VD}Q'^2_{VS} + 2Q'^3_{VS}}{(Q'_{VS} + Q'_{VD})^2} + \frac{nC'_{ox}\phi_t}{2} \right)$$
$$- \frac{LI_D}{2v_{sat}} \tag{A2.3.10}$$

$$Q_S = Q_I - Q_D \tag{A2.3.11}$$

Table A2.3.5 Capacitive coefficients

$$C_{gs} = \frac{2}{3} WL_eC'_{ox} \frac{1 + 2\alpha}{(1 + \alpha)^2} \frac{q'_{IS}}{1 + q'_{IS}} + \frac{L_e g_{ms}}{3nv_{sat}} \frac{(1 - \alpha)^2}{(1 + \alpha)^2} \tag{A2.3.12}$$

$$C_{gd} = \frac{2}{3} WL_eC'_{ox} \frac{\alpha^2 + 2\alpha}{(1 + \alpha)^2} \frac{q'_{ID}}{1 + q'_{ID}} - \frac{L_e g_{md}}{3nv_{sat}} \frac{(1 - \alpha)^2}{(1 + \alpha)^2} \tag{A2.3.13}$$

$$C_{bs(d)} = (n - 1)C_{gs(d)} \tag{A2.3.14}$$

$$C_{gb} = C_{bg} = \frac{n - 1}{n} \left(C_{ox} - C_{gso} - C_{gdo} - \frac{L_e g_{mg}}{3v_{sat}} \frac{(1 - \alpha)^2}{(1 + \alpha)^2} \right) \tag{A2.3.15}^a$$

$$C_{ds} = -\frac{4}{15} nC'_{ox}W \frac{L_e^2}{L} \frac{1 + 3\alpha + \alpha^2}{(1 + \alpha)^3} \frac{q'_{IS}}{1 + q'_{IS}} - \frac{1}{30} \frac{g_{ms}L_e^2}{v_{sat}L} \frac{(3\alpha + 7)(1 - \alpha)^2}{(1 + \alpha)^3} \tag{A2.3.16}$$

$$C_{sd} = -\frac{4}{15} nC'_{ox}W \frac{L_e^2}{L} \frac{\alpha + 3\alpha^2 + \alpha^3}{(1 + \alpha)^3} \frac{q'_{ID}}{1 + q'_{ID}} + \frac{1}{30} \frac{g_{md}L_e^2}{v_{sat}L} \frac{(3 + 7\alpha)(1 - \alpha)^2}{(1 + \alpha)^3} \tag{A2.3.17}$$

$$L_e = L - \Delta L$$

$$\alpha = \frac{Q'_{VD}}{Q'_{VS}} = \frac{Q'_{ID} - nC'_{ox}\phi_t + I_D/Wv_{sat}}{Q'_{IS} - nC'_{ox}\phi_t + I_D/Wv_{sat}} \tag{A2.3.18}$$

[a] C_{gso} and C_{gdo} are the first terms in expressions (A2.3.12) and (A2.3.13), respectively.

In saturation, usually $g_{ms} \gg g_{md}$; thus the first term in (A2.4.2) $g_{ms} - g_{md} \cong g_{ms}$. On the other hand, the second term in (A2.4.2) represents a current through a conductance connected between drain and source, which is, in most textbooks, denoted g_{ds}. Therefore, the low-frequency small-signal model of the MOSFET in saturation can be

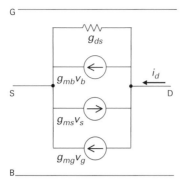

Fig. A2.4.1.   The low-frequency small-signal model of the MOSFET in saturation.

approximately represented as in Figure A2.4.1. In this textbook we have used both small-signal models, those of Figure 2.17 and Figure A2.4.1, for a MOSFET in saturation, in spite of the more general use of the latter in most textbooks and technical papers.

Problems

2.1 Considering the relationship between depletion charge and surface potential given in Appendix A2.1, derive the approximate expression for the depletion capacitance per unit area for a p-type substrate

$$C'_b \cong \frac{\sqrt{2q\varepsilon_s N_A}}{2\sqrt{\phi_s}}.$$

Calculate C'_b at $\phi_s = 2\phi_F$ for $N_A = 10^{15}\ \mathrm{cm}^{-3}$ and $N_A = 10^{17}\ \mathrm{cm}^{-3}$.

2.2 Derive expression (2.1.28) for the inversion capacitance and determine the value of the surface potential for which $C'_{ox} = C'_i$. Comment on the result.

2.3 Derive the unified charge-control model (UCCM) by using the linear relationship between inversion charge and surface potential in (2.1.55), and imposing equality between the expression for the drain current (2.1.54) and the Pao–Sah general current expression (2.1.49).

2.4 Derive expression (2.2.27) for the gate transconductance.

2.5 (a) Derive the expression

$$\frac{W}{L} = \frac{g_{mg}}{\mu C'_{ox}\phi_t} \frac{1}{\sqrt{1 + i_f} - 1}, \tag{P2.5.1}$$

which gives the aspect ratio of a saturated MOSFET in terms of its transconductance, inversion level, and technological parameters. (b) Derive the expression

$$I_D = I_{WI}\left(1 + \frac{(W/L)_{th}}{(W/L)}\right); \qquad \left(\frac{W}{L}\right)_{th} = \frac{g_{mg}}{2\mu C'_{ox}\phi_t}; \qquad I_{WI} = n g_{mg}\phi_t \tag{P2.5.2}$$

for the relationship between current consumption and the aspect ratio of an MOS transistor for a given gate transconductance. I_{WI} is the minimum current for achieving the required transconductance and $(W/L)_{th}$ is the aspect ratio for $q'_I = 2$ ($i_f = 8$).

2.6 Calculate the W/L ratio, the current consumption and V_{DSsat} for a saturated n-channel MOSFET having a (gate) transconductance of 1 mA/V operating at (a) $i_f = 0.01$, (b) $i_f = 1$, and (c) $i_f = 100$, given that $t_{ox} = 7$ nm, $n = 1.2$, $\mu_n = 400$ cm^2/V per s, and $T = 300$ K. Hint: Use the result of Problem 2.5.

2.7 Show that the total depletion charge can be written as

$$Q_B = -\frac{n-1}{n} Q'_I + Q'_{Ba} WL,$$

where Q'_{Ba} is the depletion charge for a negligible carrier charge density. Determine the expression for the gate stored charge from the charge-neutrality condition.

2.8 Find the equivalent circuit of Figure 2.24 from the general small-signal equivalent circuit of Figure 2.23. Calculate the small-signal current gain i_d/i_g for the circuit of Figure 2.24 and derive the complete expression for f_T. Derive the approximate expression (2.3.24) for f_T. Plot the curves of the approximate and complete expressions for f_T normalized with respect to $\mu_n \phi_t/(2\pi L^2)$ for both $n = 4/3$ and $n = 5/3$, and $10^{-4} < i_f < 10^4$. Comment on the resulting plots.

2.9 Consider the box-shaped resistor of Figure P2.9. Here $n_0 = 10^{15}$ cm^{-3}, $\mu_n = 500$ cm^2/V per s, $L = 100$ μm, and $A = 10$ μm^2. Calculate the current I for a voltage difference of 1 V between 0 and L. If the saturation velocity of the carriers is $v_{sat} = 100$ μm/ns, what is the maximum current through the resistor?

2.10 Show that for a saturated MOSFET operating in weak inversion the diffusion-related velocity can be written as

$$\text{velocity}_{diff} = \frac{\mu_s \phi_t}{L} \frac{Q'_I(0) - Q'_I(L)}{Q'_I(y)}$$

and interpret the short-channel parameter ζ defined in (2.4.9). Determine the channel length for which the diffusion-related velocity at the source of a saturated transistor operating in weak inversion equals the saturation velocity of the carriers. Consider $\mu_n = 400$ cm^2/V per s, $v_{sat} = 10^7$ cm/s, and $T = 300$ K. Speculate on other properties of this future "high-current weak-inversion only" technology.

2.11. Derive equation (2.4.18). Express σ in terms of the parameters of (2.4.17).

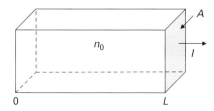

Fig. P2.9   A box-shaped semiconductor resistor.

2.12. Assume that the gate current in a MOSFET is proportional to the gate area WL. If the ratio $I_G/I_D = 10^{-4}$ for a minimum-sized device, calculate the channel length of the transistor for which $I_G = I_D$.

2.13. Consider that the moderate-inversion region is characterized by $0.22 \leq q'_I \leq 20$, as determined in Example 2.5. (a) What is the normalized forward current range of a MOSFET for which $0.22 \leq q'_{IS} \leq 20$? (b) What is the corresponding gate-voltage variation for $V_S = 0$? Assume that $n = 1.25$. Note that the moderate-inversion range extends over almost three decades of current when moderate inversion is defined as in Example 2.5.

References

[1] R. S. C. Cobbold, *Theory and Applications of Field-Effect Transistors*, New York: Wiley-Interscience, 1970.

[2] C. Galup-Montoro and M. C. Schneider, *MOSFET Modeling for Circuit Analysis and Design*, Hackensack, NJ: World Scientific, 2007.

[3] R. B. Adler, A. C. Smith, and R. L. Longini, *Introduction to Semiconductor Physics*, New York: John Wiley, 1964.

[4] M. A. Maher and C. A. Mead, "A physical charge-controlled model for MOS transistors," in *Advanced Research in VLSI*, ed. P. Losleben, Cambridge, MA: MIT Press, 1987, pp. 211–229.

[5] H. C. Pao and C. T. Sah, "Effects of diffusion current on characteristics of metal–oxide (insulator)–semiconductor transistors," *Solid-State Electronics*, vol. **9**, no. 10, pp. 927–937, Oct. 1966.

[6] J. R. Brews, "A charge sheet model for the MOSFET," *Solid-State Electronics*, vol. **21**, no. 2, pp. 345–355, Feb. 1978.

[7] C. Enz, F. Krummenacher, and E. Vittoz, "An analytical MOS transistor model valid in all regions of operation and dedicated to low-voltage and low-current applications," *Journal of Analog Integrated Circuits and Signal Processing*, vol. **8**, pp. 83–114, July 1995.

[8] E. Vittoz and J. Fellrath, "CMOS analog circuits based on weak inversion operation," *IEEE Journal of Solid-State Circuits*, vol. **12**, no. 3, pp. 224–231, June 1977.

[9] K. Lee, M. Shur, T. A. Fjeldly, and T. Ytterdal, *Semiconductor Device Modeling for VLSI*, Englewood Cliffs, NJ: Prentice Hall, 1993.

[10] A. I. A. Cunha, M. C. Schneider, and C. Galup-Montoro, "An MOS transistor model for analog circuit design," *IEEE Journal of Solid-State Circuits*, vol. **33**, no. 10, pp. 1510–1519, Oct. 1998.

[11] E. Vittoz (2003), "Weak inversion in analog and digital circuits," CCCD Workshop, Lund (available on line at http://www.es.lth.se/cccd/images/CCCD03-Weak%20inversion-Vittoz.pdf).

[12] F. Silveira, D. Flandre, and P. G. A. Jespers, "A g_m/I_D based methodology for the design of CMOS analog circuits and its application to the synthesis of a silicon-on-insulator micro-power OTA," *IEEE Journal of Solid-State Circuits*, vol. **31**, no. 9, pp. 1314–1319, Sep. 1996.

[13] S.-Y. Oh, D. E. Ward, and R. W. Dutton, "Transient analysis of MOS transistors," *IEEE Transactions on Electron Devices*, vol. **27**, no. 8, pp. 1571–1578, Aug. 1980.

[14] C. Galup-Montoro, M. C. Scheider, A. I. A. Cunha *et al.*, "The advanced compact MOSFET (ACM) model for circuit analysis and design," *IEEE Custom Integrated Circuits Conference*, pp. 519–526, Sep. 2007.

[15] Y. Tsividis, *Operation and Modeling of the MOS Transistor*, 2nd edn., New York: McGraw-Hill, 1999.

[16] C. Canali, G. Majni, R. Minder, and G. Ottaviani, "Electron and hole drift velocity measurements in silicon and their experimental relation to electric field and temperature," *IEEE Transactions on Electron Devices*, vol. **22**, no. 11, pp. 1045–1047, Nov. 1975.

[17] N. Arora, *MOSFET Models for VLSI Circuit Simulation: Theory and Practice,*" Vienna: Springer-Verlag, 1990.

[18] K. N. Ratnakumar and J. D. Meindl, "Short-channel MOST threshold voltage model," *IEEE Journal of Solid-State Circuits*, vol. **17**, no. 5, pp. 937–948, Oct. 1982.

[19] R. L. Radin, G. L. Moreira, C. Galup-Montoro, and M. C. Schneider, "A simple modeling of the Early voltage of MOSFETs in weak and moderate inversion," *Proceedings of IEEE ISCAS*, pp. 1720–1723, May 2008.

[20] R. van Langevelde, A. J. Scholten, and D. B. M. Klaassen, "Physical background of MOS Model 11," Unclassified Report 2003/00239 (available online at http://www.semiconductors. philips.com/Philips_Models/).

[21] E. Vittoz, "MOS Transistor," *Intensive Summer Course on CMOS VLSI Design*, Lausanne (EPFL), July 1989.

3 CMOS technology, components, and layout techniques

This chapter begins with a rapid overview of CMOS technology for circuit designers, including the description of a simplified deep-submicron CMOS structure and its process flow as well as the main parameters of some representative processes. The devices available in CMOS processes, including resistors, capacitors, inductors, and bipolar transistors, are then reviewed. The main electrical parameters of the passive devices are presented. Some design issues related to the use of sub-wavelength optical lithography are addressed in the introduction to the layout sections. Finally, layout topics, including design rules and layout for manufacturability, for matching, and for transistor associations, are presented.

3.1 An overview of CMOS technology

MOS integrated circuits (ICs) have been manufactured for volume delivery ever since 1970. Taking advantage of the scalability of the MOS transistors, micron technologies were developed during the 1980s and deep-submicron technologies appeared in the 1990s. In this chapter we will briefly review some deep-submicron processes that are currently (2009) employed for analog design.

3.1.1 Basic process steps in monolithic IC fabrication

Monolithic ICs are fabricated [1]–[5] on high-quality monocrystalline silicon substrates called wafers, with diameters as large as 300 mm and thicknesses up to 1250 μm. The fabrication process enables the achievement of appropriate geometries in the semiconductor and in the insulating and conducting layers deposited on the semiconductor substrate. The fabrication process consists of a series of elementary steps that allows the patterning of the semiconductor substrate and the modification of its electrical properties by selective doping as well as the deposition, removal, and patterning of conducting and insulating layers.

The distinctive property of silicon compared with other semiconductors is that the thermal oxidation of silicon produces a high-quality electrical insulator that also serves as a barrier to the diffusion of impurities in silicon. Thermal oxidation is a high-temperature process (900–1200 °C) that is employed repeatedly during the fabrication of silicon ICs.

The other fabrication steps can be classified into four groups: (i) the lithographic processes that allow the patterning of the different layers; (ii) semiconductor doping;

(iii) deposition; and (iv) removal processes. We will briefly discuss photolithography as an introduction to the section on layout.

Ion implantation is the preferred doping process. Impurities are introduced into the semiconductor by bombarding the wafer with high-energy donor and acceptor ions in a high-voltage particle accelerator (ion implanter). Ion implantation is followed by thermal annealing in furnaces or by rapid thermal annealing (RTA) to activate the implanted dopants.

Deposition processes include physical vapor deposition (PVD), chemical vapor deposition (CVD), and epitaxy. In the latter, a crystalline silicon layer is grown on the surface of a wafer, most commonly from a vapor phase.

Metal films can be deposited by evaporating metal under vacuum (PVD), and metals and insulators can be deposited by sputtering, in which a target bombarded with energetic ions loses atoms that are deposited on the wafer.

Silicon nitride (Si_3N_4), silicon dioxide (SiO_2), polysilicon, and metals can be deposited by CVD, in which the material deposited on the wafer results from a chemical reaction inside the gaseous mixture surrounding the wafer. Removal processes include chemical (wet) etching, reactive-ion etching (RIE), and chemical–mechanical polishing (CMP), which is widely employed to obtain a flat surface on the wafer before photolithography, and allows the fabrication of several interconnection layers.

3.1.2 Generic deep-submicron CMOS process flow

3.1.2.1 Transistor fabrication

The transistor process flow starts with a p-type substrate or with p^- epitaxial silicon on a p^+ substrate. The p- and n-channel devices are fabricated in an n-well and in the lightly doped p-region, respectively, formed in the epitaxial layer. The advantage of a lightly doped silicon layer, a few microns in depth, grown on a heavily doped substrate (p^+/p^- epi) is the reduction of series resistances because of the heavy doping of the bulk. The use of a thin epitaxial layer, which allows close proximity of the p^+ substrate to the surface, is the key factor for suppression of latchup [6], a phenomenon described later in this chapter.

The first fabrication steps are oxide growth and CVD of nitride. Figure 3.1(a) shows the wafer after a first lithographic step in which the active (thin oxide) areas are covered by a photoresist, which protects them from the RIE that carves the shallow trenches. After the removal of the photoresist and thick oxide deposition followed by CMP planarization, the cross section of the structure appears as in Fig. 3.1(b) with the shallow trenches filled with CVD oxide. Typical trench dimensions are 140 nm width and 400 nm depth for a 90 nm process. Figure 3.1(c) shows the structure after n-well lithography, deep phosphorous and shallow arsenic implantation, and p-well lithography and boron implantation. Figure 3.1(d) shows the structure after gate oxide growth, polysilicon deposition, lithography, and RIE of the polysilicon gates. The gate oxide can be formed by more than one layer, for example a few atomic layers of SiO_2 followed by SiON (a gate stack). The addition of nitrogen is done to increase the dielectric constant of the insulator.

Following n^+ shallow source–drain-extension (SDE) lithography and implantation and p^+ shallow SDE lithography and implantation, the structure appears as in Fig. 3.1(e). Halo implants are used to give improved short-channel characteristics for processes below 250 nm.

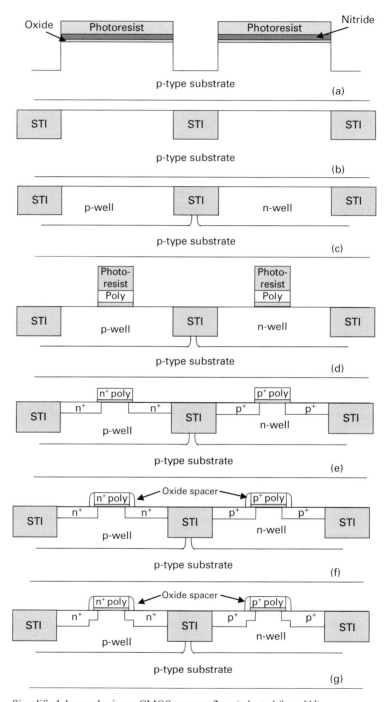

Fig. 3.1   Simplified deep-submicron CMOS process flow (adapted from [1]).

The spacer formation results from oxide or nitride deposition (CVD) and RIE. Since the oxide along the edges of the gates is thicker than elsewhere, the etching of the entire wafer results in oxide remaining around the edges of the gates, constituting the spacers, as shown in Fig. 3.1(f).

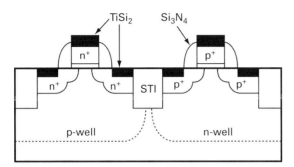

Fig. 3.2   A schematic cross section of transistors in a deep-submicron CMOS process (adapted from [7]).

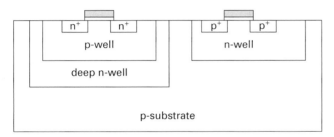

Fig. 3.3   A schematic cross section of transistors in a triple-well process.

After n$^+$ source–drain lithography and implantation and p$^+$ source–drain lithography and implantation, the structure appears as in Fig. 3.1(g). As seen in that figure, the source and drain are self-aligned with the spacers. It should be noted that the n$^+$ (p$^+$) source/drain implant also dopes the n$^+$ (p$^+$) polysilicon gate.

A refractory (high-melting-temperature) metal, such as tantalum, molybdenum, titanium, or cobalt, is then deposited over the entire wafer. Silicides are formed in the regions where the metal has contact with silicon or polysilicon. The metal over the oxide regions is then removed with a selective etch that does not remove the silicided regions.

As a result, the silicide is self-aligned with the gates and the source and drain regions (Figure 3.2). Self-aligned silicidation is called salicidation.

3.1.2.2   MOSFET structure in deep-submicron processes

Figure 3.2 illustrates the structure of the n- and p-channel MOS transistors in a generic deep-submicron CMOS process [4]–[6], [8]–[10]. In fact, this figure is a simplified representation of the structure of the transistors in the 250-, 180-, 130-, 90-, and 65-nm-technology nodes. The four main improvements of the technology illustrated in Figure 3.2, compared with older processes, are the following. Firstly, the devices are isolated by shallow trench isolation (STI) instead of by local oxidation of silicon (LOCOS). For the 250-nm node the two isolation approaches are used, whereas at or below 180 nm STI is the adopted solution. Secondly, complementary doped polysilicon gates (n$^+$ poly for the n-channel transistor and p$^+$ poly for the p-channel transistor) are used to form surface-channel n- and p-MOSFETs. Thirdly, shallow source and drain extension regions are used to reduce short-channel effects. Finally, sidewall oxide spacers allow self-aligned silicided sources, drains, and gates (TiSi$_2$, CoSi$_2$, NiSi) to give reduced effective sheet resistance of these layers.

The front-end of the line process that builds the transistor structures shown in Figure 3.1 uses six or seven masking levels. In some fabrication lines the twin-well process can be upgraded to a triple-well process, as seen in Figure 3.3, with the addition of a simple mask. The p-MOSFET is located in the (conventional) n-well, while the n-MOSFET is in the p-well located inside the deep n-well. Triple-well processes allow better noise isolation for memories and for mixed-signal circuits.

3.1.2.3   Interconnections

Aluminum is used in conventional metallization and tungsten is used as a plug to fill contact holes; CMP has enabled processes to include the formation of six or seven interconnection layers. Processes with several metallization levels use thicker and wider metals on going from the lower to the upper levels, as indicated in Table 3.1. To reduce the RC time constants of the interconnection lines, the use of low-resistivity layers and low-dielectric-constant (k) isolation layers has been adopted. The dielectric constant of silicon dioxide is $k = 3.9$. Other oxides and polymers offer dielectric constants as low as 2. Copper interconnects and low-k dielectrics are now popular. In the dual-damascene technique, a metallization level and a via are formed at the same time.

3.1.3   Main parameters in 350-, 180-, and 90-nm processes

Tables 3.2–3.4 show the main electrical parameters for three different technology nodes, 0.35 μm, 180 nm, and 90 nm [7]–[21]. It is worth noting that the areas of the six-transistor static random-access memory (RAM) cell are 20.5, 5.6, and 1 μm^2 for the 350-, 180-, and 90-nm-technology nodes, respectively [10], [7], [17].

Table 3.1 Layer pitch and thickness for a 180-nm CMOS process [7]

Layer	Pitch (nm)	Thickness (nm)
Isolation	520	530
Polysilicon	480	250
Metal 1	500	480
Metal 2, 3	640	700
Metal 4	1080	1080
Metal 5	1600	1600
Metal 6	1720	1720

Table 3.2 Main parameters for the 0.35-μm CMOS technology node [10]

	n-MOS	p-MOS
Supply voltage	3.3 V	3.3 V
Thin oxide	7 nm	7 nm
L_{gate}	0.29 μm	0.31 μm
V_T	0.50 V	− 0.65 V
I_{Dsat} (1.8 V)	540 μA/μm	300 μA/μm
I_{off}	0.5 pA/μm	0.5 pA/μm
Silicide (S, D, and poly)	8 Ω/sq	8 Ω/sq

Table 3.3 Main parameters for the six-metal-layer 180-nm CMOS technology node [7]

	n-MOS	p-MOS
Supply voltage	1.3–1.5 V	1.3–1.5 V
Thin oxide	3 nm	3 nm
L_{gate}	130 nm	150 nm
V_T	0.3 V (130 nm)	−0.24 V (150 nm)
I_{Dsat} (1.5 V)	0.94 mA/μm	0.42 mA/μm
I_{off}	3 nA/μm	3 nA/μm
g_{msat}	860 mS/mm	430 mS/mm
C_j (0 V)	0.65 fF/μm^2	0.95 fF/μm^2
Silicide (S, D, and poly)	3–5 Ω/sq	3–5 Ω/sq

Table 3.4 Main parameters for the seven-metal-layer 90-nm CMOS technology node [20]

Parameter	Logic (low power)		Analog		I/O	
	n-MOS	p-MOS	n-MOS	p-MOS	n-MOS	p-MOS
Supply voltage (V)	1.2		1.8		2.5	
Drawn gate (nm)	90		260		360	
t_{ox} (nm)^a	1.5		5.0		5.0	
V_T (mV)	420	−400	378	−338	240	−230
I_{Dsat} (mA/μm)	1.0	0.5	0.47	0.25	0.64	0.32
I_{off} (nA/μm)	15	6	2.45	0.6	3×10^{-2}	1.6×10^{-2}

^{a} Physical gate oxide thickness.

For the 180-nm node and below, halo implants of boron and arsenic are used for p-MOS and n-MOS transistors, respectively, to give improved short-channel characteristics.

The 90-nm node offers high-voltage analog and input/output (I/O) transistors with thicker gate oxide. In this node, strained silicon obtained using epitaxial SiGe to strain the silicon channel is employed to increase carrier mobility and, in turn, the drive currents of both n- and p-channel transistors.

3.2 Devices in CMOS technology

Analog circuits often require precision passive elements such as capacitors, resistors, and, sometimes, inductors. In analog processes, some extra steps allow the fabrication of these components, but in some cases an analog circuit must be built with the parasitic components available in a standard CMOS process. In this section we will review the main passive components available in analog and digital processes.

Bipolar transistors have well-known advantages compared with MOS transistors, mainly higher conductance at high currents, lower $1/f$ noise, better matching, and base–emitter voltages (V_{BE}) dependent on the semiconductor bandgap voltage. For these reasons bipolar transistors are sometimes used in analog MOS circuits, and we will review their availability in standard CMOS processes. The case of BiCMOS

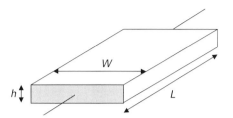

Fig. 3.4   An idealized rectangular resistor.

technologies [15], [20] with dedicated (heterostructure) bipolar transistors is outside of the scope of this book.

3.2.1   Resistors

Here, we are mostly interested in the resistance of conducting and semiconducting layers, usually with an (idealized) rectangular shape [22]–[25]. The resistance R of a uniform sample of rectangular geometry of length L, width W, and thickness h, as shown in Figure 3.4, is given by

$$R = \frac{V}{I} = \frac{FL}{hWqnv} = \frac{L}{hWqn\mu}, \tag{3.2.1}$$

where V and I are the potential drop and the current in the sample, respectively, F is the electric field, q the electronic charge, n the carrier density, v the velocity, and μ the carrier mobility.

The sheet resistance R_{SH} of the (rectangular) layer is defined by

$$R = \frac{L}{W}\left(\frac{1}{hq\mu n}\right) = \frac{L}{W}R_{SH}. \tag{3.2.2}$$

It is common to give the units of the sheet resistance in ohms per square (Ω/sq) instead of simply ohms, to emphasize that the sheet resistance is equal to the resistance of a square ($W = L$) piece of material. A resistor with $L/W = 10$ can be viewed as the series association of ten square resistors, and a resistor with $L/W = 1/10$ as the parallel association of ten square resistors.

The sheet resistance can be written in terms of the resistivity ρ of the conductor as

$$R_{SH}^{-1} = q\mu nh = \frac{h}{\rho}. \tag{3.2.3}$$

Example 3.1

Using the data in Table 3.5, find the sheet resistance of a copper layer of depth 1000 nm.

Answer

Using (3.2.3), we find that

$$R_{SH} = \frac{\rho}{h} = \frac{1.7 \times 10^{-6}\,\Omega\,\mathrm{cm}}{10^{-4}\,\mathrm{cm}} = 17\,\mathrm{m\Omega/sq}.$$

Table 3.5 Resistivities and temperature coefficients of resistivity (TCRs) of some metals

Metal (bulk)	Resistivity at 20 °C (Ω cm)	TCR (ppm/°C)
Aluminum	2.8×10^{-6}	3800
Copper	1.7×10^{-6}	4000
Gold	2.4×10^{-6}	3700

It must be observed that the resistivity of a thin metal film is usually considerably higher than that of the bulk material due to interface effects. Consequently, the above value is only a first-order estimate for the sheet resistance.

Because of the dependence of the sheet resistance of a thin film on the fabrication process (and also because of the variation in the dimensions due to photolithography inaccuracies), the absolute values of resistors are poorly controlled in IC processes. Worst-case variations of ±25% of the typical value of a sheet resistance are not uncommon. For resistors with minimum dimensions well over the minimum dimensions of the process, the variations in the resistance are due essentially to the sheet-resistance tolerances. For narrow-geometry resistors, for example, the width tolerance must be accounted for to calculate the variation in the resistance. Considering (statistically) independent variation in the variables, it follows immediately from (3.2.2) that

$$\left(\frac{\Delta R}{R}\right)^2 = \left(\frac{\Delta R_{SH}}{R_{SH}}\right)^2 + \left(\frac{\Delta W}{W}\right)^2, \tag{3.2.4}$$

where Δ represents the variation in a parameter.

The definitions above are valid for rectangular uniform samples. As a first-order approximation, the hypothesis of a uniform sample in the width and length direction is usually true, but this is not the case for the depth direction. For semiconductor layers, in most cases the doping (and carrier) density varies along the depth direction because the impurities are introduced from the surface of the sample. Fortunately, the definition of sheet resistance can be easily generalized for a non-uniform (in the depth direction) carrier concentration as

$$R_{SH}^{-1} = q \int_0^h \mu n \, dx. \tag{3.2.5}$$

The limitation of the use of the resistance formula for the rectangular geometry of the sample can also easily be overcome. For a non-rectangular geometry, an equivalent L/W can be defined and the resistance of the sample is simply given by

$$R = \left(\frac{L}{W}\right)_{eq} R_{SH}. \tag{3.2.6}$$

For the serpentine resistor shown in Figure 3.5, the equivalent L/W is calculated considering that each corner square at a bend contributes approximately half (0.56) of a square [3] due to current crowding in the region of the bend. Readers interested in the effect of layout on the resistance value can find further information in [24]. In summary,

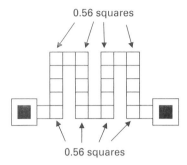

0.56 squares

0.56 squares

Fig 3.5   Top view of a serpentine resistor.

sheet resistance is universally used to characterize conducting layers in ICs because the resistance of any metal or semiconductor resistor, even with non-rectangular geometries, can very easily be determined from (3.2.6).

For most metals, the resistivity increases with temperature because of the decrease in carrier mobility with temperature. This variation is characterized by the (linear) temperature coefficient of resistivity given by

$$\mathrm{TCR} = (d\rho/\rho)/dT|_{T=T_a},\tag{3.2.7}$$

which is usually given in parts per million (ppm) per degree Celsius (°C). T_a is room temperature.

Example 3.2

Using the data in Table 3.5, find the percentage variation in the resistivity of copper for a 10 °C increase in temperature.

Answer

Using (3.2.7) we find that

$$\frac{\Delta\rho}{\rho} = \mathrm{TCR}\cdot\Delta T = 4\times 10^{-3}/°\mathrm{C}\times 10\,°\mathrm{C} \rightarrow \frac{\Delta\rho}{\rho} = 4\%.$$

Metal layers are available for interconnection and metal resistors from these layers usually have resistances in the range from tens of mΩ to several ohms; thus, their main applications are in current sensing and short-circuit protection. In some analog processes, thin-film sputtered metal resistors with high sheet resistance are available [20], but in general the best-quality resistors in CMOS processes are those made of polysilicon.

3.2.1.1   Polysilicon resistors

In deep-submicron processes the gate polysilicon is silicided [26]–[28], which results in a sheet resistance of the order of 2 Ω/sq, which is too low for the implementation of most resistors. In processes involving a silicide block mask, the n$^+$ and p$^+$ (non-silicided)

Table 3.6 Summary of properties of integrated resistors in CMOS technology

Resistor type	Sheet resistance (Ω/sq)	Temperature coefficient (ppm/°C)	Voltage coefficient (ppm/V)
n^+-Polysilicon	100	−800	50
p^+-Polysilicon	200	200	50
n^+/p^+-Polysilicon (silicided)	5		
n^+-Diffusion	50	1500	500
p^+-Diffusion	100	1500	500
n-Well	1000	2500	10000

The values of the sheet resistance and the temperature and voltage coefficients are only indicative. For a specific process the reader should consult the foundry documentation.

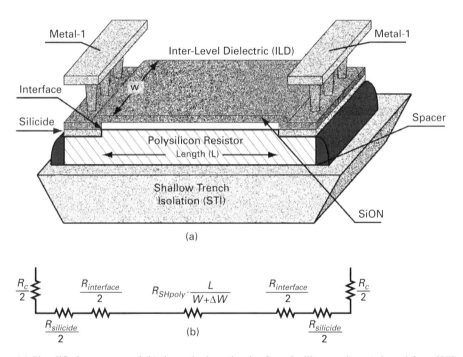

Fig 3.6   (a) Simplified structure and (b) dc equivalent circuit of a polysilicon resistor (adapted from [27]).

polysilicon gate layers, with sheet resistance usually above $100\,\Omega$/sq, as shown in Table 3.6, are available to implement precision resistors.

The simplified structure of a polysilicon resistor in a deep-submicron CMOS process is shown in Figure 3.6. As seen in this figure, the silicide block mask excludes the ends of the resistor where the contacts are located.

As shown in Figure 3.6, the total resistance R can be written as

$$R = R_C + R_{silicide} + R_{interface} + R_{SHpoly}\,\frac{L}{W + \Delta W}, \qquad (3.2.8)$$

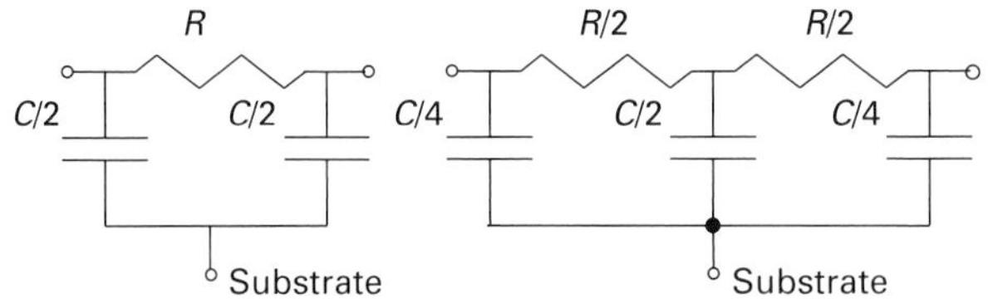

Fig. 3.7   Single-π and double-π equivalent circuits for a polysilicon resistor [23]

where R_C and $R_{silicide}$ are the effective contact and silicide resistances, R_{SHpoly} is the sheet resistance of the polysilicon, and L, W, and ΔW are the polysilicon resistor length, width, and width correction, respectively. In general, $\Delta L \ll L$ and the length correction is, therefore, negligible. Since the resistor is located on the shallow-trench isolation, the parasitic capacitance with respect to the substrate is low. Figure 3.7 shows two equivalent circuits for polysilicon resistors, including the parasitic capacitances [23].

The electrical properties of polysilicon can be easily described considering the resistance of a sample as being the series association of grains with bulk-silicon properties plus grain-boundary resistances. For resistors with polysilicon of large grain size and small boundary effects, the resistance and TCR are similar to those of equivalently doped monocrystalline silicon resistors. On the other hand, films with small grain size, larger boundaries, and/or higher trap densities can exhibit resistivities an order of magnitude higher than that of the equivalently doped monocrystalline silicon [24].

Grain-boundary resistances are highly non-linear, because of the transport mechanisms involved, including carrier hopping between traps, and crossing or tunneling through potential barriers. The non-linearity of the polysilicon film is dependent on the relative values of the grain and grain-boundary resistances and is given by the voltage coefficient in Table 3.6. To increase the linearity of the resistor, its length should be increased, with a corresponding decrease in the inner electric field.

In addition, since the grain-boundary resistances can have strong negative TCR because more carriers are available at high temperature to cross the potential barriers at grain boundaries, films for which the resistance is dominated by the grain boundary resistances have negative TCR. Polysilicon films with zero TCR can be obtained (but are not available in a standard process) for doping levels around $10^{20}\,\mathrm{cm}^{-3}$.

3.2.1.2   Implanted and diffused resistors

The best option to fabricate a large resistor in a standard CMOS process is to use the n-well, which can have a sheet resistance of 1 kΩ/sq or even higher. The main drawback of the well resistor is that it is highly non-linear. Well resistors and other implanted/diffused resistors are, in fact, junction field-effect transistors (JFETs), and they must be modeled as such. Silicon foundries furnish the JFET parameters to model resistor wells in circuit simulators.

To develop a simple model for an n-well resistor, let us consider that the well–substrate junction is a one-sided step junction with

$$N_D \gg N_A. \tag{3.2.9}$$

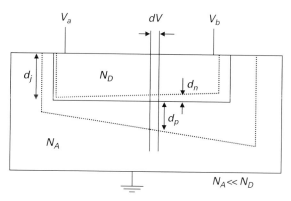

Fig. 3.8   Cross section of an implanted/diffused resistor.

Considering charge neutrality (see Figure 3.8) in the depletion region represented by the region between the dotted lines, we can write

$$N_D d_n = N_A d_p, \qquad (3.2.10)$$

which gives the following results for the depletion widths:

$$d_p \simeq \sqrt{\frac{2\varepsilon_s}{qN_A}(\phi_{bi} + V)} \qquad (3.2.11)$$

and

$$d_n \simeq \frac{N_A}{N_D}\sqrt{\frac{2\varepsilon_s}{qN_A}(\phi_{bi} + V)}, \qquad (3.2.12)$$

where $V_a \leq V \leq V_b$ is the voltage measured with respect to the substrate and ϕ_{bi} is the built-in potential. The voltage drop across a channel element of length dy is

$$dV = I_R\,dR = \frac{I_R\,dy}{W(d_j - d_n)q\mu_n N_D}, \qquad (3.2.13)$$

where d_j is the metallurgical junction depth and W is the resistor width. Integrating (3.2.13) between V_a and V_b, the voltages applied to the resistor terminals, results in the standard expression for the current in a JFET [29], which is given by

$$I_R = G_0\left\{V_b - V_a - \frac{2}{3}\frac{N_A}{N_D}\frac{1}{d_j}\sqrt{\frac{2\varepsilon_s}{qN_A}}\left[(V_b + \phi_{bi})^{3/2} - (V_a + \phi_{bi})^{3/2}\right]\right\}, \qquad (3.2.14)$$

where

$$G_0 = \frac{Wd_jq\mu_n N_D}{L} \qquad (3.2.15)$$

is the conductance of the well layer, assuming that the depletion depth $d_n = 0$.

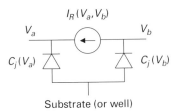

Fig. 3.9   The equivalent circuit for an implanted/diffused resistor. $I_R(V_a, V_b)$ is given by (3.2.14). Assuming that the junction area is WL, the junction capacitances are given by $C_j(V) = (WL/2)\sqrt{q\varepsilon_s N_A/[2(\phi_{bi} + V)]}$.

A more elaborate model also considers the lateral depletion regions of the well [30]. A complete model of the resistor should include the (normally reverse-biased) resistor–substrate junction, as shown in Figure 3.9. As is clear from Figures 3.8 and 3.9, the well resistor is a three-terminal device and, for this reason, two voltage coefficients characterize the non-linearity of the resistor [31] as given below:

$$R = R_0 \left[1 + \text{VCR}(V_b - V_a) + \text{BCR}\left(\frac{V_b + V_a}{2}\right)\right], \qquad (3.2.16)$$

where VCR and BCR are the voltage coefficients for the differential and common-mode voltages, respectively.

For implanted/diffused resistors, the increase in the sheet resistance is obtained by decreasing the doping and/or the depth of the layer. In either case, the non-linearity of the resistor increases, because the effect of the depletion layer modulation of the resistor depth (width) is increased. In analytical terms, from (3.2.14) it follows that

$$\text{VCR (BCR)} \propto \frac{d_p N_A}{d_j N_D}. \qquad (3.2.17)$$

Finally, concerning the temperature coefficient of implanted/diffused resistors, an analogous tradeoff between sheet resistance and linearity exists. Reducing the doping of the layer increases the variation of mobility with temperature [25], [32] and, consequently, the temperature coefficient of the resistor.

3.2.1.3   The MOS transistor as a resistor

MOS transistors can be biased either in the ohmic or in the saturation region to act as resistors. MOS transistors in the saturation region emulate large resistors, but it is difficult to predict their resistance values accurately. The MOSFET in the triode region acts as a voltage-controlled resistor, i.e. the (non-linear) resistance between source and drain is controlled by the gate voltage. The value of the ac resistance R can be calculated at $V_{DS}=0$ using the universal transconductance-to-current ratio, which gives

$$\left.\frac{1}{R}\right|_{V_{DS}=0} = \left.\frac{dI_D}{dV_D}\right|_{V_{DS}=0} = -\left.\frac{dI_D}{dV_S}\right|_{V_{DS}=0} = g_{ms} = \frac{2I_S}{\phi_t}\left(\sqrt{1+i_f}-1\right). \qquad (3.2.18)$$

The value of the inversion level i_f is dependent on the values of the source and gate voltages according to the unified current-control model (UICM). The value of the resistance can be controlled by the inversion level at the source, which, in turn, can be controlled by the gate voltage. MOS transistors, however, are inherently non-linear resistors. The non-linearities of the MOS transistor operating as a resistor, as well as the capacitive effects that limit its operation at high frequencies, will be analyzed in depth in the section of Chapter 9 on MOSFET-C filters, the MOS counterpart of active RC-filters.

Example 3.3

Verify that, in strong inversion, the equivalent resistance between source and drain of an MOS transistor at $V_{DS}=0$ is given by

$$\frac{1}{R}\bigg|_{V_{DS}=0} = g_{ms(d)} = \mu C'_{ox} n \frac{W}{L}\left(\frac{V_G - V_{T0}}{n} - V_Q\right),$$

where V_Q is the dc potential at the source.

Answer

In strong inversion, the drain current is given by

$$I_D = \mu C'_{ox} n \frac{W}{2L}\left[\left(\frac{V_G - V_{T0}}{n} - V_S\right)^2 - \left(\frac{V_G - V_{T0}}{n} - V_D\right)^2\right].$$

The drain transconductance for $V_D = V_S = V_Q$ is

$$g_{md} = \frac{dI_D}{dV_D}\bigg|_{V_{DS}=0} = \frac{1}{R}\bigg|_{V_{DS}=0} = \mu C'_{ox} n \frac{W}{L}\left(\frac{V_G - V_{T0}}{n} - V_Q\right).$$

The MOSFET conductance in strong inversion is linearly dependent on the gate voltage, and is also dependent on technological parameters, namely mobility, oxide capacitance, and threshold voltage. On the other hand, in weak inversion the conductance is an exponential function of the gate voltage.

3.2.2 Capacitors

Many analog circuits require high-quality capacitors, which are characterized by high specific capacitance, uniformity, low voltage, and low temperature coefficients. Analog processes offer options with the additional steps necessary in order to obtain components that meet these high-quality requirements. The metal–insulator–metal (MIM) and double-polysilicon options are the most frequently used for capacitors. In digital CMOS processes, capacitors are available only as parasitic elements of transistors or interconnections, but their performance tradeoffs are well known [33]. In most cases, the capacitors used in IC design have the parallel-plate structure, and use silicon oxide as the insulator.

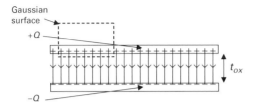

Fig. 3.10   An idealized parallel-plate capacitor showing the application of Gauss' law.

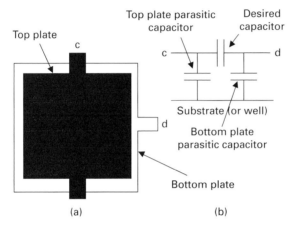

Fig. 3.11   An integrated parallel-plate capacitor: (a) simplified structure and (b) equivalent circuit.

3.2.2.1   Metal–insulator–metal (MIM) capacitors

We begin this section considering an idealized MIM capacitor and then analyze lateral flux, metal–(heavily doped) semiconductor, and transistor-gate capacitors.

Neglecting the fringing field in the parallel-plate capacitor shown in Figure 3.10, the voltage drop V is related to the stored charge Q by

$$V = Ft_{ox} = \frac{Q'}{\varepsilon_{ox}}t_{ox} = \frac{Q}{A\varepsilon_{ox}}t_{ox} = \frac{Q}{C}, \tag{3.2.19}$$

where F is the electric field in the insulator, t_{ox} the dielectric thickness, Q' the charge density per unit area, A the plate area, ε_{ox} the permittivity of the insulator, and C the capacitance given by

$$C = AC' = A\frac{\varepsilon_{ox}}{t_{ox}}. \tag{3.2.20}$$

As is clear from Figure 3.11, which illustrates an integrated parallel-plate capacitor, if we neglect the fringing field the capacitance is given by (3.2.20), with A being the overlapping area of the two plates. Since conductors are separated from adjacent conductors by dielectrics, parasitic capacitances exist between both the top plate and the bottom plate and the substrate (or well). Clearly, because of the geometry and proximity

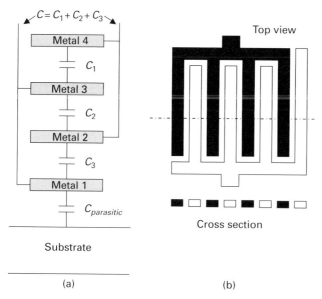

Fig. 3.12   Metal–oxide–metal (MOM) integrated capacitors: (a) vertical parallel-plate structure and (b) a lateral-flux capacitor.

to the substrate, the bottom-plate capacitance is higher than that of the top plate, but we must consider that interconnection capacitances are added in parallel with the top- and bottom-plate capacitances. The parasitic capacitances play the role of permanent ac connections between the capacitor terminals and the substrate (or well); thus, "noise" from the substrate (or well) is coupled to the capacitor and "signal" from the capacitor is coupled to the substrate (or well).

The advantage of integrated capacitors as precision analog components, compared with resistors, is their higher thermal stability. Their temperature coefficient (TCC) is given by

$$\text{TCC} = \frac{1}{C}\frac{dC}{dT} = \frac{1}{A}\frac{dA}{dT} - \frac{1}{t_{ox}}\frac{dt_{ox}}{dT} + \frac{1}{\varepsilon_{ox}}\frac{d\varepsilon_{ox}}{dT} \simeq \frac{1}{\varepsilon_{ox}}\frac{d\varepsilon_{ox}}{dT}. \tag{3.2.21}$$

Because the first and second terms, which are associated with thermal expansion, are relatively low [34], the TCC is dominated by the temperature coefficient of the dielectric constant of the silicon oxide, which is around 20 ppm/°C [34].

When no dedicated thin oxide is available for metal–metal capacitors, the available metal interconnection layers can be used to implement linear capacitors. The capacitance per unit area can be increased by placing capacitors between interconnecting layers and connecting them in parallel, as shown in Figure 3.12(a). In advanced technologies the minimum separation between metal strips is less than the spacing between metal layers; consequently, the lateral electric field is generally stronger than the vertical field. A simple structure of a capacitor that exploits this idea is shown in Figure 3.12(b), where the two terminals of the capacitor are metal lines from the same interconnecting layer. Capacitive structures with high capacitance per unit area using combinations of vertical and horizontal capacitors have been proposed [35], [36].

3.2.2.2   Metal–oxide–semiconductor (MOS) capacitors

Since the thermal (silicon) gate oxide is the thinnest available insulator in the IC structure, MOS capacitors have the highest capacitance per unit area in MOS technology. In this section we start by analyzing the voltage dependence of the general MOS structure and then consider the limit case of practical relevance of a highly doped semiconductor, which gives a highly linear capacitor. Finally, the temperature coefficient of MOS capacitances is calculated.

The total MOS capacitance (per unit area) is given by the series association of the oxide and semiconductor capacitances (per unit area) as

$$C'_{gb} = \frac{1}{\dfrac{1}{C'_c} + \dfrac{1}{C'_{ox}}}. \tag{3.2.22}$$

To calculate the voltage coefficient of the capacitance (VCC),

$$VCC = \frac{1}{C'_{gb}} \frac{dC'_{gb}}{dV_G}, \tag{3.2.23}$$

it is convenient to calculate the derivative of the total capacitance as

$$\frac{dC'_{gb}}{dV_G} = \frac{dC'_{gb}}{dC'_c} \frac{dC'_c}{d\phi_s} \frac{d\phi_s}{dV_G}, \tag{3.2.24}$$

where the first term is readily obtained from (3.2.22) and the third term is

$$\frac{d\phi_s}{dV_G} = \frac{C'_{ox}}{C'_{ox} + C'_c}, \tag{3.2.25}$$

due to the capacitive divider associated with the gate, channel, and substrate. Using (3.2.22) through (3.2.25) yields

$$VCC = \frac{C'^2_{ox}}{\left(C'_{ox} + C'_c\right)^2} \frac{1}{C'_c} \frac{dC'_c}{d\phi_s}. \tag{3.2.26}$$

The expression for the semiconductor capacitance given in Chapter 2, which is valid from accumulation to weak inversion, is

$$C'_c = \mathrm{sgn}(u_s) \frac{q p_0 \left(1 - e^{-u_s}\right)}{\sqrt{\dfrac{2 q \phi_t p_0}{\varepsilon_s}} \sqrt{\left(e^{-u_s} + u_s - 1\right)}}. \tag{3.2.27}$$

Developing (3.2.27) to first order around the flat-band condition, i.e. $u_s = 0$ (Problem 3.2), gives

$$C'_c \cong C'_{fb} \left(1 - \frac{1}{3} u_s\right), \tag{3.2.28}$$

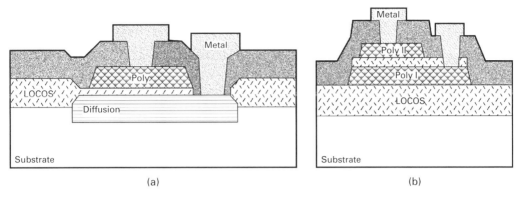

Fig. 3.13   (a) Poly–semiconductor and (b) poly–poly capacitors (adapted from [37]).

where

$$C'_{fb} = \sqrt{\frac{q\varepsilon_s p_0}{\phi_t}} = \sqrt{\frac{q^2\varepsilon_s p_0}{kT}} \qquad (3.2.29)$$

is the flat-band capacitance.

Finally, from (3.2.26), (3.2.28), and (3.2.29) it follows that

$$\text{VCC} \cong -\frac{C'^2_{ox}}{3q\varepsilon_s N_A}. \qquad (3.2.30)$$

Equation (3.2.30) gives VCC at flat band if we consider that $C'_c \gg C'_{ox}$, which is valid for heavily doped semiconductors. The VCC values at flat band are similar to those for highly doped substrates at the bias-voltage values which are usually applied, these being typically around 100 ppm/V (Problem 3.4).

The implementation of MOS capacitors with a heavily doped semiconductor (Figure 3.13(a)) in a self-aligned MOS process requires an extra mask, since in a conventional process the heavily doped source and drain implants are masked by the silicon gate structure.

Another capacitor structure sometimes available in CMOS processes is the highly linear poly-to-poly capacitor illustrated in Figure 3.13(b) (in Problem 3.5, the analysis of the dependence of the capacitance on the gate voltage is extended to these poly–poly capacitors), which is often used in switched-capacitor filters.

The temperature coefficient of the MOS capacitance is defined as

$$\text{TCC} = \frac{1}{C_{gb}}\frac{dC_{gb}}{dT} = \frac{C'_{ox} + C'_c}{AC'_{ox}C'_c}\frac{d}{dT}\left(\frac{AC'_{ox}C'_c}{C'_{ox} + C'_c}\right), \qquad (3.2.31)$$

where A is the capacitor area. Considering that $C'_c \gg C'_{ox}$, it can be shown that [34]

$$\text{TCC} = \left[\frac{1}{A}\frac{dA}{dT} - \frac{1}{t_{ox}}\frac{dt_{ox}}{dT}\right] + \frac{C'_{ox}}{C'^2_c}\frac{dC'_c}{dT} + \frac{1}{\varepsilon_{ox}}\frac{d\varepsilon_{ox}}{dT} \cong \frac{1}{\varepsilon_{ox}}\frac{d\varepsilon_{ox}}{dT}. \qquad (3.2.32)$$

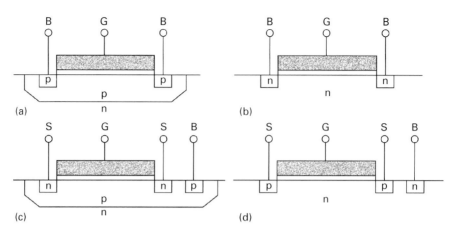

Fig. 3.14　(a)–(d) Gate capacitors in a p-well CMOS technology (adapted from [33]).

Since the terms in square brackets, which correspond to the thermal expansion and thermal coefficient of the semiconductor capacitance, respectively, can be neglected [34], the TCC is dominated by the thermal coefficient of the dielectric constant, which is very low, of the order of 20 ppm/°C.

3.2.2.3　MOSFET gate capacitors

For reasons of cost, it is highly desirable to implement the analog part of a mainly digital system on a chip with standard VLSI processes. One way to avoid the need for non-standard processes to implement linear capacitors is to use the MOSFET gate structure, which is the intrinsic element of any MOS technology. Compared with double-poly capacitors, thin-oxide capacitors present a larger capacitance per unit area and have better matching properties [38].

The thin-oxide capacitive structures available in a p-well CMOS technology are summarized in Figure 3.14. The first two, Figures 3.14(a) and (b), must operate in accumulation, whereas the last two can be biased in either accumulation or strong inversion. The device in Figure 3.14(a) can act as a floating capacitor, but the one in Figure 3.14(b) is an ac grounded capacitor since the n substrate is connected to the power supply. The device in Figure 3.14(c) can be shielded from substrate noise by applying a fixed potential to the well, whereas the one in Figure 3.14(d) can act as a floating capacitor when biased in strong inversion.

The MOS gate capacitor is a highly non-linear device, but, by properly biasing this structure, in either accumulation or strong inversion, capacitors presenting weak non-linearities can be obtained. The experimental and theoretical $C-V$ characteristics of a test capacitor in a 2-μm n-well CMOS technology are shown in Figure 3.15 [33]. When the capacitor is constrained to operate in either strong inversion ($V_{GB} < -1$ V) or accumulation ($V_{GB} > 1$ V), the capacitance is almost insensitive to the gate voltage.

The model of the MOS transistor with $V_{DS} = 0$, which is valid for low- and medium-frequency operation, is shown in Figure 3.16. The approximate expressions for the intrinsic gate capacitance are as follows. In strong inversion

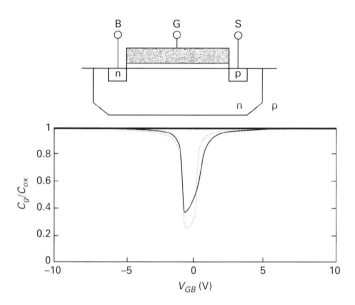

Fig. 3.15   A test capacitor in an n-well CMOS technology and its corresponding gate capacitance for $V_{BS}=0$. Solid and dotted lines represent experiment and theory, respectively (adapted from [33]).

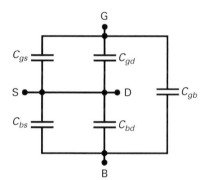

Fig. 3.16   Intrinsic capacitances of the MOS transistor at $V_{DS}=0$. For operation as a weakly non-linear capacitor, the gate is the top plate whereas either the source for strong-inversion operation or the bulk for accumulation can be the bottom plate (adapted from [33]).

$$C_{gs} = C_{gd} \cong \frac{1}{2}C_{ox},$$

$$C_{gb} \cong 0, \tag{3.2.33}$$

whereas in accumulation

$$C_{gs} = C_{gd} \cong 0,$$

$$C_{gb} \cong C_{ox}, \tag{3.2.34}$$

where C_{ox} is the thin-oxide intrinsic capacitance. The capacitances C_{bs} and C_{bd} are non-linear dependent on V_{BS}. For an actual device, the overlap capacitances are in parallel with C_{gs} and C_{gd}, while the junction capacitances are in parallel with C_{bs} and C_{bd}. A third junction parasitic capacitance (well-to-substrate) must be considered if the device is inside a well. In order to obtain a nearly voltage-independent capacitance, a voltage bias must be applied to the gate terminal to keep the structure in either accumulation or strong inversion. This is easy to carry out for grounded or virtually grounded capacitors, as will be shown in Chapter 9 for typical OTA-C systems and in Chapter 10 for SC filters.

To calculate the voltage coefficient of the MOS gate capacitance in accumulation, we approximate the semiconductor charge by retaining the dominant exponential term [33] only, which gives

$$-Q'_C \cong -\sqrt{2q\phi_t\varepsilon_s p_0}e^{-u_s/2}. \tag{3.2.35}$$

Consequently, the semiconductor capacitance is given by

$$C'_c = -\frac{dQ'_C}{d\phi_s} = \frac{Q'_C}{2\phi_t} = -\frac{C'_{ox}(V_G - V_{FB} - \phi_s)}{2\phi_t}. \tag{3.2.36}$$

The total gate capacitance (per unit area), which is the series association of the oxide and semiconductor capacitances (per unit area), is given by

$$C'_{gb} = C'_{ox}\left(1 - \frac{2\phi_t}{2\phi_t + V_{FB} - V_G + \phi_s}\right) \cong C'_{ox}\left(1 - \frac{2\phi_t}{2\phi_t + V_{FB} - V_G}\right), \tag{3.2.37}$$

which is valid for

$$V_{FB} - V_G \gg 2\phi_t. \tag{3.2.38}$$

Since (3.2.37) is valid in accumulation, we consider that $\phi_s \cong 0$. The voltage coefficient of the MOS gate capacitance in accumulation thus follows directly from (3.2.37) (Problem 3.4), resulting in

$$VCC = -\frac{2\phi_t}{(V_{FB} - V_G)^2}. \tag{3.2.39}$$

In inversion, a similar expression (Problem 3.4) holds:

$$VCC = \frac{2\phi_t}{(V_G - V_T)^2}. \tag{3.2.40}$$

It should be observed that, even when biased in accumulation or in strong inversion, gate capacitors with thin oxide present non-negligible non-linearities, as shown in the example that follows.

Example 3.4
Determine the voltage capacitance coefficient for a MOSFET capacitor in strong inversion with a gate-to-source voltage of 2.5 V. The threshold voltage is 0.6 V.

Answer
Using (3.2.40) we find that

$$\text{VCC} = \frac{2\phi_t}{(V_G - V_T)^2} = \frac{2 \times 26 \times 10^{-3}}{(2.5 - 0.6)^2} = 14.4 \times 10^{-3} \, \text{V}^{-1}$$

or 14.4 kppm/V.

So far, the gate capacitor has been modeled as a lumped element. For high-frequency operation, the distributed nature of the gate capacitor must be taken into account, since neither the channel nor the gate resistance is zero. In the following analysis, we assume that the gate resistance is significantly lower than the channel resistance; thus, to model the gate capacitor we assume that the gate area is equipotential. Under small-signal operation the gate structure can be modeled as a uniformly distributed RC transmission line. For a gate structure in which the drain is open, the equivalent impedance between gate and source is [39]

$$Z(s) = \frac{R_t}{\sqrt{s\tau}} \coth \sqrt{s\tau}, \tag{3.2.41}$$

where s is the complex frequency, $R_t = R_{SH}L/W$ (R_{SH} is the channel sheet resistance, and W and L are the channel width and length, respectively) is the total channel resistance, and $\tau = R_t C_t$, with $C_t = C'_{ox}WL$ the total gate capacitance. For the gate structure shown in Figure 3.14(c), expression (3.2.41) for the gate impedance still holds, but with $\tau = R_t C_t/4$, since the drain and source terminals are short-circuited. If the operating frequency is much lower than $1/\tau$, the transmission-line nature of the gate structure in Figure 3.14(c) can be modeled with the approximate impedance parameter

$$Z(s) \cong \frac{1}{sC_t} + \frac{1}{12}R_t. \tag{3.2.42}$$

Consequently, for frequencies under, say, $0.2/\tau$, the MOSFET gate structure acts as a capacitor with a quality factor given by

$$Q \cong \frac{12}{L^2 \omega R_{SH} C'_{ox}}. \tag{3.2.43}$$

Therefore, for high-frequency operation, short-channel capacitors are required in order to avoid an excessive phase shift due to their parasitic resistance.

It is worth making a final comment about the use of the MOS capacitor (in either inversion or accumulation) as a varactor (variable reactor) in electronically tuned circuits, the most common application being the voltage controlled oscillator (VCO). Varactors suitable for operation in the range 300 MHz–3 GHz with quality factors over 100 in 0.5-μm CMOS technology have been presented [40].

To conclude this subsection, Table 3.7 summarizes some relevant data on capacitors integrated in CMOS technologies.

3.2.3 Inductors

Inductors are the least attractive passive devices in IC technologies. Integrated inductors are implemented either as bond wires or as planar spirals [41]. The former have very

Table 3.7 Summary of capacitor properties in CMOS technology

Capacitor type	Capacitance per unit area (aF/μm^2)	Temperature coefficient (ppm/°C)	Voltage coefficient (ppm/V)
MOM	150	20	10
MOM (combined lateral and vertical structure)	200	20	10
MOS gate (biased)	5000	200	10000
MOS (heavily doped Si option)	1000	20	10
MIM (thin-oxide option)	1000	20	10
Poly–poly	1000	20	10

The values of capacitance per unit area and the temperature and voltage coefficients are only indicative. For a specific process, the reader should consult the foundry documentation.

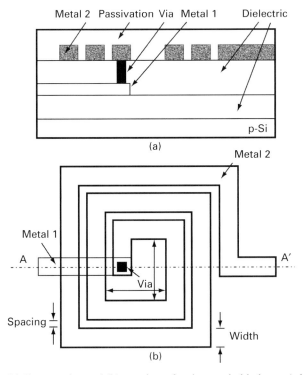

Fig. 3.17   (a) Cross section and (b) top view of a planar spiral inductor (adapted from [42]).

constrained values and are very sensitive to production fluctuations [41]. Thus, the most commonly used structure for inductors is the planar spiral, one example of which is shown in Figure 3.17.

The spiral inductor is commonly implemented in the topmost metal layer because this gives two advantages. The topmost layer is usually the thickest and, therefore, the one with the lowest sheet resistance. It is at the greatest distance from the substrate, thus presenting the lowest parasitic capacitance to the substrate. The underpass connection to

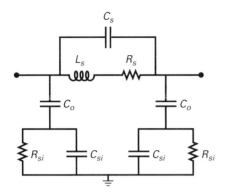

Fig. 3.18   A lumped inductor model (adapted from [41]).

the center of the spiral is built in a lower metal level, and the interconnection between the two layers is generally through several vias to reduce the series resistance. The equivalent circuit of the inductor is shown in Figure 3.18. L_s and R_s represent the inductance and series resistance of the spiral and the underpass. The overlap between the spiral and the underpass originates a capacitive coupling between the two terminals of the inductor, which is represented by C_s. The oxide capacitance between the spiral and the silicon substrate is represented by C_o. Finally, R_{si} and C_{si} model the silicon substrate.

The inductance of a spiral inductor is a complicated function of its geometry, and numerous publications have presented approximation methods [35], [41] to avoid the need for numerical (tridimensional) magnetic field solvers. Here, we will review first-order estimation formulas only.

For a simple circular loop, the exact expression for the inductance is given in terms of elliptic integrals. If the wire radius a is much smaller than the loop radius r ($a \ll r$), the inductance L of the loop [43] is given by

$$L = \mu_0 r \left[\ln\left(\frac{8r}{a}\right) - 2 \right],$$ (3.2.44)

where $\mu_0 = 4\pi \times 10^{-7}$ H/m is the vacuum permeability. For an ideal planar circular coil of n turns with a circular cross section of radius a, the inductance is given by (3.2.44) multiplied by n^2, i.e.

$$L = n^2 \mu_0 r \left[\ln\left(\frac{8r}{a}\right) - 2 \right].$$ (3.2.45)

A very simple estimate of the inductance of a spiral inductor is given [35] by

$$L \approx n^2 \mu_0 r,$$ (3.2.46)

which corresponds to a planar circular inductor with $r/a \cong 2.5$.

Example 3.5

Give a rough estimate of the inductance of a five-turn spiral inductor with an average radius of 50 µm.

Answer

A rough approximation of the inductance is, from (3.2.46),

$$L \approx n^2 \mu_0 r = 5^2 \times 4\pi \times 10^{-7} \times 50 \times 10^{-6} \cong 1.6\,\text{nH}.$$

Spiral inductors are very expensive in silicon real estate and, for this reason, values of on-chip inductances are commonly under 10 nH [35].

For the design of an inductor one must consider the effects of the parasitic components shown in Figure 3.18. Because of the parasitic capacitances the spiral inductor has a self-resonance frequency, which gives an upper boundary to its useful frequency range. The calculation of the quality factor Q of the inductor is involved because not only losses in the series resistances and contacts but also those associated with the induced (eddy) currents in the substrate must be considered. The interested reader can refer to [35] and the references therein for more details on inductor modeling.

3.2.4   Bipolar transistors

The bipolar junction transistor (BJT) has several advantages over the MOS transistor for analog applications, the two most important being that it brings less $1/f$ noise and has better matching characteristics. Additionally, the BJT offers exponential current–voltage characteristics at higher current densities than the MOSFET and can be used for bandgap references. BiCMOS technologies were developed to combine the best characteristics of MOSFETs and BJTs, but they are not always available, their cost can be high, and the MOS transistor performance in the BiCMOS process is inferior to that in the CMOS process. For these reasons, whenever a BJT is required, it is highly desirable to use the parasitic BJTs in a conventional CMOS process since they are always available at no extra cost.

As seen in Figure 3.19, the p-MOS transistor inside the n-well has three associated parasitic bipolar devices, consisting of one lateral and two vertical p–n–p structures. The

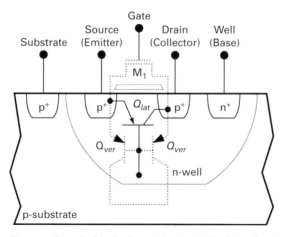

Fig. 3.19   Cross section of a bipolar lateral device in CMOS technology (adapted from [44]).

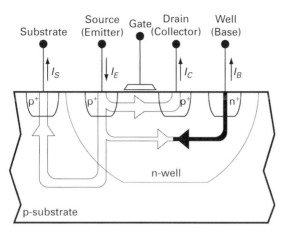

Fig. 3.20   Flow of carriers in the CMOS-compatible BJT (adapted from [44]).

use of the vertical transistors is limited since their collectors are at the substrate; thus, they are suitable only for common-collector topologies. Lateral devices have received much more attention for building analog circuits, such as bandgap references, accurate current mirrors, and low-noise amplifiers [45]. To conveniently operate a parasitic bipolar device, the MOS transistor must remain off, and thus a sufficiently positive (for a transistor inside the n-well) gate-to-well voltage must be applied.

The carrier flow for a lateral bipolar transistor operating in the forward active region (emitter–base junction forward biased and collector–base junction reverse biased) is shown in Figure 3.20. The holes injected by the emitter into the base (n-well) diffuse to the lateral collector as desired; however, they also diffuse to the p-substrate, because there is no n^+ buried layer that would prevent hole injection into the substrate. Thus, the lateral and vertical transistors operate in conjunction, the emitter current I_E being split into a lateral collector current I_C, a substrate collector current I_S, and a base current I_B. As a consequence, neither of the common-base current gains $\alpha_{lat} = I_C/I_E$ and $\alpha_{vert} = I_S/I_E$ is close to the ideal value of unity. On the other hand, due to the high emitter efficiency and reduced recombination in the base region, both the common-emitter current gains, $\beta_{lat} = I_C/I_B$ and $\beta_{vert} = I_S/I_B$, are usually sufficiently large for circuit applications. For common-collector topologies, the lateral and vertical (substrate) collectors can be connected, but for the other cases the lateral device current has to be higher than that of the vertical device. To maximize the ratio I_C/I_S, the emitter area must be minimized and the lateral collector must surround the emitter area, because in this way the efficiency of the vertical collector is reduced and that of the lateral collector is maximized. Additionally, the channel length of the MOSFET, which corresponds to the base width of the lateral transistor, must be set at its minimum value in order to increase the lateral BJT performance.

If a triple-well CMOS process is available, it offers the additional advantage of a vertical parasitic n–p–n BJT, as shown in Figure 3.21. Compared with the lateral p–n–p transistor, the vertical n–p–n transistor has the advantages of better uniformity, higher drive capability, and characteristics that are closer to the ideal ones. It should be observed

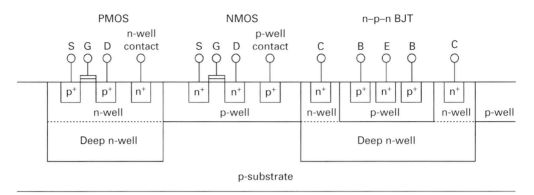

Fig. 3.21   Cross section of the triple-well CMOS technology (adapted from [46]).

that, although lateral BJTs have been used mainly in low-frequency applications, parasitic vertical n–p–n BJTs are currently being used in RF mixers for the operating-frequency range 1–2 GHz in a 0.18-μm technology [46].

3.3   Latchup

CMOS circuits can develop a low-resistance path between the power rails, which can cause catastrophic meltdown. This phenomenon, called latchup, occurs when the parasitic bipolar transistors inherent to the conventional CMOS process turn on. The parasitic bipolar transistors are shown in Figure 3.22, where only the source (or drain) of an n-channel and the source (or drain) of a p-channel MOSFET are shown. The two parasitic BJTs are a lateral n–p–n and a vertical p–n–p transistor. The parasitic resistances in the bulk and well regions are also indicated in Figure 3.22. The equivalent circuit of the two cross-coupled transistors is shown in Figure 3.22(b). This n–p–n–p structure constitutes a bistable silicon-controlled rectifier (SCR) or a thyristor. In normal operation the two bipolar transistors are off, but the structure can turn on if, for example, a large enough transient current causes a voltage drop across R_{nwell} that turns on the p–n–p BJT. The collector current of the p–n–p transistor, which flows through R_{sub}, can turn on the n–p–n transistor. The collector current of the n–p–n transistor lowers the base voltage of the p–n–p transistor, creating a positive feedback that can lead to a runaway process, known as the latchup of the structure. The value of the current between the power rails is limited only by the parasitic resistances. Latchup results in the failure of the circuit, which remains until the power is removed or can even cause the destruction of the chip. It should be noted that there is a parasitic n–p–n–p structure for each pair of n-channel and p-channel transistors, thus causing a chip to be very susceptible to latchup if measures to avoid it are not taken. Latchup is minimized by a combination of technology and design techniques. The use of a lightly doped epitaxial layer over a heavily doped substrate reduces the substrate parasitic resistances, making the turning on of the n–p–n BJTs less likely. The foundries furnish specific design rules to minimize latchup, including rules (of maximum spacing) for well and substrate contacts, and placement rules for the

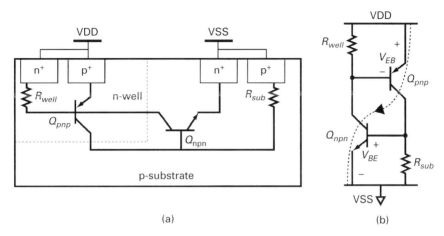

Fig. 3.22 Parasitic bipolar transistors in CMOS technology, which may lead to latchup: (a) cross section of the CMOS structure and (b) the equivalent circuit of the parasitic bipolar transistors and resistors (adapted from [47]).

transistors (avoiding intertwining of n-channel and p-channel MOSFETs). Input and output structures are the most prone to latchup because external voltages can ring above V_{DD} or below ground, forward biasing the base–emitter junctions and turning on the parasitic BJTs. Specific rules for the I/O, such as the use of guard rings as well as pre-designed I/O cells resistant to latchup, are furnished by the foundry.

3.4 Analog layout issues

3.4.1 Optical lithography

Photolithography is the process that allows the transfer of a pattern from a mask to the surface of the silicon wafer [4], [48]–[51]. Photolithography is currently a very sophisticated and costly technique, but the basic processing steps, illustrated in Figure 3.23, have remained the same since the development of the planar technology. Photoresist, a light-sensitive viscous liquid polymer, is applied to the top surface of the wafers. The wafers are then exposed to ultraviolet light through a mask (or reticle) that contains the geometric pattern to be transferred to the wafer. In current systems the reticle geometry is in large scale (e.g. 5× or 10×) and a reduction lens system projects the image of the reticle onto the photoresist. Before exposure, each mask following the first must be precisely aligned with the wafer. The alignment of each new mask level to one of the previous levels is achieved through the use of alignment marks on the wafer and mask.

For positive photoresists, the ultraviolet light breaks bonds, making the exposed regions more soluble in the developer. For negative photoresists, the exposed regions are less soluble because they polymerize under illumination. Finally, depending on the type of photoresist, a positive or negative image relative to the mask pattern is obtained after developing the photoresist. After hardening of the photoresist, the wafer can be etched to transfer the geometric pattern of the photoresist onto the underlying layer on the wafer.

Table 3.8 Wavelengths used for optical lithography [49]

Source	Wavelength (nm)	Intended resolution (nm)	Year of introduction
G-line[a]	436	1000	
I-line[a]	365	500	1984
KrF laser	248	250	1989
ArF laser	193	100	2001
F$_2$ laser	157	65	None[b]

[a] Filtered spectral components of high-pressure Hg or Hg–rare-gas discharge lamps.
[b] The technology was abandoned.

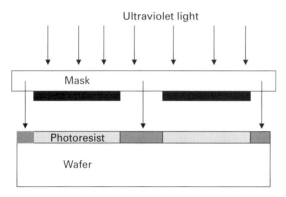

Fig. 3.23   A simplified representation of the basic steps of photolithography.

DESIGN   MASK WITH OPC   RESULT ON WAFER

Fig. 3.24   Optical proximity correction (OPC) counteracts lithography distortions (adapted from [48]).

As summarized in Table 3.8, there are only a few available wavelengths for lithography and the scaling of the wavelength of the light sources has not followed the scaling of the minimum dimensions of transistors. In the middle of the 1990s, the minimum dimensions of chip features approached the light wavelength used for lithography, requiring new techniques, known generically as resolution-enhancement techniques (RETs), to maintain control over the minimum dimensions, which are shorter than the wavelength. It should be noted that for the 90 (45) nm process node the critical dimension is slightly above one half (quarter) of the wavelength of the ArF laser, which is the shortest available.

The first RET was optical proximity correction (OPC), illustrated in Figure 3.24. In OPC, predictable distortions in the printed geometry are corrected by applying an inverse

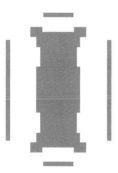

Fig. 3.25 Optical proximity correction (OPC) and SRAFs applied to a layout (adapted from [51]).

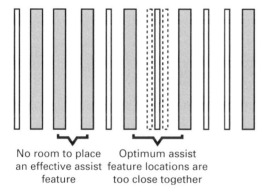

No room to place   Optimum assist
an effective assist feature locations are
feature   too close together

Fig. 3.26 Layout space widths to be avoided in the use of assist features (adapted from [48]).

distortion to the mask geometry. Corner rounding, for example, is compensated for by adding additional shapes (serifs) to outer corners or removing shapes from inner corners. The first generation of OPC was rule-based. The OPC rules were developed on the basis of simulation and experimental results. More advanced OPC algorithms are also available.

Another technique to enhance the printing of sub-wavelength geometries is the use of scatter bars, also called sub-resolution assist features (SRAFs), illustrated in Figure 3.25. These are narrow lines placed on the mask to enhance the image of adjacent figures. The resolution of the small geometry is enhanced by additional light interference produced by the surrounding scatter bars. The scatter bars themselves should not be printed on the wafer and, therefore, they must be smaller than the resolution limit of the lithography [48]. There must be space available around a geometry for it to be surrounded by scatter bars. As illustrated in Figure 3.26, layouts must avoid space widths that preclude the use of SRAFs.

One of the main techniques of sub-wavelength lithography is the use of a phase-shift mask (PSM), a technique invented in 1982. Interference is used beneficially in photo-lithography to improve image resolution. In alternating PSMs, the thickness of the clear region on the mask is modulated. The clear patterns have recessed and unrecessed areas. In the recessed areas the thickness is adjusted such that the phase of transmitted light shifts 180° relative to light in the unrecessed areas. Interference between the light waves

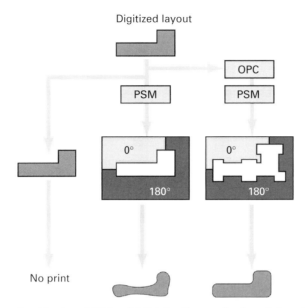

Fig. 3.27   Combination of OPC and PSM for 65-nm technology (adapted from [51]).

coming from adjacent regions creates a notch in the light intensity that allows a resolution appropriate for very fine lines. The phase assignment for the different regions presents some challenges, which are discussed in the literature [4], [48], [51]. The combination of PSM and OPC is illustrated in Figure 3.27.

3.4.1.1   Design for manufacturability

Deep-submicron manufacturability is a big issue and in this regard there is a strong focus on best practices for layout. A brief summary follows.

Design for manufacturability (DFM) [4] is strongly dependent on circuit layouts because of the dependence of the critical dimensions (CDs), such as poly width, on the poly density and pitch. If the transistors are all oriented in the same direction, correction of the CDs using RET is much easier. On the other hand, the simpler the shape of a polygon, the simpler the OPC. Limiting the degrees of freedom in a layout, such as requiring all transistors to be oriented in the same direction, can considerably improve process control and optimization.

The use of multiple contacts and vias also has a major impact on yield. Design uniformity and the use of folded devices improve matching, which is desirable in many analog, and even digital, circuits.

In Section 3.4.3 on MOSFET layout, we will develop analog layout techniques that are compatible with the best practices for DFM.

3.4.2   Mask layout and design rules

The patterns of the different layers constituting the integrated circuit are provided by the designer for the manufacturer (foundry) as a set of binary masks (dark, transparent),

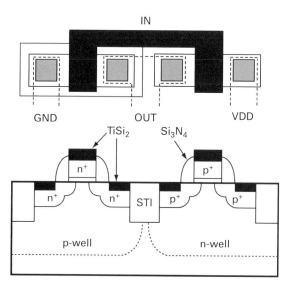

Fig. 3.28   Mask layout and cross section of a CMOS inverter (adapted from [7]). The n-well and p-well contacts are not shown. Dashed lines represent metal connections.

such that each mask determines the regions in which the corresponding layer is built [52]–[57]. Besides the masks furnished by the designer, other masks dependent on the designed masks, which are necessary for the processing, are generated by the manufacturer. The basic set of masks for a standard CMOS process (Figure 3.28) is the following.

(1) Wells are deep diffusions on which MOSFETs are built. In the basic CMOS process, the doping of the wells is the opposite to that of the substrate. Twin-well processes provide a well for each type of transistor, and triple-well technologies enable local bias for n- and p-wells, as explained in Section 3.1.2.2.

(2) The active regions are those which do not have a thick (field) oxide layer over them. Active regions are the source and drain of the transistors and the contact regions of the substrate and wells. The intersection of the active region with the (poly) gate region defines the intrinsic MOS transistor. The type of doping that an active region receives is determined by using a select mask or two active masks, n^+ active and p^+ active.

(3) Polysilicon, or simply poly, is the layer from which the gates of the transistors are built. If a silicide blocking mask is available, polysilicon can also be used to implement high-resistivity resistors.

(4) Various metal layers are available for interconnection.

(5) Contacts are holes in the insulation layer that allow the metal-1 layer to connect the active and poly regions under it.

(6) Via 1 connects metal 1 with metal 2, via 2 connects metal 2 with metal 3, etc.

(7) A thick insulator layer protects the entire circuit except the regions where electrical connections with the outside, the so-called bonding pads, are needed. The passivation or overglass mask defines the cuts in the overglass for bonding-pad positioning.

The non-idealities of a technology are taken into account as a series of geometric restrictions that must be obeyed in the design of a layout and are called design rules.

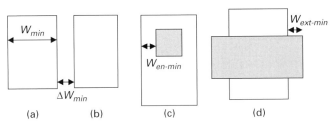

Fig. 3.29   The four main types of design rule: (a) minimum width, (b) minimum spacing, (c) minimum enclosure, and (d) minimum extension.

The design rules are related to the process steps involved in CMOS fabrication, including lithography, etching, implantation, and diffusion. Some of the physical phenomena that are at the origin of the design rules are mask (mis)alignment, optical resolution, etching undercut, and lateral diffusion.

Most of the design rules fall into one of the following four categories (see Figure 3.29).

Minimum element width (W_{min})	The width and length of the pattern drawn on a mask must exceed minimum values as a consequence of the lithography and process characteristics. Geometries with minimum dimensions under the allowed limits could present catastrophic failures, such as open circuits, or, alternatively, strong fluctuations in electrical parameters, such as resistance, which can be considered as parametric failures.
Minimum spacing (ΔW_{min})	To avoid shorts, geometries on the same mask, or in some cases on different masks, must be separated by a minimum spacing.
Minimum enclosure (W_{en-min})	Two layers to be connected must overlap even in the worst case of misalignment. A contact between two layers, which must remain within the two layers, is one example of the minimum-enclosure rule. A transistor within a well is another example of this rule.
Minimum extension ($W_{ext-min}$)	The polysilicon gate must extend beyond the edge of the active area. In this case the rule avoids the possibility that, due to mask misalignments and other effects, a source/drain diffusion bypasses the gate region of the transistor making the transistor unable to cut off.

In some cases, such as for contacts, the design rule can establish mandatory fixed dimensions.

Another kind of design rule is related to reliability issues and/or to failure mechanisms; the interested reader should consult the design kit of the specific process. Since design rules apply to every mask level and also to the relation between different layers, and their number is quite large, well over 100 layout rules are specified in design kits of modern CMOS processes.

Because of the importance of complying with design rules, a specific computer tool, the design-rule checker, is included in computer packages for layout generation.

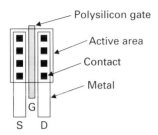

Polysilicon gate

Active area

Contact

Metal

Fig. 3.30   Typical layout of a transistor (the substrate contact is not shown).

3.4.3   MOSFET layout

A typical layout of a transistor, which is neither wide nor of minimum width, is shown in Figure 3.30 [54]–[59]. The active-area regions not protected by the polysilicon gate become the source and drain regions; the channel is the region under the gate. The polysilicon gate must extend beyond the edge of the active area according to a design rule. In Figure 3.30 only the source/drain contacts are shown; the substrate and gate contacts are not indicated. The use of multiple contacts, as shown in Figure 3.30, is typical in MOSFET layouts, and there are several reasons for this. Multiple contacts reduce series resistance and increase yield (the failure of one contact is not catastrophic), and the alternative of large-area contacts is usually not available, since the process is optimized for minimum-area contacts.

Layout styles aimed at both improving the matched behavior between (ideally identical) devices and reducing parasitic capacitances will be presented next.

3.4.3.1   Layout for matching

Analog circuits such as current mirrors and differential pairs rely on the concept of matched behavior between identically designed devices. Time-independent variations between identically designed transistors, called mismatch, affect the performance of most analog (and even digital) MOS circuits. Small MOSFET dimensions and reduced supply voltage make matching limitations even more important. Mismatch results from either systematic or stochastic (random) effects [61], [62]. Systematic effects originate from either poor layout or uncontrollable variation during IC fabrication. Systematic mismatch can originate from equipment-induced non-uniformities such as temperature gradients and differences in photomask size across the wafer. Systematic effects are important for large distances, but appropriate layout techniques can minimize them. Random mismatch refers to local variation in parameters such as doping concentration, mobility, oxide thickness, and polysilicon granularity. Random mismatch, which is dominant compared with systematic mismatch for short distances (that is, distances of the same order as the transistor size as opposed to the die size), will be covered in Chapter 4.

Matching of critical devices may be improved by following the rules summarized in Table 3.9, which are aimed at minimizing systematic mismatch. To reduce random mismatch the devices should be sized adequately to account for the spatial averages of the fluctuating quantities. A statistical model relating the standard deviation of the drain

Table 3.9 Rules for minimizing systematic mismatch of integrated devices [60]

Rule no.	Rule
1	Same structure
2	Same shape, same size
3	Same orientation
4	Same surroundings
5	Minimum distance
6	Common-centroid geometries
7	Same temperature

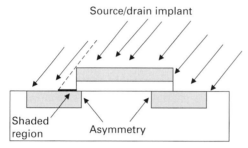

Fig. 3.31   Diagonal shift in the source/drain regions of a transistor due to a tilted implant (adapted from [59]).

current to transistor size is developed in Chapter 4. Here, we will address only the reduction of systematic mismatches.

The above rules are applicable to all IC technologies, but the quantitative importance of each rule depends on the process under consideration.

Some of the rules are self-evident, but a few comments are worthy of note.

(1) Devices to be matched must have the same structure, that is, a polysilicon resistor cannot be matched with a well resistor.

(2) They must have the same shape and size. For example, matched transistors must have not simply the same aspect ratio W/L, but also the same channel length L and the same channel width W.

(3) Matched devices must have the same orientation for the current flow in order to avoid the effect of non-isotropy of the substrate, including mechanical-stress-induced anisotropies. The current flow must be parallel and in the same direction to avoid the effect of non-symmetry between contacts, as illustrated in Figure 3.31.

(4) Matched devices should have the same surroundings in order to avoid there being different edge effects on the (ideally) matched devices during the etching or implantation processes. Dummy devices should be placed on either side of matched devices, as indicated in Figure 3.32. Unconnected dummies are prone to accumulate electrostatic charge since they are electrically unconnected. Connection of dummies to ground or a low-impedance node eliminates the possibility of electrostatic modulation of adjacent resistors [24].

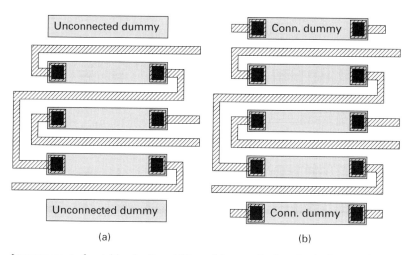

Fig. 3.32   Improvement of matching by the addition of dummy devices for the layout of two resistors with a resistance ratio of 2 : 1: (a) unconnected dummy resistors and (b) connected dummy resistors (adapted from [24] and [58]).

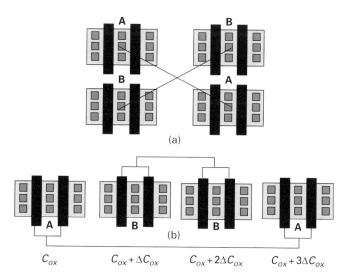

Fig. 3.33   Mock layouts of some possible common-centroid geometries for improved matching (adapted from [59]). Transistors with the same label are connected in parallel. It can be shown that, under certain conditions, $I_{DA} = I_{DB}$ (Problem 3.7).

(5) Minimizing the distance between devices reduces the effects of variation in parameters due to global mismatch.

(6) Common-centroid geometries, examples of which are illustrated in Figure 3.33, ideally cancel out the effect of constant gradients of parameters.

(7) Since the semiconductor device characteristics are strongly dependent on temperature, matched devices must be placed on the same isotherm. If the chip layout has

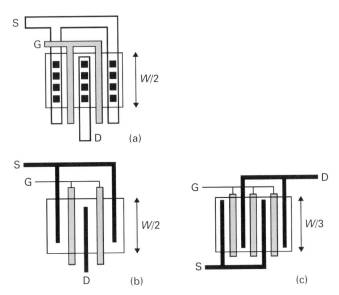

Fig. 3.34   (a) A transistor with an even number of folds. (b) Mock layout of the transistor with an even number of folds. (c) Mock layout of the transistor with an odd number of folds.

some symmetry with respect to the most dissipative devices, the isotherm location can be derived from symmetry considerations.

It is important to observe that device contacts must be considered when complying with rules (1) and (2), namely that devices be of the same structure and shape, and that the interconnections must be considered in relation to rule (4), concerning the need for matched devices to have the same surroundings.

In the next section, a very regular layout style that can easily comply with the matching rules indicated previously will be presented.

3.4.3.2   Folded layout

Folding the MOSFET (the transistor is split into two parallel connected pieces), as shown in Figure 3.34, potentially halves the source/drain diffusion areas because of the sharing of the diffusion strips by two subdevices [53], [58]. Consequently, the junction capacitances and leakage currents are also reduced, potentially being halved. As shown in Figure 3.34, use of an even number of folds places either the drain or source at both ends. Use of an odd number of folds places the drain at one end and the source at the other. An additional advantage of the use of an even number of folds is that the series resistance of the source or drain is insensitive to contact-mask misalignment.

As shown in Figure 3.35(a), a folded layout is very convenient for a differential pair. The inclusion of a third device, as shown in Figure 3.35(b), maintains symmetric surroundings in the layout of the differential pair.

A further reduction in the source/drain areas can be obtained by sharing them on the four sides. The resulting waffle-shaped layout [52], [56] is seldom used because of the non-rectangular shape of the channel, which makes it difficult to control the equivalent aspect ratio.

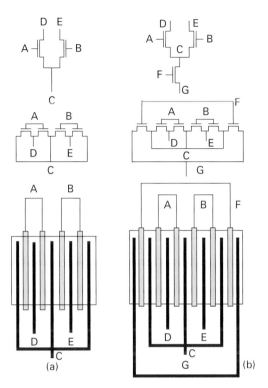

Fig. 3.35 (a) A differential pair with folded layout. (b) A third device is added without degrading the symmetry of the layout.

3.4.3.3 Interdigitated layout

Mismatch between folded transistors can be minimized using the interdigitated layouts of Figures 3.36 and 3.37 [53]. The simultaneous use of folded and interdigitated arrays allows a compact and symmetric layout to be achieved, which improves both speed and matching. With regard to matching, folded layouts obey rules (1) and (2) from Table 3.9 and interdigitated layouts additionally obey rules (3), (4), and (5). Rule (6) for obtaining a common-centroid differential pair can be satisfied using the layout of Figure 3.37.

3.4.3.4 Series association of transistors

In this section we demonstrate that the series association of two transistors is dc-equivalent to a single transistor. The gate and bulk terminals are common to both transistors, as shown in Figure 3.38. In the triode region, using the gradual-channel approximation, the drain current of an MOS transistor can be written as

$$I_D = \frac{W}{L}[g(V_G, V_S) - g(V_G, V_D)], \tag{3.4.1}$$

where all voltages are referred to the substrate, emphasizing the symmetry between source and drain. W and L are the channel width and length and $g(V_G, V)$ is a function that describes the dc behavior of the MOSFET, including the body effect and mobility reduction.

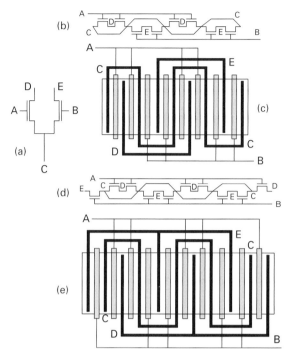

Fig. 3.36   A differential pair. (a) A simplified schematic diagram. (b) A schematic diagram of a case with an even number of folds. (c) Mock layout of an interdigitated even number of folds. (d) A schematic diagram of a case with an odd number of folds. (e) Mock layout of an interdigitated odd number of folds. Compared with the differential pair in (c), the one in (e) has a fold of transistor BEC added on the left and a fold of transistor ADC added on the right.

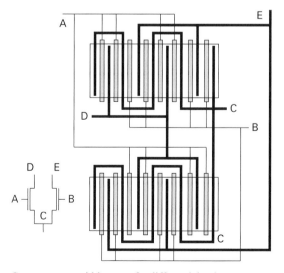

Fig. 3.37   Common-centroid layout of a differential pair.

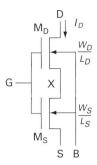

Fig. 3.38 Series association of transistors.

Assuming that V_D is such that M_D is biased in the triode region, from the equality of the drain currents of M_S and M_D in Figure 3.38 it follows that

$$g(V_G, V_X) = \frac{(W/L)_S g(V_G, V_S) + (W/L)_D g(V_G, V_D)}{(W/L)_S + (W/L)_D}. \qquad (3.4.2)$$

From (3.4.1) and (3.4.2), the drain current of the series association of transistors is

$$I_D = \left(\frac{W}{L}\right)_{eq} [g(V_G, V_S) - g(V_G, V_D)], \qquad (3.4.3)$$

where

$$\left(\frac{W}{L}\right)_{eq} = \frac{(W/L)_S (W/L)_D}{(W/L)_S + (W/L)_S} \qquad (3.4.4)$$

The case $W_D > W_S$ has applications for high-gain and high-frequency circuits [63]; here we will emphasize the simple case $W_D = W_S$, for which (3.4.4) reduces to

$$L_{eq} = L_S + L_D. \qquad (3.4.5)$$

It is worth emphasizing that the equivalence of a series association of transistors with the same width and a single transistor with the combined length given by (3.4.5) is valid considering the body effect, mobility degradation, and even velocity saturation (Problem 3.10).

Combining the series and parallel associations of transistors allows us to implement compact layouts for equivalent transistors having very different aspect ratios but high accuracy in current ratios [64]. As an example, the case of a current mirror exhibiting a current attenuation factor of 16 is shown in Figure 3.39. The input transistor is the parallel association of four unit transistors and the output transistor is the series association of four unit transistors. As shown in Figure 3.39, the layout can be very compact. Improved matching is possible, using interdigitated transistors, but at the price of a more cumbersome layout.

The layout techniques presented in the previous sections comply with the best practices of design for manufacturability summarized in Section 3.4.1.1. A detailed

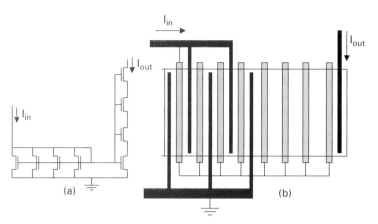

Fig. 3.39   A current mirror with an attenuation factor of 16: (a) schematic diagram and (b) mock layout.

coverage of layout issues extends over entire books [24] and courses [52], [53], [65], to which the interested reader is referred.

Problems

3.1 Determine the voltage coefficients VCR and BCR defined in (3.2.16) for an implanted resistor for which the current–voltage relation is given by (3.2.14). Calculate the numerical values of VCR and BCR for the case $N_A = 10^{16}\,\text{cm}^{-3}$, $N_D = 10^{18}\,\text{cm}^{-3}$, $d_j = 100\,\text{nm}$, and $(V_a + V_b)/2 = 1\,\text{V}$.

3.2 Derive expressions (3.2.28) and (3.2.30).

3.3 Derive expression (3.2.32).

3.4 Derive the general expression for the voltage coefficient VCC of an MOS capacitor [34], [37]:

$$\text{VCC} = \frac{C_{ox}'^2}{\varepsilon_s q N_D} \frac{e^{2u_s} - 2u_s e^{u_s} - 1}{(e^{u_s} - 1)^3},$$

where $u_s = \phi_s/(kT/q)$ is the normalized surface potential and N_D is the donor concentration. Determine the asymptotic limits of VCC for a heavily inverted/accumulated semiconductor. Plot VCC(flat band), VCC($V_G = -10\,\text{V}$), and VCC($V_G = 10\,\text{V}$) versus N_D for (a) $10^{17}\,\text{cm}^{-3} \le N_D \le 10^{19}\,\text{cm}^{-3}$ with $t_{ox} = 20\,\text{Å}$ and (b) $20\,\text{Å} \le t_{ox} \le 200\,\text{Å}$, applying the following scaling rule:

$$N_D\,(\text{cm}^{-3}) = 10^{16}\,[200/t_{ox}\,(\text{Å})]^2.$$

3.5 As shown in Figure P3.5, the capacitance of a poly-to-poly capacitor can be considered as the series association of the two space-charge capacitances and an oxide capacitance, i.e.

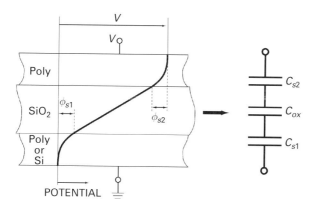

Fig. P3.5   A model for a poly-to-Si or poly-to-poly capacitor showing two space-charge capacitances (adapted from [34]).

$$\frac{1}{C_t} = \frac{1}{C_{ox}} + \frac{1}{C_{s1}} + \frac{1}{C_{s2}}.$$

Neglecting ϕ_{s1} when calculating ϕ_{s2} and vice versa, show that VCC = VCC$_1$ − VCC$_2$ [34]. Calculate VCC(−10 V), VCC(0 V), and VCC(10 V) for the case in which the poly doping is 10^{20} cm^{-3} for the two poly layers.

3.6 Using Ampère's law, calculate the magnetic field for each point along the axis of a circular loop of wire carrying a direct current I_0 (see, for example, [43]). Calculate the inductance of the loop assuming that the flux density at the center of the loop is half the average flux in the plane [35] and compare the result with Equation (3.2.46).

3.7 Assume that (i) the transistors labeled A in Figure 3.33(b) are connected in parallel and that the same is true for the two transistors labeled B; (ii) all transistors operate with the same applied voltages (V_G, V_S, V_D, and V_B). Neglecting the effect of the variation of oxide capacitance in the slope factor n, show that $I_{DA} = I_{DB}$.

3.8 Draw a common-centroid interdigitated layout with five folds for a differential pair (Figure 3.37 shows a common-centroid layout with an even number of folds).

3.9 Derive Equations (3.4.2) through (3.4.5).

3.10 The current law for a MOSFET (Chapter 4 in [66]) is given by

$$I_D = \frac{W}{L}\frac{\mu(V_G)}{1 + \alpha(Q'_{ID} - Q'_{IS})}\left[f(Q'_{IS}) - f(Q'_{ID})\right],$$

where the denominator $1 + \alpha(Q'_{ID} - Q'_{IS})$ models the effect of the saturation velocity on the drain current and $1/\alpha = nC'_{ox}LF_C$, where F_C is a constant critical field, $n = n(V_G)$ is the slope factor, and $Q'_{IS(D)}$ is the source (drain) carrier charge density. Consider the series association of two transistors, with the same channel width W and channel lengths L_S and L_D, and modeled by the current law shown above and the same technological parameters. Prove that the series association is equivalent to a single transistor with $L = L_S + L_D$.

References

[1] Y. Taur and T. H. Ning, *Fundamentals of Modern VLSI Devices*, Cambridge: Cambridge University Press, 1998.

[2] P. R. Gray, P. J. Hurst, S. H. Lewis, and R. G. Meyer, *Analysis and Design of Analog Integrated Circuits*, 4th edn., New York: John Wiley & Sons, 2001.

[3] R. C. Jaeger, *Introduction to Microelectronic Fabrication*, 2nd edn., Upper Saddle River, NJ: Prentice Hall, 2002.

[4] B. P. Wong, A. Mittal, Y. Cao, and G. Starr, *Nano-CMOS Circuit and Physical Design*, Hoboken, NJ: Wiley-Interscience, 2005.

[5] R. Jacob Baker, *CMOS Circuit Design, Layout, and Simulation*, 2nd edn., Hoboken, NJ: IEEE Press and Wiley-Interscience, 2005.

[6] Y. Taur, W. H. Chang, and R. H. Dennard, "Characterization and modeling of a latchup free 1-μm CMOS technology," *IEDM Technical Digest*, 1984, pp. 398–401.

[7] S. Yang, S. Ahmed, B. Arcot *et al.*, "A high performance 180 nm generation logic technology," *IEDM Technical Digest*, 1998, pp. 197–200.

[8] W.-H. Chang, B. Davari, M. R. Wordeman *et al.*, "A high-performance 0.25-μm CMOS technology: I – Design and characterization," *IEEE Transactions on Electron Devices*, vol. **39**, no. 4, pp. 959–966, Apr. 1992.

[9] B. Davari, W.-H. Chang, K. E. Petrillo *et al.*, "A high-performance 0.25-μm CMOS technology: II – Technology," *IEEE Transactions on Electron Devices*, vol. **39**, no. 4, pp. 967–975, Apr. 1992.

[10] M. Bohr, S. U. Ahmed, L. Brigham *et al.*, "A high performance 0.35 μm logic technology for 3.3 and 2.5 V operation," *IEDM Technical Digest*, 1994, pp. 273–276.

[11] T. C. Holloway, G. A. Dixit, D. T. Grider *et al.*, "0.18 μm CMOS technology for high-performance, low-power, and RF applications," *Symposium on VLSI Technical Digest*, 1997, pp. 13–14.

[12] S. Ikeda, Y. Yoshida, K. Shoji *et al.*, "A highly manufacturable 0.18 μm generation LOGIC technology," *IEDM Technical Digest*, 1999, pp. 675–678.

[13] C. Wann, J. Harrington, R. Mih *et al.*, "CMOS with active well bias for low-power and RF/analog applications," *Symposium on VLSI Technical Digest*, 2000, pp. 158–159.

[14] A. Chatterjee, D. Mosher, S. Sridhar *et al.*, "Analog integration in a 0.35 μm Cu metal pitch, 0.1 μm gate length, low-power digital CMOS technology," *IEDM Technical Digest*, 2001, pp. 211–214.

[15] J. Kirchgessner, S. Bigelow, F. K. Chai *et al.* "A 0.18 μm SiGe : C RFBiCMOS technology for wireless and gigabit optical communication applications," *IEEE BCTM Technical Digest*, 2001, pp. 151–154.

[16] N. Yanagiya, S. Matsuda, S. Inaba *et al.*, "65 nm CMOS technology (CMOS5) with high density embedded memories for broadband microprocessor applications," *IEDM Technical Digest*, 2002, pp. 57–60.

[17] S. Thompson, N. Anand, M. Armstrong *et al.*, "A 90 nm logic technology featuring 50 nm strained silicon channel transistors, 7 layers of Cu interconnects, low *k* ILD, and 1 μm² SRAM cell," *IEDM Technical Digest*, 2002, pp. 61–64.

[18] C. C. Wu, Y. K. Leung, C. S. Chang *et al.*, "A 90-nm CMOS device technology with high-speed, general-purpose, and low-leakage transistors for system on chip applications," *IEDM Technical Digest*, 2002, pp. 65–68.

[19] Y. W. Kim, C. B. Oh, Y. G. Ko *et al.*, "50 nm gate length logic technology with 9-layer Cu interconnects for 90 nm node SoC applications," *IEDM Technical Digest*, 2002, pp. 69–72.

[20] K. Kuhn, M. Agostinelli, S. Ahmed *et al.*, "A 90 nm communication technology featuring SiGe HBT transistors, RF CMOS, precision *R–L–C* RF elements and 1 μm^2 6-T SRAM cell," *IEDM Technical Digest*, 2002, pp. 73–76.

[21] MOSIS Integrated Circuits Fabrication Service, www.mosis.org.

[22] J. F. Shackelford, *Introduction to Materials Science for Engineers*, 4th edn., Upper Saddle River, NJ: Prentice Hall, 1996.

[23] D. J. Hamilton and W. G. Howard, *Basic Integrated Circuit Engineering*, Tokyo: McGraw-Hill Kogakusha, 1975.

[24] A. Hastings, *The Art of Analog Layout*, Upper Saddle River, NJ: Prentice Hall, 2001.

[25] D. J. Allstot and W. C. Black Jr., "Technological design considerations for monolithic MOS switched-capacitor filtering systems," *Proceedings of the IEEE*, vol. **71**, pp. 967–986, Aug. 1983.

[26] W. A. Lane and G. T. Wrixon, "The design of thin-film polysilicon resistors for analog IC applications," *IEEE Transactions on Electron Devices*, vol. **36**, no. 4, pp. 738–744, Apr. 1989.

[27] W.-C. Liu, K.-B. Thei, H.-M. Chuang *et al.*, "Characterization of polysilicon resistors in sub-0.25 μm CMOS ULSI applications," *IEEE Electron Device Letters*, vol. **22**, no. 7, pp. 318–320, July 2001.

[28] H.-M. Chuang, K.-B. Thei, S.-F. Tsai, and W.-C. Liu, "Temperature-dependent characteristics of polysilicon and diffused resistors," *IEEE Transactions on Electron Devices*, vol. **50**, no. 5, pp. 1413–1415, May 2003.

[29] A. S. Grove, *Physics and Technology of Semiconductor Devices*, New York: John Wiley & Sons, 1967.

[30] R. V. H. Booth and C. C. McAndrew, "A 3-terminal model for diffused and ion-implanted resistors," *IEEE Transactions on Electron Devices*, vol. **44**, no. 5, pp. 809–814, May 1997.

[31] A. M. Niknejad and B. E. Boser, EECS 240 Analog Integrated Circuits, http://rfic.eecs.berkeley.edu/~niknejad/ee240sp06.

[32] S. M. Sze, *Physics of Semiconductor Devices*, New York: John Wiley & Sons, 1969.

[33] A. T. Behr, M. C. Schneider, S. Noceti Filho, and C. Galup Montoro, "Harmonic distortion caused by capacitors implemented with MOSFET gates," *IEEE Journal of Solid-State Circuits*, vol. **27**, no. 10, pp. 1470–1475, Oct. 1992.

[34] J. L. McCreary, "Matching properties, and voltage and temperature dependence of MOS capacitors," *IEEE Journal of Solid-State Circuits*, vol. **16**, no. 6, pp. 608–616, Dec. 1981.

[35] T. H. Lee, *The Design of CMOS Radio-Frequency Integrated Circuits*, 2nd edn., New York: Cambridge University Press, 2004.

[36] R. Aparicio and A. Hajimiri, "Capacity limits and matching properties of lateral flux integrated capacitors," *CICC Technical Digest*, 2001, pp. 365–368.

[37] D. B. Slater Jr. and J. J. Paulos, "Low-voltage coefficient capacitors for VLSI processes," *IEEE Journal of Solid-State Circuits*, vol. **24**, no. 1, pp. 165–173, Feb. 1989.

[38] K. R. Lakshmikumar, R. H. Hadaway, and M. A. Copeland, "Characterization and modeling of mismatch in MOS transistors for precision analog design," *IEEE Journal of Solid-State Circuits*, vol. **21**, no. 6, pp. 1057–1066, Dec. 1986.

[39] L.-J. Pu and Y. Tsividis, "Transistor-only frequency-selective circuits," *Proceedings of IEEE ISCAS*, 1988, pp. 2851–2854.

[40] A.-S. Porret, *Design of a Low-Power and Low-Voltage UHF Transceiver Integrated in a CMOS Process*, PhD Thesis, EPFL, Lausanne, 2002.

[41] S. S. Mohan, M. del Mar Hershenson, S. P. Boyd, and T. H. Lee, "Simple accurate expressions for planar spiral inductances," *IEEE Journal of Solid-State Circuits*, vol. **34**, no. 10, pp. 1419–1424, Oct. 1999.

[42] M. Park, S. Lee, C. S. Kim, H. K. Yu, and K. S. Nam, "The detailed analysis of high Q CMOS-compatible microwave spiral inductors in silicon technology," *IEEE Transactions on Electron Devices*, vol. **45**, no. 9, pp. 1953–1959, Sep. 1998.

[43] S. Ramo, J. R. Whinnery, and T. van Duzer, *Fields and Waves in Communication Electronics*, New York: John Wiley & Sons, 1965.

[44] D. MacSweeney, K. G. McCarthy, A. Mathewson, and B. Mason, "A SPICE compatible subcircuit model for lateral bipolar transistors in a CMOS process," *IEEE Transactions on Electron Devices*, vol. **45**, no. 9, pp. 1978–1984, Sep. 1998.

[45] E. A. Vittoz, "MOS transistors operated in the lateral bipolar model and their application in CMOS technology," *IEEE Journal of Solid-State Circuits*, vol. **18**, no. 3, pp. 273–279, June 1983.

[46] I. Nam, Y. J. Kim, and K. Lee, "Low $1/f$ noise and DC offset RF mixer for direct conversion receiver using parasitic vertical NPN bipolar transistor in deep N-well CMOS technology," *Symposium on VLSI Circuits Digest*, 2003, pp. 223–226.

[47] M.-D. Ker, H.-C. Jiang, J.-J. Peng, and T.-L. Shieh, "Automatic methodology for placing the guard ring into chip layout to prevent latchup in CMOS IC's," *Proceedings of IEEE ICECS*, 2001, pp. 113–116.

[48] M. L. Rieger, J. P. Mayhew, and S. Panchapakesan, "Layout design methodologies for sub-wavelength manufacturing," *Proceedings of DAC*, 2001, pp. 85–88.

[49] L. W. Liebmann, "Layout impact of resolution enhancement techniques: impediment or opportunity?," *Proceedings of the International Symposium on Physical Design*, 2003, pp. 110–117.

[50] J. Kawa, C. Chiang, and R. Camposano, "EDA challenges in nano-scale technology," *CICC Technical Digest*, 2006, pp. 845–851.

[51] J. Kawa and C. Chiang, "DFM issues for 65 nm and beyond," *Proceedings of GLSVLSI*, 2007, pp. 318–322.

[52] E. Vittoz, "Layout techniques for analog circuits," *Intensive Summer Course on CMOS VLSI Design*, EPFL, Lausanne, July 1989.

[53] T. Schmerbeck, "Analog and mixed signal circuit layout techniques and floorplanning," *Advanced Engineering Course on Practical Aspects in Analog and mixed ICs*, EPFL, Lausanne, July 1994.

[54] C. Mead, *Analog VLSI and Neural Systems*, Reading, MA: Addison-Wesley, 1989.

[55] S.-C. Liu, J. Kramer, G. Indiveri, T. Delbrück, and R. Douglas, *Analog VLSI: Circuits and Principles*, Cambridge, MA: MIT Press, 2002.

[56] F. Maloberti, *Analog Design for CMOS VLSI Systems*, Boston, MA: Kluwer Academic Publishers, 2001.

[57] U. Gatti and F. Maloberti, "Analog and mixed analog-digital layout," in *Analog VLSI – Signal and Information Processing*, ed. M. Ismail and T. Fiez, New York: McGraw-Hill, 1994, pp. 699–726.

[58] Y. Tsividis, *Mixed Analog-Digital VLSI Devices and Technology*, New York: McGraw-Hill, 1996.

[59] B. Razavi, *Design of Analog CMOS Integrated Circuits*, New York: McGraw-Hill, 2001.

[60] E. Vittoz, "The design of high-performance analog circuits on digital CMOS chips," *IEEE Journal of Solid-State Circuits*, vol. **20**, no. 3, pp. 657–665, June 1985.

[61] M. J. M. Pelgrom, A. C. J. Duinmaijer, and A. Welbers, "Matching properties of MOS transistors," *IEEE Journal of Solid-State Circuits*, vol. **24**, no. 5, pp. 1433–1440, Oct. 1989.

[62] C. Galup-Montoro, M. C. Schneider, H. Klimach, and A. Arnaud, "A compact model of MOSFET mismatch for circuit design," *IEEE Journal of Solid-State Circuits*, vol. **40**, no. 8, pp. 1649–1657, Aug. 2005.

[63] C. Galup-Montoro, M. Cherem Schneider, and I. J. B. Loss, "Series-parallel association of FET's for high gain and high frequency applications," *IEEE Journal of Solid-State Circuits*, vol. **29**, no. 98, pp. 1094–1101, Sep. 1994.

[64] A. Arnaud, R. Fiorelli, and C. Galup-Montoro, "Nanowatt, sub-nS OTAs, with sub-10-mV input offset, using series-parallel current mirrors", *IEEE Journal of Solid-State Circuits*, vol. **41**, no. 9, pp. 2009–2018, Sep. 2006.

[65] T. Schmerbeck, "Mechanisms and effects of noise coupling in analog and mixed signal ICs," *Advanced Engineering Course on Practical Aspects in Analog and mixed ICs*, EPFL, Lausanne, July 1994.

[66] C. Galup-Montoro and M. C. Schneider, *MOSFET Modeling for Circuit Analysis and Design*, Singapore: World Scientific, 2007.

4 Temporal and spatial fluctuations in MOSFETs

This chapter deals with temporal and spatial fluctuations in electronic devices, with strong emphasis on MOSFETs. The spontaneous fluctuations over time of the current and voltage inside a device, which are basically related to the discrete nature of electric charge, are called electrical noise. Noise imposes minimum values for the input signals of amplifiers and other analog circuits.

Time-independent variations between identically designed devices in an integrated circuit due to the spatial fluctuations in the technological parameters and geometries are called mismatch. Since digital and analog integrated circuits often rely on the matched behavior between identically designed devices, mismatch affects the performance of most integrated circuits.

Mismatch (spatial fluctuation) and noise (temporal fluctuation) are accuracy-limiting factors, both depending on the fabrication process, device dimensions, temperature, and bias. The shrinkage of the MOSFET dimensions and the reduction in the supply voltage of advanced technologies have made the consideration of matching and noise even more important for analog design. Consequently, we have included a detailed presentation of mismatch and noise so that they can be considered in the subsequent study of the basic circuits and building blocks. This chapter begins with a short summary of the various sources of noise. The general model for drain-current fluctuations in MOSFETs is then introduced, and the fundamental thermal- and flicker-noise models are developed. Small-dimension and high-frequency effects on thermal noise are considered. Design-oriented models for thermal and flicker noise are then presented. In the second part of the chapter, the main models for random mismatch are summarized and the dependence of mismatch on bias, dimensions, and technology is discussed. Finally, the main analysis methods regarding matching for analog circuits are briefly reviewed.

4.1 Types of noise

4.1.1 Thermal noise

Thermal noise is due to the random thermal motion of electrons and holes, and is independent of the direct current in the device if the drift velocity is much less than the thermal velocity, as is the case for resistors and transistors operating under normal conditions. The thermal noise, calculated as a mean-square current or voltage, is directly

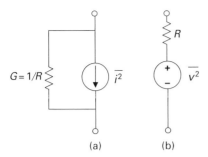

Fig. 4.1   (a) Norton and (b) Thévenin equivalent circuits of a real (noisy) resistor. The resistor in the
equivalent circuits is an ideal (noiseless) resistor.

proportional to the absolute temperature since the mean-square value is a measure of the
energy and the thermal kinetic energy of the carriers is proportional to the absolute
temperature T.

The mean-square value of the thermal-noise current of a resistance R $(=1/G)$ [1], [2] is

$$\overline{i^2} = 4kT\frac{1}{R}\,\Delta f = 4kTG\,\Delta f, \qquad (4.1.1)$$

where $k = 1.38 \times 10^{-23}$ J/K is the Boltzmann constant and Δf is the bandwidth in hertz
over which the noise is measured. Since the noise current has a mean-square value that is
proportional to the bandwidth, a noise current spectral density $\overline{i^2}/\Delta f$ that is independent
of frequency can be defined. Noise with a constant spectral density is called white noise.
Two equivalent circuits can represent the thermal noise of a resistor, as shown in
Figure 4.1. The Norton (shunt) equivalent circuit is directly represented by Equation
(4.1.1), whereas the Thévenin (series) equivalent circuit corresponds to

$$\overline{v^2} = 4kTR\,\Delta f. \qquad (4.1.2)$$

Example 4.1
Calculate the root mean square (rms) value of the thermal-noise current per $\mathrm{Hz}^{1/2}$ of a
1-mA/V conductance at $T = 300$ K.

Answer
Using (4.1.1) we find that

$$\sqrt{\overline{i^2}/\Delta f} = \sqrt{4kTG} = \sqrt{4 \times 1.38 \times 10^{-23} \times 300 \times 10^{-3}} \cong 4\,\mathrm{pA/Hz}^{1/2}.$$

On the other hand, the rms value of the thermal noise voltage per $\mathrm{Hz}^{1/2}$ is

$$\sqrt{\overline{v^2}/\Delta f} = \sqrt{4kTR} = \sqrt{4 \times 1.38 \times 10^{-23} \times 300 \times 10^{3}} \cong 4\,\mathrm{nV/Hz}^{1/2}.$$

4.1.2   Shot noise

Shot noise is associated with the random flow of carriers across a potential barrier. This noise is present in vacuum tubes, diodes, and bipolar transistors. The mean-square value of the fluctuation of a current I around its average value I_{DC} can be written [1], [2] as

$$\overline{i^2} = \overline{(I - I_{DC})^2}. \tag{4.1.3}$$

Shot noise is given by the Schottky formula [1], [2] as

$$\overline{i^2} = 2qI_{DC}\,\Delta f, \tag{4.1.4}$$

where q is the electronic charge and Δf is the bandwidth in hertz over which the noise is measured. As in the case of thermal noise, shot noise is white.[1] Thermal noise and shot noise are related phenomena, both being due to the thermal agitation of carriers. Conventionally, bipolar-transistor noise has been modeled as shot noise and FET noise as thermal noise. In some cases, for example FETs operating in weak inversion, the fundamental noise can be calculated either as thermal noise or as shot noise. The two calculations give exactly the same result, thus adding the thermal to the shot noise effectively means counting the same thing twice.

Example 4.2

Calculate the rms value of the shot noise per $Hz^{1/2}$ in a p–n junction diode for a direct current of 1 µA.

Answer

Using (4.1.4) we find that

$$\sqrt{\overline{i^2}/\Delta f} = \sqrt{2qI_{DC}} = \sqrt{2 \times 1.6 \times 10^{-19} \times 10^{-6}} \cong 0.57\,\mathrm{pA}/\mathrm{Hz}^{1/2}.$$

4.1.3   Flicker noise

All active devices, and some passive devices such as carbon resistors, present excess noise at low frequencies, in addition to thermal or shot noise, which is called flicker noise. Flicker noise is low-frequency noise that occurs when a direct current is flowing [1], [2], thus flicker noise can be regarded as being produced by a fluctuation in conductance. The power spectral density (PSD) of flicker noise varies with frequency in the form

$$\frac{\overline{i^2}}{\Delta f} = \frac{K}{f^{\mathrm{EF}}}, \tag{4.1.5}$$

with K and EF being constants. Since in most cases $\mathrm{EF} \cong 1$, flicker noise is also called $1/f$ noise.

[1] Here, we assume that the operating frequency is limited to values at which shot noise can be considered white.

This $1/f$ noise is as ubiquitous as thermal noise, but, unlike for thermal noise, there is no universal mechanism fully responsible for it. Even if flicker noise in MOSFETs is associated with interface traps, there is no known procedure to determine $1/f$-noise parameters other than from noise measurements. On the other hand, thermal (or shot) noise can be estimated from dc measurements. There is an extensive literature concerning flicker noise in MOSFETs since high $1/f$ noise has plagued MOS transistors ever since the beginning of the technology.

Example 4.3

Assume that $EF = 1$ and $K = 10^{-20}\,\mathrm{A}^2$. Calculate the mean-square value of the flicker-noise current over frequency ranges of (a) $[10\,\mathrm{Hz},\ 10\,\mathrm{kHz}]$, (b) $[10\,\mathrm{Hz},\ 10\,\mathrm{MHz}]$, and (c) $[(1\ (\text{million years})^{-1},\ 1\ \text{year}^{-1}]$.

Answer

Using (4.1.5) we find that

$$\overline{i^2} = \int_{f_{min}}^{f_{max}} \frac{K}{f}\,df = K\ln\left(\frac{f_{max}}{f_{min}}\right),$$

which leads to the following results:

(a) $\overline{i^2} = 10^{-20}\ln 10^3 \cong 7 \times 10^{-20}\,\mathrm{A}^2$,
(b) $\overline{i^2} = 10^{-20}\ln 10^6 \cong 14 \times 10^{-20}\,\mathrm{A}^2$,
(c) $\overline{i^2} = 10^{-20}\ln 10^6 \cong 14 \times 10^{-20}\,\mathrm{A}^2$.

Note that the results of (b) and (c) are equal even though the bandwidth in (c) is around 315×10^{12} narrower than that in (b). Of course, if the flicker noise is given by (4.1.5) and $EF = 1$, the noise current is dependent on the number of frequency decades, not on the bandwidth.

4.2. Modeling the drain-current fluctuations in MOSFETs

The fluctuation of the drain current in a MOSFET is a consequence of the internal (microscopic) fluctuations along the channel. Thus, a physics-based model of fluctuations must consider the distributed nature of the transistor. Some empirical models of mismatch and noise attribute drain-current variations to the fluctuation of a specific parameter of the transistor, e.g. the threshold voltage or the flat-band voltage. Even if the simplicity of the empirical models of mismatch and noise is appealing, the fact that they are not based on physics means that these models do not consistently represent the series association of transistors. Thus, we will recall the calculation of the drain-current fluctuation of a MOSFET considering the distributed nature of the channel. This calculation is carried out in three steps. Firstly, the (microscopic) fluctuations are calculated in a generic elementary region of the device and a local current or voltage generator is selected to represent the local fluctuations. Secondly, the transfer from the local current

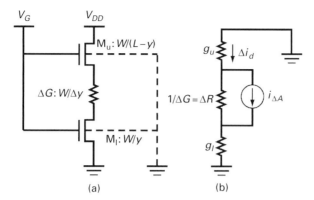

Fig. 4.2   (a) An MOS transistor represented as the series association of three elements: an upper transistor, a lower transistor, and a channel element in between. (b) The small-signal circuit used to calculate the drain-current fluctuation due to the local fluctuation in the channel element ΔA.

or voltage generator to the output current of the device is determined. Finally, the contributions of the elementary local sources to the fluctuation of the output current must be combined.

To determine the transfer function from an elementary channel segment to the drain current, the transistor is split into three series elements: the upper transistor, the lower transistor, and a small channel element of length Δy and area $\Delta A = W \Delta y$, as in Figure 4.2(a). On denoting the local current fluctuation produced by the channel element as $i_{\Delta A}$ (Figure 4.2(b)), small-signal analysis allows one to calculate the resulting effect of $i_{\Delta A}$ on the drain current. Current division between the channel element and the equivalent small-signal resistance of the rest of the channel gives

$$\Delta i_d = \frac{\Delta R}{1/g_u + 1/g_l} i_{\Delta A}. \tag{4.2.1}$$

Small-signal analysis can be carried out considering the general expression for the source (drain) (trans)conductance given in (2.2.25). Thus,

$$g_u = -\mu \frac{W}{L - y} Q'_I, \tag{4.2.2}$$

$$g_l = -\mu \frac{W}{y} Q'_I, \tag{4.2.3}$$

where Q'_I is the inversion charge density in the channel element of length Δy and μ is the effective mobility. The equivalent resistance ΔR of a small element of the channel of length Δy is

$$\Delta R = \Delta V / I_D = -\Delta y / (\mu W Q'_I). \tag{4.2.4}$$

It should be noted that (4.2.4) is a consequence of the general expression for the drain current (2.1.49) including drift and diffusion, and thus it is valid from weak to strong inversion.

The substitution of (4.2.2) through (4.2.4) into (4.2.1) yields

$$\Delta i_d = (\Delta y/L)i_{\Delta A}. \tag{4.2.5}$$

The current division coefficient is simply given by a geometric ratio because the three (trans)conductances involved are proportional to the inversion charge density at the same point in the channel.

Finally, the mean-square value of the total drain-current fluctuation is

$$\overline{i_d^2} = \lim_{\Delta y \to 0} \sum_{channel} \overline{(\Delta i_d)^2} = \lim_{\Delta y \to 0} \sum \overline{\left(\frac{\Delta y}{L} i_{\Delta A}\right)^2} = \frac{1}{L^2} \int_0^L \Delta y \overline{(i_{\Delta A})^2} dy. \tag{4.2.6}$$

On calculating (4.2.6) we have assumed that local current fluctuations along the channel are uncorrelated. Equation (4.2.6) is a very general expression, valid for long-channel transistors, which can be applied to calculate thermal noise, flicker noise, and current mismatch due to the fluctuation in the technological parameters.

4.3 Thermal noise in MOSFETs

4.3.1 Channel thermal noise

Since the conductance of a MOSFET channel element of length Δy is

$$\frac{1}{\Delta R} = \frac{W}{\Delta y}\mu(-Q_I'), \tag{4.3.1}$$

the mean-square thermal noise of the channel element, $\Delta A = W\Delta y$, calculated through (4.1.1) is

$$\frac{\overline{(i_{\Delta A})^2}}{\Delta f} = 4kT\frac{W}{\Delta y}\mu(-Q_I'). \tag{4.3.2}$$

Substituting (4.3.2) into (4.2.6) yields

$$\frac{\overline{i_d^2}}{\Delta f} = \frac{1}{L^2}\int_0^L 4kTW\mu(-Q_I')dy = -4kT\mu\frac{Q_I}{L^2}, \tag{4.3.3}$$

where the rightmost term in (4.3.3) holds for the simple case of constant mobility. Q_I is the total channel charge given by Equation (2.3.16) in terms of the inversion charge densities at source and drain. Equation (4.3.3) gives the channel thermal noise for FETs in all operating regions, from weak to strong inversion, and from the linear to the saturation region.

Since the expression of the total charge in terms of the applied bias is rather cumbersome, design-oriented expressions for thermal noise in terms of the transistor transconductances will be presented in Section 4.5.2.

4.3.2   Short-channel effects on channel thermal noise

To account for the effect of high electric-field strengths on channel thermal noise we approximate the dependence of the mobility on the field strength F by

$$\mu = \frac{\mu_s}{1 + F/F_c}, \tag{4.3.4}$$

where μ_s is the low longitudinal-field mobility and F_c a critical field strength.

The expression for the drain current using (4.3.4) in the Pao–Sah expression given by (2.1.49) is

$$I_D = -W \frac{\mu_s}{1 + F/F_c} Q_I' \frac{dV_C}{dy}. \tag{4.3.5}$$

Using (4.3.5), the contribution of the channel element to the drain current is given [3] by

$$\Delta i_d = \frac{\Delta y}{L_e} \frac{1 + F/F_c}{1 + \dfrac{\phi_{sL} - \phi_{s0}}{L_e F_c}} i_{\Delta A}, \tag{4.3.6}$$

where $L_e = L - \Delta L$ is the electric channel length, since for short-channel transistors the channel-length modulation (CLM) must be taken into account. The result in (4.3.6) is a generalization of the current-division principle previously shown as (4.2.5) for a long-channel device, taking into account the saturation-velocity phenomenon. In the case of (4.3.6), as compared with (4.2.5), the part of the elementary current $i_{\Delta A}$ directed to the drain is higher for values of the electric field above its mean value (closer to the drain) and reduced for those below its mean value (closer to the source).

We can now proceed to determining the total drain-current noise by adding the contributions of all channel elements, i.e.

$$\overline{i_d^2} = \overline{\left(\sum_{channel} \Delta i_d \right)^2}. \tag{4.3.7}$$

If the elementary thermal-noise sources are assumed to be uncorrelated, substitution of (4.3.6) into (4.3.7) gives

$$\overline{i_d^2} = \sum_{channel} \left(\frac{\Delta y}{L_e} \frac{1 + F/F_c}{1 + \dfrac{\phi_{sL} - \phi_{s0}}{L_e F_c}} \right)^2 \overline{i_{\Delta A}^2}. \tag{4.3.8}$$

The thermal noise for a channel element located between y and $y + \Delta y$ [3] is[2]

$$\overline{i_{\Delta A}^2} = -4qDQ_I'(y) \frac{W}{\Delta y} \Delta f, \tag{4.3.9}$$

---

[2] For low electric fields $D = \mu \phi_t$ and (4.3.9) is equivalent to (4.3.2).

where D is the carrier diffusion coefficient. For the sake of compactness in the derivation of the noise model, we will assume simply, as in [3], that the diffusion coefficient is independent of the electric field. The use of a constant diffusion coefficient in (4.3.9) implicitly accounts for the hot-carrier case in which the carriers are "heated" by the electric field.

Substituting expression (4.3.9) into (4.3.8) and taking the limit $\Delta y \rightarrow 0$ leads to

$$\frac{\overline{i_d^2}}{\Delta f} = -\frac{4qDW \int_0^{L_e} \left(1 + \dfrac{F}{F_c}\right)^2 Q_I' \, dy}{\left(1 + \dfrac{\phi_{sL} - \phi_{s0}}{L_e F_c}\right)^2 L_e^2}. \tag{4.3.10}$$

Using the linear relationship between surface potential and inversion charge density it is possible to calculate (4.3.10), but the final expression is rather cumbersome [4], [5]. A compact analytical expression is obtained on approximating the electric-field-dependent term in the numerator by its mean value, thus

$$\left(1 + \frac{F}{F_c}\right)^2 \cong \left(1 + \frac{\phi_{sL} - \phi_{s0}}{L_e F_c}\right)^2. \tag{4.3.11}$$

Using the Einstein relationship between the diffusion coefficient D and the low transversal field mobility μ_s and substituting (4.3.11) into (4.3.10), we obtain

$$\frac{\overline{i_d^2}}{\Delta f} = -4kT\mu_s \frac{Q_I}{L_e^2}, \tag{4.3.12}$$

which is formally the same expression as we derived for long-channel transistors. Note that μ_s is dependent only on the gate-to-bulk voltage, while L_e models CLM. Although (4.3.12) looks very simple, the explicit expression for the total channel charge Q_I of a short-channel transistor (Appendix A2.2) is fairly complicated. For this reason, a simplified explicit expression for the thermal noise of short-channel transistors will be given in Section 4.5.2. An alternative approach to the calculation of thermal noise in short-channel MOSFETs is developed in [6] as a generalization of the Klaassen–Prins equation [7].

4.3.3 Induced gate noise

The origin of the gate noise of a MOSFET is the thermal noise of the conducting channel which is capacitively coupled to the gate. Noise voltages develop along the entire channel and, because of the capacitive coupling between channel and gate, a noise current will flow through the transistor gate. This noise is referred to as induced gate noise. In Figure 4.3 the MOSFET channel is represented as an active transmission line. Each channel element gives rise to channel noise, which is represented by a current source $i_{\Delta A}$ in parallel with the channel element. Figure 4.3 shows one such current source. The superposition of the gate currents of all the small-channel elements gives the total induced gate noise [4] as shown below:

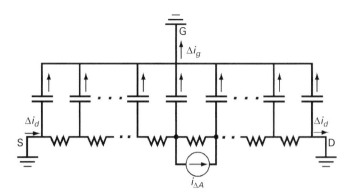

Fig. 4.3   A distributed ac MOSFET model used for calculation of the induced gate noise.

$$\frac{\overline{i_g^2}}{4kT\Delta f} = \frac{16}{135} \frac{\omega^2 (C'_{ox}L)^3 W\phi_t}{\mu Q_F'^2}$$

$$\times \left(\frac{Q'_F}{Q'_{IP}} \frac{1 + 5\alpha + 10.5\alpha^2 + 5\alpha^3 + \alpha^4}{(1+\alpha)^5} - \frac{15}{8} \frac{1 + 5\alpha + 5\alpha^2 + \alpha^3}{(1+\alpha)^5} \right) \quad (4.3.13)$$

Expression (4.3.13) gives the mean-square value of the induced gate noise as a function of both the inversion charge density at the source ($Q'_F = Q'_{IS} - nC'_{ox}\phi_t$) and the channel linearity index α for a long-channel device in quasi-static operation. In strong inversion saturation ($Q'_F/Q'_{IP} \gg 1$, $\alpha \ll 1$), expression (4.3.13) can be approximated by

$$\frac{\overline{i_g^2}}{4kT\Delta f} \cong -\frac{16}{135} \frac{\omega^2 C'^2_{ox} WL^3}{\mu Q'_{IS}} = \frac{16}{135} \frac{\omega^2 (C'_{ox}WL)^2}{g_{ms}} = \frac{4}{15} \frac{\omega^2 C^2_{gs}}{g_{ms}}, \quad (4.3.14)$$

which fully agrees with the result given by van der Ziel in [1] as

$$\frac{\overline{i_g^2}}{4kT\Delta f} = \delta g_g \quad (4.3.15)$$

with[3]

$$g_g = \frac{\omega^2 C^2_{gs}}{5g_{ms}}, \qquad \delta = \frac{4}{3}. \quad (4.3.16)$$

One can also note that, in strong inversion in the linear region, for $V_{DS} \to 0$, or equivalently $\alpha \to 1$, the induced gate noise is given by

$$\frac{\overline{i_g^2}}{4kT\Delta f} \cong -\frac{1}{12} \frac{\omega^2 C'^2_{ox} WL^3}{\mu Q'_{IS}} = \frac{1}{3} \frac{\omega^2 C^2_{gs}}{g_{ms}}, \quad (4.3.17)$$

---

[3] The drain conductance for $V_{DS} = 0$ is $g_{do} = -\mu(W/L)Q'_{IS} = g_{ms}$.

which is slightly lower than the gate noise in saturation. In weak inversion, the mean-square value of the gate-current noise [4] is

$$\frac{\overline{i_g^2}}{4kT\Delta f} = -\frac{1}{12}\frac{\omega^2 WL^3}{\mu n^2 \phi_t^2}\left(\frac{Q_{IS}' + Q_{ID}'}{2}\right) = \frac{1}{12}\frac{\omega^2 \left(C_{gs} + C_{gd}\right)^2}{\left(g_{ms} + g_{md}\right)/2} \tag{4.3.18}$$

both for linear and for saturation regions. In weak inversion, the PSD of the gate current increases proportionally with the channel inversion charge owing to the increase in the elementary noise current with an increase in the charge. On the other hand, in strong inversion, the fluctuation in voltage drop (and in charge density) along the channel is less sensitive to the noise current owing to the lower resistance of the channel. Thus, the gate current is inversely proportional to the inversion charge of the channel.

Example 4.4

Using the simplified high-frequency model of Figure 2.22, demonstrate that in strong inversion the admittance y_{gs} of a saturated MOSFET is $y_{gs} = j\omega C_{gs} + g_{gs}$, where $g_{gs} = (\omega C_{gs})^2/(5g_{ms}/n)$.

Answer

The gate-to-source admittance is

$$y_{gs} = \frac{j\omega C_{gs}}{1 + j\omega(\tau_1 - \tau_2)} \cong j\omega C_{gs} + \omega^2 C_{gs}(\tau_1 - \tau_2).$$

The approximation in the above equation is valid when $\omega(\tau_1 - \tau_2) \ll 1$. In strong inversion saturation (see Example 2.9 and Table 2.2), $i_f \gg 1$ and $\alpha \to 0$, and thus we have

$$\tau_1 = 2\tau_2 \cong \frac{4}{15}\frac{L^2}{\mu\phi_t}\frac{1}{\sqrt{i_f}} \cong \frac{4}{15}\frac{L^2}{\mu\phi_t}\frac{2I_S}{g_{ms}\phi_t}; \qquad \tau_1 - \tau_2 \cong \frac{1}{5}\frac{C_{gs}}{g_{ms}/n}.$$

In the derivation above we used the universal transconductance-to-current ratio and the approximation $C_{gs} = (2/3)C_{ox}$. Finally, the gate-to-source conductance is written as

$$g_{gs} = \frac{\omega^2 C_{gs}^2}{5g_{ms}/n} = \frac{\omega^2 C_{gs}^2}{5g_m}.$$

Note that $g_{gs} = ng_g$, but deep in strong inversion $n \to 1$.

The correlation coefficient c between gate and drain noise is defined as

$$c = \frac{\overline{i_g i_d^*}}{\sqrt{\overline{i_g^2}}\sqrt{\overline{i_d^2}}}. \tag{4.3.19}$$

To obtain c, one needs to calculate the cross-spectral intensity $\overline{i_g i_d^*}$ from

$$\overline{i_g i_d^*} = \left(\sum_{channel}\Delta i_g\right)\left(\sum_{channel}\Delta i_d\right)^*. \tag{4.3.20}$$

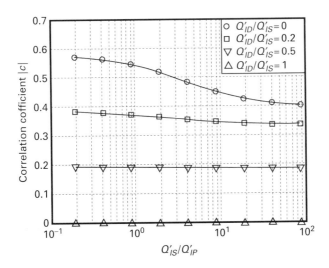

Fig. 4.4   The correlation coefficient of gate and drain noise versus the normalized inversion charge density at source, from the linear region, $Q'_{ID}/Q'_{IS} = 1$, to the saturation region, $Q'_{ID}/Q'_{IS} = 0$.

The assumption that the noise sources along the channel are uncorrelated [4] leads to

$$\frac{\overline{i_g i_d^*}}{4kT\Delta f} = \frac{1}{9}j\omega\left(C'_{ox}WL\right)\left[\frac{(1-\alpha)(1+4\alpha+\alpha^2)}{(1+\alpha)^3}\right]. \tag{4.3.21}$$

Using (4.3.3), (4.3.13), and (4.3.21) we obtain the expression for the correlation coefficient between gate and drain thermal noise, which is plotted in Figure 4.4.

As shown in Figure 4.4, for very low V_{DS}, $\alpha \to 1$ and thus $|c| \to 0$. In saturation, $|c| \to 0.4$ in strong inversion and $|c| \to 0.6$ in weak inversion.

4.4   Flicker noise in MOSFETs

Flicker or $1/f$ noise in resistors and transistors is due to fluctuations in the conductance of the devices and, thus, manifests itself only with a direct current flow. In effect, with no direct current flowing, conductance fluctuations do not produce voltage or current variations. Fluctuations in the number of carriers or in their mobility, or these two effects combined, can be at the origin of conductance fluctuations and, consequently, of flicker noise [8]. We will summarize here only the basic carrier-number-fluctuation theory, which is the most accepted approach to modeling the $1/f$ noise and requires only one specific noise parameter. More elaborate models including mobility fluctuations are available [8], but the model equations require several noise parameters that are not easy to extract from low-frequency noise measurements.

In the field of semiconductor devices, the McWhorter model [9] has successfully been used to explain low-frequency noise. This model considers $1/f$ noise as the summation of many random telegraph signals (RTSs) having a Lorentzian spectrum [1] with characteristic time τ and statistical distribution given by (4.4.1). In the MOSFET, each RTS noise can be explained by the existence of a trap inside the gate oxide that can be occupied by an electron from the channel. Electrons in the channel may tunnel to and back from this trap in a random process, making the drain current fluctuate.

Let us consider a small channel element of area ΔA and designate by $N'_{t\Delta A}$ the number of occupied traps per unit area in the whole oxide volume above the channel element ΔA. Assuming that oxide traps have a uniform spatial distribution and that the probability of an electron penetrating into the oxide decreases exponentially with the distance from the interface, it can be shown [1] that the trapping time constants have a probability distribution $g(\tau)$ of the form

$$g(\tau) = C/\tau \tag{4.4.1}$$

with the normalization constant C given by

$$C = 1/\ln(\tau_{max}/\tau_{min}). \tag{4.4.2}$$

In [9], the authors describe the simulation of $1/f$ waveforms generated as the summation of various RTSs with time constants having the probability distribution given by (4.4.1), with $\tau_{min} = 1$ and $\tau_{max} = 2^{12}$. The results for the PSD (refer to [9] for details) averaged over a large number of realizations are shown in Figure 4.5(a). The spectrum

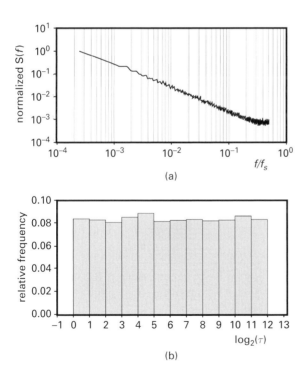

(a)

(b)

Fig. 4.5 (a) The normalized spectral density of a simulated waveform obtained by superimposing many random telegraph signals with single time constants having τ values generated according to (4.4.1) (see details in [9]). f_s is the sampling frequency. (b) The distribution of the relaxation times τ (adapted from [9]).

follows a $1/f$ dependence. Figure 4.5(b) shows the histogram of the time constants τ on a log scale. Note that the number of time constants within each octave interval is almost constant, which is consistent with the probability distribution $g(\tau)$ of (4.4.1).

For frequency values between $1/\tau_{max}$ and $1/\tau_{min}$, the PSD of the fluctuation (in time) $\delta N'_{t\Delta A}$ is given [4] by the very simple expression

$$\frac{\overline{(\delta N'_{t\Delta A})^2}}{\Delta f} = \frac{1}{\Delta A}\frac{N_t(E)kT}{\lambda}\frac{1}{f} = \frac{N_{ot}}{\Delta A}\frac{1}{f},$$ (4.4.3)

where λ (cm^{-1}) is the attenuation coefficient of the electron wave function in the oxide, $N_t(E)$ $(\text{eV}^{-1}/\text{cm}^3)$ is the density of oxide traps per unit volume and unit energy, and the parameter N_{ot} (cm^{-2}) is the equivalent density of oxide traps defined by

$$N_{ot} = \frac{kTN_t(E)}{\lambda}.$$ (4.4.4)

Thus, the time constants associated with the oxide traps must be distributed uniformly on a log scale to obtain a $1/f$ spectrum, as shown in Figure 4.5.

To obtain the PSD of the drain current, the contribution of each local noise current $i_{\Delta A}$ must be accounted for. The value of $i_{\Delta A}$ is dependent on the fluctuation in the carrier density which, in turn, is dependent on the fluctuation in the density of occupied traps (for details on the derivation of $i_{\Delta A}$ see [4] and [10]). Integrating the contributions of all channel segments to the total current using (4.2.6) results in

$$\frac{\overline{i_d^2}}{I_D^2\,\Delta f} = \frac{q^2 N_{ot}\mu}{L^2 n C'_{ox}I_D\,f}\frac{1}{\ln}\left(\frac{nC'_{ox}\phi_t - Q'_{IS}}{nC'_{ox}\phi_t - Q'_{ID}}\right).$$ (4.4.5)

An expression similar to (4.4.5) is used in BSIM to model strong inversion noise due to charge trapping, but it must be emphasized that (4.4.5) is valid for any inversion level, including moderate inversion.

Approximate expressions can be found for the different operating regions. In weak inversion, $-Q'_{ID}, -Q'_{IS} \ll nC'_{ox}\phi_t$. On making a first-order series expansion, it is possible to rewrite (4.4.5) in a more concise manner as

$$\frac{\overline{i_d^2}}{I_D^2\,\Delta f} = \frac{N_{ot}}{WLN^{*2}}\frac{1}{f},$$ (4.4.6)

where $N^* = nC'_{ox}\phi_t/q$. Equation (4.4.6) is the same as the one used in BSIM3 and BSIM4 to model flicker noise in weak inversion ($N_{ot} = AkT/\lambda$).

In the linear region in strong inversion, $Q'_{ID} \cong Q'_{IS} \cong -C'_{ox}(V_G - V_T)$, and the first-order expansion of (4.4.5) leads to

$$\frac{\overline{i_d^2}}{I_D^2\,\Delta f} = \frac{q^2 N_{ot}}{WLC'^2_{ox}(V_G - V_T)^2}\frac{1}{f}.$$ (4.4.7)

Summarizing, the bias dependence of the PSD of the flicker noise is as follows: in subthreshold operation the PSD increases with I_D^2 according to (4.4.6), whereas in strong-inversion saturation the PSD increases linearly with I_D as predicted by (4.4.5), if one neglects the variation in the logarithmic term.

4.5.   Design-oriented noise models

4.5.1   Consistency of noise models

A noise model is consistent regarding series or parallel associations if the composition of the noise contributions from the individual series (or parallel) elements is the same as the noise from the series (or parallel) equivalent. The thermal-noise model for a resistor given by Equation (4.1.2), namely, repeated below for convenience,

$$\overline{v^2} = 4kTR\,\Delta f, \tag{4.5.1}$$

is consistent, as expected. In effect, for two series elements R_1 and R_2 (Figure 4.6), the total noise voltage introduced into the circuit $\overline{v^2} = 4kT(R_1 + R_2)\Delta f$ can be obtained by composing the individual noise sources or using (4.5.1) to calculate the noise of the equivalent resistor $R_{eq} = R_1 + R_2$. The analysis can be extended to MOS transistors because, for these devices, series and parallel equivalents are clearly defined.

The flicker-noise model of (4.4.5) as well as the thermal-noise model of (4.3.3) were deduced from expression (4.2.6), which gives the combined effect of the local fluctuations integrated along the transistor channel on the drain current, thus resulting in an inherently consistent model for the series and parallel associations of transistors. However, some popular $1/f$-noise models are inconsistent, as pointed out in [11].

4.5.2   The thermal noise excess factor

As expected, in the linear region, for $V_{DS} \rightarrow 0$, the transistor is equivalent to a resistor and (4.3.3) reduces to the expression of the thermal noise of a resistor

$$\frac{\overline{i_d^2}}{\Delta f} = -4kT\mu\frac{WLQ'_{IS}}{L^2} = 4kTg_{ms}, \tag{4.5.2}$$

where $g_{ms}\ (=g_{md})$ is the equivalent conductance of the transistor. It is important to observe that outside the linear region the general expression (4.3.3) does not give the thermal noise of an equivalent resistor and, thus, there is an excess noise factor (with respect to an equivalent resistor) to be considered.

In weak inversion, $\left|Q'_{IS(D)}\right| \ll nC'_{ox}\phi_t$, and (4.3.3) becomes

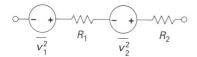

Fig. 4.6   A circuit for the calculation of the total noise produced by two resistors in series.

$$\frac{\overline{i_d^2}}{\Delta f} \cong -4kT\mu \frac{W}{L}\frac{Q'_{IS} + Q'_{ID}}{2} = 4kT\frac{g_{ms} + g_{md}}{2}. \qquad (4.5.3)$$

For a saturated transistor ($g_{ms} \gg g_{md}$) in weak inversion

$$\frac{\overline{i_d^2}}{\Delta f} = 2kTg_{ms}. \qquad (4.5.4)$$

In general, the channel thermal noise is written as

$$\frac{\overline{i_d^2}}{\Delta f} = 4kT\gamma g_{ms}, \qquad (4.5.5)$$

where g_{ms} is the source transconductance at $V_{DS}=0$. Expression (4.5.5) is convenient for first-order calculations since the thermal noise, rather than being expressed as in (4.3.3), is written as the noise of a simple resistor with conductance equal to g_{ms}, modified by a correction factor[4] γ with a value close to unity. This γ is named the excess-noise factor and its value is 2/3 for a long-channel saturated transistor in strong inversion.

Considering the effect of saturation velocity on the thermal noise as given by (4.3.12), the excess-noise factor for a short-channel transistor is given by

$$\gamma_{short} = \frac{L_e Q_I}{L_{esat}^2 WQ'_I}, \qquad (4.5.6)$$

where L_e and L_{esat} are the electric lengths of the channel in the linear and saturation regions, respectively. Considering that $L_{esat}=L_e-\Delta L$, where ΔL is the channel shortening due to CLM, we can write

$$\gamma_{short} \cong \left(1 + \frac{2\,\Delta L}{L_e}\right)\frac{Q_I}{WL_e Q'_{IS}}. \qquad (4.5.7)$$

Thus, for short-channel transistors it is possible that $\gamma > 1$ [3] due to the CLM effect.

4.5.3   Flicker noise in terms of inversion levels

A useful alternative expression for the flicker-noise formula (4.4.5) is obtained if the charge densities at source (drain) are expressed in terms of the normalized forward and reverse currents i_f and i_r, respectively. Using the relationship between normalized charges and currents of (2.2.3), expression (4.4.5) can be rewritten as

$$\frac{\overline{i_d^2}}{I_D^2 \Delta f} = \frac{N_{ot}}{WLN^{*2}}\frac{1}{f}\frac{1}{i_f - i_r}\ln\left(\frac{1+i_f}{1+i_r}\right). \qquad (4.5.8)$$

From weak to strong inversion in the linear region, $i_f \approx i_r$ and (4.5.8) reduces to

---

[4] We use the symbol γ to represent both the body-effect factor and the thermal-noise excess factor, following the customary notation.

$$\frac{\overline{i_d^2}}{I_D^2 \Delta f} = \frac{N_{ot}}{WLN^{*2}} \frac{1}{f} \frac{1}{1 + i_f}.$$ (4.5.9)

In weak inversion, $i_f \ll 1$ and $i_r \ll 1$. The first-order series expansion of (4.5.8) leads to

$$\frac{\overline{i_d^2}}{I_D^2 \Delta f} = \frac{N_{ot}}{WLN^{*2}} \frac{1}{f}.$$ (4.5.10)

Sometimes, designers prefer to represent the channel noise as an equivalent noise-voltage source in series with the gate rather than as a noise-current source connected between drain and source. In this case, the PSD of the noise-voltage source in series with the gate is

$$\frac{\overline{v_g^2}}{\Delta f} = \frac{1}{g_m^2} \frac{\overline{i_d^2}}{\Delta f}.$$ (4.5.11)

The current-to-transconductance ratio in terms of the inversion level for a transistor operating in saturation, given by (2.2.28), is repeated below as

$$\frac{I_F}{n\phi_t g_m} = \frac{1 + \sqrt{1 + i_f}}{2}.$$ (4.5.12)

Thus, from (4.5.8), (4.5.11), and (4.5.12) it follows that

$$\frac{\overline{v_g^2}}{\Delta f} = \frac{q^2 N_{ot}}{WLC_{ox}'^2} \frac{1}{f} \psi(i_f),$$ (4.5.13)

where

$$\psi(i_f) = \left(\frac{1 + \sqrt{1 + i_f}}{2} \right)^2 \frac{\ln(1 + i_f)}{i_f}.$$ (4.5.14)

Because $\psi(i_f)$ shows very small variation with i_f, as can be seen in Figure 4.7, the following empirical model results if we consider ψ equal to 1:

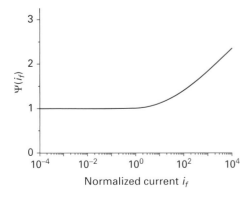

Fig. 4.7 The function $\psi(i_f)$ versus the normalized forward current [10].

$$\frac{\overline{v_g^2}}{\Delta f} \cong \frac{q^2 N_{ot}}{WLC_{ox}'^2} \frac{1}{f}. \tag{4.5.15}$$

Expression (4.5.15) is sometimes presented as

$$\frac{\overline{v_g^2}}{\Delta f} \cong \frac{K_F}{WLC_{ox}'} \frac{1}{f}, \tag{4.5.16}$$

where the flicker-noise constant $K_F = q^2 N_{ot}/C_{ox}'$.

The physical meaning of (4.5.16) is very simple; it models the fluctuations distributed along the channel by a lumped fluctuation in the gate voltage (or equivalently by a fluctuation of the threshold voltage). Owing to its simplicity, the empirical model (4.5.16) is very convenient for hand calculations. Moreover, in current designs, the inversion level is seldom higher than 10^2, thus $\psi \cong 1$ is a fair approximation. Even though the empirical model of (4.5.16) gives a good estimate of the flicker noise of a transistor in saturation, it is not consistent for modeling the series association of transistors. In effect, the empirical model does not consider the distributed nature of the MOSFET, because it represents noise as a gate-voltage source independent of the bias condition. Additionally, it should be noted that, for high-current applications such as those in RF circuits, Equation (4.5.16) can give large errors. For a gate overdrive $V_{GS} - V_T$ of 1.5 V, we have $i_f \approx 2000$ and $\psi \approx 2$.

4.5.4   The corner frequency

The corner frequency f_c, defined as the frequency at which the flicker-noise and thermal-noise PSDs have the same value, can be calculated directly in terms of Q_{IS}' and Q_{ID}' from Equations (4.3.3) and (4.4.5). However, simpler results are obtained determining f_c in saturation ($Q_{ID}' = 0$)

$$f_c = \frac{\alpha g_m}{WLC_{ox}'} \frac{N_{ot}}{N^*} \cong \frac{\pi}{2} \frac{N_{ot}}{N^*} f_T \tag{4.5.17}$$

with $\alpha = 1/2$ in weak inversion and $\alpha \cong 9/16$ in strong inversion. The approximation in (4.5.17) is acceptable since α is almost insensitive to the inversion level. Note that the corner frequency in (4.5.17) is proportional to the transition frequency f_T of the transistor, which is a useful approximation for designers.

The total noise in a frequency band ($f_2 - f_1$) resulting from the contributions of thermal and flicker noise can be calculated as an equivalent gate mean-square voltage. For a saturated transistor operating in weak inversion, the integration of both (4.5.4) and (4.5.10), yields

$$\overline{v_g^2} = \frac{2nkT}{g_m} \left[(f_2 - f_1) + f_c \ln\left(\frac{f_2}{f_1}\right) \right]. \tag{4.5.18}$$

For strong inversion, an analogous formula with slightly different coefficients exists.

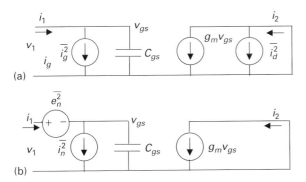

Fig. 4.8 Two equivalent representations of the noise in a saturated MOSFET with $V_{SB} = 0$. (a) The MOSFET small-signal equivalent circuit with noise-current generators. (b) The MOSFET small-signal equivalent circuit with equivalent input-noise generators.

4.5.5 Two-port noise models

In the previous sections, the noise of an MOS transistor is represented by noise-current sources, as shown in Figure 4.8(a). In this figure, the MOS transistor is represented as a two-port network, which is an appropriate model for noise analysis in common circuits. In fact, the two-port equivalent circuit is strictly valid if the four-terminal transistor has short-circuited bulk and source terminals. The channel thermal noise and the $1/f$ noise are represented by the noise-current source i_d connected between source and drain. The noise-current source connected to the transistor gate i_g represents the noise due to the high-frequency capacitive coupling between the channel thermal noise and the gate input. There is an alternative and useful representation of noise in two-port networks consisting of the use of two equivalent input noise generators. The noise introduced by the MOSFET can be represented by the voltage noise generator e_n and the current noise generator i_n connected to the input port of the transistor, as shown in Figure 4.8(b). If an ideal voltage source is connected to the input, the current noise generator is shorted and the input noise-voltage generator only will produce noise at the output. Thus, the input noise-voltage source e_n models the noise introduced by the two-port network when it is driven by an ideal voltage generator. Similarly, in the case of an ideal current source connected to the input, the input noise-voltage source does not produce output noise. Thus, the input noise-current source i_n models the noise introduced by the two-port network when it is driven by an ideal current generator. Clearly, in the common case of an input generator with internal resistance R_S, both e_n and i_n contribute to the output noise. It is important to note that the two input noise generators are correlated in the general case because both are dependent on the internal noise sources of the device.

4.5.5.1 The correlation admittance

When dealing with correlated current and voltage noise sources, it is useful to express the noise current as the sum of two components, one fully correlated with the voltage generator and the other uncorrelated. Thus,

$$i_n = i_u + i_c \qquad (4.5.19)$$

and by definition

$$\overline{i_u e_n^*} = 0. \tag{4.5.20}$$

The fully correlated current noise is proportional to the voltage noise, thus

$$i_c = Y_c e_n, \tag{4.5.21}$$

where the constant Y_c is the correlation admittance. To determine Y_c we can calculate the average value using (4.5.20) as

$$\overline{i_n e_n^*} = \overline{(i_u + i_c)e_n^*} = Y_c \overline{|e_n|^2} \tag{4.5.22}$$

and, consequently,

$$Y_c = \frac{\overline{i_n e_n^*}}{\overline{|e_n|^2}}. \tag{4.5.23}$$

The mean-square value of the current source is

$$\overline{|i_n|^2} = \overline{(i_u + Y_c e_n)(i_u + Y_c e_n)^*} = \overline{|i_u|^2} + |Y_c|^2 \overline{|e_n|^2}. \tag{4.5.24}$$

4.5.5.2   MOS-transistor equivalent input noise generators

As shown in Sections 4.3.3 and 4.5.2, the thermal-noise model of the saturated MOSFET can be represented by the two current sources i_d and i_g, the mean-square values of which are given by

$$\overline{i_d^2} = 4kT\gamma g_{ms}\,\Delta f \tag{4.5.25}$$

and

$$\overline{i_g^2} = 4kT\delta g_g\,\Delta f, \tag{4.5.26}$$

where, in strong-inversion saturation,

$$g_g = \frac{\omega^2 C_{gs}^2}{5g_{ms}} \quad \text{and} \quad \delta = \frac{4}{3}. \tag{4.5.27}$$

The gate and drain noise-current generators are correlated, with the correlation factor defined by expression (4.3.19) as

$$c = \frac{\overline{i_g i_d^*}}{\sqrt{\overline{i_g^2}}\sqrt{\overline{i_d^2}}} = j|c|. \tag{4.5.28}$$

The correlation factor is purely imaginary, as given in (4.3.21), its absolute value varying from approximately 0.6 in weak inversion to 0.4 in strong-inversion saturation, as shown in Figure 4.4.

To determine the values of the equivalent input noise generators of Figure 4.8(b), we equate the short-circuit output currents of the circuits (a) and (b) considering inputs firstly short-circuited and secondly open-circuited.

On short-circuiting the input of each circuit in Figure 4.8 and equating the output short-circuit current i_2, we obtain

$$i_d = -g_m e_n, \tag{4.5.29}$$

which gives the value of the input noise voltage

$$e_n = -\frac{i_d}{g_m}. \tag{4.5.30}$$

Considering now the two circuits with open-circuited inputs, equating short-circuited output currents gives

$$i_n = i_g - \frac{j\omega C_{gs}}{g_m} i_d. \tag{4.5.31}$$

To determine the correlation admittance using (4.5.23) we calculate the mean value of the product $i_n e_n^*$ using (4.5.30) and (4.5.31), which gives

$$\overline{i_n e_n^*} = -\frac{\overline{i_g i_d^*}}{g_m} + \frac{j\omega C_{gs}}{g_m^2} \overline{i_d i_d^*}. \tag{4.5.32}$$

Using the definition of the correlation coefficient of (4.5.28), we can rewrite (4.5.32) as

$$\overline{i_n e_n^*} = -j|c| \frac{\sqrt{\overline{i_g^2}}\sqrt{\overline{i_d^2}}}{g_m} + \frac{j\omega C_{gs}}{g_m^2} \overline{i_d i_d^*}. \tag{4.5.33}$$

Finally, from (4.5.23), (4.5.30), and (4.5.33) we obtain

$$Y_c = \frac{g_m^2}{|i_d|^2} \overline{i_n e_n^*} = -j|c|g_m \sqrt{\frac{|i_g|^2}{|i_d|^2}} + j\omega C_{gs} \tag{4.5.34}$$

or, using the mean-square values for i_g and i_d given by (4.5.25) and (4.5.26) in (4.5.34), we obtain a useful expression for the correlation admittance [12] as

$$Y_c = j\omega C_{gs}\left(1 - \frac{|c|}{n}\sqrt{\frac{\delta}{5\gamma}}\right). \tag{4.5.35}$$

Thus, the correlation admittance is not equal to the input admittance $j\omega C_{gs}$, although it is proportional to it. The final input noise parameter which needs to be calculated is the input-current noise generator. The mean-square value of the input-current noise calculated using (4.5.31) is

$$\overline{|i_n|^2} = \overline{|i_g|^2} + \frac{\omega^2 C_{gs}^2}{g_m^2} \overline{|i_d|^2} + \frac{j\omega C_{gs}}{g_m}\left(\overline{i_g i_d^*} - \overline{i_d i_g^*}\right). \tag{4.5.36}$$

Using the definition of the correlation coefficient of (4.5.28) in (4.5.36), we obtain a useful expression for the input-current noise generator as

$$\overline{|i_n|^2} = \overline{|i_g|^2} + \frac{\omega^2 C_{gs}^2}{g_m^2} \overline{|i_d|^2} - \frac{2|c|\omega C_{gs}}{g_m} \sqrt{\overline{|i_g|^2}}\sqrt{\overline{|i_d|^2}}. \tag{4.5.37}$$

The uncorrelated part of the input current noise can be determined from (4.5.24), (4.5.30), (4.5.34), and (4.5.37) as

$$\overline{|i_u|^2} = \overline{|i_g|^2}\left(1 - |c|^2\right).$$

(4.5.38)

If the frequency is low enough ($\omega \ll \omega_T$) for the gate-current noise generator to be neglected, the equivalent input-current noise generator reduces to

$$\overline{|i_n|^2} = \frac{\omega^2 C_{gs}^2}{g_m^2}\overline{|i_d|^2}.$$

(4.5.39)

Considering the thermal and $1/f$ noise components of noise, (4.5.39) can be rewritten in weak inversion saturation as

$$\overline{|i_n|^2} = \frac{\omega^2 C_{gs}^2}{g_m^2}\left(\frac{N_{ot}I_D^2}{WLN^{*2}}\frac{\Delta f}{f} + 2kTng_m\,\Delta f\right).$$

(4.5.40)

In strong inversion, the complete expression for the flicker noise (4.5.8) in terms of the inversion level at the source must be employed and the corresponding expression is more complicated.

At low frequencies $i_n \to 0$, and in this case the MOSFET does not introduce noise if it is driven by a very-high-impedance signal source since, with the input open-circuited (in small-signal operation), the e_n generator does not produce output noise (see Figure 4.8). Thus, the MOS transistor has excellent noise performance when driven by a high source resistance at low and moderate frequencies. At high frequencies, the current noise generator at the input cannot be neglected. In this case, the value of the input admittance that minimizes the noise introduced by the MOSFET can be easily determined in terms of the correlation admittance, as shown in Problem 4.7.

4.6   Systematic and random mismatch

Mismatch is the name given to the time-independent differences between identically designed and identically used devices. The performance of most analog, or even digital, circuits relies on the concept of matched behavior between identically designed and used devices. In analog circuits, unwanted differences between the effective values for equally designed components, such as threshold voltage differences of millivolts or less, can critically reduce the performance and/or yield of a circuit. Even for digital circuits, transistor mismatches can lead to propagation-delay differences in clock trees, reducing the robustness of the circuit. The shrinkage of the MOSFET dimensions and the reduction in the supply voltage make matching limitations even more important.

Manufacturing variations result in device parameter variations from lot to lot, wafer to wafer, die to die, and device to device. Variations in parameters are the result of either systematic or stochastic (random) effects. Systematic errors are mainly due to lithographic and chemical effects during the fabrication process. One of the origins of systematic mismatch is equipment-induced non-uniformities such as temperature gradients and photo-mask size differences across the wafer. Proximity effects on the line width, due to diffused light from neighboring structures, or in the implantation dose, due to ion scattering into neighboring structures, are other kinds of systematic error. Systematic errors are important for large distances and can be addressed using appropriate layout techniques [13]. The proximity effects are minimized by using dummy structures and the gradient effects by using common centroid structures as explained in Section 3.4.3. The layout strategies summarized in that section are intended to minimize systematic errors in matched structures.

Random component mismatches are due to local variations that occur in the fabrication process. Polysilicon granularity, which produces edge roughness in polysilicon lines, is one of the causes of random errors. It is called peripheral fluctuation because it scales with the device perimeter. Since impurity atoms and other fixed charges, such as oxide charges, are not manipulated in an atom-by-atom controlled manner, random errors result from the location of the fixed charges. Doping fluctuations are called areal fluctuations because they scale with device area. Random mismatch predominates over systematic mismatch for small distances. Random mismatch is dependent on the device dimensions (area, width, and length) and bias, and the designer needs a probabilistic model by which to include the mismatch error in the design process.

The purpose of the next sections is to develop simple design-oriented models for random mismatch in bulk CMOS transistors. We will mainly focus on mismatch due to random fluctuations in the dopant concentration [14], [15], which is nowadays recognized as the main cause of mismatch in bulk CMOS transistors.

4.6.1   Pelgrom's model of mismatch

The number of atoms per unit volume in silicon is $N_{Si} = 5 \times 10^{22}\,\mathrm{cm}^{-3}$ and the doping concentrations for the MOS substrates are typically below $10^{18}\,\mathrm{cm}^{-3}$. For the sake of simplicity, let us assume that only acceptor atoms are present. For an acceptor doping density N_A, the probability[5] p of having an acceptor atom in the place of a silicon atom in the crystal lattice is

$$p = \frac{N_A}{N_{Si}} \ll 1. \tag{4.6.1}$$

The fluctuation in the number of acceptor atoms under the gate of an n-channel transistor can be determined considering the number N of crystal atoms in the depletion region under the gate given by

$$N = (WLx_d)N_{Si}, \tag{4.6.2}$$

where W, L, and x_d are the transistor width and length and the depletion depth under the gate, respectively. N can be regarded as the number of trials of the experiment consisting of determining whether an atom in the lattice is silicon or an acceptor atom. Considering the binomial distribution, the N-trial mean and mean-square deviation (variance) are given [16] by

---

[5] We follow in this section the standard probability notation for the binomial distribution. In the rest of the book p represents the hole density.

$$\mu_N = Np \qquad \text{and} \qquad \sigma_N^2 = Np(1-p). \qquad (4.6.3)$$

The mean number of acceptor atoms in the depletion region under the gate of a transistor of area WL is obtained by substituting (4.6.1) and (4.6.2) into (4.6.3) as

$$\mu_N = Np = (WLx_d)N_A. \qquad (4.6.4)$$

Since $p \ll 1$, the doping concentration follows a Poisson distribution and the variance is given by

$$\sigma_N^2 = Np(1-p) \cong Np = \mu_N = (WLx_d)N_A. \qquad (4.6.5)$$

Thus, the Poisson distribution is fully determined by a single value, the mean value.

Let us now consider the variation in the threshold voltage of an MOS transistor of length L and width W with the number of ionized acceptor atoms in the depletion region under the gate. Since the Fermi potential has a logarithmic dependence on the substrate doping, it can be considered constant. Thus, considering only the fluctuation in the depletion charge term in the expression for the threshold voltage (2.1.63), we can define the average value for the threshold voltage over the area WL as

$$\langle V_{T0}\rangle = V_{FB} + 2\phi_F + q\frac{\text{number of acceptors under gate}}{WLC'_{ox}}. \qquad (4.6.6)$$

Assuming a Poisson distribution, the mean of the average threshold voltage values is given by substituting (4.6.4) into (4.6.6):

$$\mu_{WL}(\langle V_{T0}\rangle) = V_{FB} + 2\phi_F + \frac{q(WLx_dN_A)}{WLC'_{ox}}. \qquad (4.6.7)$$

Thus, the conventional expression for the threshold voltage can be regarded as the mean value for the average threshold over the gate area. In (4.6.7) we use the sub-index WL to emphasize that the trial number N is proportional to the gate area WL. The square root of the variance, or standard deviation, is obtained from (4.6.5) and (4.6.6) as

$$\sigma_{WL}(\langle V_{T0}\rangle) = q\frac{\sigma_{WL}(\text{number of acceptors under gate})}{WLC'_{ox}}$$
$$= q\sqrt{WLx_dN_A}/(WLC'_{ox}). \qquad (4.6.8)$$

In most applications, the matching between pairs of transistors is the main concern. The standard deviation of the difference between the threshold voltages of two identical transistors $(\Delta V_{T0} = V_{T1} - V_{T2})$ is given [17] by

$$\sigma_{WL}(\langle \Delta V_{T0}\rangle) = \sqrt{2}\sigma_{WL}(\langle V_{T0}\rangle) = \frac{q\sqrt{2x_dN_A}}{C'_{ox}\sqrt{WL}} = \frac{A_{VT}}{\sqrt{WL}}, \qquad (4.6.9)$$

where

$$A_{VT} = \frac{q\sqrt{2x_dN_A}}{C'_{ox}} \qquad (4.6.10)$$

is called the threshold (mis)matching coefficient.

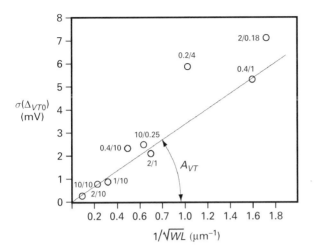

Fig. 4.9   The standard deviation of the n-channel MOSFET threshold voltage versus the inverse square root of the gate area for a 0.18-μm process (adapted from [18]). *W/L* values are indicated close to the circles.

Figure 4.9 shows experimental results corroborating the proportionality of the standard deviation of the threshold voltage to the inverse square root of the transistor area, except for narrow ($W = 0.2$ μm) and short ($L = 0.18$ μm) transistors. In Figure 4.9 and in the rest of the text we simplify the notation to

$$\sigma_{WL}(\langle \Delta V_{T0} \rangle) = \sigma(\Delta V_{T0}) \tag{4.6.11}$$

according to common practice. Writing the depletion depth in terms of the technological parameters as (see Appendix A2.1)

$$x_d = \sqrt{\frac{2\varepsilon_s(2\phi_F)}{qN_A}} \tag{4.6.12}$$

and substituting (4.6.12) into (4.6.10) yields

$$A_{VT} = \frac{q}{C'_{ox}} \sqrt[4]{\frac{8\varepsilon_s(2\phi_F)N_A}{q}} = \frac{qt_{ox}}{\varepsilon_{ox}} \sqrt[4]{\frac{8\varepsilon_s(2\phi_F)N_A}{q}}. \tag{4.6.13}$$

The experimental results in Figure 4.10 confirm the linear dependence of the threshold matching coefficient on the oxide thickness present in (4.6.13), for technologies above 0.25 μm. In fact, A_{VT} is not a linear function of t_{ox} since transistor-scaling practice forces N_A to increase [18]. Assuming, however, that A_{VT} is a linear function of t_{ox}, from Figure 4.10 the approximate relationship

$$A_{VT} \text{ (mV μm)} \cong t_{ox} \text{ (nm)} \tag{4.6.14}$$

follows.

The current-factor differences between pairs of transistors $\Delta\beta$ ($\beta = \mu C'_{ox} W/L$) are modeled in the same way as the threshold voltage differences [17]. Thus,

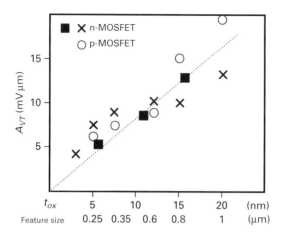

Fig. 4.10 The threshold-voltage matching coefficient over process generations (adapted from [18]).

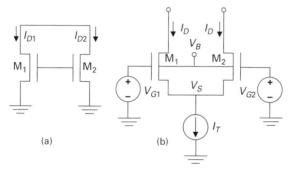

Fig. 4.11 Two basic MOSFET topologies: (a) a current mirror and (b) a differential pair. The drain currents of M_1 and M_2 in the differential pair are assumed to be equal for the calculation of the offset voltage.

$$\frac{\sigma(\Delta\beta)}{\beta} = \frac{A_\beta}{\sqrt{WL}}. \tag{4.6.15}$$

Typical values of A_β, which do not change significantly with technology-node scaling [19], range from 1% μm to 3% μm. Although V_T and β show some correlation, because both are dependent on some common technological parameters, C'_{ox} for instance, they are assumed to be independent random variables in order to obtain simple design formulas.

Assuming that V_{T0} and β are uncorrelated, the normalized variance of the drain current of saturated transistors having the same gate and source voltages as in the current mirror of Figure 4.11(a) is given by

$$\frac{\sigma^2(\Delta I_D)}{I_D^2} = \frac{\sigma^2(\Delta\beta)}{\beta^2} + \left(\frac{g_m}{I_D}\right)^2 \sigma^2(\Delta V_{T0}). \tag{4.6.16}$$

The variance of the gate-voltage mismatch of two saturated transistors having the same drain current and the same source voltage as in the differential pair of Figure 4.11(b) is given by

$$\sigma^2(\Delta V_G) = \sigma^2(\Delta V_{T0}) + \left(\frac{I_D}{g_m}\right)^2 \frac{\sigma^2(\Delta\beta)}{\beta^2}. \tag{4.6.17}$$

The g_m/I_D ratio for a saturated MOSFET is given by (2.2.27) and (2.2.28) as

$$\frac{g_m}{I_D} = \frac{1}{n\phi_t} \frac{2}{1 + \sqrt{1 + i_f}}. \tag{4.6.18}$$

For a current mirror, it can be seen from Equations (4.6.16) and (4.6.18) that biasing the transistor in weak inversion ($i_f \ll 1$, and thus with the maximum g_m/I_D) results in the largest current mismatch, which is given by

$$\frac{\sigma(\Delta I_D)}{I_D} \cong \frac{\sigma(\Delta V_{T0})}{n\phi_t}, \tag{4.6.19}$$

whereas the minimum current mismatch, which is set by $\sigma(\Delta\beta)/\beta$, is obtained with the transistors operating deep in strong inversion.

On the other hand, for a differential pair, for example, operation of the transistors in weak inversion results in the minimum mismatch between the gate voltages, equal to the threshold-voltage mismatch $\sigma(\Delta V_{T0})$ as follows from (4.6.17).

It should be noted that, in general, the design for best matching of a current mirror or a differential pair is more complicated than in the analysis above. In the design problem we must compare solutions corresponding to transistors with different geometries and biases. If we compare a current mirror with transistors operating in strong inversion with another having the transistors operating in moderate inversion, which one will have the better current matching? If the area of the transistors operating in moderate inversion is large enough, their current matching can be even better than that of the pair biased in strong inversion. The examples that follow illustrate the effects of transistor dimensions on matching.

Example 4.5

A unity-gain current mirror operates at a constant input current equal to I_B. Using Pelgrom's model, evaluate the current mismatch for the following conditions: (a) constant aspect ratio and (b) constant channel length.

Answer

The transconductance-to-current ratio is

$$\frac{g_m}{I_B} = \frac{1}{n\phi_t} \frac{2}{1 + \sqrt{1 + I_B/(I_{SH}W/L)}}.$$

(a) For constant $S = W/L$ the transconductance-to-current ratio does not change and, thus, the normalized variance of the current is

$$\frac{\sigma^2(\Delta I_D)}{I_B^2} = \frac{A_\beta^2}{SL^2} + \left(\frac{1}{n\phi_t} \frac{2}{1 + \sqrt{1 + I_B/(SI_{SH})}}\right)^2 \frac{A_{VT}^2}{SL^2}.$$

In this case, the standard deviation of the current is simply inversely proportional to the channel length (or width).

(b) For constant L, the transconductance-to-current ratio is not constant. The normalized variance of the current is

$$\frac{\sigma^2(\Delta I_D)}{I_B^2} = \frac{A_\beta^2}{WL} + \left(\frac{1}{n\phi_t} \frac{2}{1 + \sqrt{1 + I_B/(I_{SH}W/L)}}\right)^2 \frac{A_{VT}^2}{WL}.$$

In this case, the variance of the current is inversely proportional to W when the term in parentheses is independent of W, i.e. when the transistors are so wide that they operate in weak inversion. For narrower channels the MOS transistors tend to operate in strong inversion, for which the transconductance-to-current ratio is proportional to the square root of W. Therefore, in strong inversion the variance of the current deviation caused by threshold mismatch becomes independent of W, whereas the term due to the specific current mismatch is inversely proportional to the channel width. For very high inversion levels (narrow channels) the error caused by A_β can be even more significant than that caused by A_{VT}.

Example 4.6

The tail current of the differential pair in Figure 4.11(b) is 6 μA. Using Pelgrom's model, evaluate the standard deviation of the offset for the following transistor dimensions: (a) $W = 20$ μm, $L = 2$ μm; (b) $W = 2$ μm, $L = 20$ μm. Assume that $I_{SH} = 100$ nA, $n\phi_t = 30$ mV, $A_\beta = 2\%$ μm, and $A_{VT} = 10$ mV μm.

Answer

(a) The specific current is $I_S = I_{SH}W/L = 0.1 \times 20/2 = 1$ μA and $i_f = I_T/(2I_S) = 3$. The current-to-transconductance ratio is

$$\frac{I_D}{g_m} = n\phi_t \frac{1 + \sqrt{1 + i_f}}{2} = 30 \times 10^{-3} \times 1.5 = 45\,\text{mV}.$$

The standard deviation of the offset voltage calculated from (4.6.17) is

$$\sigma^2(V_{OS}) = \frac{A_{VT}^2}{WL} + \left(\frac{I_D}{g_m}\right)^2 \frac{A_\beta^2}{WL} = \frac{(10^{-8})^2}{40 \times 10^{-12}} + (0.045)^2 \frac{(2 \times 10^{-8})^2}{40 \times 10^{-12}},$$

$$\sigma(V_{OS}) \cong 1.58\,\text{mV}.$$

(b) The specific current is $I_S = I_{SH}W/L = 0.1 \times 2/20 = 0.01$ μA and $i_f = I_T/(2I_S) = 300$. The current-to-transconductance ratio is

$$\frac{I_D}{g_m} = 30 \times 10^{-3} \frac{1 + \sqrt{1 + 300}}{2} \cong 275\,\text{mV}.$$

The standard deviation of the offset voltage is

$$\sigma^2(V_{OS}) = \frac{A_{VT}^2}{WL} + \left(\frac{I_D}{g_m}\right)^2 \frac{A_\beta^2}{WL} = \frac{(10^{-8})^2}{40 \times 10^{-12}} + (0.275)^2 \frac{(2 \times 10^{-8})^2}{40 \times 10^{-12}},$$

$$\sigma(V_{OS}) \cong 1.80\,\mathrm{mV}.$$

4.6.2 (Mis)matching energy

To compare the effects of noise and mismatch on the achievable precision of a circuit, let us calculate the mean-square gate voltage representing the channel thermal noise, over a bandwidth equal to the transition frequency. The channel thermal noise of a saturated MOSFET can be modeled approximately by a drain-current noise generator or a gate-voltage noise generator as indicated below:

$$\overline{i_d^2} = \frac{2}{3} 4kTg_m\,\Delta f, \qquad \overline{v_g^2} = \frac{2}{3} 4kT\frac{1}{g_m}\,\Delta f. \qquad (4.6.20)$$

Approximating the transition frequency as

$$f_T \cong \frac{g_m}{2\pi\frac{1}{2}C_{ox}'WL} = \frac{g_m}{\pi C_{ox}'WL} = \Delta f \qquad (4.6.21)$$

yields

$$\overline{v_g^2} = \frac{2}{3} 4kT\frac{1}{g_m}\frac{g_m}{\pi C_{ox}'WL} \cong \frac{kT}{C_{ox}'WL}. \qquad (4.6.22)$$

To compare the relative effects of thermal noise and mismatch (we assume that the number of impurities is the dominant mismatch component) on a circuit, using (4.6.9) and (4.6.20), we can calculate the ratio [19]

$$\frac{\sigma^2(\Delta V_{T0})}{\overline{v_g^2}} = \frac{C_{ox}'A_{VT}^2/2}{kT/2}. \qquad (4.6.23)$$

The denominator in (4.6.23) is the thermal energy for a system with one degree of freedom while the numerator can be regarded as indicating the (mis)match energy

$$E_{match} = C_{ox}'A_{VT}^2/2 = (C_{ox}'WL)\sigma^2(\Delta V_{T0})/2. \qquad (4.6.24)$$

Thus, the (mis)match energy is the energy stored in the gate capacitance ($C_{ox}'WL$) by a voltage equal to the standard deviation of the threshold voltage and is independent of the transistor geometry as shown below:

$$E_{match} = \frac{q^2(2x_d N_A)}{2C_{ox}'}. \qquad (4.6.25)$$

Figure 4.12 shows the ratio of the matching energy to the thermal energy for various technology nodes. The effect of mismatch is dominant over thermal noise by two orders

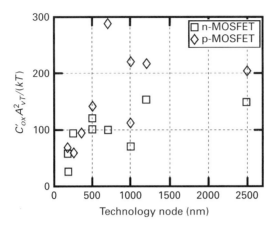

Fig. 4.12   The ratio of "mismatch energy" to thermal energy for several MOS technology nodes (adapted from [19]).

of magnitude for several generations of technology. A reduction in the mismatch energy for advanced technologies is apparent in Figure 4.12. For deep-submicron technologies the effect of the gate current must be considered when one needs to assess mismatch, as developed in [20].

4.6.3   The number-fluctuation mismatch model

In Pelgrom's model the effect of the (atomistic) doping fluctuations under the gate of the transistor is modeled as a fluctuation of the threshold voltage, and the dc model of the transistor is then used to determine the drain-current fluctuation. Thus, Pelgrom's model considers average fluctuations over the entire channel and represents them with a lumped parameter. The drawback of this approach is that the impact of the local doping fluctuations is not considered. Since the amplitude of the stochastic fluctuations is proportional to $1/\sqrt{\text{volume}}$, local fluctuations are much greater than the global fluctuation and, thus, their effects should be considered [21]. Fortunately, the formalism needed to include local fluctuations is already available in flicker ($1/f$)-noise modeling, namely, carrier-number-fluctuation theory. Mismatch (spatial fluctuation) and noise (temporal fluctuation) are similar phenomena, both being dependent on the process, device dimensions, temperature, and bias. Put simply, mismatch can be regarded and modeled as "dc noise."

In the derivation of the mismatch model that follows, we calculate the fluctuations in the drain current around its nominal value resulting from the sum of all the tiny contributions from local fluctuations along the channel, in the same way as for the noise calculations using expression (4.2.6) repeated below:

$$\overline{\Delta I_D^{\,2}} = \lim_{\Delta y \to 0} \sum \overline{\left(\frac{\Delta y}{L} i_{\Delta A}\right)^2} = \frac{1}{L^2} \int_0^L \overline{\Delta y (i_{\Delta A})^2}\, dy, \qquad (4.6.26)$$

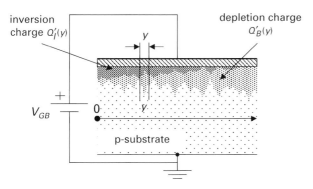

Fig. 4.13   Cross section of an MOS transistor showing the (greatly exaggerated) fluctuations in both inversion and depletion charge densities due to local dopant fluctuations (adapted from [22]).

with a slight change in the notation.[6] The local current fluctuation ($i_{\Delta A}$) is assumed to be a zero-mean stationary random process dependent on the variable y.

Local current fluctuations arise from three independent physical origins, namely fluctuations in both channel and polysilicon gate doping, surface state density, and gate oxide thickness. Note that, since $i_{\Delta A}$ is related to local fluctuations calculated in the area $W\Delta y$, its variance is proportional to $1/(W\Delta y)$. As in Pelgrom's model, we have assumed that fluctuation of channel doping is the main factor that determines local current fluctuations. Figure 4.13 shows the fluctuations in the inversion charge density Q_I' due to local dopant fluctuations. Note that both depletion and inversion charge densities, Q_B' and Q_I', change as a result of the variation in the number of impurity atoms along the y-axis.

In order to derive the mismatch model, we have adopted the following asumptions and/ or approximations:

(1) the capacitive model of the MOS transistor, assuming the depletion capacitance to be dependent on the gate voltage only and the inversion capacitance to be proportional to the inversion charge density;

(2) the fluctuation of the impurity concentration in the depletion layer as the main source of mismatch;

(3) Poisson distribution of impurity atoms;

(4) uncorrelated local impurity fluctuations;

(5) the charge-based model of the channel current.

The application of assumptions/approximations (1) through (5) yields [23]

$$\frac{\sigma_{I_D}^2}{I_D^2} = \frac{N_{oi}}{WLN^{*2}} \frac{1}{i_f - i_r} \ln\left(\frac{1 + i_f}{1 + i_r}\right) \tag{4.6.27}$$

with

$$N^* = \frac{nC_{ox}'\phi_t}{q} \tag{4.6.28}$$

---

[6] Here ΔI_D represents the difference between the actual drain current in a specific transistor and the mean value of the drain current of that transistor.

and

$$N_{oi} = \frac{N_A x_d}{3}.$$ (4.6.29)

Equation (4.6.27) presents the dependences of mismatch on geometry (W and L), bias (i_f and i_r), and technology (N^* and N_{oi}), which are the three degrees of freedom used by circuit designers.

The result in (4.6.27) is essentially the same as that derived for flicker noise in MOS transistors in Equation (4.5.8). This similarity results from the physical origin of both matching and 1/f noise; while the former is related to (spatial) fluctuations in fixed charges, the latter results from (temporal) fluctuations in the occupancy of localized states along the channel. As in Pelgrom's model, expression (4.6.27) indicates that the ratio of mismatch power to dc power is inversely proportional to the gate area WL.

Expression (4.6.27) can be simplified under specific conditions. In weak inversion, i_f, $i_r \ll 1$; thus, the first-order series expansion of (4.6.27) leads to

$$\frac{\sigma_{I_D}^2}{I_D^2} = \frac{N_{oi}}{WLN^{*2}}.$$ (4.6.30)

Therefore, in weak inversion, the normalized mismatch is not sensitive to the current level, for either saturation ($i_f \gg i_r$) or the linear region ($i_f \cong i_r$).

From weak to strong inversion in the linear region, $i_f \cong i_r$ and (4.6.27) reduces to

$$\frac{\sigma_{I_D}^2}{I_D^2} = \frac{N_{oi}}{WLN^{*2}} \frac{1}{1 + i_f}.$$ (4.6.31)

Expression (4.6.31) indicates once again that the normalized mismatch is not sensitive to the inversion level in weak inversion ($i_f \ll 1$) and is inversely proportional to i_f in strong inversion ($i_f \gg 1$). Finally, in saturation ($i_r \to 0$), expression (4.6.27) can be written as

$$\frac{\sigma_{I_D}^2}{I_D^2} = \frac{N_{oi}}{WLN^{*2}} \frac{\ln(1 + i_f)}{i_f}.$$ (4.6.32)

Expressions (4.6.31) and (4.6.32) indicate that, under strong inversion, the current mismatch decreases when the inversion level increases, this behavior being more accentuated in the linear than in the saturation region.

4.6.4   The dependence of mismatch on bias, dimensions, and technology

Figure 4.14 presents the mismatch power normalized with respect to the dc power for drain-to-source voltages ranging from 10 mV (within the linear region) to 2 V (saturation) for medium-sized NMOS devices. Mismatch was measured for six inversion levels (0.01, 0.1, 1, 10, 100, and 1000), with $V_{SB} = 0$.

In weak inversion ($i_f = 0.01$ and 0.1), mismatch is almost constant from the linear region to the saturation region and independent of the inversion level, as predicted by (4.6.30). Measured and simulated curves for weak inversion are almost coincident, being hardly distinguishable.

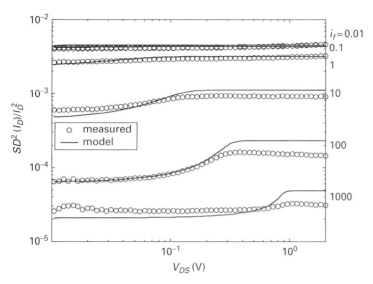

Fig. 4.14 Normalized current mismatch power for the medium-sized ($W = 3\,\mu m$, $L = 2\,\mu m$) n-MOS transistor array (adapted from [22]).

From moderate ($i_f = 1$ and 10) to strong ($i_f = 100$) inversion, the simulated and measured curves present similar behavior, increasing on going from the linear to the saturation region, where a plateau is reached. Differences between measured and simulated curves at saturation may be associated with a spatially non-uniform concentration of dopant atoms.

The parameter N_{oi} was estimated from measurements in weak inversion, using Equation (4.6.30). The same values of N_{oi}, i.e. $1.8 \times 10^{12}\,\text{cm}^{-2}$ for the NMOS devices and $7 \times 10^{12}\,\text{cm}^{-2}$ for the PMOS devices, were obtained both for the large ($W = 12\,\mu m$, $L = 8\,\mu m$) and for the medium-sized ($W = 3\,\mu m$, $L = 2\,\mu m$) transistor.

In order to account for mismatch factors other than doping fluctuations, one can include the random errors due to the specific sheet current $I_{SH} = \mu C'_{ox} n \phi_t^2 / 2$, which results in a modification of expression (4.6.27), yielding

$$\frac{\sigma_{I_D}^2}{I_D^2} = \frac{1}{WL}\left[\frac{N_{oi}}{N^{*2}}\frac{1}{i_f - i_r}\ln\left(\frac{1 + i_f}{1 + i_r}\right) + A_{ISH}^2\right]. \qquad (4.6.33)$$

A_{ISH} is the parameter which accounts for mismatch in I_{SH}. For high inversion levels, mismatch levels out at a minimum value determined by A_{ISH}, a result that is corroborated by experimental data.

A_{ISH} was estimated from measurements in strong inversion in the linear region, using (4.6.33). A_{ISH} values of the order of $0.89\%\,\mu m$ and $0.71\%\,\mu m$ resulted for NMOS and PMOS devices, respectively, both for large and for medium-sized devices. The simulated curves presented in Figures 4.14 and 4.15 are based on the values extracted for both N_{oi} and A_{ISH}, for either NMOS or PMOS transistors.

Figure 4.15 shows the measured and simulated dependences of current matching on inversion level (or bias current I_B) for the linear and saturation regions, for three sizes of

(a)

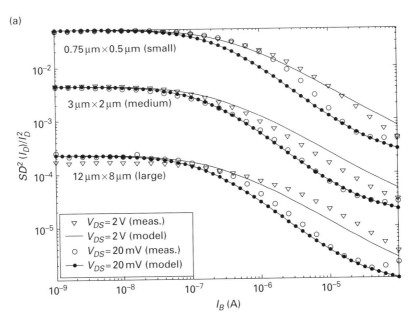

(b)

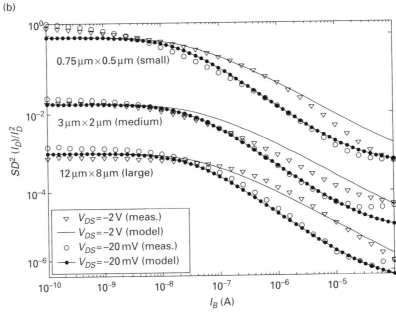

Fig. 4.15   The dependence of current matching on inversion level (bias current I_B) in the linear ($|V_{DS}| = 20$ mV) and saturation ($|V_{DS}| = 2$ V) regions for the large, medium, and small n-MOS (a) and p-MOS (b) transistor arrays (adapted from [22]).

NMOS and PMOS transistors. From Figure 4.15, one can see that larger transistors follow the "area rule," as shown in our model. We used the same A_{ISH} value to model the matching both of the large and of the medium-sized devices. The small transistors do not follow this rule, presenting a mismatch 55% lower (NMOS) or 80% higher (PMOS) than the model estimates obtained using the same N_{oi} as that for the large and medium-sized transistors. However, values of N_{oi} for the small transistors different from those measured for the *large* transistors were chosen in order to obtain better fitting of the curves.

For the dies we characterized, small transistors presented an unpredictable N_{oi}. In fact, electrical characteristics of short-channel devices are very sensitive to fluctuations due to a greater dependence on edge effects. This high sensitivity of short-channel devices is one of the main reasons for the difficulties found in modeling mismatch, particularly in today's very complex submicron technologies. Also, for minimum-length devices, drain and source doped regions are very close to each other, affecting strongly the shape of the depletion layer below the channel.

The curves presented in Figure 4.15 have a similar behavior to that seen in $1/f$-noise characterization, showing that both phenomena, mismatch and $1/f$ noise, arise from the same mechanism, namely carrier-density fluctuation, although the first is related to random fluctuations in doping density and the second to random trapping–detrapping of carriers at the channel oxide–substrate interface.

4.6.5 Matching analysis of analog circuits

Owing to the random variations in the device characteristics, from lot to lot, wafer to wafer, die to die, and device to device, a statistical approach to analyzing circuit performance is required. The most common way of estimating the statistical distributions of circuit performance is Monte Carlo analysis, in which hundreds, or even thousands, of circuit simulations are run to obtain reliable estimations of the statistics. Since the Monte Carlo method is computationally very costly, faster methods have been developed for the determination of common random circuit parameters, such as the offset voltage of an amplifier/comparator.

4.6.5.1 Monte Carlo simulation

The ultimate goal of Monte Carlo analysis is to obtain simulation results for which the probability distribution is statistically the same as that of the actual measurements of the fabricated circuits. To obtain these kinds of results, a series of simulations is performed, with the parameters of the devices varying according to specified probability distributions.

Usually, all parameter variations are defined using device models. Two types of parameter variations are allowed: (individual) device variations, which represent device mismatch; and (as a group) lot variations, which represent process variations [24]. Either type of variation, or both types, may be specified when describing parameter variations for Monte Carlo analysis. To run a Monte Carlo simulation the basic specifications are the total number of simulations, the type of analysis to be done (dc, ac, transient, etc.), the probability distributions (uniform, normal (or Gaussian), etc.) of the parameters, and the kind and amount of outputs generated. Figure 4.16 shows the outputs of the Monte Carlo

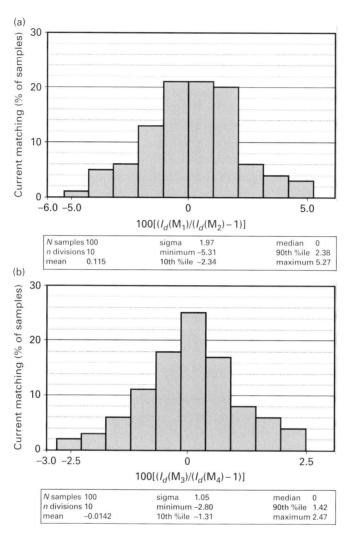

Fig. 4.16   Monte Carlo simulations of the current matching of two current mirrors, for transistors with (a) $W/L = 10/4$ and (b) $W/L = (2 \times 10)/(2 \times 4)$. The main statistical metrics are given below the histogram: N samples is the number of extracted data values, n divisions is the number of bins in the histogram, and mean and sigma are the arithmetic mean and standard deviation of the extracted values, respectively, 10th (90th) %ile is the tenth (ninetieth) percentile that describes the portion of the distribution below which 10% (90%) of the extracted values lie (adapted from [25]).

simulations for two current mirrors [25]. The standard deviation of the current mismatch in the bottom histogram is half of that in the top histogram, since the area of the transistors of the current mirror represented by the bottom histogram is four times larger than that of the top one.

Since in many cases manufacturers do not provide a Monte Carlo library, designers must build themselves the mismatch libraries. For resistors, the expressions giving the resistance value in terms of the geometrical and technological parameters are relatively simple and the elaboration of Monte Carlo simulation netlists is well documented [26]. Thus, it is easy for the designer, using the basic information from the manufacturer, to develop model cards and determine the parameters with which to simulate the process variations and the mismatch between resistors. With transistors the situation is much more complicated because the transistor models provided by the manufacturers consist of complicated non-linear equations involving numerous parameters for which statistical data are not available. Since a complete statistical model is clearly not possible, two approximative approaches are useful. In the first approach [27], the built-in MOSFET control cards *delvtho* and *delu0* of SPICE are used to give an independent random Gaussian distribution for the threshold voltage and for the current gain factor β of every transistor on the netlist. The parameters of *delvtho* and *delu0* cards are obtained from Pelgrom's model. In the second approach, a (random) current generator is connected between the source and drain of each transistor of the netlist. The value of the current is calculated with the help of Equation (4.6.27) of the carrier-number-fluctuation model. The use of Pelgrom's model or the number-fluctuation model is up to the user since the parameters of the two models are related as given below:

$$A_{VT}^2 = \frac{6q^2}{C_{ox}'^2} N_{oi},$$
(4.6.34)

$$A_{\beta}^2 = 2A_{ISH}^2.$$
(4.6.35)

Equation (4.6.34) results from (4.6.10) and (4.6.29). The fact that A_{β} and A_{ISH} differ by a factor of 2 is because the number-fluctuation parameters represent the variations in an individual device and Pelgrom's model represents the current variation between two devices, thus the variances in Pelgom's model are twice those in the carrier-number-fluctuation model. Summarizing, the parameters of the carrier-number-fluctuation model are used directly for simulations, whereas the Pelgrom-model parameters for the variances must be divided by 2.

4.6.5.2 Small-signal analysis of the mismatch sensitivity of a circuit

Since Monte Carlo analysis requires typically hundreds of circuit simulations, more rapid simplified methods of analysis have been developed. By representing device mismatches as voltage or current (error) sources, the calculation of the sensitivity of a circuit to device mismatches can be reduced to a small-signal analysis. The standard deviation of a current or voltage in a circuit is then easily calculated. Expressions (4.6.16) and (4.6.17), which give the current standard deviation of a current mirror and the input-voltage standard deviation of a differential pair, respectively, are examples of such analysis. The same procedure can be applied to more complicated cases, such as the calculation of the input random offset of an operational transconductance amplifier [28]. Small-signal analysis can even be applied to determine mismatch effects on circuit dynamics or transient response, modeling mismatch as an equivalent ac pseudo-noise generator [29]. Thus, it is possible to take advantage of the strong analogies between noise and mismatch even at the circuit-simulation level.

Problems

4.1 Calculate the rms values of the noise current and voltage due to thermal noise of a 1-kΩ resistor at room temperature in a bandwidth of (a) 1 Hz and (b) 1 MHz.

4.2 The two (noisy) resistors in the figure below are at the same temperature. The voltage generators in Figure P4.2 represent the thermal noise of the resistors. Calculate the power P_{12} (P_{21}) that the generator v_1 (v_2) delivers to resistance R_2 (R_1). Assuming no knowledge of the thermal-noise formula (4.1.1), what property of the thermal noise can you derive from the power balance $P_{12} = P_{21}$? Derive the condition of maximum power transfer between the resistors. Use expression (4.1.1) to calculate the so-called *available noise power* $P_{12} = P_{21}$, which is the maximum thermal noise that can be extracted from a resistor.

4.3 Assume that the ideal capacitance C_2 in Figure P4.3 generates a thermal noise voltage v_2. Considering a power balance between the resistor and capacitor in equilibrium, prove that $\overline{v_2^2} = 0$.

4.4 In the *RC* circuit of Figure P4.4, calculate the mean-square value of the noise voltage of the capacitor, assuming that the power spectral density $\overline{v_1^2}/\Delta f$ is independent of frequency. Derive the Nyquist expression for the thermal noise (4.1.2) from the application of the energy-equipartition principle to the circuit in Figure P4.4. (Reminder of the *energy-equipartition principle*: in thermal equilibrium the mean thermal energy per degree of freedom is $(1/2)kT$, where k is the Boltzmann constant and T is the absolute temperature.)

4.5 Consider the virtual cut of a transistor that slices it into two series elements as in Figure P4.5. Show that the relation among the thermal-noise current sources associated with the lower, upper, and whole transistor is

$$\frac{\overline{i_d^2}}{\Delta f} = \left(\frac{a}{1+a}\right)^2 \frac{\overline{i_{dl}^2}}{\Delta f} + \left(\frac{1}{1+a}\right)^2 \frac{\overline{i_{du}^2}}{\Delta f}, \tag{P4.5.1}$$

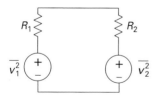

Fig. P4.2   Two (noisy) resistors connected in parallel.

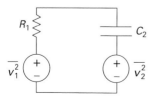

Fig. P4.3   An *RC* circuit assuming a "noisy" capacitor.

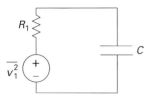

Fig. P4.4   An *RC* circuit.

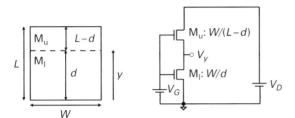

Fig. P4.5   A virtual cut of a transistor into two slices and its representation as a series association of transistors.

where $a = g_{msu}/g_{mdl}$, and g_{msu} and g_{mdl} are the source and drain transconductances of transistors M_u and M_l, respectively. Show that (P4.5.1) can be rewritten as

$$\frac{\overline{i_d^2}}{\Delta f} = \left(\frac{d}{L}\right)^2 \frac{\overline{i_{dl}^2}}{\Delta f} + \left(\frac{L-d}{L}\right)^2 \frac{\overline{i_{du}^2}}{\Delta f}. \tag{P4.5.2}$$

Using the expression for the thermal noise in terms of the inversion charge, prove that

$$\frac{\overline{i_d^2}}{\Delta f} = -4kT\mu\left[\frac{Q_{ll}}{d^2}\left(\frac{d}{L}\right)^2 + \frac{Q_{lu}}{(L-d)^2}\left(\frac{L-d}{L}\right)^2\right] = \frac{-4kT\mu Q_I}{L^2}, \tag{P4.5.3}$$

where Q_{ll}, Q_{lu}, and Q_I are the total inversion charge in the channel of the lower, upper, and whole transistor, respectively. Comment on the results.

4.6 The following formulas are used to represent the PSD of the MOSFET flicker noise in the SPICE simulator:

$$\frac{\overline{i_d^2}}{\Delta f} = \frac{K_F I_D^{AF}}{C'_{ox} L^2}\frac{1}{f}, \tag{P4.6.1}$$

$$\frac{\overline{i_d^2}}{\Delta f} = \frac{K_F I_D^{AF}}{C'_{ox} WL}\frac{1}{f}, \tag{P4.6.2}$$

$$\frac{\overline{i_d^2}}{\Delta f} = \frac{K_F g_{mg}^2}{C'_{ox} WL}\frac{1}{f^{EF}}, \tag{P4.6.3}$$

for **noiselevel** $= 0$, 1, and 2/3, respectively. Verify whether these formulas are consistent for the series and parallel association of transistors and specify the value of

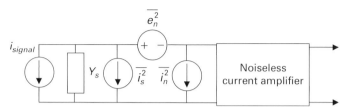

Fig. P4.7   Equivalent input noise representation for the calculation of the noise factor.

AF required for consistency. Also determine the asymptotic value of the normalized PSD $\overline{i_d^2}/(I_D^2\,\Delta f)$ for deep weak inversion (for a discussion on the consistency of $1/f$-noise models, see [11]).

4.7 Using the equivalent circuit in Figure P4.7, show that the noise factor F, defined as [12]

$$F = \frac{\text{total output noise power}}{\text{output noise power due to input source}}, \tag{P4.7.1}$$

can be written as

$$F = \frac{\overline{i_s^2} + \overline{|i_n + Y_s e_n|^2}}{\overline{i_s^2}} = 1 + \frac{\overline{i_u^2} + |Y_c + Y_s|^2\overline{e_n^2}}{\overline{i_s^2}}. \tag{P4.7.2}$$

The noise current i_n is the sum of two components, one fully correlated (i_c) with the input noise voltage generator and the other (i_u) uncorrelated,

$$i_n = i_u + i_c \tag{P4.7.3}$$

with

$$i_c = Y_c e_n, \tag{P4.7.4}$$

where Y_c is the correlation admittance.

By considering the three independent noise sources of Figure P4.7 as being produced by equivalent resistances or conductances

$$\overline{e_n^2} = 4kT\Delta f\, R_n,$$
$$\overline{i_u^2} = 4kT\Delta f\, G_u, \tag{P4.7.5}$$
$$\overline{i_s^2} = 4kT\Delta f\, G_s,$$

and decomposing each admittance into a sum of a conductance G and a susceptance B, prove that

$$F = 1 + \frac{G_u + \left[(G_c + G_s)^2 + (B_c + B_s)^2\right]R_n}{G_s}. \tag{P4.7.6}$$

Show, using expression (P4.7.6), that the noise factor is minimized for

$$B_s = -B_c = B_{opt},$$

$$G_s = \sqrt{\frac{G_u}{R_n} + G_c^2} = G_{opt},$$

(P4.7.7)

and the minimum noise factor is given by

$$F_{min} = 1 + 2R_n\left(G_{opt} + G_c\right) = 1 + 2R_n\left(\sqrt{\frac{G_u}{R_n} + G_c^2} + G_c\right).$$

(P4.7.8)

Verify that for $\overline{i_n^2} = 0$ the optimum noise factor is $F = 1$.

The above results, (P4.7.2) through (P4.7.8) are valid not only for current amplifiers but also for two-port devices for which the noise is represented by the current and noise generators at the input port as in Figure P4.7.

4.8 Show that for the common-source amplifier of Figure 4.8

$$R_n = \frac{\gamma n}{g_m}, \qquad G_u = \frac{\delta\omega^2 C_{gs}^2\left(1 - |c|^2\right)}{5g_{ms}},$$

(P4.8.1)

when the channel thermal noise is given by (4.5.5). Using the results of Problem 4.7, show that

$$B_{opt} = -\omega C_{gs}\left(1 - \frac{|c|}{n}\sqrt{\frac{\delta}{5\gamma}}\right) \qquad G_{opt} = \frac{\omega C_{gs}}{n}\sqrt{\frac{\delta\left(1 - |c|^2\right)}{5\gamma}}$$

(P4.8.2)

and

$$F_{min} = 1 + \frac{2\gamma\omega C_{gs}}{g_m}\sqrt{\frac{\delta\left(1 - |c|^2\right)}{5\gamma}} \cong 1 + \frac{2\omega}{\omega_T}\sqrt{\frac{\gamma\delta\left(1 - |c|^2\right)}{5}}.$$

(P4.8.3)

Calculate, at 1 GHz, the minimum noise factor for a $2L_{min}$ n-channel MOSFET common-source amplifier operating in strong inversion with $i_f = 1000$ in the 1-μm and 0.25-μm technologies with the parameters given in the table of Problem 4.9 below. Comment on the results.

4.9 Assume that the bulk concentration and the oxide thickness of n-MOSFETs in a 1-μm and in a 0.25-μm technology are those given in the table below.

Technology	1 μm	0.25 μm
N_A	2.5×10^{16}	10^{17}
t_{ox} (nm)	20	5

Considering the real transistor dimensions to be equal to the nominal ones and considering $W_{min} = 2L_{min}$, determine the mean and the standard deviation of the number of impurity atoms under the gate of a minimum-dimension n-channel transistor in 1-μm and 0.25-μm technologies. Consider that the depletion depth x_d

under the gate is constant and corresponds to a surface potential value of $2\phi_F$. Determine the threshold matching coefficient A_{VT} for both technologies. Assuming that $A_\beta = 2\%$ μm for both technologies, determine the minimum dimensions of the transistors of a current mirror that allow a current mismatch of less than 1% for devices operating in weak inversion ($i_f = 1$) and in strong inversion ($i_f = 1000$).

4.10 The mismatch parameters for the p-channel transistors of a 0.35-μm and a 0.18-μm technology are reported in the table below.

Technology	A_{VT} (mV μm)	A_β (% μm)
0.35 μm	7	0.8
0.18 μm	4	0.6

Design a unity-gain p-MOS current mirror using minimum-length transistors with current mismatch less than 1%, with the transistors operating in weak ($i_f = 1$) and in moderate inversion ($i_f = 100$) in both technologies. Consider the real transistor dimensions to be equal to the nominal ones. Determine the unity-gain frequency of the transistors in the four cases and comment on the results.

4.11 Consider a transistor represented by two series elements as in Figure P4.11. Show that the variance of the drain current of the whole transistor is related to the variances of the transistors M_1 and M_2 by

$$\sigma_{I_D}^2 = \left(\sigma_{I_{D1}} \frac{g_{md2}}{g_{ms1} + g_{md2}} \right)^2 + \left(\sigma_{I_{D2}} \frac{g_{ms1}}{g_{ms1} + g_{md2}} \right)^2. \tag{P4.11.1}$$

Using the expressions for $g_{ms(d)}$ of a long-channel transistor, prove that

$$L^2 \sigma_{I_D}^2 = L_1^2 \sigma_{I_{D1}}^2 + L_2^2 \sigma_{I_{D2}}^2. \tag{P4.11.2}$$

Show that the number-fluctuation model of mismatch given by (4.6.27), repeated as (P4.11.3),

$$\frac{\sigma_{I_D}^2}{I_D^2} = \frac{N_{oi}}{WLN^{*2}} \frac{1}{i_f - i_r} \ln\left(\frac{1 + i_f}{1 + i_r} \right), \tag{P4.11.3}$$

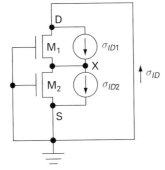

Fig. P4.11   Representation of a transistor as a series association of two transistors.

verifies the consistency condition in (P4.11.2), but that Pelgrom's model given by (P4.11.4),

$$\frac{\sigma_{I_D}^2}{I_D^2} = \frac{A_{VT}^2}{WL}\left(\frac{g_m}{I_D}\right)^2, \qquad (P4.11.4)$$

does not.

References

[1] A. van der Ziel, *Noise in Solid State Devices and Circuits*, New York: Wiley, 1986.

[2] P. R. Gray and R. G. Meyer, *Analysis and Design of Analog Integrated Circuits*, New York: Wiley, 1984.

[3] K. Han, H. Shin, and K. Lee, "Analytical drain thermal noise current model valid for deep submicron MOSFETs," *IEEE Transactions on Electron Devices*, vol. **51**, no. 8, pp. 261–269, Aug. 2004.

[4] C. Galup-Montoro and M. C. Schneider, *MOSFET Modeling for Circuit Analysis and Design*, Singapore: World Scientific, 2007.

[5] C. C. Enz and E. A. Vittoz, *Charge-Based MOS Transistor Modeling. The EKV Model for Low-Power and RF IC Design*, Chichester: Wiley, 2006.

[6] J. C. Paasschens, A. J. Scholten, and R. van Langevelde, "Generalization of the Klaassen–Prins equation for calculating the noise of semiconductor devices," *IEEE Transactions on Electron Devices*, vol. **52**, no. 11, pp. 2463–2472, Nov. 2005.

[7] F. M. Klaassen and J. Prins, "Thermal noise of MOS transistors", *Philips Research Reports*, vol. **22**, pp. 505–514, 1967.

[8] K. K. Hung, P. K. Ko, C. Hu, and Y. C. Cheng, "A physics-based MOSFET noise model for circuit simulators," *IEEE Transactions on Electron Devices*, vol. **37**, no. 5, pp. 1323–1333, May 1990.

[9] F. Principato and G. Ferrante, "1/*f* noise decomposition in random telegraph signals using the wavelet transform," *Physica A*, vol. **380**, no. 7, pp. 75–97, July 2007.

[10] A. Arnaud, *Very Large Time Constant G_m–C Filters*, Ph.D. Thesis, Universidad de la Republica, Uruguay, 2004, http://eel.ufsc.br/~lci/work_doct.html.

[11] A. Arnaud and C. Galup-Montoro, "Consistent noise models for analysis and design of CMOS circuits," *IEEE Transactions on Circuits and Systems I*, vol. **51**, no. 10, pp. 1909–1915, Oct. 2004.

[12] T. H. Lee, *The Design of CMOS Radio-Frequency Integrated Circuits*, 2nd edn., New York: Cambridge University Press, 2004.

[13] A. Hastings, *The Art of Analog Layout*, Upper Saddle River, NJ: Prentice Hall, 2001.

[14] K. R. Lakshmikumar, R. A. Hadaway, and M. A. Copeland, "Characterization and modeling of mismatch in MOS transistors for precision analog design," *IEEE Journal of Solid-State Circuits*, vol. **21**, no. 6, pp. 1057–1066, Dec. 1986.

[15] M. J. M. Pelgrom, A. C. J. Duinmaijer, and A. P. G. Welbers, "Matching properties of MOS transistors," *IEEE Journal of Solid-State Circuits*, vol. **24**, no. 5, pp. 1433–1440, Oct. 1989.

[16] V. Ambegaokar, *Reasoning about Luck: Probability and Its Uses in Physics*, Cambridge: Cambridge University Press, 1996.

[17] M. J. M. Pelgrom, "Low-power CMOS data conversion," in *Low-Voltage/Low-Power Integrated Circuits and Systems*, edited by E. Sánchez-Sinencio and A. Andreou, New York: IEEE Press, 1999, pp. 432–484.

[18] M. J. M. Pelgrom, H. P. Tuinhout, and M. Vertregt, "Transistor matching in analog CMOS applications," *IEDM Technical Digest*, pp. 915–918, Dec. 1998.

[19] P. R. Kinget, "Device mismatch: an analog design perspective," *Proceedings of IEEE ISCAS*, 2007, pp. 1245–1248.

[20] A.-J. Annema, B. Nauta, R. Van Langevelde, and H. Tuinhout, "Analog circuits in ultra-deep-submicron CMOS," *IEEE Journal of Solid-State Circuits*, vol. **40**, no. 1, pp. 132–143, Jan. 2005.

[21] H. Yang, V. Macary, J. L. Huber *et al.*, "Current mismatch due to local dopant fluctuations in MOSFET channel," *IEEE Transactions on Electron Devices*, vol. **50**, no. 11, pp. 2248–2254, Nov. 2003.

[22] H. Klimach, C. Galup-Montoro, M. C. Schneider, and A. Arnaud, "MOSFET mismatch modeling: a new approach," *IEEE Design and Test of Computers*, vol. **23**, no. 1, pp. 20–29, Jan.–Feb. 2006.

[23] C. Galup-Montoro, M. C. Schneider, H. Klimach, and A. Arnaud, "A compact model of MOSFET mismatch for circuit design," *IEEE Journal of Solid-State Circuits*, vol. **40**, no. 8, pp. 1649–1657, Aug. 2005.

[24] P. W. Tuinenga, *SPICE: A Guide to Circuit Simulation and Analysis using PSpice*, 3rd edn., Englewood Cliffs, NJ: Prentice Hall, 1995.

[25] F. Krummenacher, "Matching analysis of analog circuits," MOS Modeling and Parameter Extraction Group Meeting, Lausanne, May 2000, http://legwww.epfl.ch/ekv/mos-ak/lausanne/index.html.

[26] D. O'Riordan, "Recommended Spectre Monte Carlo modeling methodology," Version 1, Dec. 17, 2003, http://www.designers-guide.org/Modeling/.

[27] K. Papathanasiou, "A designer's approach to device mismatch: theory, modeling, simulation techniques, scripting, applications and examples," *Journal of Analog Integrated Circuits and Signal Processing*, vol. **48**, no. 8, pp. 95–106, Aug. 2006.

[28] P. R. Kinget, "Device mismatch and tradeoffs in the design of analog circuits," *IEEE Journal of Solid-State Circuits*, vol. **40**, no. 6, pp. 1212–1224, June 2005.

[29] J. Kim, K. D. Jones, and M. A. Horowitz, "Fast, non-Monte-Carlo estimation of transient performance variation due to device mismatch," *Proceedings of DAC*, 2007, pp. 440–443.

5 Current mirrors

The current mirror is one of the most useful building blocks for analog integrated circuits. It is largely employed as a biasing element and as a load device for amplifier stages. It can also find other uses such as arrays of current sources in D/A converters and current amplifiers in current-mode filters. Basically, a current mirror is a circuit that copies a current flowing through one active device (input) to another active device (output) of a circuit, keeping the output current independent of loading. Current mirrors, from their simplest version to more elaborate circuits, together with analysis of their characteristics, are the subject of this chapter.

5.1 A simple MOS current mirror

5.1.1 The ideal current mirror

The simplest configuration of the ideal current mirror is shown in Figure 5.1. Ideally, the current gain A_I is independent of the input frequency, and the output current is independent of the output voltage; in other words, the output impedance is infinite [1], [2]. Additionally, the input impedance is zero, i.e., the voltage drop across the input device is zero for any input current. The ideal current mirror is equivalent to a two-port current-controlled current source having a common terminal that connects the input and output ports.

In the section that follows, we will present a first-order analysis of the current mirror implemented with a pair of MOS transistors.

5.1.2 The two-transistor current mirror

The basic current mirror is shown in Figure 5.2. It consists of a diode-connected input transistor and an output transistor that share the same gate, source, and substrate voltages. For typical technologies and not-too-low currents, the input transistor operates in saturation (see Problem 5.2).

For a first-order analysis, let us assume that the dependence of the drain current on the voltages for a transistor in saturation is given by

$$I_D = \frac{W}{L} G(V_G, V_S),$$

(5.1.1)

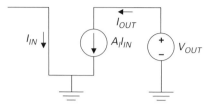

Fig. 5.1   The ideal current mirror.

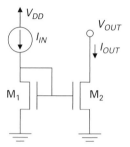

Fig. 5.2   A basic MOS current mirror.

where G describes the dependence of the current on the gate and source voltages. Let us also assume, for the time being, that the two transistors are perfectly identical and that the output transistor also operates in saturation. Additionally, let us assume that the gate currents are negligible. Since the identical transistors M_1 and M_2 are connected to the same gate, source, and bulk nodes and are in saturation, we conclude that $I_{OUT} = I_{IN}$. If the transistors have different aspect ratios, then

$$I_{OUT} = A_I I_{IN} = \frac{(W/L)_2}{(W/L)_1} I_{IN}. \tag{5.1.2}$$

The unity-gain current mirror is the most commonly used of the current mirrors. If a current gain of two is required, then two of the possible topologies for achieving the required gain from the association of identically designed transistors are those shown in Figure 5.3.

If a current gain equal to a rational number N/M is required, the current mirror can be implemented by M diode-connected transistors in parallel at the input and N parallel-connected transistors at the output.

At this point one may ask whether, e.g., a gain-of-two current mirror can be implemented by simply choosing the output transistor with the same channel length but twice the channel width of the input transistor. The reason for not doing this is that the effective transistor width differs from the drawn channel width due to edge effects, e.g., the transition from the thin oxide to the thick oxide, a phenomenon referred to as bird's beak [3] in some technologies. Using a simplified model for the effect of the bird's beak on the transistor width, Example 5.1 illustrates how the current-mirror gain can be affected by an inappropriate design.

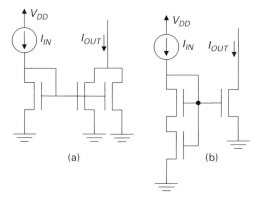

Fig. 5.3   Gain-of-two current mirrors using identical transistors: (a) parallel-connected transistors at the output; (b) series-connected transistors at the input. The series-connected transistors share a common substrate.

Example 5.1

A gain-of-two current mirror, as shown in Figure 5.2, is designed with the following drawn lengths and widths: $L_2 = L_1$ and $W_2 = 2W_1$. Suppose that W_1 is the minimum width allowed for this technology. Calculate the relative error gain if the effective channel width is $W_{eff} = W_{drawn} - \Delta W$. What is the relative error in the gain if the typical $\Delta W = W_1/5$? What is the fluctuation in the relative error if the fluctuation of $\Delta W/W_1$ around the typical value is $\pm 25\%$?

Answer

The relative error $\cong \Delta W/(2W_1)$; 10%; $\pm 2.5\%$.

5.1.3   Error caused by difference between drain voltages

So far, we have assumed that the output transistor of the current mirror is operating in saturation, a condition that is fulfilled for an n-channel transistor if

$$V_{DS} > V_{DSsat} = \phi_t\left(\sqrt{1 + i_f} + 3\right).$$

We have assumed not only that the output transistor operates in saturation, but also that its output conductance is negligible. Now let us assume that the drain voltages of the input and output transistors are different and the output conductances of both transistors are non-negligible. In this case, the current mirror in Figure 5.2 presents the following ratio of the output current to the input current:

$$\frac{I_{OUT}}{I_{IN}} = A_I \frac{1 + V_{DS2}/V_{A2}}{1 + V_{DS1}/V_{A1}} = A_I \frac{1 + V_{OUT}/V_{A1}}{1 + V_{IN}/V_{A2}}, \tag{5.1.3}$$

where V_{A1} and V_{A2} are the Early voltages of the input and output transistors. If the input and output transistors have the same channel length, then $V_{A1} = V_{A2}$. If, in addition, we assume that the errors in the numerator and denominator of (5.1.3) are much less than 1, then we can write

$$\frac{I_{OUT}}{I_{IN}} \cong A_I\left(1 + \frac{V_{OUT} - V_{IN}}{V_A}\right), \tag{5.1.4}$$

where $V_A = V_{A1} = V_{A2}$. As shown in (5.1.4), a simple way to reduce the error due to different drain voltages is to increase the Early voltage through an increase in the channel length, which can be an appropriate solution as long as the mirror does not operate at high frequencies (recall that the transistor unity-gain frequency is inversely proportional to the square of the channel length).

Example 5.2

A unity-gain current mirror as shown in Figure 5.2 is designed to operate at around $I_{IN} = 50\,\mu\text{A}$. Assume that $n = 1.25$, $V_{T0} = 0.4\,\text{V}$, $I_{SH} \cong 50\,\text{nA}$, the Early voltage per unit channel length $V_E \cong 5\,\text{V}/\mu\text{m}$, and the transistor channel width and length are $W = 40\,\mu\text{m}$ and $L = 4\,\mu\text{m}$. (a) Calculate the saturation voltage of the output transistor. (b) What is the percentage error in the output current relative to the input current if the output voltage varies between the saturation voltage and 3.3 V?

Answer

(a) The specific current of both transistors is $I_S = I_{SH}W/L = 0.5\,\mu\text{A}$ and the inversion level $i_f = I_{IN}/I_S = 100$. Therefore, the saturation voltage $V_{DSsat} \cong \phi_t(\sqrt{1 + i_f} + 3)$ is around 0.34 V at room temperature. (b) The Early voltage is $V_A = V_E L = 20\,\text{V}$. Using the UICM

$$V_P = \phi_t\left[\sqrt{1 + i_f} - 2 + \ln\left(\sqrt{1 + i_f} - 1\right)\right],$$

we find $V_P \cong 0.27\,\text{V}$. The approximation $V_P = (V_G - V_{T0})/n$ gives $V_{IN} = V_G \cong 0.73\,\text{V}$. Using (5.1.4), we find that the percentage error for $0.34\,\text{V} < V_{OUT} < 3.3\,\text{V}$ and $V_{IN} = 0.73\,\text{V}$ is within the range -1.95% to 12.85%.

5.1.4   Error caused by transistor mismatch

Let us now calculate the inaccuracy of a unity-gain current mirror due to mismatch between M_1 and M_2, identically designed transistors subjected to the same bias voltages, as shown in Figure 5.4.

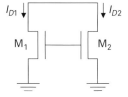

Fig. 5.4   A unity-gain current mirror for mismatch calculation.

In the following, we will derive expressions for the statistics of the error of the drain current that are based on the random fluctuations of both the specific current and the threshold voltage (or substrate doping concentration).

Assuming that the mismatch of components is due to local variations only, Pelgrom's model described in Chapter 4 gives the squared standard deviation of the threshold voltage relative to the average threshold voltage as approximately

$$\sigma^2(V_{T0}) = \frac{A_{VT}^2}{WL}. \tag{5.1.5}$$

As previously mentioned in Chapter 4, the very simplified model of (5.1.5) assumes that (i) the fluctuations in doping concentration are the major cause of mismatch of the threshold voltage; and (ii) the transistor is represented as a lumped element rather than a distributed one. On the other hand, the standard deviation of the specific current is characterized by

$$\frac{\sigma^2(I_S)}{I_S^2} \simeq \frac{A_{ISH}^2}{WL}. \tag{5.1.6}$$

We emphasize here that expressions (5.1.5) and (5.1.6) are good approximations as long as the transistors are neither very short nor very narrow; otherwise, expressions (5.1.5) and (5.1.6) must include length- and width-dependent terms in addition to the area-dependent term. A list of mismatch parameters A_{VT} and A_{ISH}, the latter represented by A_{KP} in the list, for several technologies is presented in [4].

Writing the drain current as

$$I_D = I_S(i_f - i_r), \tag{5.1.7}$$

the variation in the current due to the variations in both the normalized current and the threshold voltage is given by

$$\Delta I_D = \Delta I_S(i_f - i_r) + I_S \frac{\partial(i_f - i_r)}{\partial V_{T0}} \Delta V_{T0}. \tag{5.1.8}$$

Considering that

$$I_S \frac{\partial(i_f - i_r)}{\partial V_{T0}} \Delta V_{T0} = \frac{\partial I_D}{\partial V_{T0}} \Delta V_{T0} = -\frac{\partial I_D}{\partial V_G} \Delta V_{T0} = -g_m \Delta V_{T0}, \tag{5.1.9}$$

the normalized variation of the drain current is

$$\frac{\Delta I_D}{I_D} = \frac{\Delta I_S}{I_S} - \frac{g_m}{I_D} \Delta V_{T0} = \frac{\Delta I_S}{I_S} - \frac{2}{n\phi_t\sqrt{1 + i_f} + 1} \Delta V_{T0}. \tag{5.1.10}$$

If one assumes the errors in the threshold voltage and in the specific current to be uncorrelated, then from (5.1.10) the normalized variance of the drain current is

$$\frac{\sigma^2(I_D)}{I_D^2} = \frac{\sigma^2(I_S)}{I_S^2} + \left(\frac{g_m}{I_D}\right)^2 \sigma^2(V_{T0}). \tag{5.1.11}$$

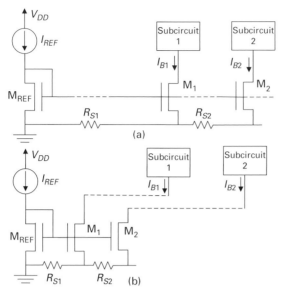

Fig. 5.5   Replication of the reference current using (a) voltage-routing and (b) current-routing techniques (adapted from [2]).

Using Equations (5.1.5) and (5.1.6) of Pelgrom's model and the expression for the transconductance over the drain current in terms of the inversion level for a saturated transistor, we obtain

$$\frac{\sigma^2(I_D)}{I_D^2} \cong \frac{2}{WL}\left\{A_{ISH}^2 + \left(\frac{A_{VT}}{n\phi_t}\right)^2\left[\frac{2}{(\sqrt{1+i_f}+1)}\right]^2\right\}. \tag{5.1.12}$$

Since (5.1.5) and (5.1.6) give the normalized mismatch relative to the average value of the current, a factor of 2 was included in (5.1.12) to account for the uncorrelated statistical errors of the input and output transistors, which are added quadratically. Roughly speaking, the statistics of the drain current in a current mirror are represented by a Gaussian (bell-shaped) characteristic centered on I_D, the average drain current, with a standard deviation calculated from Equation (5.1.12).

Typically, the error in the drain current is dominated by the term in the threshold-voltage variation (ΔV_{T0}), but the error in the specific current prevails in deep strong inversion [5].

One of the most important applications of current mirrors is the biasing of analog integrated circuits. Basically, the biasing technique commonly employed in analog circuits is the replication of amplified and/or attenuated copies of a reference current to the various subcircuits of the integrated circuit, as shown in the two schemes in Figure 5.5 [2]. The voltage-routing technique, illustrated in Figure 5.5(a), consists of wiring the voltage generated at the gate of the reference transistor to other parts of the chip. The voltage routing has two significant disadvantages [2]: (i) the reference transistor and the output transistors of the current mirrors can be far away from each other, increasing the mismatch due to global variations; and (ii) the bias currents of the subcircuits are sensitive to the series resistances R_{S1} and R_{S2} of the ground bus. Of course, problem (ii) becomes worse the further the subcircuits are from the reference transistor.

In the current-routing technique illustrated in Figure 5.5(b), the output transistors are close to the reference transistor and the bias currents are routed to the blocks that require them. The disadvantages of the current-routing technique are (i) the need for one node to be routed for each bias signal, which can increase tremendously the area consumption if the number of output subcircuits is large; and (ii) that the parasitic capacitance at the drains of M_1 and M_2 can greatly increase and thus impose some high-frequency limitations on the blocks connected to M_1 and M_2.

Example 5.3

A unity-gain current mirror, as shown in Figure 5.2, is designed to operate in the range $50\,\text{nA} < I_{IN} < 500\,\mu\text{A}$. For the given technology assume that $n = 1.25$, $V_{T0} = 0.4\,\text{V}$, $I_{SHN} \cong 50\,\text{nA}$, the Early voltage is infinite, and the transistor channel width and length are $W = 40\,\mu\text{m}$ and $L = 4\,\mu\text{m}$. (a) Calculate the gate-voltage range. (b) What is the variation in the percentage error of the output current relative to the input current if the relative error of the specific current is 4% and $\Delta V_{T0} = -5\,\text{mV}$?

Answer

(a) The specific current is $I_S = 0.5\,\mu\text{A}$ while the inversion level is $0.1 < i_f < 1000$ for $50\,\text{nA} < I_{IN} < 500\,\mu\text{A}$. Using the UICM results in $0.27\,\text{V} < V_G < 1.47\,\text{V}$. (b) Using (5.1.10), we find that the percent error in the drain current lies within the range $5\% < \Delta I_D/I_D < 20\%$.

As previously pointed out in Chapter 4, the mismatch model given by (5.1.12) assumes that mismatch can be characterized by lumped parameters, which is inconsistent with the distributed nature of the MOSFET. The mismatch model using a distributed model of the MOS transistor, which was derived in [6] and summarized in Chapter 4, rewritten here for the sake of convenience, is

$$\frac{\sigma_{I_D}^2}{I_D^2} = \frac{1}{WL}\left[\frac{N_{oi}}{N^{*2}}\frac{1}{i_f - i_r}\ln\left(\frac{1 + i_f}{1 + i_r}\right) + A_{ISH}^2\right]. \tag{5.1.13}$$

We recall here that the relationship between N_{oi} and A_{VT0} is

$$\frac{N_{oi}}{N^{*2}} = \left(\frac{A_{VT}}{n\phi_t}\right)^2. \tag{5.1.14}$$

Since (5.1.13) gives the normalized mismatch relative to the average value of the current, a factor of 2 must be included in (5.1.13) to account for the mismatch between the input and output transistors. Thus, the application of (5.1.13) to a current mirror with both the input and output transistors in saturation ($i_r \to 0$) gives

$$\frac{\sigma_{I_D}^2}{I_D^2} = \frac{2}{WL}\left[\left(\frac{A_{VT}}{n\phi_t}\right)^2 \frac{1}{i_f}\ln(1 + i_f) + A_{ISH}^2\right]. \tag{5.1.15}$$

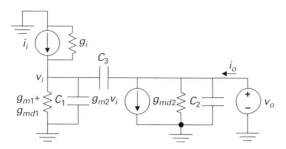

Fig. 5.6   The small-signal equivalent circuit of the current mirror.

Example 5.4

A unity-gain current mirror, as shown in Figure 5.2, is designed to operate in the range $50\,\text{nA} < I_{IN} < 500\,\mu\text{A}$. For the given technology assume that $n = 1.25$, $V_{T0} = 0.4\,\text{V}$, $I_{SH} = 50\,\text{nA}$, the Early voltage is infinite, $A_{VT} = 10\,\text{mV}\,\mu\text{m}$ and $A_{ISH} = 2\%\,\mu\text{m}$. The transistor channel width and length are $W = 40\,\mu\text{m}$ and $L = 4\,\mu\text{m}$. Calculate the expected value of the mismatch between the input and output currents in the specified range of the input current using (a) Equation (5.1.12) and (b) Equation (5.1.15).

Answer

As calculated in Example 5.3, the inversion level $0.1 < i_f < 1000$. (a) Using (5.1.12), we find that the normalized standard deviation of the output current relative to the input current lies within the range $0.31\% < \sigma(I_D)/I_D < 3.6\%$. (b) Equation (5.1.15) gives, for the normalized standard deviation, $0.37\% < \sigma(I_D)/I_D < 3.6\%$. Note that the difference between the models associated with Equations (5.1.12) and (5.1.15) becomes important in strong inversion only.

Even though we have calculated the mismatch between the input and output currents in the particular case of a unity-gain current mirror, the results we have obtained can be extrapolated to non-unity-gain current mirrors.

5.1.5   Small-signal characterization and frequency response

The frequency response of the current mirror in Figure 5.2 is determined from the small-signal equivalent circuit in Figure 5.6. To calculate the small-signal parameters and frequency response we use the following approximations: (i) the operating frequency is such that $\omega\tau_1 \ll 1$ or, in other words, the quasi-static MOSFET model is valid;[1] (ii) both the input and the output transistors operate in saturation; therefore, C_{gd}, C_{bd}, and C_{sd}, the intrinsic capacitances in the quasi-static model which are associated with the variation in the drain voltage, are negligible.

---

[1] If $\omega\tau_1 \ll 1$, it follows from Equation (2.3.21) that the component of the drain current with magnitude equal to $\omega C_m v_i$ is much smaller than $g_m v_i$.

In Figure 5.6, conductance g_i is associated with the input current source whereas $g_{m1} + g_{md1}$ is the conductance of the diode-connected input transistor; g_{md2} is the output conductance of transistor M_2. The capacitance C_1 between the gate node and ground can be written as

$$C_1 = C_{gs1} + C_{gb1} + C_{gs2} + C_{gb2} + \Delta C_1, \tag{5.1.16}$$

where ΔC_1 is the sum of some of the extrinsic capacitances of both M_1 and M_2, namely the drain-to-bulk junction capacitance of M_1, the field oxide capacitance from gate to substrate, the source overlap capacitance of both M_1 and M_2, and other capacitances such as that between the interconnection line and substrate and the output capacitance of the current source. C_2 is the drain-to-bulk junction capacitance plus other parasitic capacitances connected to the output node. Finally, C_3 is the gate-to-drain overlap capacitance of transistor M_2.

The nodal equations for the circuit in Figure 5.6 yield

$$A_I(s) = \frac{I_O}{I_I}(s)\bigg|_{V_O=0} = \frac{g_{m2} - sC_3}{g_{m1} + g_{md1} + g_i + s(C_1 + C_3)}, \tag{5.1.17}$$

where the low-frequency gain A_{I0} is

$$A_{I0} \cong \frac{g_{m2}}{g_{m1}}\left(1 - \frac{g_{md1} + g_i}{g_{m1}}\right) = \frac{(W/L)_2}{(W/L)_1}\left(1 - \frac{g_{md1} + g_i}{g_{m1}}\right). \tag{5.1.18}$$

To write (5.1.18), we assume that $(g_{md1} + g_i)/g_{m1} \ll 1$. To give the reader an insight into the meaning of (5.1.17), we assume that capacitance C_1 is composed predominantly of the intrinsic gate-to-source and gate-to-bulk capacitances of M_1 and M_2. Also, let us assume that $C_1 \gg C_3$, two conditions that are usually fulfilled in practical situations. In this case, we can write (5.1.17) as

$$A_I(s) \cong \frac{g_{m2}}{g_{m1}}\frac{1 - sC_3/g_{m2}}{1 + sC_1/g_{m1}} \cong \frac{A_{I0}}{1 + s(1 + A_{I0})/\omega_{T1}}, \tag{5.1.19}$$

where ω_{T1} is the intrinsic cutoff frequency of M_1. According to (5.1.19), the gain–bandwidth (gain A_{I0}, bandwidth $\omega_{T1}/(A_{I0} + 1)$) product of the current mirror is approximately equal to $A_{I0}/(A_{I0} + 1)$ times the intrinsic cutoff frequency of the input transistor. A unity-gain current mirror has a bandwidth of the order of half the intrinsic cutoff frequency. Therefore, to avoid significant attenuation in the current gain at higher frequencies, unity-gain current mirrors must operate at frequencies that do not exceed one tenth of ω_{T1}. Even though at frequencies below $\omega_{T1}/(1 + A_{I0})$ the attenuation in the magnitude of the current gain is not significant, the phase shift of the output current relative to the input current is important in some applications such as current-mode filters and differential pairs with current mirror load.

The input admittance can readily be calculated as

$$Y_I = \frac{I_I}{V_I}\bigg|_{V_O=0} = g_{m1} + g_{md1} + s(C_1 + C_3) \cong g_{m1}\left[1 + \frac{s(1 + A_{I0})}{\omega_{T1}}\right]. \tag{5.1.20}$$

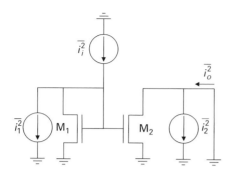

Fig. 5.7   A schematic diagram of the current mirror, showing the noise sources: $\overline{i_i^2}$ represents the noise associated with the input source, whereas $\overline{i_1^2}$ and $\overline{i_2^2}$ represent the channel noise of M_1 and M_2, respectively.

As can be seen in (5.1.20), the input admittance increases with frequency due to the capacitors associated with the input node. This increase in the admittance at higher frequencies is the reason for the decreasing mirror gain. On the other hand, the output admittance is given by

$$Y_O = \left.\frac{I_O}{V_O}\right|_{I_I=0} = g_{md2} + sC_2 + sC_3\left(1 + \frac{g_{m2}}{g_{m1}}\right)\frac{1 + sC_1/(g_{m1} + g_{m2})}{1 + s(C_1 + C_3)/g_{m1}}$$

$$\cong g_{md2} + s[C_2 + C_3(1 + A_{I0})]. \tag{5.1.21}$$

The output admittance of the current mirror is equal to the output conductance of M_2 in parallel with a capacitance equal to C_2 plus the Miller capacitance associated with C_3. To arrive at the approximation in (5.1.21), we have assumed that the maximum operating frequency of the current mirror is sufficiently low for us to neglect the frequency dependence of the rational function in the last term of (5.1.21).

5.1.6   Noise

The output noise current of the current mirror can be calculated from the circuit in Figure 5.7, where $\overline{i_i^2}$ represents the noise associated with the input current signal, and $\overline{i_1^2}$ and $\overline{i_2^2}$ represent the channel noise of transistors M_1 and M_2, respectively. In the general case, for the calculation of the power-spectral density (PSD) of the output noise one has to include the transistor capacitances as well as other parasitic capacitances. However, for the sake of simplicity, we will restrict our analysis to the case of low to moderate frequencies, for which these capacitances are assumed to have a negligible effect on the output noise. Therefore, the small-signal circuit for noise analysis is the same as that shown in Figure 5.6, except for the capacitances, which have been removed. Once again, we assume $(g_{md1} + g_i)/g_{m1} \ll 1$. Under these conditions and given the fact that the noise sources are uncorrelated, we can write

$$\overline{i_o^2} = A_{I0}^2\left(\overline{i_i^2} + \overline{i_1^2}\right) + \overline{i_2^2}, \tag{5.1.22}$$

where A_{I0} is the low-frequency current-mirror gain. The PSDs of the thermal and $1/f$ noise of a MOSFET are given by

$$\left.\frac{\overline{i_d^2}}{\Delta f}\right|_{th} = 4kT\gamma g_{ms} \quad \text{with } \gamma = \frac{2}{3}\frac{\sqrt{1+i_f}+1/2}{\sqrt{1+i_f}+1}, \tag{5.1.23}$$

$$\left.\frac{\overline{i_d^2}}{\Delta f}\right|_{1/f} = I_S^2 i_f \frac{N_{ot}}{WLN^{*2}}\frac{\ln(1+i_f)}{f}. \tag{5.1.24}$$

In weak inversion, the expressions above simplify to

$$\left.\frac{\overline{i_d^2}}{\Delta f}\right|_{th} = 2kTng_m = 2qI_F, \qquad \left.\frac{\overline{i_d^2}}{\Delta f}\right|_{1/f} = \frac{N_{ot}}{WLN^{*2}}\frac{I_F^2}{f}, \tag{5.1.24}$$

whereas in strong inversion the thermal-noise excess factor is $\gamma = 2/3$ and the complete expression (5.1.24) must be employed for the $1/f$ noise.

The inversion level i_f is the same for both M_1 and M_2. The application of expressions (5.1.23) and (5.1.24) to both M_1 and M_2, and their subsequent substitution into (5.1.22), yields

$$\frac{\overline{i_o^2}}{\Delta f} = A_{I0}^2\frac{\overline{i_i^2}}{\Delta f} + (A_{I0}+1)4kT\gamma g_{ms2}$$

$$+ I_{D2}I_{S2}\left(\frac{1}{W_1L_1}+\frac{1}{W_2L_2}\right)\frac{N_{ot}}{N^{*2}}\frac{\ln(1+i_f)}{f}, \tag{5.1.25}$$

where I_{D2} is the direct current of M_2. The relative contribution of M_1 and M_2 to the thermal noise is proportional to their aspect ratios, whereas their relative contribution to $1/f$ noise is inversely proportional to their areas.

5.2   Cascode current mirrors

One of the main problems with a simple current mirror is its relatively low output impedance, which can be insufficient for some applications. A simple way to increase the output impedance of a transistor is to increase the channel length. If the transistor width is increased by the same factor as the channel length in order to keep the inversion level constant, the output impedance increases (approximately) proportionally to the channel length,[2] whereas the transition frequency will decrease proportionally to the reciprocal of the square of the channel length, which, in some cases, leads to a transition frequency that is not high enough for the application.

---

[2] In technologies with pocket implants, the output impedance increases sub-linearly with the channel length.

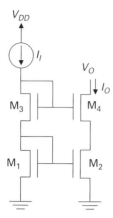

Fig. 5.8   A self-biased cascode current mirror.

5.2.1   Self-biased cascode current mirrors

One way to increase the output impedance of a current mirror is through a change in the circuit topology to a self-biased cascode configuration, which is shown in Figure 5.8. The concept of the self-biased cascode current mirror is to use feedback to ensure that the drain voltage of the mirror transistor follows the drain voltage of the input transistor. If transistors M_1–M_4 are identical and all of them operate in saturation, then $I_O \cong I_I$ and, consequently, $V_{S3} \cong V_{S4}$, which means that M_1 and M_2 operate with approximately the same set of voltages. This means that the sensitivity of the output current, which is equal to the current flowing through M_2, to the output voltage is very low, since the drain voltage of M_2 is almost unaffected by the output voltage as long as M_4 operates in saturation. The operation of M_4 in saturation requires that

$$V_{DS4} = V_O - V_{S4} > V_{DSsat} = \phi_t(\sqrt{1 + i_f} + 3), \tag{5.2.1}$$

where i_f is the inversion level[3] of transistor M_4. Since the currents flowing through M_4 and M_3 are approximately the same, we have $V_{S4} \cong V_{G1}$. The gate voltage V_{G1} can be readily calculated using the UICM. Once V_{G1} is known, one can immediately find the minimum output voltage V_{Omin} required to keep M_4 in saturation through

$$V_{Omin} = V_{G1} + \phi_t(\sqrt{1 + i_f} + 3). \tag{5.2.2}$$

Using the small-signal transistor model, the output conductance G_O can readily be calculated as

$$\frac{1}{G_O} = \frac{1}{g_{md4}} + \frac{1}{g_{md2}\,g_{md4}}\frac{g_{ms4}}{} \cong \frac{1}{g_{md2}\,g_{md4}}\frac{g_{ms4}}{}. \tag{5.2.3}$$

---

[3] In fact, since the four transistors have the same W and L, their inversion levels are the same as long as the small variations in the specific currents due to the different gate voltages are neglected.

The approximation in (5.2.3) is easily interpreted; the output resistance of the cascode current mirror is equal to the output resistance of a simple current mirror, which is equal to $1/g_{md2}$, multiplied by the factor g_{ms4}/g_{md4}, the gain of the common-gate amplifier, which will be analyzed in the next chapter.

Example 5.5

Assume that the cascode current mirror in Figure 5.8 is built with transistors with $L = 0.5\,\mu m$ and that the Early voltage per unit length of all transistors is equal to $10\,V/\mu m$. Calculate the increase in the output resistance compared with that of a simple current mirror for inversion levels spanning the range 0.01 to 80.

Answer

The increase in the output resistance of the cascode current mirror over that of a simple current mirror is equal to g_{ms}/g_{md}, according to (5.2.3). Using the simplified model for the output conductance $g_{md} = I_D/(V_E L)$ and the expression for the source transconductance of a long-channel transistor in saturation, we find that

$$\frac{g_{ms}}{g_{md}} = \frac{I_D}{\phi_t}\frac{2}{\sqrt{1+i_f}+1}\bigg/ \frac{I_D}{V_E L} = \frac{V_E L}{\phi_t}\frac{2}{\sqrt{1+i_f}+1}.$$

For an inversion level between 0.01 and 80, the output resistance of the self-biased cascode current mirror is between 200 and 40 times higher than that of a simple current mirror, for $\phi_t = 25\,mV$.

5.2.2   High-swing cascode current mirrors

The main drawback of self-biased cascode current mirrors is a very serious loss of signal swing, which is especially critical for low supply voltages. As previously discussed, the minimum output voltage required to keep M_4 in saturation is given by (5.2.2), where $V_{G1} \cong V_{DS2}$ is usually relatively larger than V_{DSsat2}. Thus, the minimum output voltage necessary to keep M_4 in saturation can be reduced if a scheme for decreasing V_{DS2} is devised. Such a scheme, shown in Figure 5.9, was presented in [7] and [8] for current mirrors operating in strong and weak inversion, respectively, and later in [9] and [10] for a general inversion regime. The basic idea behind the circuit shown in Figure 5.9 is to generate a voltage V_{DS2} that is just enough to bias the drain-to-source voltage of M_2 close to V_{DSsat2}. The problem to be solved is that of how to design the bias network (M_5 and I_{B5}) or, as we will see later in this section, how to determine the inversion level of M_5 to bias M_2 on the verge of saturation. The equations that follow are derived according to [10]. We note here that the body of M_4 is tied to ground. In this case, the pinch-off voltages of M_4 and M_5, which are the same, can be written as

$$V_{P5} = \phi_t\left[\sqrt{1+i_{f5}} - 2 + \ln\left(\sqrt{1+i_{f5}} - 1\right)\right], \tag{5.2.4}$$

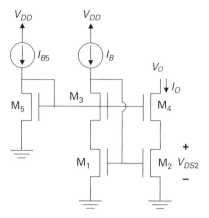

Fig. 5.9   A high-swing cascode current mirror.

$$V_{P4} = V_{S4} + \phi_t \left[\sqrt{1 + i_{f4}} - 2 + \ln\left(\sqrt{1 + i_{f4}} - 1\right) \right]. \tag{5.2.5}$$

The source voltage of M_4 must ensure that M_2 is in saturation. Thus,

$$V_{S4} = V_{DS2} = V_{DSsat2} + \Delta V, \tag{5.2.6}$$

where ΔV is a safety margin that is included in the design equations to ensure that the drain voltage of M_2 is slightly above the drain saturation voltage. This safety margin prevents M_2 from operating in the linear region due to component mismatch and/or inaccurate dc modeling.

Let us now choose $\Delta V = \alpha \phi_t$, where α is, for practical purposes, around unity. Using (5.2.4), (5.2.5), and (5.2.6) results in the following expression for i_{f5}:

$$\left[\sqrt{1 + i_{f5}} - 1 + \ln\left(\sqrt{1 + i_{f5}} - 1\right) \right] = \\ \left[\sqrt{1 + i_{f4}} + 3 + \alpha \right] + \left[\sqrt{1 + i_{f4}} - 1 + \ln\left(\sqrt{1 + i_{f4}} - 1\right) \right]. \tag{5.2.7}$$

In (5.2.7) we use the approximation $i_{f2} \cong i_{f4}$ since we assume that the specific currents of M_2 and M_4 are about the same. We also assume that $V_{DSsat} = \phi_t(\sqrt{1 + i_f} + 3)$. Expression (5.2.7) gives the value of the inversion level i_{f5} required in order to bias M_2 close to the edge of saturation.

Example 5.6

The high-swing cascode current mirror in Figure 5.9 is to be laid out by connecting single transistors from an array of identical transistors. The specific current of a single transistor is $I_S = 0.25\ \mu A$ and the current I_B is equal to $10\ \mu A$. (a) Find an implementation for transistors M_1–M_4 such that transistor M_4 operates in saturation for output voltage higher than $13\phi_t$. (b) Discuss some possible solutions for designing the bias generator. (c) Calculate the drain-to-source voltage of M_3 for $n = 1.25$ and $V_{T0} = 0.4\ V$.

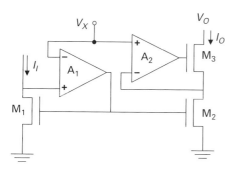

Fig. 5.10   An active regulated cascode current mirror.

Answer

(a) We will assume, for the sake of simplicity, that M_1 through M_4 are identical and that their specific currents are independent of the gate voltages. We decided to choose a safety margin $\Delta V = \phi_t$; thus, since $V_{Omin} = 13\phi_t$, the saturation voltages of both M_2 and M_4 are equal to $6\phi_t$. Using the approximate expression for the saturation voltage $V_{DSsat} = \phi_t(\sqrt{1 + i_f} + 3)$, we find that $i_{f2} = i_{f4} = 8$. Since the current flowing through the mirror is 10 µA, and the inversion level is 8, the specific current of each transistor of the set M_1–M_4 is $10/8 = 1.25$ µA. The specific current of a unity transistor is 0.25 µA; thus, $1.25/0.25 = 5$ unity transistors, connected in parallel, are required to make up each of the transistors of the current mirror. (b) Using (5.2.7) we find $i_{f5} = 74$. A possible solution for the bias generator is $I_{B5} = 1$ µA and M_5 as the series association of 19 unity transistors (in this case, $i_{f5} = 76$). If N is the number of series-connected transistors that constitute M_5, then $NI_{B5} = i_{f5}I_S$, $74 \times 0.25 = 18.5$ µA. If the aim is to save on silicon, one choice would be $I_{B5} = 3.7$ µA and $N = 5$. (c) Using the UICM we can write

$$V_{G1} = V_{T0} + n\phi_t\left[\sqrt{1 + i_{f1}} - 2 + \ln\left(\sqrt{1 + i_{f1}} - 1\right)\right],$$

which gives $V_{G1} \cong 0.46$ V. Since $V_{G1} = V_{DS1} + V_{DS3} = 7\phi_t + V_{DS3}$, we find $V_{DS3} = 0.31$ V, which is higher than $V_{DSsat} \cong 0.15$ V, the drain-to-source saturation voltage for an inversion level of 8.

5.3   Advanced current mirrors

Besides the simple and cascode current mirrors that are largely employed to enhance accuracy and/or output impedance, several others can be found in the technical literature. In recent years, a plethora of articles on MOS current mirrors has been published, especially in the context of low-voltage integrated circuits [11]–[17]. One of them, the active regulated cascode current mirror [11], is shown in Figure 5.10.

The purpose of the two operational amplifiers (or differential amplifiers) is to keep the drains of both the input and the output transistor at the same voltage V_X. This current mirror can operate with both M_1 and M_2 either in the saturation or in the ohmic region, since they are biased by (almost) the same set of voltages. Clearly, if M_1 and M_2 operate in the triode region, the accuracy of the current mirror will generally diminish due to the offset voltages of the operational amplifiers, with, however, the benefit that the output voltage can then swing closer to the negative supply. Note that the signal fed back to A_1 is the output voltage of a common-source amplifier, which is an inverting amplifier, whereas the signal fed back to A_2 is the output voltage of a source follower, a non-inverting amplifier.

Example 5.7
Determine the input and output conductances of the current mirror in Figure 5.10.

Answer

$$G_i = g_{md1} + A_1 g_{m1} \cong A_1 g_{m1},$$

$$G_o = \frac{g_{md2} g_{md3}}{g_{md2} + g_{md3} + g_{ms3} + A_2 g_{m3}} \cong \frac{g_{md2}}{A_2 (g_{m3}/g_{md3})}.$$

The input conductance is equal to that of a simple current mirror multiplied by the amplifier voltage gain, whereas the output conductance is approximately equal to that of a simple current mirror divided by the product of the amplifier gain and the intrinsic gain of transistor M_3.

5.4   Class-AB current mirrors

So far, we have studied class-A current mirrors, for which bias currents are supplied at the input and output due to the bidirectional nature of the current flow. In class-A current mirrors, the bias current must be greater than the signal current in order to ensure that the input transistor does not turn off. The bias current usually leads to a significant offset error due to transistor mismatch [18], [19], and also an increase in the power budget. On the other hand, class-AB current mirrors are able to deal with currents several times larger than the bias current. As a result, class-AB current mirrors have a small offset current at the output and are considerably more power-efficient than class-A current mirrors.

The basic class-AB CMOS current mirror is shown in Figure 5.11. The leftmost branch is the bias circuit, composed of M_3 and M_4, two current sources, and a bias voltage V_{Bias}, which has to be judiciously designed to keep transistors M_1 and M_2 and the current sources in the saturation region. In the analysis that follows, let us assume that the current-mirror gain is unity. If the input current I_I is equal to zero, a copy of the current I_B flows through M_1 and M_2. For a non-zero input current, the output current is equal to the difference between the currents flowing through M_2 and M_1, which, in turn, is equal to the input current. Note that the gate voltages of M_1 and M_2 are constant and that a positive variation in the input current leads to an increase in the voltage at the node common to the sources of M_1 and M_2.

One of the main drawbacks of the class-AB topology shown in Figure 5.11 is the relatively high supply voltage required due to the diode-connected (M_5 and M_6) drive transistors of the mirrors. A solution to alleviate this problem is reported in [18].

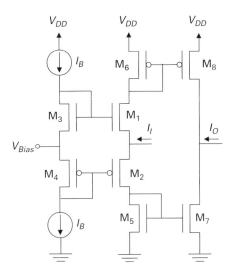

Fig. 5.11   A class-AB current mirror.

Appendix

A.5.1   Harmonic distortion

Current mirrors are key components of current-processing circuits [1]. In these circuits, the input current for the mirrors can, sometimes, change by one order of magnitude or even more. Ideally, the output current must be a replica of the input current with a gain determined by the aspect ratios and independent of the value of the input current. Mismatch between transistors can cause not only a gain error but also harmonic distortion [20]. It is the purpose of this subsection to evaluate the effects of mismatch on the harmonic distortion at low frequencies, i.e. at frequencies at which the capacitive effects of the transistors are negligible. Differences in the drain-to-source voltages of the input and output transistors can also cause harmonic distortion; however, in the analysis that follows we assume that differences in the drain-to-source voltages of the input and output transistors have been reduced by using appropriate circuit techniques such as cascoding [20]. For the sake of simplicity, we will analyze the effects of both the threshold voltage and the specific current mismatch on the topology of Figure A5.1.1, where the drain-to-source voltages are assumed to be equal.

In order to calculate the harmonic distortion at the output, we first calculate the dc mismatch between the direct currents I_{B1} and I_{B2}. According to (5.1.10), the current mismatch is

$$\frac{\Delta I_B}{I_B} = \frac{I_{B2} - I_{B1}}{\left(\dfrac{I_{B2} + I_{B1}}{2}\right)} \simeq \frac{\Delta I_S}{I_S} - \frac{2}{\sqrt{1 + i_f} + 1}\frac{\Delta V_{T0}}{n\phi_t}, \qquad (A5.1.8)$$

where

$$\frac{\Delta I_S}{I_S} = \frac{I_{S2} - I_{S1}}{\left(\dfrac{I_{S2} + I_{S1}}{2}\right)}, \qquad \Delta V_{T0} = V_{T02} - V_{T01}, \qquad \text{and} \qquad i_f = \frac{I_B}{I_S}.$$

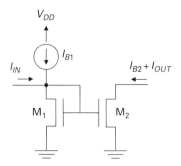

Fig. A5.1.1   A simple MOS current mirror and symbols used to analyze harmonic distortion. The drain-to-source voltages of M_1 and M_2 are assumed to be equal.

Now, the relationship between the output current and the input current is derived from the UICM, written for both M_1 and M_2:

$$\frac{V_{P1}}{\phi_t} = \sqrt{1 + \frac{I_{B1} + I_{IN}}{I_{S1}}} - 2 + \ln\left(\sqrt{1 + \frac{I_{B1} + I_{IN}}{I_{S1}}} - 1\right), \qquad \text{(A5.1.9)}$$

$$\frac{V_{P2}}{\phi_t} = \sqrt{1 + \frac{I_{B2} + I_{OUT}}{I_{S2}}} - 2 + \ln\left(\sqrt{1 + \frac{I_{B2} + I_{OUT}}{I_{S2}}} - 1\right). \qquad \text{(A5.1.10)}$$

Taking the difference between expressions (A5.1.9) and (A5.1.10), and linearizing the pinch-off voltage around the threshold voltage, yields

$$\frac{V_{P1} - V_{P2}}{\phi_t} = \frac{\Delta V_{T0}}{n\phi_t} = \sqrt{1 + \frac{I_{B1} + I_{IN}}{I_{S1}}} - \sqrt{1 + \frac{I_{B2} + I_{OUT}}{I_{S2}}}$$

$$+ \ln\left(\sqrt{1 + \frac{I_{B1} + I_{IN}}{I_{S1}}} - 1\right) - \ln\left(\sqrt{1 + \frac{I_{B2} + I_{OUT}}{I_{S2}}} - 1\right).$$

$$\text{(A5.1.11)}$$

To calculate the harmonic distortion, we will make use of the power series of the output current in terms of the input current. The coefficients of the power series are calculated from the successive derivatives of the output current with respect to the input current at the quiescent point, i.e. for $I_{OUT} = I_{IN} = 0$. Using (A5.1.11), we find that

$$\left.\frac{dI_{OUT}}{dI_{IN}}\right|_0 = 1 + \frac{\Delta I_S}{I_S} - \frac{1}{\sqrt{1 + i_f}}\frac{\Delta V_{T0}}{n\phi_t} \cong 1, \qquad \text{(A5.1.12)}$$

$$\left.\frac{d^2 I_{OUT}}{dI_{IN}^2}\right|_0 = \frac{I_{S2}}{I_{S1}^2}\frac{\Delta V_{T0}}{2n\phi_t}\frac{1}{\left(\sqrt{1 + i_f}\right)^3} \cong \frac{\Delta V_{T0}}{2n\phi_t I_S}\frac{1}{\left(\sqrt{1 + i_f}\right)^3}, \qquad \text{(A5.1.13)}$$

$$\left.\frac{d^3 I_{OUT}}{dI_{IN}^3}\right|_0 = -\frac{I_{S2}}{I_{S1}^3}\frac{3}{4}\frac{\Delta V_{T0}}{n\phi_t}\frac{1}{\left(\sqrt{1 + i_f}\right)^5} \cong -\frac{3}{4}\frac{\Delta V_{T0}}{n\phi_t I_S^2}\frac{1}{\left(\sqrt{1 + i_f}\right)^5}. \qquad \text{(A5.1.14)}$$

The power-series expansion of the output current in terms of the input current is written as

$$I_{OUT} = \frac{dI_{OUT}}{dI_{IN}}\bigg|_0 I_{IN} + \frac{1}{2!}\frac{d^2 I_{OUT}}{dI_{IN}^2}\bigg|_0 I_{IN}^2 + \frac{1}{3!}\frac{d^3 I_{OUT}}{dI_{IN}^3}\bigg|_0 I_{IN}^3 + \cdots. \qquad (A5.1.15)$$

Assuming the input current to be $I_{IN} = I_M \cos(\omega_0 t)$, we can readily calculate the magnitudes of the fundamental, second-harmonic, and third-harmonic components of the output current, I_{MO1}, I_{MO2}, and I_{MO3}. The output current is written as

$$I_{OUT} = I_{MO1} \cos(\omega_0 t) + I_{MO2} \cos(2\omega_0 t) + I_{MO3} \cos(3\omega_0 t) + \cdots. \qquad (A5.1.16)$$

On substituting $I_{IN} = I_M \cos(\omega_0 t)$ into (A5.1.15) and using trigonometric identities and the approximations in (A5.1.12), (A5.1.13), and (A5.1.14), we find that

$$I_{MO1} = I_M \frac{dI_{OUT}}{dI_{IN}}\bigg|_0 \cong I_M,$$

$$I_{MO2} = \frac{I_M^2}{4}\frac{d^2 I_{OUT}}{dI_{IN}^2}\bigg|_0 \cong I_M \frac{I_M}{8 I_B}\frac{\Delta V_{T0}}{n\phi_t}\frac{i_f}{\left(\sqrt{1+i_f}\right)^3}, \qquad (A5.1.17)$$

$$I_{MO3} = \frac{I_M^3}{24}\frac{d^3 I_{OUT}}{dI_{IN}^3}\bigg|_0 \cong -\frac{I_M}{32}\left(\frac{I_M}{I_B}\right)^2 \frac{\Delta V_{T0}}{n\phi_t}\frac{i_f^2}{\left(\sqrt{1+i_f}\right)^5}.$$

On defining the nth-order harmonic distortion HD_n as the power ratio $(I_{MOn}/I_{MO1})^2$, we finally find that

$$\sqrt{HD_2} = \left|\frac{I_{MO2}}{I_M}\right| = \frac{I_M}{8 I_B}\frac{|\Delta V_{T0}|}{n\phi_t}\frac{i_f}{\left(\sqrt{1+i_f}\right)^3},$$

$$\sqrt{HD_3} = \left|\frac{I_{MO3}}{I_M}\right| = \frac{1}{32}\left(\frac{I_M}{I_B}\right)^2 \frac{|\Delta V_{T0}|}{n\phi_t}\frac{i_f^2}{\left(\sqrt{1+i_f}\right)^5}. \qquad (A5.1.18)$$

The expressions in (A5.1.18) show that, for a first-order calculation, the mismatch in the specific currents does not cause harmonic distortion, as could be expected. Mismatch in the specific current causes only a gain error. On the other hand, threshold-voltage mismatch gives rise to harmonic distortion, which is dependent on both the inversion level and the bias current. We can also see that (A5.1.18) provides the classical linear and quadratic dependence of the second- and third-harmonic distortion on the signal magnitude, which is valid for low distortion levels.

Problems

5.1 The transistors of the unity-gain current mirror in Figure P5.1 are assumed to have the following parameters: $V_{T0} = -0.4$ V, $I_S = 1\,\mu$A, and $n = 1.25$. (a) Calculate the range of the resistance values for an output current in the range 0.5–$50\,\mu$A. (b) Calculate the maximum output voltage at which the output transistor remains in saturation. (c) Assuming the Early voltage to be 5 V, calculate the error in the

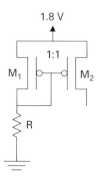

1.8 V

1:1

M_1 M_2

R

Fig. P5.1   A unity-gain current mirror.

specified output current range for an output voltage equal to 0 V. (d) Now, assuming
the Early voltage to be negligible, calculate the error in the output current for the
specified current range when $V_{T1} = V_{T0}$ and $V_{T2} = V_{T0} \pm 10$ mV.

5.2 Assume that an n-channel MOSFET connected as a diode operates in weak inversion. Using the linearized expression for the pinch-off voltage and considering the
drain-to-source saturation voltage of a MOS transistor in weak inversion equal to
$4\phi_t$, verify that the diode-connected transistor is in saturation when

$$I_D \geq 2 I_S \exp\left(\frac{-V_{T0} + (4+n)\phi_t}{n\phi_t}\right).$$

Calculate the minimum inversion level at the source such that the transistor is in
saturation when $V_{T0} = 0.4$ V, $n = 1.2$, and $T = 290$ K.

5.3 Using the expression for the UICM, demonstrate that, for small deviations in the
specific current and in the threshold voltage, the drain-current error in a saturated
transistor is given by

$$\frac{\Delta I_D}{I_D} \cong \frac{\Delta I_S}{I_S} - \frac{2}{n\phi_t\left(\sqrt{1+i_f}+1\right)}\Delta V_{T0}.$$

5.4 Design a gain-of-25 current mirror employing identically designed transistors with
the minimum number of transistors.

5.5 (a) Design a 25 : 1 (or 24 : 1) current mirror employing identically designed transistors. Using series/parallel association of transistors at the input and output, find
all the combinations possible for a 25 (or 24) : 1 current mirror. The maximum
number of transistors of your implementation cannot exceed 26.

(b) Assume now that the input current can be either as high as $1000 I_{SH}$ or as low as
$0.01 I_{SH}$. Also assume that the aspect ratio of the transistors is 1. Comment on the
input impedance, output impedance, matching (assume that the matching
depends on the inverse of the area), noise, inversion levels at the input and
output, input voltage drop, and output saturation voltage for the combinations
you have found in (a).

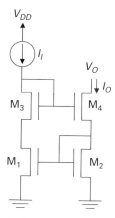

Fig. P5.6   An improved Wilson current mirror.

5.6 Figure P5.6 shows the scheme of the improved Wilson current mirror. The transistors have the same dimensions and the input current is 6 μA. Assume that, for the technology employed, the specific current, measured for a unit transistor with $W/L = 3$ μm/ 0.5 μm is 1 μA, the threshold voltage $V_{T0} = 0.5$ V, and $n = 1.2$ is independent of V_G. (a) Using series/parallel association of transistors, determine the number of unit transistors required to bias the transistors at an inversion level of 3. What is the gate voltage of M_2 in this case, and what is the minimum output voltage at which M_4 remains in saturation? (b) Repeat the calculations of (a) for $i_f = 10$. (c) Show that, if the output conductances of the transistors are much smaller than their transconductances, then the output conductance of the improved Wilson current mirror is, approximately $G_O \cong g_{md1}(g_{md4}/g_{ms4})$, the same value as that obtained for the self-biased cascode current mirror.

5.7 Assume that the high-swing cascode current mirror in Figure 5.9 is to be laid out by connecting either single transistors from an array of identical transistors or customized transistors with any aspect ratio. Both the specific current of a transistor from the array and the sheet normalization current are equal to 0.25 μA. The current I_B is equal to 2 μA. (a) Find an implementation for transistors M_1–M_4 such that transistor M_4 operates in saturation when the output voltage is higher than $13\phi_t$. Use a safety margin of $\Delta V = \phi_t$. Use either the identically designed transistors or customized transistors. (b) Discuss some possible solutions for designing the bias generator (current source I_{B5} and equivalent transistor M_5), using either the identically designed transistors or customized transistors.

5.8 Verify that the small-signal output current of the high-swing cascode current mirror in Figure 5.9 depends on the input current and the output voltage according to

$$i_o = \alpha i_{in} + G_o v_o,$$

where
$$G_o = \frac{g_{md2}g_{md4}}{g_{md2} + g_{ms4}} \cong \frac{g_{md2}g_{md4}}{g_{ms4}},$$

$$\alpha = \frac{g_{m2}}{g_{m1}} \frac{1 + g_{md1}/g_{ms3}}{1 + g_{md2}/g_{ms4}} \frac{1}{1 + (g_{md1}/g_{m1})(g_{md3}/g_{ms3})}.$$

References

[1] B. Gilbert, "Bipolar current mirrors," in *Analogue IC Design: The Current-Mode Approach*, ed. C. Toumazou, F. J. Lidgey, and D. G. Haigh, London: Peter Peregrinus Ltd., 1990, pp. 239–296.

[2] P. R. Gray, P. J. Hurst, S. H. Lewis, and R. G. Meyer, *Analysis and Design of Analog Integrated Circuits*, 4th edn., New York: John Wiley & Sons, 2001.

[3] A. Hastings, *The Art of Analog Layout*, Upper Saddle River, NJ: Prentice Hall, 2001.

[4] D. M. Binkley, B. J. Blalock, and J. M. Rochelle, "Optimizing drain current, inversion level, and channel length in analog CMOS design," *Journal of Analog Integrated Circuits and Signal Processing*, vol. **47**, pp. 137–163, May 2006.

[5] E. Vittoz, "Elementary building blocks," *Intensive Summer Course on CMOS VLSI Design: Analog & Digital*, Lausanne, 1989.

[6] C. Galup-Montoro, M. C. Schneider, H. Klimach, and A. Arnaud, "A compact model of MOSFET mismatch for circuit design," *IEEE Journal of Solid-State Circuits*, vol. **40**, no. 8, pp. 1649–1657, Aug. 2005.

[7] T. C. Choi, R. T. Kaneshiro, R. W. Brodersen *et al.*, "High-frequency CMOS switched-capacitor filters for communications application," *IEEE Journal of Solid-State Circuits*, vol. **18**, no. 6, pp. 652–664, Dec. 1983.

[8] E. Vittoz, "Micropower techniques," in *Design of Analog-Digital VLSI Circuits for Telecommunications and Signal Processing*, 2nd edn., ed. J. E. Franca and Y. Tsividis, Englewood Cliffs, NJ: Prentice Hall, 1994.

[9] V. C. Vincence, C. Galup-Montoro, and M. C. Schneider, "A high-swing MOS cascode bias circuit," *IEEE Transactions on Circuits and Systems II*, vol. **47**, no. 11, pp. 1325–1328, Nov. 2000.

[10] P. Aguirre and F. Silveira, "Bias circuit design for low-voltage cascode transistors," *Proceedings of SBCCI 2006*, pp. 94–98, Sep. 2006.

[11] T. Serrano and B. Linares-Barranco, "The active-input regulated-cascode current mirror," *IEEE Transactions on Circuits and Systems II*, vol. **41**, no. 6, pp. 464–467, June 1994.

[12] H. C. Yang and D. J. Allstot, "An active-feedback cascode current source," *IEEE Transactions on Circuits and Systems*, vol. **37**, no. 5, pp. 644–646, May 1990.

[13] E. Säckinger and W. Guggenbühl, "A high-swing, high-impedance MOS cascode circuit," *IEEE Journal of Solid-State Circuits*, vol. **25**, no. 1, pp. 289–298, Feb. 1990.

[14] B. B. Blalock and P. Allen, "A low-voltage, bulk-driven MOSFET current mirror for CMOS technology," *Proceedings of IEEE ISCAS*, 1995, vol. **3**, pp. 1972–1975.

[15] V. I. Prodanov and M. M. Green, "CMOS current mirrors with reduced input and output voltage requirements," *Electronics Letters*, vol. **32**, no. 2, pp. 104–105, Jan. 18, 1996.

[16] F. You, S. H. K. Embabi, J. F. Duque-Carrillo, and E. Sánchez-Sinencio, "An improved current source for low voltage applications," *IEEE Journal of Solid-State Circuits*, vol. **32**, no. 8, pp. 1173–1180, Aug. 1997.

[17] J. Ramírez-Angulo, R. G. Carvajal, and A. Torralba, "Low-supply-voltage high-performance CMOS current mirror with low input and output voltage requirements," *IEEE Transactions on Circuits and Systems II*, vol. **51**, no. 3, pp. 124–129, Mar. 2004.

[18] G. Palumbo and S. Pennisi, "A Class AB CMOS current mirror with low-voltage capability" *Proceedings of ICECS*, 1999, pp. 891–894.

[19] S. Kawahito and Y. Todokoro, "CMOS class-AB current mirrors for precision current-mode analog-signal processing elements," *IEEE Transactions on Circuits and Systems II*, vol. **43**, no. 12, pp. 843–845, Dec. 1996.

[20] E. Bruun, "Analytical expressions for harmonic distortion at low frequencies due to device mismatch in CMOS current mirrors," *IEEE Transactions on Circuits and Systems II*, vol. **46**, no. 7, pp. 937–941, July 1999.

6   Current sources and voltage references

The design of dc current and voltage sources internal to the chip is an essential step in monolithic-integrated-circuit design [1]. The performance of an analog circuit is essentially dependent on its dc operating conditions, which are set by the bias circuit, which is most often a current source. Voltage or current references are required in applications such as D/A and A/D converters, and voltage regulators. This chapter presents some basic CMOS building blocks for implementing current sources, voltage references, and proportional-to-absolute-temperature (PTAT) voltage references.

6.1   A simple MOS current source

A simple structure for implementing a current source is shown in Figure 6.1. It is composed of a resistor and a current mirror. Ideally, the output current I_{OUT} is a copy of the input current I_{REF} as long as M$_2$ remains in saturation. The graph shown in Figure 6.1(b) represents the current–voltage characteristics of the diode-connected transistor and the load line. The combination of Ohm's law for the load and the unified current-control model (UICM) for the MOS transistor yields

$$RI_{REF} \cong V_{DD} - V_{T0} - n\phi_t \left[\sqrt{1 + \frac{I_{REF}}{I_S}} - 2 + \ln\left(\sqrt{1 + \frac{I_{REF}}{I_S}} - 1 \right) \right]. \qquad (6.1.1)$$

Expression (6.1.1) allows the calculation of the resistance in terms of the transistor parameters for a specified value of the reference current.

Example 6.1
A current source as shown in Figure 6.1 is designed for $I_{REF} = 1\,\mu\text{A}$. Assume that $V_{DD} = 3.3$ V and the technology parameters are $V_{T0} = 0.4$ V, $I_{SH} = 100\,\text{nA}$, and $n = 1.2$. (a) Calculate the approximate value of the load resistance for transistor aspect ratios equal to 100, 1, and 0.01. (b) Recalculate the load resistance for the same transistor aspect ratios and $I_{REF} = 10\,\text{nA}$.

Answer
From (6.1.1) we find (a) 3.02, 2.84, and 1.91 MΩ for aspect ratios 100, 1, and 0.01, respectively. (b) R is of the order of 300 MΩ for the three aspect ratios, a value that is quite demanding in terms of silicon area!

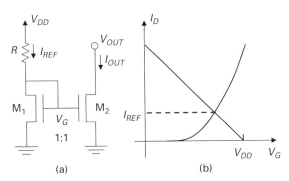

Fig. 6.1   (a) A simple MOS current source. (b) The current–voltage characteristic of the input transistor and load line.

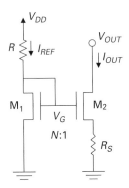

Fig. 6.2   The MOS Widlar current source.

The current-source implementation in Figure 6.1 has several drawbacks. The reference current is strongly dependent on the supply voltage as well as on the process parameters. Moreover, the generation of low currents requires a large silicon area owing to the large resistance values needed. A possible way to reduce the value of the overall resistance to generate the low bias currents often needed for low-power design is to insert a resistor in series with the source of the output transistor, as shown in the next section.

6.2   The Widlar current source

The bipolar Widlar current source was proposed in [2] for generating very small currents using resistors of moderate value. The MOS counterpart to the bipolar Widlar current source is shown in Figure 6.2. Even though one can use the Widlar current source for the generation of small currents, the designer of MOS circuits has an additional degree of freedom for design, namely the possibility of connecting transistors in series and/or in parallel. It is generally more practical to employ parallel-connected transistors at the input and series-connected transistors at the output for lowering the input current. Example 6.2 illustrates how to generate an output current equal to $I_{REF}/100$.

Example 6.2

Using the data and results of Example 6.1, design a current source for $I_{REF} = 1\,\mu A$ and $I_{OUT} = 10\,nA$ using (a) a $1:1$ current mirror and a source degeneration resistance R_S; and (b) an $N:1$ current mirror and $R_S = 0$.

Answer

Two possible solutions to this problem are (a) $R = 2.84\,M\Omega$, and a $1:1$ current mirror (transistors with aspect ratios of 1 at both the input and the output) and $R_S = 18.4\,M\Omega$; and (b) $R = 2.84\,M\Omega$, $R_S = 0$ and a $100:1$ current mirror composed of 10 identical parallel-connected transistors at the input and 10 identical series-connected transistors at the output, all of them with aspect ratios of 0.1.

The current sources we have described so far provide currents that are highly dependent on process, supply voltage, and temperature. For most applications, the performance of the subcircuits and, consequently, of the overall system relies on the stability of the bias current. In many cases, neither the supply voltage nor the temperature is constant over the lifetime of the equipment or from one sample to another. Thus, more appropriate schemes must be devised to generate more stable currents. The next sections of this chapter describe current and voltage sources that have low sensitivity to the supply voltage and/or temperature.

6.3 Self-biased current sources (SBCSs)

The circuit shown in Figure 6.3 [3]–[6] generates a supply-independent output current. Both the original MOS version [4] of the circuit and its bipolar equivalent were conceived in order to generate a PTAT voltage, which is then converted into a current through resistor R_S.

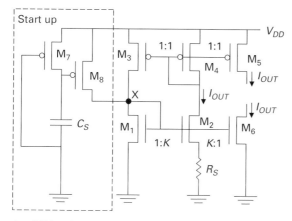

Fig. 6.3 A self-biased current source.

The core cell of the current source is made up of M_1, M_2, resistor R_S, and a unity-gain current mirror (M_3–M_4). The relative aspect ratios of the transistors are given in Figure 6.3, with $K > 1$. Since the currents through M_1 and M_2 are the same, for $K > 1$ the difference $V_{GS1} - V_{GS2} > 0$ is applied across resistor R_S. For $K > 1$, the SBCS is stable. To derive the design equations, let us assume that all transistors operate in saturation. Thus, using the UICM for M_1, we can write

$$\frac{V_{P1}}{\phi_t} = \left[\sqrt{1 + \frac{I_{OUT}}{I_{S1}}} - 2 + \ln\left(\sqrt{1 + \frac{I_{OUT}}{I_{S1}}} - 1 \right) \right], \tag{6.3.1}$$

whereas for M_2

$$\frac{V_{P2} - V_{S2}}{\phi_t} = \left[\sqrt{1 + \frac{I_{OUT}}{KI_{S1}}} - 2 + \ln\left(\sqrt{1 + \frac{I_{OUT}}{KI_{S1}}} - 1 \right) \right]. \tag{6.3.2}$$

Since $V_{S2} = R_S I_{OUT}$ and $V_{P2} = V_{P1}$ we can write, from (6.3.1) and (6.3.2), the following design equation:

$$\frac{R_S I_{OUT}}{\phi_t} = \left[\sqrt{1 + \frac{I_{OUT}}{I_{S1}}} - \sqrt{1 + \frac{I_{OUT}}{KI_{S1}}} + \ln\left(\frac{\sqrt{1 + I_{OUT}/I_{S1}} - 1}{\sqrt{1 + I_{OUT}/(KI_{S1})} - 1} \right) \right]. \tag{6.3.3}$$

It is interesting to note that, for $I_{OUT}/I_{S1} \ll 1$, (6.3.3) becomes

$$I_{OUT} = \frac{\phi_t}{R_S} \ln K, \tag{6.3.4}$$

a result that was previously presented in [4]. On the other hand, for operation of both M_1 and M_2 in strong inversion, the output current, as shown in [5], [6], is given by

$$I_{OUT} = \frac{\phi_t^2}{R_S^2 I_{S1}} \left(1 - \sqrt{\frac{1}{K}} \right)^2 = \frac{2}{R_S^2 \mu_n C_{ox}' n(W/L)_1} \left(1 - \sqrt{\frac{1}{K}} \right)^2. \tag{6.3.5}$$

The temperature coefficient TC of the SBCS, that is,

$$TC = \frac{1}{I_{OUT}} \frac{dI_{OUT}}{dT},$$

can be readily calculated from (6.3.3) for any operating region, but, for a simple interpretation, we derive the temperature coefficient for the asymptotic cases of weak and strong inversion using (6.3.4) and (6.3.5), respectively, which give

$$TC|_{WI} = \frac{1}{T} - \frac{1}{R_S} \frac{dR_S}{dT}, \tag{6.3.6}$$

$$TC|_{SI} = -\frac{1}{\beta} \frac{d\beta}{dT} - \frac{2}{R_S} \frac{dR_S}{dT}, \tag{6.3.7}$$

where $\beta = \mu_n C_{ox}' n$. It can be clearly seen in (6.3.6) that a zero temperature coefficient for weak-inversion operation requires a PTAT resistance. However, this is not the case in

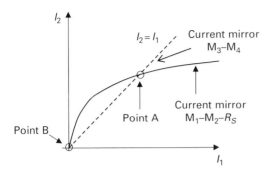

Fig. 6.4   Transfer characteristics of the current mirrors of the core cell in Figure 6.3 (adapted from [3]).

strong inversion, and (6.3.7) defines the relationship required for the temperature coefficient to equal zero. In general, it is difficult to obtain a temperature coefficient of the current source close to zero since the resistor's temperature coefficient cannot be chosen at will due to the scarcity of available materials for the resistor in a low-cost CMOS process.

In many self-biased circuits, which operate with positive feedback, there are two operating points. To bias the circuit at the desired operating point, a start-up circuit is often needed [1], [3], [5], [6]. To better understand why the core circuit in Figure 6.3 has two operating points, we have recourse to Figure 6.4, which shows the current transfer characteristics of the two subcircuits that comprise the core cell, namely the current mirror M_3–M_4 and the degenerate current mirror M_1–M_2–R_S. As can be seen in Figure 6.4, two operating points, A and B, satisfy simultaneously both transfer characteristics. To see that the core circuit in Figure 6.3 operates with positive feedback, let us cut the connection between transistors M_2 and M_4, drive transistor M_4 with a small-signal current i_{in}, and examine the open-loop current gain. Suppose first that the operating point of the circuit is point B; in this case, the small-signal current of M_2 is Ki_{in} since the gain of mirror M_3–M_4 is equal to unity and the effect of the voltage drop across R_S is negligible for low currents. Note that the output current is in phase with the input current; therefore, the closed-loop cell operates with positive feedback and the open-loop gain is equal to K, which is greater than unity. This simplified analysis shows that point B is unstable due to the loop gain being greater than unity. On the other hand, for point A, the loop gain, which is given by the derivative of the transfer characteristic of M_1–M_2–R_S, is smaller than unity; thus, we can conclude that point A is stable. However, as pointed out in [3], point B is frequently a stable operating point (as is the case for SBCSs based on bipolar transistors) because the currents in the transistors at this point are very low. At such low current levels, leakage currents cause a reduction in the current gain of the mirrors, usually causing the loop gain to be less than unity [3].

As regards the circuit in Figure 6.3, it is self-starting (point B is unstable) as long as the leakage of M_2 is greater than that of M_1 [7]. In this case, the stable operating point is A only. However, as mentioned in [8], in some systems, such as passive RFID tags and low-power sensor nodes, all circuits must "wake up" as soon as possible after power-up. Even if point B is unstable, it can take a very long time (of the order of seconds) for the system to converge to point A because the currents of the circuit immediately after power-up are extremely low, resulting in a very long charge-up time for the internal node capacitances. Therefore, the inclusion of a start-up circuit to accelerate the convergence of the self-biased circuit to the desired operating point is highly recommended. An example of a start-up circuit convenient for low-power applications is shown inside the dashed box in Figure 6.3 [9]. Assume initially that the power supply is off and that all node capacitances are discharged. When the power supply is switched on, M_8 turns on and starts to charge the capacitance associated with node X. The drive current of M_8 is maximal on power-up and gradually starts to decrease as time passes, since M_7 charges capacitor C_S. After a short time, which depends on the drive current of both M_7 and M_8, C_S charges up to V_{DD} and completely turns off M_8 except for a small leakage current at $V_{GS8} = 0$. Thus, the start-up circuit shown in Figure 6.3 does not introduce extra static power into the SBCS.

To conclude this section, we calculate the noise associated with the current generated at the core cell of the SBCS in Figure 6.3. We calculate the noise power-spectral density (PSD) for frequencies lower than the -3 dB cut-off frequency of the current mirrors. The final result shows the correlation of the noise PSD with the direct current and the transistor parameters. To start with, let us represent the noise associated with each transistor by a current source i_{nj} between the source and drain of transistor M_j. The resistor noise is represented by a current source i_s in parallel with the resistor. The block diagram used to calculate the SBCS noise is shown in Figure 6.5. Each block represents the low-frequency short-circuit current gain. The sum of the current sources gives

$$i_{out} = A_i \left(i_{n1} + \frac{i_{n2} + i_s g_{ms2} R_S}{1 + g_{ms2} R_S} + i_{n3} + i_{n4} \right), \tag{6.3.8}$$

where A_i is the closed-loop gain given by

$$A_i = \frac{g_{m2}/g_{m1}}{1 + g_{ms2} R_S - g_{m2}/g_{m1}} = \frac{g_{ms2}/g_{ms1}}{1 + g_{ms2} R_S - g_{ms2}/g_{ms1}}. \tag{6.3.9}$$

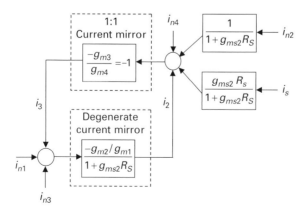

Fig. 6.5   A block diagram for the calculation of the SBCS noise. The formulas inside the blocks represent the short-circuit current gains.

The equality $g_{m2}/g_{m1} = g_{ms2}/g_{ms1}$ holds because both M_1 and M_2 operate in saturation.

Recalling that, for a saturated transistor $g_{ms}\phi_t/I_D = 2/(\sqrt{1+i_f}+1)$, observing that the currents through M_1 and M_2 are the same, and taking into account (6.3.3), one can write

$$A_i = \frac{\sqrt{1 + i_{f1}} + 1}{\sqrt{1 + i_{f1}} - \sqrt{1 + i_{f1}/K} + 2 \ln \left(\dfrac{\sqrt{1 + i_{f1}} - 1}{\sqrt{1 + i_{f1}/K} - 1} \right)}. \tag{6.3.10}$$

Each of the noise currents in (6.3.8) can be represented by its PSD as given in Chapter 4 for both thermal and $1/f$ noise. Since the noise sources in Figure 6.5 are uncorrelated, we can write (6.3.8) as[1]

$$\frac{\overline{i_{out}^2}}{\Delta f} = A_i^2 \left[\frac{\overline{i_{n1}^2}}{\Delta f} + \frac{\overline{i_{n2}^2}/\Delta f + (g_{ms2}R_S)^2 \overline{i_s^2}/\Delta f}{(1 + g_{ms2}R_S)^2} + \frac{\overline{i_{n3}^2}}{\Delta f} + \frac{\overline{i_{n4}^2}}{\Delta f} \right]. \tag{6.3.11}$$

To illustrate the usefulness of (6.3.11), we calculate the PSD of the current in the core cell assuming that all transistors operate in weak inversion. In this case, we have

$$\frac{\overline{i_{out,th}^2}}{\Delta f} = \frac{2qI_{OUT}}{(\ln K)^2} \left(3 + \frac{1 + 2\ln K}{(1 + \ln K)^2} \right) \tag{6.3.12}$$

for the thermal noise and

$$\frac{\overline{i_{out,1/f}^2}}{\Delta f} = \frac{I_{OUT}^2}{(WL)_1 (\ln K)^2} \left[\frac{N_{otN}}{N_N^{*2}} \left(1 + \frac{1}{K(1 + \ln K)^2} \right) + 2 \frac{N_{otP}}{N_P^{*2}} \frac{(WL)_1}{(WL)_3} \right] \frac{1}{f} \tag{6.3.13}$$

for the flicker noise. In (6.3.13) we have assumed that $W_2 = KW_1$, $L_2 = L_1$, and the flicker noise of R_S is negligible.

A drawback of the circuit in Figure 6.3 is the use of a resistor. For some very-low-power applications, a current of the order of nA or even less would require extremely high values of resistance, which, in turn, would be very costly in terms of silicon area. In the next section we describe a MOSFET-only self-biased current source.

6.4   A MOSFET-only self-biased current source

The basic idea behind the MOSFET-only SBCS presented in [10] is to replace the resistor R_S of Figure 6.3 with a transistor operating in the linear region. The circuit of the MOSFET-only self-biased current source shown in Figure 6.6 gives a current proportional to the specific current and can operate at low supply voltages. For the sake of simplicity, the start-up circuit is not shown.

---

[1] $\overline{i_{out}^2}$ is the noise power at the drain of transistor M_2. To calculate the noise at the drain of M_6, for example, we must include the noise generated by M_6 itself.

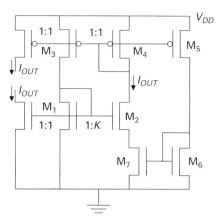

Fig. 6.6   A MOSFET-only self-biased current source (adapted from [10]).

To derive the design equations, we use (6.3.1) and (6.3.2) to give

$$\frac{V_{S2}}{\phi_t} = \left[\sqrt{1 + i_{f1}} - \sqrt{1 + \frac{i_{f1}}{K}} + \ln\left(\frac{\sqrt{1 + i_{f1}} - 1}{\sqrt{1 + i_{f1}/K} - 1} \right) \right] \cong \ln K. \tag{6.4.1}$$

The approximation in (6.4.1) is valid if $i_{f1} \ll 1$, i.e. both M_1 and M_2 operate deep in weak inversion. Choosing the operating point deep in weak inversion is convenient not only owing to the simplicity of equations but also for having low-voltage circuits. We denote the aspect ratio of transistor M_i as $S_i = (W/L)_i$. Thus, $K = S_2/S_1$ and, as in the SBCS of Figure 6.3, $K > 1$. To derive the equations to be used for the current source design, we note that

$$I_{D6} = I_{D5} = \frac{S_5}{S_4} I_{D1}, \tag{6.4.2}$$

$$I_{D7} = I_{S7}(i_{f7} - i_{r7}) = I_{D1} = I_{S1} i_{f1}, \tag{6.4.3}$$

$$i_{f7} = i_{f6} = \frac{I_{D6}}{I_{S6}}. \tag{6.4.4}$$

Now we can apply the UICM to express voltage $V_{S2} \cong \phi_t \ln K$ in terms of the inversion levels at source and drain of M_7, yielding

$$\ln K \cong \frac{V_{S2}}{\phi_t} = \sqrt{1 + i_{f7}} - \sqrt{1 + i_{r7}} + \ln\left(\frac{\sqrt{1 + i_{f7}} - 1}{\sqrt{1 + i_{r7}} - 1} \right). \tag{6.4.5}$$

To derive the next expression, we assume that the specific current is not dependent on the gate voltage. Using (6.4.2), (6.4.3), and (6.4.4), we can write

$$i_{f7} = \frac{S_5 S_1}{S_4 S_6} i_{f1}, \tag{6.4.6}$$

$$i_{r7} = \left(1 - \frac{S_4 S_6}{S_5 S_7}\right) i_{f7}. \tag{6.4.7}$$

The combination of Equations (6.4.5), (6.4.6), and (6.4.7) leads to a single solution for i_{f1}, which is dependent only on the relative aspect ratios and is independent of temperature. As a result, the output current, which is a copy of the core cell current, will be a replica of the specific current of the n-channel transistors. Thus, this SBCS can also be called a specific current generator, among many others [11]–[13] that have previously been reported in the technical literature. The current source described in [11] is included in the problems (see Problem 6.4).

There is no closed form for the output current of the circuit in Figure 6.6; however, since transistor M_7 operates in moderate or strong inversion, we assume, for the purpose of obtaining an approximate expression for the output current, that both i_{f7} and i_{r7} are much greater than unity. In this case, we can approximate (6.4.5) by

$$\ln K \cong \sqrt{i_{f7}} - \sqrt{i_{r7}} = \sqrt{i_{f7}}\left(1 - \sqrt{1 - \frac{S_4 S_6}{S_5 S_7}}\right). \tag{6.4.8}$$

Recalling that $I_{OUT} = I_{SH} S_1 i_{f1}$, and using (6.4.6) through (6.4.8), yields

$$I_{OUT} = \mu_n C'_{ox} n \frac{\phi_t^2}{2} S_7 \left[2J - 1 + 2\sqrt{J(J-1)}\right] \ln^2 K, \tag{6.4.9}$$

where $J = S_5 S_7/(S_4 S_6)$. Even though (6.4.9) has been derived using some approximations, the output current will be, in any case, a scaled copy of the sheet-specific current of n-channel transistors. Likewise, the output current will be a replica of the sheet-specific current of p-channel devices for the complementary cell. The sheet-specific current is proportional to the mobility, slope factor, oxide capacitance, and squared absolute temperature. The temperature dependence of the mobility is, in general, dominant over those arising from the other technological parameters. Since the mobility decreases with the temperature following a T^{-m} dependence, the specific current depends on T as T^{2-m}. Some authors have reported specific currents that are almost PTAT [8], [11] or proportional to $T^{0.4}$ [10].

Example 6.3
Assume that the technology parameters are $V_{T0N} = 0.4$ V, $V_{T0P} = -0.5$ V, $I_{SHN} = 100$ nA, $I_{SHP} = 40$ nA, and $n_N = n_P = 1.2$. Design a current source for $I_{OUT} = 10$ nA using the circuit in Figure 6.6. Calculate the minimum power-supply voltage for proper operation of the current source.

Answer
The reader is referred to [10] and [11] for some design guidelines. We will make the following choices: $i_{f1} = 0.01$ and $i_{f7} = 10$. The operation of the core cell in weak inversion is more appropriate for low-voltage design and allows us to use the approximation in (6.4.1). Transistor M_7 must typically operate in moderate or strong inversion. A moderate

Table 6.1 Aspect ratios of transistors for the circuit in Figure 6.6 with the specifications of Example 6.3

S_1	S_2	S_3	S_4	S_5	S_6	S_7
10	7.39	10	10	100	0.1	1.49×10^{-2}

inversion level for M_7 is also appropriate for low supply voltages. From (6.4.6) we find that $(S_5/S_4)(S_1/S_6) = 1000$. Reasonable values for V_{S2} are in the range 40–80 mV. We have chosen $V_{S2}/\phi_t = 2$, which gives $K = \exp 2 \cong 7.39$. Now, using (6.4.5) we find that $i_{r7} \cong 3.3$ and from (6.4.7) we calculate $(S_4/S_5)(S_6/S_7) = 0.67$. Table 6.1 gives the aspect ratios for our design.

Assuming that the transistors are implemented as an association of unit transistors with $W/L = 1 \, \mu m/1 \, \mu m$ we find the following combinations of unit transistors for implementing M_1–M_7. M_1, M_3, and M_4 are implemented using the parallel association of 10 transistors, whereas M_5 is the parallel association of 100 transistors. M_6 and M_7 can be implemented as the series association of 10 and 67 unit transistors, respectively, whereas a possible choice for M_2 is approximately realized as the parallel association of 7.33 transistors. The "0.33 transistor" is the series association of three unit transistors.

Note that this circuit will provide an output current equal to $I_{S1}i_{f1}$. Since i_{f1} is independent of temperature in the MOSFET-only SBCS, the output current will follow the temperature variations of the specific current. The minimum supply voltage can be calculated from the following expression:

$$V_{DD} = \max\{(-V_{GB4} + V_{DSsat2} + V_{S2}), (V_{SDsat5} + V_{GB6})\}.$$

We first note that $i_{f3} = i_{f4} = 0.025$. From the UICM we find that

$$-V_{GB4} = -V_{T0P} + n\phi_t \left[\sqrt{1 + i_{f4}} - 2 + \ln\left(\sqrt{1 + i_{f4}} - 1\right)\right] \cong 0.34 \, \text{V}.$$

Since both M_2 and M_5 operate in weak inversion,

$$V_{DSsat2} = V_{SDsat5} \cong 4\phi_t = 100 \, \text{mV}.$$

Finally, using once again the UICM, we find that

$$V_{GB6} = V_{T0N} + n\phi_t \left[\sqrt{1 + i_{f6}} - 2 + \ln\left(\sqrt{1 + i_{f6}} - 1\right)\right] \cong 0.47 \, \text{V}.$$

Therefore,

$$V_{DD} > \max\{(0.34 + 0.1 + 0.05), (0.1 + 0.47)\}; \qquad V_{DD} > 0.57 \, \text{V}.$$

6.5   Bandgap voltage references

Many integrated circuits require stable reference voltages with very low temperature coefficients. The most popular implementation of a voltage reference for integrated circuits is the bandgap reference. The main idea on which the bandgap reference is

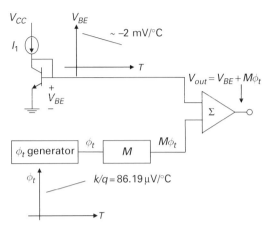

Fig. 6.7   A block diagram of the operating principle of the bandgap reference (adapted from [3]).

based was first presented in 1963 [14] for discrete components and later further developed for application in monolithic integrated circuits in bipolar technology [15]–[17].

6.5.1   The operating principle of the bandgap reference

Bandgap-voltage-reference circuits operate on the principle of adding two voltages having equal and opposite temperature coefficients. The bandgap reference is obtained by adding the base–emitter junction voltage of a forward-biased transistor, which is complementary to absolute temperature (CTAT), to the positive temperature coefficient of the (scaled) thermal voltage [1], as illustrated in Figure 6.7 [3]. To analyze the principle of the bandgap reference, we assume that the base current of the bipolar junction transistor in Figure 6.7 is negligible. The collector current is then expressed as

$$I_1 = I_C = I_{sat} \exp\left(\frac{V_{BE}}{\phi_t}\right). \tag{6.5.1}$$

The saturation current I_{sat} of the bipolar transistor is approximately given by

$$I_{sat} = \frac{q A n_i^2 \overline{\mu_n} \phi_t}{G_B}. \tag{6.5.2}$$

In (6.5.2), the intrinsic concentration n_i can be expressed approximately [3] as

$$n_i^2 = DT^3 \exp\left(-\frac{E_G}{kT}\right), \tag{6.5.3}$$

where $D \approx 7.7 \times 10^{32} \, \mathrm{K}^{-3} \, \mathrm{cm}^{-6}$ is a constant term and E_G is the silicon bandgap extrapolated to $0 \, \mathrm{K}$, which is around $1.206 \, \mathrm{eV}$ [18]. We assume that the dependence of the average mobility of electrons in the base region on temperature is described as

$$\overline{\mu_n} = \mu_{n0} (T/T_0)^{-m}. \tag{6.5.4}$$

We also assume that the area A and Gummel number $G_B = \int_0^{W_B} N_A \, dx$, which is equal to the areal density of impurities in the base, are temperature-independent. The dependence of the current source on the temperature is assumed to be

$$I_1 = I_{C0}(T/T_0)^\alpha. \tag{6.5.5}$$

Using (6.5.2)–(6.5.5) in (6.5.1) yields

$$V_{BE} = \frac{T}{T_0} V_{BE0} + \phi_t \left[\left(\frac{E_G}{k} \right) \left(\frac{1}{T} - \frac{1}{T_0} \right) - (4 - m - \alpha) \ln \left(\frac{T}{T_0} \right) \right]. \tag{6.5.6}$$

The subscript "0" in a parameter of (6.5.6) indicates that the parameter is evaluated at temperature T_0. Finally, the temperature dependence of the reference voltage V_{OUT} is

$$\frac{dV_{OUT}}{dT} = \frac{d(V_{BE} + M\phi_t)}{dT}$$

$$= \frac{V_{BE0}}{T_0} + \frac{k}{q} \left[-(4 - m - \alpha) \left(1 + \ln \left(\frac{T}{T_0} \right) \right) - \frac{E_G}{kT_0} + M \right]. \tag{6.5.7}$$

In order to have a zero temperature coefficient at $T = T_0$, the value of M is, according to (6.5.7), given by

$$M = -\frac{V_{BE0}}{kT_0/q} + \frac{E_G}{kT_0} + (4 - m - \alpha). \tag{6.5.8}$$

Substitution of (6.5.8) into $V_{OUT} = V_{BE} + M\phi_t$ results in

$$V_{OUT} = \frac{E_G}{q} + \frac{kT}{q}(4 - m - \alpha) \left[1 + \ln \left(\frac{T_0}{T} \right) \right]$$

$$\cong V_{OUT}|_{T=T_0} - \frac{kT}{2q}(4 - m - \alpha) \left(\frac{T - T_0}{T} \right)^2. \tag{6.5.9}$$

At $T = T_0$ the reference voltage is equal to the bandgap voltage plus an additional term whose value is around 2–3 times the thermal voltage. Thus, the scheme in Figure 6.7 generates reference voltages slightly higher than the extrapolated bandgap voltage. The approximation in (6.5.9) is valid for small temperature fluctuations around the reference temperature T_0. If, for example, $m = \alpha = 1$, then a temperature variation of ± 30 K around 300 K will cause a fluctuation in the reference voltage of approximately 260 μV. Finally, from (6.5.9), the temperature coefficient of the reference voltage is

$$\frac{dV_{OUT}}{dT} = \frac{k}{q}(4 - m - \alpha) \ln \left(\frac{T_0}{T} \right). \tag{6.5.10}$$

6.5.2   CMOS bandgap references

In this subsection we will briefly describe some circuits commonly used to implement bandgap references in CMOS technology. The principles of CMOS bandgap references are similar to those of bipolar technologies [19], i.e. a PTAT voltage is added to the

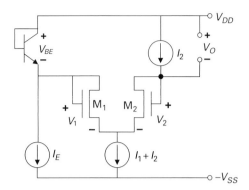

Fig. 6.8   A concept developed in [20] for the generation of a temperature-independent voltage reference. Transistors M_1 and M_2 are biased in weak inversion.

base–emitter voltage of a bipolar transistor to compensate for the temperature dependence of the base–emitter voltage.

The concept proposed in [20] to generate the bandgap reference voltage is shown in Figure 6.8. The bipolar transistor is a substrate transistor. When M_1 and M_2 operate in weak inversion, the output voltage V_O is

$$V_O = V_{BE} + V_1 - V_2 = V_{BE} + n \frac{kT}{q} \ln \left(\frac{I_1 (W/L)_2}{I_2 (W/L)_1} \right). \tag{6.5.11}$$

Both the ratio of currents and the ratio of aspect ratios can be adjusted to compensate for the negative temperature coefficient of the base–emitter voltage. One of the drawbacks of the circuit in Figure 6.8 for more advanced CMOS technologies is the high minimum supply voltage, of the order of 2 V, required to bias the circuit appropriately. Also, the slope factor n is dependent on temperature and, thus, the last term in (6.5.11) is not a PTAT voltage. Last, but not least, the MOSFETs operate on quite different gate voltages, which lead to different slope factors. Consequently, (6.5.11) is not strictly valid.

The PTAT core cell in [21], shown in Figure 6.9, is a pair of gate-connected transistors biased in weak inversion. A PTAT voltage V_{R1} given by

$$V_{R1} = V_{S3} - V_{S1} = \frac{kT}{q} \ln \left(\frac{S_3 S_2}{S_1 S_4} \right) \tag{6.5.12}$$

develops across R_1. In (6.5.12) S_i is the aspect ratio of transistor M_i. The maximum practical value obtainable for V_{R1} is of the order of 100 mV [21].

The output voltage of the PTAT core cell is

$$V_O = \left[1 + \frac{R_2}{R_1} \left(1 + \frac{S_2}{S_4} \right) \right] \frac{kT}{q} \ln \left(\frac{S_3 S_2}{S_1 S_4} \right), \tag{6.5.13}$$

which can be properly adjusted to compensate for the base–emitter voltage of a lateral bipolar transistor used in the bandgap reference of [21].

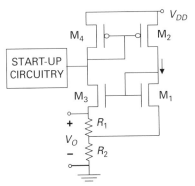

Fig. 6.9   The PTAT voltage generator. Transistors M_1 and M_3 operate in weak inversion (adapted from [21]).

Example 6.4

Assume that the technology parameters for the circuit in Figure 6.9 are $V_{T0N}=0.4$ V, $V_{T0P}=-0.5$ V, $I_{SHN}=100$ nA, $I_{SHP}=40$ nA, and $n_N=n_P=1.2$. Assume that the base–emitter voltage to be compensated is equal to 0.6 V at room temperature and that its temperature coefficient is around -2 mV/K. Design the PTAT generator in Figure 6.9 to give a PTAT voltage that compensates for the temperature coefficient of the bipolar transistor. Determine the reference voltage $V_{REF}=V_{BE}+V_O$. Calculate the inversion levels of all transistors. Calculate the minimum power-supply voltage for proper operation of the PTAT generator.

Answer

Assuming $\phi_t=25.9$ mV, a possible solution is $R_1=1$ MΩ, $R_2=1.03$ MΩ, $S_3=208$, $S_1=208/5$, $S_2=1.3$, and $S_4=1.3/5$. In this case $V_{REF}\cong1.2$ V. The inversion levels are $i_{f1}=0.1$, $i_{f3}=0.004$, and $i_{f2}=i_{f4}=8$. The minimum power-supply voltage for the core cell in Figure 6.9 is

$$V_{DD} = \max\{(-V_{GB4} + V_{DSsat3} + V_O), (V_{SDsat2} + V_{GB1})\},$$
$$V_{DD} = \max\{(0.5 + 0.1 + 0.6), (0.15 + 0.89)\} = 1.2 \text{ V}.$$

Another bandgap reference is shown in Figure 6.10 [22]. Q_1 and Q_2 are parasitic vertical bipolar n–p–n transistors, with the n-substrate acting as the collector, the p-well as the base, and the n-diffusion volume as the emitter. Since the currents flowing through Q_1 and Q_2 are equal, the voltage across R_1 is given by

$$V_{R1} = V_{E2} - V_{E1} = \frac{kT}{q} \ln N, \qquad (6.5.14)$$

where N is the ratio of the emitter areas of Q_2 and Q_1. The reference voltage is given by

$$V_{REF} = -\left[V_{BE2} + \left(1 + \frac{R_2}{R_1}\right)\frac{kT}{q}\ln N\right]. \qquad (6.5.15)$$

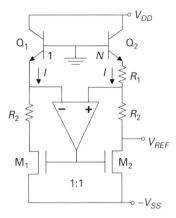

Fig. 6.10 A circuit for bandgap voltage generation (adapted from [22]).

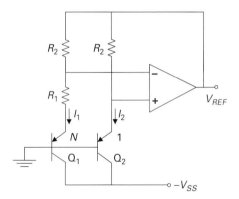

Fig. 6.11 A schematic circuit for generation of a bandgap reference [23], [24].

The analysis for the derivation of (6.5.15) assumes that the operational amplifier is ideal. In the real case, the offset voltage of the operational amplifier contributes to the spread of the reference voltage values, which can be unacceptably high for some applications [22].

The bandgap references presented in [23], [24] make use of the simplified schematic circuit of Figure 6.11. Assuming that the area of Q_1 is N times larger than that of Q_2 and that the operational amplifier is ideal, it can be readily deduced that

$$V_{REF} = V_{EB2} + \frac{R_2}{R_1} \frac{kT}{q} \ln N. \tag{6.5.16}$$

The most significant error sources in the bandgap reference of Figure 6.11 are the operational amplifier's offset voltage and the sometimes relatively poor performance of CMOS-compatible bipolar transistors. The reader is referred to [23], [24] for techniques to implement precision bandgap references.

The use of CMOS-compatible lateral bipolar transistors for the implementation of bandgap references is reported in [25], [26]. The lateral bipolar transistor is accompanied

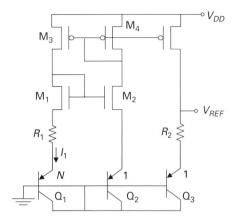

Fig. 6.12   A schematic circuit for generation of a bandgap reference [3], [27].

by the parasitic action of substrate bipolar and MOS transistors (see Section 3.2.4 on CMOS-compatible bipolar transistors). The MOSFET action must be turned off to avoid degradation of the $I_C - V_{BE}$ characteristic of the bipolar transistor. A simplified circuit diagram of the bandgap reference of [25] is presented in Problem 6.5.

Another bandgap-reference implementation in CMOS technology is shown in Figure 6.12 [3], [27]. A PTAT current I_1 is generated due to the difference between the emitter voltages of Q_1 and Q_2 across resistance R_1. Current I_1 is mirrored to the output and causes a PTAT voltage drop across resistance R_2, which, added to the CTAT emitter–base voltage of Q_3, produces the (almost) temperature-independent reference voltage.

Example 6.5

Assuming that the current mirrors have unity gain and that transistors M_1 and M_2 are matched, calculate V_{REF} in the circuit of Figure 6.12.

Answer

The voltage drop across R_1 is given by

$$V_{R1} = V_{EB2} - V_{EB1} = \phi_t \left[\ln \left(\frac{I_2}{I_{Sat2}} \right) - \ln \left(\frac{I_1}{I_{Sat1}} \right) \right] = \phi_t \ln N,$$

since $I_1 = I_2$ and $I_{Sat1} = N I_{Sat2}$. The reference voltage is

$$V_{REF} = V_{EB3} + R_2 I_3 = V_{EB3} + R_2 \frac{V_{R1}}{R_1} = V_{EB3} + \frac{R_2}{R_1} \frac{kT}{q} \ln N.$$

6.5.3   A CMOS bandgap reference with sub-1-V operation

The bandgap-reference circuits that we have shown so far generate an output voltage that is around 1.25 V, a value that is a little higher than the extrapolated bandgap voltage. The

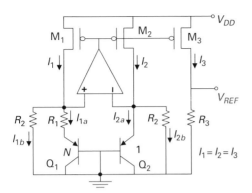

Fig. 6.13 A circuit for generating a sub-1-V bandgap reference.

minimum supply voltage required for the appropriate operation of such circuits is at least equal to the reference voltage plus some additional voltage drop between the output node and the supply voltage. Therefore, supply voltages for the bandgap references described in this chapter should be at least 1.5 V.

An interesting strategy for the generation of bandgap-reference circuits with sub-1-V operation was first presented in [28]; the prototype of the bandgap reference was fabricated in a flash-memory process in which a native n-MOS transistor with $V_T = -0.2$ V was available in addition to conventional enhancement p- and n-channel transistors. This implementation was later modified for fabrication in either BiCMOS [29] or conventional CMOS [30], [31] technologies.

The basic idea behind these sub-bandgap references is to convert both the base–emitter voltage of a diode and the PTAT voltage into two currents and convert the sum of these two currents into a reference voltage that is a fraction of the bandgap voltage [28]. A conceptual scheme of the sub-bandgap reference is shown in Figure 6.13.

For a first-order analysis of the circuit in Figure 6.13, let us assume the operational amplifier to be ideal and the currents flowing through the p-channel devices to be the same. The voltage across resistance R_1 can be readily calculated as

$$V_{R1} = R_1 I_{1a} = V_{EB2} - V_{EB1} = \frac{kT}{q} \ln N. \tag{6.5.17}$$

The reference voltage is given by

$$V_{REF} = R_3(I_{1a} + I_{1b}) = \frac{R_3}{R_2}\left(V_{EB2} + \frac{R_2}{R_1}\frac{kT}{q} \ln N\right). \tag{6.5.18}$$

According to (6.5.18), the output reference voltage can be a fraction of the bandgap voltage of conventional bandgap references. It can be seen that in the circuit of Figure 6.13 the minimum power-supply voltage required for adequate operation of the circuit is equal to the base–emitter voltage of Q_2 plus a saturation voltage of the p-channel transistor as long as the reference voltage is equal to or less than V_{EB} of Q_2. One should note that stringent requirements are imposed on the operational amplifier, since it must be

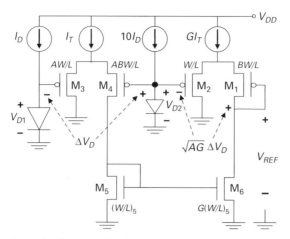

Fig. 6.14   A circuit of a resistorless bandgap reference [32].

able to handle a common-mode input voltage that is very close to that of the positive supply voltage and an output voltage that is very close to that of the negative supply.

6.5.4   A resistorless CMOS bandgap reference

All bandgap references that we have shown so far employ resistors as the scaling elements of the PTAT voltage. In some cases, in particular for low-power design, the resistance values are extremely high, demanding a large silicon area. A possible scheme to implement a CMOS bandgap reference without resistors is shown in Figure 6.14 [32]. The circuit consists of a pair of diodes, a set of current sources, two differential pairs M_1–M_2 and M_3–M_4, and a current mirror M_5–M_6. The voltage difference between the two diodes is a PTAT voltage the value of which is dependent on the currents flowing through them and on the ratio of their areas (A_1/A_2) according to

$$\Delta V_D = V_{D2} - V_{D1} = \frac{kT}{q} \ln\left(10\frac{A_1}{A_2}\right). \tag{6.5.19}$$

The differential amplifier M_3–M_4 converts the voltage difference between the two diodes into a current that flows through M_5 and is amplified by M_6. Finally, the current in M_1, which is equal to that in M_6, is converted back into a differential voltage across the differential pair M_1–M_2, which is an amplified copy of the PTAT voltage difference between the two diodes. The PTAT voltage difference across M_1–M_2 is added to V_{D2} to generate the bandgap reference. The main drawback of the circuit in Figure 6.14 is that transistors M_1 to M_4 operate in strong inversion. Thus, the minimum power-supply voltage for this application is the bandgap reference plus the source-to-gate voltage of M_1 (which is higher than the threshold voltage due to operation in strong inversion) plus the voltage across the terminals of the current source GI_T required for its proper operation.

For the analysis of the circuit in Figure 6.14 we make the following simplifying assumptions: (i) transistors M_1–M_4 operate in strong inversion; (ii) the slope factor is the same for M_1–M_4; (iii) all transistors operate in saturation; and (iv) the simplified expression of the UICM in strong inversion $V_P/\phi_t \cong (V_{GB} - V_{T0})/(n\phi_t) \cong \sqrt{i_f}$ for $V_{SB} = 0$ (with the source connected to the local substrate) holds. Under these assumptions, one can derive the following expressions:

$$\frac{\Delta V_D}{n\phi_t} = \frac{V_{G4} - V_{G3}}{n\phi_t} \cong \sqrt{i_{f3}} - \sqrt{i_{f4}} = \sqrt{\frac{I_T - I_{F4}}{I_{S3}}} - \sqrt{\frac{I_{F4}}{BI_{S3}}}, \tag{6.5.20}$$

$$\frac{V_{G1} - V_{G2}}{n\phi_t} \cong \sqrt{i_{f2}} - \sqrt{i_{f1}} = \sqrt{\frac{GI_T - I_{F1}}{I_{S2}}} - \sqrt{\frac{I_{F1}}{BI_{S2}}}. \tag{6.5.21}$$

Since $I_{S2} = I_{S3}/A$ and $I_{F1} = GI_{F4}$, we can rewrite (6.5.21) as

$$\frac{V_{G1} - V_{G2}}{n\phi_t} = \sqrt{AG}\sqrt{\frac{I_T - I_{F4}}{I_{S3}}} - \sqrt{\frac{I_{F4}}{BI_{S3}}}. \tag{6.5.22}$$

The comparison between (6.5.22) and (6.5.20) allows us to write

$$V_{G1} - V_{G2} = \sqrt{AG}\,\Delta V_D. \tag{6.5.23}$$

The reference voltage is then given by the sum of a CTAT voltage, V_{D2}, and a PTAT voltage given by (6.5.23). Note that, to reduce the source-to-gate voltage of both M_1 and M_2, the aspect ratios of M_1 and M_4 are higher than those of M_2 and M_3, respectively, by a factor of B. The prototype of the voltage reference implemented in [32] was calibrated through the trimming of factor A by adjustment of the aspect ratios of M_3 and M_4 under digital control.

6.6 CMOS voltage references based on weighted V_{GS}

A voltage reference is necessary for the design of low-dropout linear regulators (LDOs) [33]. It must have low dependence on both supply voltage and temperature. Most of the voltage references in CMOS standard technologies use a bandgap reference, as previously described, and others are based either on weighted gate-source voltages of MOS transistors [33], [34] or on compensating a PTAT-based voltage with a MOSFET gate-to-source voltage in the subthreshold region [9]. Despite not having a technology-independent value as the bandgap reference, voltage references based on weighted V_{GS} values can be used in LDOs as long as they have low sensitivity to supply voltage and temperature. The core circuit of the voltage reference presented in [33] is reproduced in Figure 6.15; a similar circuit is presented in [34] with an n-channel transistor replacing the p-channel device. The current source I_B is generated through a circuit identical to that in Figure 6.3. According to the scheme in Figure 6.15, the reference voltage is

$$V_{REF} = \left(1 + \frac{R_1}{R_2}\right)V_{GSn} - |V_{GSp}|. \tag{6.6.1}$$

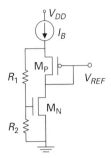

Fig. 6.15   The core circuit of the voltage reference with weighted gate-to-source voltages [33], [34].

In order to achieve a reference voltage with negligible sensitivity to temperature, it is required that the weighted temperature coefficient of V_{GSn} be equal to the temperature coefficient of V_{GSp}, i.e.

$$\left(1 + \frac{R_1}{R_2}\right) \frac{dV_{GSn}}{dT} = \frac{d|V_{GSp}|}{dT}. \tag{6.6.2}$$

It is not an easy task to design this kind of voltage reference since the gate-to-source voltages are sensitive not only to the bias current but also to technological parameters such as threshold voltage and mobility. Moreover, the bias current is also dependent on process parameters and temperature. In order to improve the power-supply rejection ratio, the n-well of M_P is connected to its source.

6.7   A current-calibrated CMOS PTAT voltage reference

We have previously shown that the difference between the base–emitter voltages of two isothermal bipolar transistors is proportional to the absolute temperature, the proportionality factor being dependent on the emitter areas and on the currents through the devices. We will now describe an approach that can be applied to MOS transistors to derive a PTAT voltage. A core cell that can be used as a PTAT voltage generator is shown in Figure 6.16.

To analyze the cell in Figure 6.16 we first assume that both M_1 and M_2 operate in saturation. Then we can write

$$\frac{V_{P1}}{\phi_t} = \sqrt{1 + i_{f1}} - 2 + \ln\left(\sqrt{1 + i_{f1}} - 1\right) = F(i_{f1}) \tag{6.7.1}$$

and

$$\frac{V_{P2} - V_{S2}}{\phi_t} = \sqrt{1 + i_{f2}} - 2 + \ln\left(\sqrt{1 + i_{f2}} - 1\right) = F(i_{f2}). \tag{6.7.2}$$

Since M_1 and M_2 share common gates and substrates, it follows that $V_{P1} = V_{P2}$. Using (6.7.1) and (6.7.2) we can then write

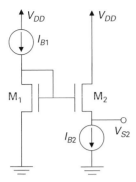

Fig. 6.16   The PTAT voltage generator. Bias currents I_{B1} and I_{B2} are proportional to the sheet normalization current. M_1 and M_2 share a common grounded bulk.

$$V_{S2} = \phi_t \left[F(i_{f1}) - F(i_{f2}) \right].$$ (6.7.3)

Thus, if the term in square brackets in (6.7.3) is independent of temperature, the source voltage of M_2 is PTAT. Since $i_{f1} = I_{B1}/I_{S1}$ and $i_{f2} = I_{B2}/I_{S2}$, in order to make both inversion levels i_{f1} and i_{f2} independent of temperature, both currents I_{B1} and I_{B2} must be proportional to the specific currents I_{S1} and I_{S2}. This can be easily achieved with the circuit in Figure 6.6 [10] or other specific-current generators [11]–[13]. The temperature coefficient of the PTAT can be adjusted by changing the bias currents and/or the aspect ratios of the transistors.

Problems

6.1 Using (6.3.3), calculate the temperature coefficient of the current source in Figure 6.3. Assume that the temperature coefficients of the sheet normalization current and of the resistor are equal to TC_{ISH} and TC_{RS}, respectively. Find the asymptotic values for the temperature coefficient of the current source in the asymptotic cases of weak and strong inversion.

6.2 (a) Verify that the block diagram in Figure 6.5 corresponds to the circuit in Figure 6.3. (b) Verify the correctness of block gains and of the closed-loop current gain given by (6.3.10). (c) Calculate the expressions for thermal noise and flicker noise in strong inversion. (d) Design a network such as the one in Figure 6.3 for an output current equal to $0.1\,\mu A$ at room temperature. Use $K = 10$ and operation in weak inversion. Determine the aspect ratios of all transistors for inversion levels equal to 0.01 for M_1 and 0.1 for both M_3 and M_4. What is the minimum supply voltage for the proper operation of the circuit? What are the PSD values for flicker and thermal noise? The parameters are $V_{T0N} = -V_{T0P} = 0.4\,V$, $I_{SHN} = 100\,nA$, $I_{SHP} = 40\,nA$, $n_P = 1.25$, and $n_N = 1.2$, $t_{ox} = 40\,\text{Å}$, $N_{ot,N} = N_{ot,P} = 2 \times 10^7\,cm^{-2}$. Use $L \geq 1\,\mu m$ for this design.

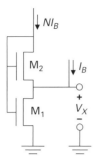

Fig. P6.3   A PTAT voltage generator.

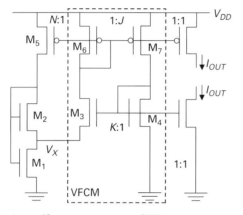

Fig. P6.4   A specific-current generator [11].

6.3 The association of transistors in Figure P6.3 can act as a PTAT voltage generator. (a) Demonstrate that voltage V_X at the intermediate node of the association is given by

$$\frac{V_X}{\phi_t} = \left[\sqrt{1 + MN\frac{I_B}{I_{S2}}} - \sqrt{1 + N\frac{I_B}{I_{S2}}} + \ln\left(\frac{\sqrt{1 + MNI_B/I_{S2}} - 1}{\sqrt{1 + NI_B/I_{S2}} - 1}\right) \right],$$

where $M = 1 + (S_2/S_1)/[(N+1)/N]$. (b) If the bias current I_B is a replica of the specific current, i.e. $I_B = \alpha I_{S2}$, where α is independent of temperature, verify that voltage V_X is PTAT. (c) If $\alpha = 1$ and $N = 1$, calculate the ratio S_2/S_1 for temperature coefficients of the PTAT voltage equal to 200 μV/K and 1 mV/K. (d) If $\alpha = 0.05$ and $N = 1$, calculate the ratio S_2/S_1 for temperature coefficients of the PTAT voltage equal to 200 μV/K and 1 mV/K. (e) What are the corresponding gate voltages for the aspect ratios you calculated in (c) and (d)? Assume $V_{T0} = 0.4$ V and $n = 1.15$.

6.4 A specific current generator is shown in Figure P6.4 [11]. It is composed of the transistor association M_1–M_2 as in the previous problem and a voltage-following current mirror (VFCM), a name coined by B. Gilbert [35], composed of M_3, M_4, M_6, and M_7. The purpose of the VFCM is to provide a voltage at node

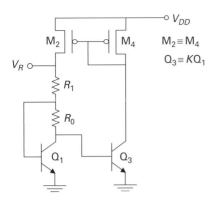

Fig. P6.5   The bandgap reference circuit presented in [25].

V_X that is a shifted copy of the voltage at the source of M$_4$, at 0 V in this case. (a) Demonstrate that, if both M$_3$ and M$_4$ operate in weak inversion, then $V_X = \phi_t \ln(JK)$. (b) Verify that the relationship between the forward inversion levels of M$_1$ and M$_2$ is given by $i_{f1} = i_{f2}[1 + (S_2/S_1)(N+1)/N]$. (c) Find the relationship between V_X and the forward inversion level i_{f2}. (Hint: use the UICM expression for the sources of M$_1$ and M$_2$). (d) Assume that in this problem you have chosen $i_{f2} = 3$ and $i_{f1} = 15$. What is the value required for V_X and the corresponding value of the product JK? (e) Using $N = 0.25$ and the technological parameters provided below, give a list of transistor sizes for a current through M$_6$ equal to 10 nA. The minimum size for both W and L is 4 μm. Choose inversion levels of p-channel devices between 0.1 and 1 and use 0.1 for the inversion level of M$_4$. Calculate the minimum supply voltage for the proper operation of the circuit. Assume the following parameters: $V_{T0N} = -V_{T0P} = 0.4$ V, $I_{SHN} = 100$ nA, $I_{SHP} = 40$ nA, $n_P = 1.25$, and $n_N = 1.2$.

6.5 The circuit shown in Figure P6.5 is a bandgap reference introduced in [25].

(a) Neglecting the base currents of the bipolar transistors, show that the current and the bandgap-reference voltage are given by

$$I_{C1} = \frac{\phi_t}{R_0} \ln K \qquad \text{and} \qquad V_R = V_{BE1} + \frac{R_1}{R_0} \phi_t \ln K.$$

(b) Assume that measurements for Q$_1$ at room temperature showed that for operation in the active mode at $V_{BE} = 0.5$ V the collector current is equal to 1 μA. The temperature coefficient of the base–emitter voltage is around −2.4 mV/°C. Design the bandgap-reference circuit for a collector current $I_{C1} = 1$ μA. What is the expected value of the reference voltage? What is the expected variation of the current for a variation of ±50 °C around room temperature? Assume that the temperature coefficient of the resistors is negligible.

6.6 The circuit of the resistorless bandgap reference shown in Figure 6.14 [32] has been fabricated in a 0.5-μm CMOS technology. The nominal sizes of the devices are given in Table P6.6. The area of diode D$_1$ is eight times larger than that of diode D$_2$.

Table P6.6 Device sizes for Figure 6.13

Device	W (μm)	L (μm)
M_1	80	15
M_2	20	15
M_3	80	15
M_4	320	15
M_5	25	4
M_6	150	4

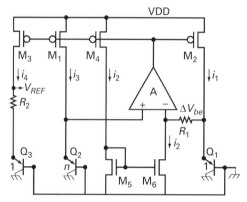

Fig. P6.7   A schematic diagram of the circuit for the CMOS voltage bandgap reference with improved headroom [36].

(a) Given that the value of V_{REF} is around 1.160 V, calculate the diode voltages V_{D1} and V_{D2} at room temperature. (b) Assuming that for operation in strong inversion the source-to-gate voltage of the p-channel devices must exceed the threshold voltage by at least 400 mV, estimate the minimum supply voltage required for the circuit. Assume $V_{TOP} = -0.65$ V.

6.7 Consider the following patent abstract [36]: "A voltage bandgap reference voltage circuit is provided. The circuit includes an amplifier having a first and second transistor coupled to the inputs of the amplifier. The circuit is adapted to operate with lower headroom by effecting a subtraction of a voltage substantially equivalent to ΔV_{be} of the first and second transistors from the voltage applied to the common input of the amplifier." Describe the operation of the bandgap-reference circuit shown in Figure P6.7.

References

[1] A. B. Grebene, *Bipolar and MOS Analog Integrated Circuit Design*, Hoboken, NJ: Wiley Interscience, 2003.

[2] R. J. Widlar, "Some circuit design techniques for linear integrated circuits," *IEEE Transactions on Circuit Theory*, vol. **12**, no. 4, pp. 586–590, Dec. 1965.

[3] P. R. Gray, P. J. Hurst, S. H. Lewis, and R. G. Meyer, *Analysis and Design of Analog Integrated Circuits*, 4th edn., New York: John Wiley & Sons, 2001.

[4] E. Vittoz and J. Fellrath, "CMOS analog integrated circuits based on weak inversion operation," *IEEE Journal of Solid-State Circuits*, vol. **12**, no. 3, pp. 224–231, June 1977.

[5] B. Razavi, "*Design of Analog CMOS Integrated Circuits*," Boston, MA: McGraw-Hill, 2001.

[6] R. Jacob Baker, "*CMOS Circuit Design, Layout, and Simulation*," 2nd edn., Hoboken, NJ: Wiley Interscience, 2005.

[7] E. Vittoz, "Weak inversion in analog and digital circuits", CCCD Workshop, Lund, 2003 (available on line at http://www.es.lth.se/cccd/images/CCCD03-Weak%20inversion-Vittoz.pdf).

[8] S. Mandal, S. Arfin, and R. Sarpeshkar, "Fast startup CMOS current references," *Proceedings of IEEE ISCAS*, 2006, pp. 2845–2848.

[9] G. Giustolisi, G. Palumbo, M. Criscione, and F. Cutri, "A low-voltage low-power voltage reference based on subthreshold MOSFETs," *IEEE Journal of Solid-State Circuits*, vol. **38**, no. 1, pp. 151–154, Jan. 2003.

[10] H. J. Oguey and D. Aebischer, "CMOS current reference without resistance," *IEEE Journal of Solid-State Circuits*, vol. **32**, no. 7, pp. 1132–1135, July 1997.

[11] E. M. Camacho-Galeano, C. Galup-Montoro, and M. C. Schneider, "A 2-nW 1.1-V self-biased current reference in CMOS technology" *IEEE Transactions on Circuits and Systems* II, vol. **52**, no. 2, pp. 61–65, Feb. 2005.

[12] E. A. Vittoz and C. C. Enz, "CMOS low-power analog circuit design", *Proceedings of IEEE ISCAS*, 1996, pp. 79–133.

[13] F. Serra-Graells and J. L. Huertas, "Sub-1-V CMOS proportional-to-absolute temperature references," *IEEE Journal of Solid-State Circuits*, vol. **38**, no. 1, pp. 84–88, Jan. 2003.

[14] D. F. Hilbiber, "A new semiconductor voltage standard," *International Solid-State Circuits Conference Digest Technical Papers*, 1964, pp. 32–33.

[15] R. J. Widlar, "New developments in IC voltage regulators," *IEEE Journal of Solid-State Circuits*, vol. **6**, no. 1, pp. 2–7, Jan. 1971.

[16] K. E. Kuijk, "A precision reference voltage source," *IEEE Journal of Solid-State Circuits*, vol. **8**, no. 3, pp. 222–226, June 1973.

[17] A. P. Brokaw, "A simple three-terminal IC bandgap reference," *IEEE Journal of Solid-State Circuits*, vol. **9**, no. 6, pp. 388–393, Dec. 1974.

[18] M. A. Green, "Intrinsic concentration, effective densities of states, and effective mass in silicon," *Journal of Applied Physics*, vol. **67**, no. 6, pp. 2944–2954, Mar. 15, 1990.

[19] G. C. M. Meijer, G. Wang, and F. Fruett, "Temperature sensors and voltage references implmented in CMOS technology," *IEEE Sensors Journal*, vol. **1**, no. 3, pp. 225–234, Oct. 2001.

[20] Y. P. Tsividis and R. W. Ulmer, "A CMOS voltage reference," *IEEE Journal of Solid-State Circuits*, vol. **13**, no. 6, pp. 774–778, Dec. 1978.

[21] E. A. Vittoz and O. Neyroud, "A low-voltage CMOS bandgap reference," *IEEE Journal of Solid-State Circuits*, vol. **14**, no. 3, pp. 573–577, June 1979.

[22] R. Gregorian, G. A. Wegner, and W. E. Nicholson, Jr., "An integrated single-chip PCM voice codec with filters," *IEEE Journal of Solid-State Circuits*, vol. **16**, no. 4, pp. 322–333, Aug. 1981.

[23] B. S. Song and P. Gray, "A precision curvature compensated CMOS bandgap reference," *IEEE Journal of Solid-State Circuits*, vol. **18**, no. 6, pp. 634–643, Dec. 1983.

[24] J. Michejda and S. K. Kim, "A precision CMOS bandgap reference," *IEEE Journal of Solid-State Circuits*, vol. **19**, no. 6, pp. 1014–1021, Dec. 1984.

[25] E. Vittoz, "MOS transistors operated in the lateral bipolar mode and their application in CMOS technology," *IEEE Journal of Solid-State Circuits*, vol. **18**, no. 3, pp. 273–279, June 1983.

[26] M. G. R. Degrauwe, O. N. Leuthold, E. A. Vittoz, H. J. Oguey, and A. Descombes, "CMOS voltage references using lateral bipolar transistors," *IEEE Journal of Solid-State Circuits*, vol. **20**, no. 6, pp. 1151–1157, Dec. 1985.

[27] P. K. T. Mok and K. N. Leung, "Design considerations of recent advanced low-voltage low-temperature-coefficient CMOS bandgap voltage reference," *Proceedings IEEE CICC*, 2004, pp. 635–642.

[28] H. Banba, H. Shiga, A. Umezawa *et al*, "A CMOS bandgap reference circuit with sub-1-V operation," *IEEE Journal of Solid-State Circuits*, vol. **34**, no. 5, pp. 670–674, May 1999.

[29] P. Malcovati, F. Maloberti, C. Fiocchi, and M. Pruzzi, "Curvature-compensated BiCMOS bandgap with 1-V supply voltage," *IEEE Journal of Solid-State Circuits*, vol. **36**, no. 7, pp. 1076–1081, July 2001.

[30] K. N. Leung and P. K. T. Mok, "A sub-1-V 15-ppm/°C CMOS bandgap voltage reference without requiring low threshold voltage device," *IEEE Journal of Solid-State Circuits*, vol. **37**, no. 4, pp. 526–530, Apr. 2002.

[31] J. Doyle, Y. J. Lee, Y.-B. Kim, H. Wilsch, and F. Lombardi, "A CMOS subbandgap reference circuit with 1-V power supply voltage," *IEEE Journal of Solid-State Circuits*, vol. **39**, no. 1, pp. 252–255, Jan. 2004.

[32] A. E. Buck, C. L. McDonald, S. H. Lewis, and T. R. Viswanathan, "A CMOS bandgap reference without resistors," *IEEE Journal of Solid-State Circuits*, vol. **37**, no. 1, pp. 81–83, Jan. 2002.

[33] K. N. Leung and P. K. T. Mok, "A CMOS voltage reference based on weighted ΔV_{GS} for CMOS low-dropout linear regulators," *IEEE Journal of Solid-State Circuits*, vol. **38**, no. 1, pp. 146–150, Jan. 2003.

[34] G. de Vita and G. Iannaccone, "An ultra-low-power, temperature compensated voltage reference generator," *Proceedings of IEEE CICC*, 2005, pp. 751–754.

[35] B. Gilbert, "Current-mode, voltage-mode, or free mode? A few sage suggestions," *Analog Integrated Circuits and Signal Processing*, vol. **38**, pp. 83–101, Feb. 2004.

[36] US Patent 6,885,178, April 2005, CMOS voltage bandgap reference with improved headroom; Stefan Marinca, *IEEE Journal of Solid-State Circuits*, vol. **40**, no. 11, p. 2345, Nov. 2005; http://www.freepatentsonline.com/6885178.html.

7 Basic gain stages

This chapter presents elementary amplifier stages widely used in analog integrated circuits. We start with the simplest amplifier topologies, namely common-source, common-gate, and source follower, and derive their dc characteristics, noise, and frequency response. We then introduce the cascode amplifier, a combination of the common-source and common-gate amplifiers. After that, we present the differential amplifier, the ubiquitous block at the input of operational amplifiers, together with its transfer characteristics, noise analysis, and small-signal response, and the effects of transistor mismatch on its performance. We then apply the MOSFET model equations for sizing and biasing transistors for analog design and introduce MOSVIEW, a transistor-level design tool. Finally, we describe a methodology for the migration of an analog design to a scaled-down technology.

7.1 Common-source amplifiers

7.1.1 Resistive load

The scheme of the common-source amplifier with the load being a resistor is shown in Figure 7.1, together with the transistor's output characteristics and voltage transfer curve. For low input voltages, the drain current is small and the voltage drop across R is negligible; thus, the output voltage remains constant at V_{DD}. As the input voltage increases, so do the drain current and the voltage drop across R, thus decreasing the output voltage, which is given by

$$v_o = V_{DD} - Ri_D. \qquad (7.1.1)$$

Assuming the transistor output resistance to be much greater than R, the small-signal voltage gain is

$$\frac{v_o}{v_i} = -Rg_m = -\frac{RI_D}{n\phi_t}\frac{2}{\sqrt{1+i_f}+1}, \qquad (7.1.2)$$

where the transconductance has been replaced with its value in saturation. An inspection of the circuit in Figure 7.1 reveals that the maximum voltage drop across the resistor is limited by the supply voltage and by the minimum drain-to-source voltage required to keep M_1 in saturation, i.e.

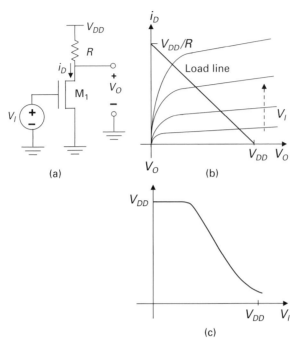

Fig. 7.1   (a) A common-source amplifier with a resistive load. (b) MOSFET output characteristics and load line. (c) The voltage transfer characteristic.

$$RI_D \leq V_{DD} - V_{DSsat}.$$
(7.1.3)

The maximum voltage gain is obtained in weak inversion, for which the transconductance-to-current ratio is maximal and the MOSFET saturation voltage is minimal, around 100 mV. Assuming that the supply voltage is much higher than the saturation voltage, the maximum gain becomes

$$\left|\frac{v_o}{v_i}\right|_{max} \cong \frac{V_{DD}}{n\phi_t}.$$
(7.1.4)

For a 1.8-V supply and $n\phi_t = 30$ mV, the maximum gain of the resistively loaded common-source amplifier is 60. In addition to the low voltage gain, the structure of Figure 7.1 requires a resistor, which can be a disadvantage in many applications. Assume, for example, that, due to low-power design constraints, the direct current of the amplifier is 10 nA and the ac requirements of the amplifier impose a voltage drop across R of 1 V. In this case, a resistance R of 100 MΩ, a prohibitively large value for standard CMOS technologies, would be required.

7.1.2   Diode-connected load

The common-source amplifier with a diode load is shown in Figure 7.2 [1]. The load line associated with the diode-connected transistor M_2 is shown together with the output characteristics of M_1. The resulting voltage transfer characteristic, also given in

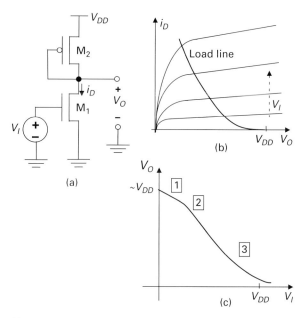

Fig. 7.2 (a) A common-source amplifier with a diode-connected load. (b) MOSFET output characteristics and load line. (c) The voltage transfer characteristic.

Figure 7.2, has a high-gain region, where both transistors operate in saturation. The small-signal voltage gain is

$$\frac{v_o}{v_i} = -\frac{g_{m1}}{g_{m2} + g_{ds1} + g_{ds2}} \cong -\frac{g_{m1}}{g_{m2}}. \tag{7.1.5}$$

The approximation in (7.1.5) is valid for cases in which $g_m \gg g_{ds}$, a condition that is generally applicable. Using the universal transconductance-to-current ratio, (7.1.5) can be rewritten as

$$\frac{v_o}{v_i} \cong -\frac{g_{m1}}{g_{m2}} = -\frac{n_2}{n_1} \frac{\sqrt{1 + i_{f2}} + 1}{\sqrt{1 + i_{f1}} + 1}, \tag{7.1.6}$$

where $i_f = I_D/I_S$ for both M_1 and M_2 in saturation.

For both transistors in weak inversion, region 1 in the voltage transfer characteristic of Figure 7.2, the voltage gain is equal to $-n_2/n_1$, which is close to -1. For both transistors in strong inversion, (7.1.6) gives

$$\frac{v_o}{v_i} \cong -\sqrt{\frac{\mu_n n_2 (W/L)_1}{\mu_p n_1 (W/L)_2}} \tag{7.1.7}$$

as long as M_1 is in saturation. Therefore, in strong inversion, region 3, the gain is determined by the ratio of the aspect ratio of M_1 to that of M_2. In the intermediate region of Figure 7.2, M_1 operates in weak/moderate inversion while M_2 operates in moderate/strong inversion when the aspect ratio of M_1 is much higher than that of M_2.

As can be noted from (7.1.6), in order to achieve a high gain the drive transistor must operate at an inversion level considerably lower than that of the diode-connected load. This means that the aspect ratio of M_1 must be considerably higher than that of M_2 to achieve a high gain and also that M_2 must operate in very strong inversion. Since the voltage gain of the amplifier is proportional to the square root of the inversion level, which, in turn, is approximately proportional to the gate-to-source voltage, the higher the voltage gain, the higher the power supply voltage needed to achieve the required gain. Summarizing, the topologies shown in Figures 7.1 and 7.2 are not appropriate for achieving high gain in low-voltage circuits.

Example 7.1

For the circuit in Figure 7.2, $W_1 = 100\,\mu m$, $L_1 = 4\,\mu m$, $W_2 = 10\,\mu m$, $L_2 = 4\,\mu m$, $V_{DD} = 3.3\,V$, $V_{T0N} = 0.5\,V$, $V_{T0P} = -0.7\,V$, $\mu_N = 400\,cm^2/V$ per s, and $\mu_P = 160\,cm^2/V$ per s. For the sake of simplicity, assume that $n_1 = n_2 = 1.25$. (a) Calculate the voltage at which $V_O = V_I$. (b) Calculate the maximum input voltage at which M_1 remains in saturation. (c) What is the variation in voltage gain when the input voltage is changed from 0.5 V to the voltage calculated in (b)? Assume that $T = 300\,K$ and the output conductance of M_1 is negligible.

Answer

(a) For the data given, we have $I_{S1} = 25I_{S2}$, which implies that, in saturation, $i_{f2} = 25i_{f1}$. Using the UICM for both M_1 and M_2 in saturation and equating the input and output voltages, we find that $i_{f1} \cong 110$, which gives $V_I \cong 0.85\,V$. (b) Recalling that the saturation voltage $V_{DSsat1} \cong \phi_t\left[\sqrt{1 + i_{f1}} + 3\right]$ and that $V_O = V_{DSsat1}$ on the edge of saturation, we apply once again the UICM for transistor M_2 and find that $i_{f1} \cong 170$, which, substituted into the UICM for transistor M_1, gives $V_I \cong 0.94\,V$. (c) We note that, for $V_I = V_{T0} = 0.5\,V$, $i_{f1} = 3$ and $i_{f2} = 75$, whereas for the condition in (b), $i_{f1} = 170$ and $i_{f2} = 4250$. On using these two sets of inversion levels in (7.1.6), we find that $v_o/v_i \cong -3.24$ and $v_o/v_i \cong -4.7$, respectively.

7.1.3   The intrinsic gain stage

Before studying the common-source amplifiers that are usually employed in integrated circuits, we will analyze the so-called "intrinsic gain stage" [2], which consists of a single transistor loaded by an ideal current source and a load capacitance as shown in Figure 7.3(a).

Figure 7.3(b) shows the output characteristics of M_1 together with the characteristic of the current source that delivers a direct current equal to I_B. The dc input voltage is equal to V_{GQ}, an appropriate value at which to bias the transistor in the saturation region.

Before analyzing the topology in Figure 7.3(a) as a voltage amplifier, we first review the fundamental characteristic of the MOS transistor as a voltage-to-current converter. To

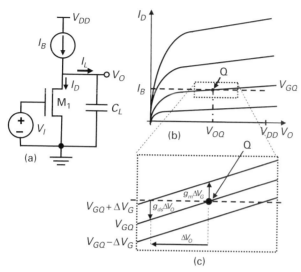

Fig. 7.3   (a) A common-source amplifier with an ideal current source load. (b) MOSFET output characteristics and load line. (c) A zoom of the quiescent point region.

this end, let us assume that a constant voltage source with a value of V_{OQ} is connected to the output. Now, let us assume that the input voltage varies from V_{GQ} to $V_{GQ}+\Delta V_G$. In this case, the output current $I_L = I_B - I_D$, where $I_D = I_B + \Delta I_D$. The value of the variation in drain current in response to a small variation in the input voltage is

$$\Delta I_D \cong \left.\frac{dI_D}{dV_G}\right|_{V_O=V_{OQ}} \Delta V_G = g_m\, \Delta V_G. \tag{7.1.8}$$

Figure 7.3(c), a zoom of the dotted rectangle in Figure 7.3(b), shows the current variation $g_m \Delta V_G$ at a constant output voltage of V_{OQ}. On the other hand, for constant input voltage, a variation in the output voltage produces a variation in the drain current equal to

$$\Delta I_D \cong \left.\frac{dI_D}{dV_O}\right|_{V_G=V_{GQ}} \Delta V_O = g_{ds}\, \Delta V_O. \tag{7.1.9}$$

Figure 7.3(c) also shows the variation in the drain current in response to a negative variation in the output voltage.

The operation of the circuit in Figure 7.3(a) as a voltage amplifier is analyzed as follows. When the input voltage increases by ΔV_G, the output voltage will change by an amount such that

$$\Delta I_D = g_m\, \Delta V_G + g_{ds}\, \Delta V_O = 0, \tag{7.1.10}$$

since for an open-circuit output the transistor drain current is not allowed to change due to the constant biasing source. A_{V0}, the low-frequency voltage gain of the intrinsic-gain stage, can be readily interpreted from the graph in Figure 7.3(c) and from (7.1.10) as

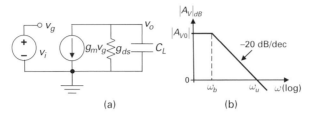

Fig. 7.4   (a) The small-signal equivalent circuit of the common-source amplifier loaded with an ideal current source. (b) The asymptotic frequency-response magnitude of the voltage gain.

follows. Assume that the input voltage has increased by an amount equal to ΔV_G. The output current would then increase by an amount equal to $g_m \Delta V_G$. However, the drain current cannot change, since it is imposed by the constant current source. Thus, the output voltage has to change by an amount equal to ΔV_O in order to keep the drain current unchanged, as (7.1.10) shows. The voltage gain A_{V0}, calculated from (7.1.10) is

$$A_{V0} = \frac{\Delta V_O}{\Delta V_I} = -\frac{g_m}{g_{ds}}. \qquad (7.1.11)$$

Note that the voltage gain is the negative of the ratio of the transconductance to the output conductance. This useful result will be used intensively throughout this text; in most cases the voltage gain will be determined by computing the ratio of the short-circuit transconductance to the output impedance.

Next, we compute the frequency response of the voltage gain using the small-signal equivalent circuit of Figure 7.4. The frequency-dependent voltage gain is

$$A_V(s) = \frac{\Delta V_O}{\Delta V_I} = \frac{v_o}{v_i} = -\frac{g_m}{g_{ds}} \frac{1}{1 + sC_L/g_{ds}}. \qquad (7.1.12)$$

The amplifier bandwidth is $\omega_b = g_{ds}/C_L$ (rad/s). The unity-gain frequency at which the voltage gain equals unity is $\omega_u = g_m/C_L$ (rad/s). We can note that $\omega_u = |A_{V0}|\omega_b$, that is, the unity-gain frequency of the amplifier is equal to the product of the open-loop voltage gain and the −3 dB bandwidth. For this reason, the unity-gain frequency is often called the gain–bandwidth product.

Example 7.2

Consider the common-source amplifier in Figure 7.3. (a) Determine the open-loop low-frequency voltage gain and the gain–bandwidth product in terms of the bias current. Assume that the output conductance is proportional to the channel length. (b) Calculate the maximum product $A_{V0}\omega_u$, assuming that the load capacitance must also include a parasitic capacitance C_P associated with the MOS transistor, and that the parasitic capacitance is roughly proportional to the channel width, i.e. $C_P = C_W W$, where C_W is a process-dependent parameter.

Answer

(a) The universal relationship between transconductance and current and the expression of the output conductance give, respectively,

$$g_m = \frac{2I_S}{n\phi_t}\left(\sqrt{1+\frac{I_B}{I_S}}-1\right) \qquad \text{and} \qquad g_{ds} = \frac{I_B}{V_E L}.$$

The open-loop gain and the unity-gain frequency are

$$-A_{V0} = \frac{g_m}{g_{ds}} = \frac{2V_E L}{n\phi_t}\bigg/\left(\sqrt{1+\frac{I_B}{I_S}}+1\right) \qquad \text{and} \qquad \omega_u = \frac{2I_S}{n\phi_t C_L}\left(\sqrt{1+\frac{I_B}{I_S}}-1\right).$$

(b) Using the previous two formulas and including the parasitic capacitance, we find that

$$|A_{V0}\omega_u| = \frac{2V_E \mu C'_{ox}}{n(C_L/W + C_W)}\frac{\sqrt{1+I_B/I_S}-1}{\sqrt{1+I_B/I_S}+1}$$

The maximum of the product $A_{V0}\omega_u$ is reached in strong inversion for $C_L/W \ll C_W$. In this case $|A_{V0}\omega_u|_{\max} = 2V_E\mu C'_{ox}/(nC_W)$, a result similar to that presented in [3].

7.1.4 Current source load

The voltage gain of the common-source amplifier can be increased when a transistor operating in the saturation region is used for the load. The amplifier configuration, in this case, is shown in Figure 7.5. I_B is a direct current that is replicated at the output branch via current mirror M_3–M_2 to provide the bias current of M_1.

Figure 7.5(b) shows the output characteristics of M_1 together with the load characteristic, which is similar to that of a current source, provided that M_2 operates in saturation. For very low input voltages, the drain current is close to zero, M_2 operates in the triode region, and $V_O \cong V_{DD}$. For higher input voltages, M_1 operates in saturation whereas M_2 is still in the triode region (region 1 in the graphs). A higher input voltage drives M_2 into the saturation region, labeled 2 in the graphs. At an even higher input voltage we enter region 3, where M_1 is no longer in saturation. In this region, the transconductance of M_1 drops significantly, whereas its output conductance increases considerably, resulting in a voltage gain that decreases substantially for higher voltages, as reflected in the voltage transfer characteristic.

For operation as an amplifier, the maximum and minimum values of the output voltage should be such that both M_1 and M_2 remain in saturation, i.e.

$$V_{DSsat1} \le V_O \le V_{DD} - V_{DSsat2}, \qquad (7.1.13)$$

where

$$\frac{V_{DSsat1(2)}}{\phi_t} = \sqrt{1+\frac{I_B}{I_{S1(2)}}}+1. \qquad (7.1.14)$$

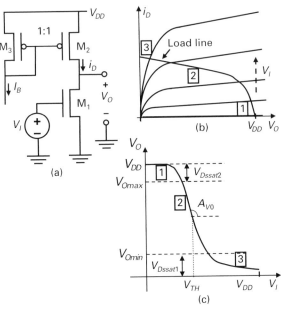

Fig. 7.5   (a) A common-source amplifier with a current-source load. (b) MOSFET output characteristics and load line. (c) The voltage transfer characteristic.

Note that we have used I_B, the approximate value for the output current in the high-gain region, for the calculation of the saturation voltages. For a high-gain amplifier, the dc input voltage V_{TH} for which both transistors are in saturation is, according to the UICM, approximately given by

$$V_{TH} = V_{T0N} + n\phi_t \left[\sqrt{1 + \frac{I_B}{I_{S1}}} - 2 + \ln\left(\sqrt{1 + \frac{I_B}{I_{S1}}} - 1 \right) \right]. \qquad (7.1.15)$$

The common-source amplifier's current capability is shown in Figure 7.6. It operates as a class-A amplifier with a maximum current-sourcing capability of I_B. On the other hand, the current-sinking capability is, in general, much higher than I_B. As a result, the falling slew rate is, in general, greater than the rising slew rate, which, in turn, is limited by I_B, i.e.

$$C_L \frac{dV_O}{dt} \leq I_B. \qquad (7.1.16)$$

The small-signal equivalent circuit of the common-source amplifier is shown in Figure 7.6(c). Here C_g ($=C_{gs} + C_{gb}$) is the gate capacitance of M_1, C_{gd} is the gate-to-drain overlap capacitance of M_1, C_L is the output capacitance composed of both the load capacitance and other capacitances associated with the output node, and g_o is equal to the sum of the output conductances of M_1 and M_2. Using the small-signal equivalent circuit in Figure 7.6, circuit analysis leads to

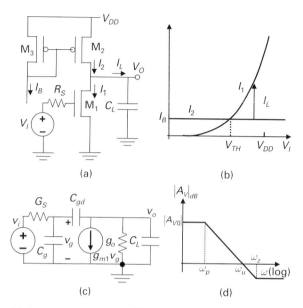

Fig. 7.6   (a) A common-source amplifier. (b) Current sourcing and sinking capabilities. (c) The small-signal
equivalent circuit. (d) The magnitude of the voltage gain versus frequency.

$$A_V(s) = \frac{V_O}{V_I} = -\frac{G_S(g_{m1} - sC_{gd})}{D(s)},$$

$$D(s) = s^2[(C_L + C_{gd})C_g + C_L C_{gd}]$$
$$+ s[G_S(C_L + C_{gd}) + g_{m1}C_{gd} + g_o(C_g + C_{gd})] + G_S g_o. \qquad (7.1.17)$$

The low-frequency gain of the common-source amplifier is

$$A_{V0} = -\frac{g_{m1}}{g_o} = -\frac{(1/V_{A1} + 1/V_{A2})^{-1}}{n\phi_t} \frac{2}{\sqrt{1 + I_B/I_{S1}} + 1}. \qquad (7.1.18)$$

In (7.1.18) we have used both the transconductance-to-current ratio and the approximation $g_{ds} = I_D/V_A$ to calculate the output conductances g_{ds1} and g_{ds2}.

In a well-designed amplifier, an acceptable phase margin requires that one of the poles
be dominant and also that the frequency of the secondary pole be higher than the unity-
gain frequency. Applying the dominant-pole approximation[1] to the denominator of the
voltage transfer function (see Problem 7.2) in (7.1.17) yields

---

[1] Let $D(s)$ be a second-order polynomial in s. If p_1 and p_2 are widely spaced (i.e. $-p_1 \ll -p_2$) negative real roots
of $D(s)$, then

$$D(s) = (1 - s/p_1)(1 - s/p_2) = 1 - s(1/p_1 + 1/p_2) + 1/(p_1 p_2)$$
$$\approx 1 - s/p_1 + 1/(p_1 p_2).$$

$$p_1 \cong -\frac{G_S g_o}{G_S(C_L + C_{gd}) + g_{m1}C_{gd} + g_o(C_g + C_{gd})}. \tag{7.1.19}$$

If we assume that the voltage source is ideal ($G_S \to \infty$), the voltage gain in (7.1.17) becomes

$$A_V(s) = \frac{v_o}{v_i} = \frac{-g_{m1}}{g_o}\frac{1 - s/\omega_z}{1 + s/\omega_p}. \tag{7.1.20}$$

For the transfer function in (7.1.20) we have

$$\omega_p = \frac{g_o}{C_L + C_{gd}}, \qquad \omega_z = \frac{g_{m1}}{C_{gd}}, \qquad \omega_u = |A_{V0}|\omega_p = \frac{g_{m1}}{C_L + C_{gd}}.$$

The value calculated for the extrapolated unity-gain frequency ω_u holds if the low-frequency gain is much greater than unity. Figure 7.6(d) illustrates the approximate magnitude response of the amplifier for the case of source resistance equal to zero.

At this point, we should advise the reader that the zero frequency ω_z is greater than the transistor's intrinsic transition frequency ω_t. For operation in strong inversion, we have previously demonstrated that the quasi-static small-signal model shown in Figure 7.6 is valid only for frequencies below the intrinsic transition frequency. Thus, the frequency response depicted in Figure 7.6(d) is not valid for frequencies approaching ω_z in the case of strong inversion, since the model in Figure 7.6(c) is not valid for such high frequencies.

The input admittance of the common-source amplifier calculated from the small-signal circuit in Figure 7.6(c) is

$$Y_i \cong s\left[C_g + C_{gd}(1 - A_V(s))\right], \tag{7.1.21}$$

where A_V is given by (7.1.20). For frequencies below the pole frequency ω_p the signal source sees an input capacitance equal to the gate capacitance plus the gate-to-drain capacitance multiplied by the low-frequency gain [4]. This modification to the capacitance C_{gd}, which arises from the voltage gain across C_{gd}, is referred to as the Miller effect [1]. For frequencies greater than the pole frequency and less than the unity-gain frequency, the contribution of the second term on the right-hand side of (7.1.21) becomes a conductance equal to $g_{m1}[C_{gd}/(C_L + C_{gd})]$, a fraction of the transistor's transconductance.

The input-referred noise sources i_n and e_n shown in Figure 7.7 can be calculated from the channel noise associated with each transistor and from the small-signal equivalent circuit. For the sake of simplicity we assume that, for the calculation of the noise referred to the input, the channel noise of M_3 is completely transferred to M_2. Therefore, the equivalent current noise source between the output node and ground is the sum of the individual noise currents of M_1, M_2, and M_3. For frequencies below the unity-gain frequency of M_1, we can write the following expressions for the PSDs of e_n and i_n:

$$\frac{\overline{e_n^2}}{\Delta f} = \frac{1}{g_{m1}^2}\left(\frac{\overline{i_{n1}^2}}{\Delta f} + 2\frac{\overline{i_{n2}^2}}{\Delta f}\right) \tag{7.1.22}$$

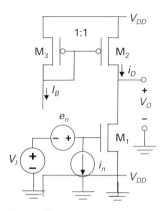

Fig. 7.7 Input-referred noise sources e_n and i_n.

and

$$\frac{\overline{i_n^2}}{\Delta f} = \frac{\omega^2 \left(C_{gs} + C_{gb} + C_{gd}\right)^2}{g_{m1}^2} \left(\frac{\overline{i_{n1}^2}}{\Delta f} + 2\frac{\overline{i_{n2}^2}}{\Delta f}\right). \tag{7.1.23}$$

The terms $\overline{i_{n1,2}^2}/\Delta f$ represent the PSD of the channel noise current. We will compute only the input-referred noise voltage for both thermal and flicker noise, since the value of the noise voltage allows one to compute the value of the noise current directly, as can be seen from (7.1.22) and (7.1.23). Recalling that the thermal noise of a saturated long-channel MOS transistor is given by

$$\frac{\overline{i_{n1}^2}}{\Delta f} = \frac{8}{3} k T g_{m1} n_1 \frac{\sqrt{1 + i_{f1}} + 1/2}{\sqrt{1 + i_{f1}} + 1} \approx 4 k T g_{m1}, \tag{7.1.24}$$

expression (7.1.22) can be written approximately as

$$\frac{\overline{e_n^2}}{\Delta f} \simeq \frac{4kT}{g_{m1}} \left(1 + 2\frac{g_{m2}}{g_{m1}}\right) = \frac{4kT}{g_{m1}} \left(1 + 2\frac{\sqrt{1 + I_B/I_{S1}} + 1}{\sqrt{1 + I_B/I_{S2}} + 1}\right) \tag{7.1.25}$$

for the thermal noise.

Note that the approximation in (7.1.24) was used in (7.1.25) as a means for first-order calculations.

The use in (7.1.22) of the approximate expression (4.5.16) for the flicker noise presented in Chapter 4 allows us to write

$$\frac{\overline{e_n^2}}{\Delta f} \simeq \left(\frac{K_{F,n}}{W_1 L_1 C_{ox}'} + \frac{2 K_{F,p}}{W_2 L_2 C_{ox}'} \frac{g_{m2}^2}{g_{m1}^2}\right) \frac{1}{f}. \tag{7.1.26}$$

The values of g_{m1} and g_{m2} needed for the calculation of the thermal and flicker noise can be calculated through the universal transconductance-to-current relationship of MOSFETs.

Example 7.3

For the circuit in Figure 7.6, $W_1 = 10\,\mu m$, $W_2 = W_3 = 20\,\mu m$, $L_1 = L_2 = L_3 = 2.5\,\mu m$, $V_{DD} = 3.3\,V$, $I_B = 4.5\,\mu A$, $V_{T0N} = 0.5\,V$, $V_{T0P} = -0.7\,V$, $C'_{ox} = 5\,fF/\mu m^2$, $\mu_N = 400\,cm^2/V$ per s, $\mu_P = 200\,cm^2/V$ per s, and $C_L = 2\,pF$. For the sake of simplicity, assume that $n_1 = n_2 = 1.2$, $\phi_t = 25\,mV$, and $V_{A1} = V_{A2} = 9\,V$. (a) Calculate the voltage at which $V_O = V_I$. (b) Calculate the output voltage range such that both M_1 and M_2 remain in saturation. (c) What is the voltage gain? (d) What is the input-voltage variation such that the output voltage remains within the voltage range calculated in (b)? (e) What are the approximate maximum and minimum rates of change of the output? (f) What is the PSD of the input-referred thermal noise? (g) What is the input-referred flicker noise when $N_{otp} = N_{otn} = 3 \times 10^7\,cm^{-2}$? (h) What is the corner frequency?

Answer

(a) For the data given, we have $I_{S1} = I_{S2} = 300\,nA$, which implies that, for the two transistors in saturation, $i_{f2} = i_{f1} = 15$. Using the UICM for $i_{f1} = 15$, we find that $V_I = 593\,mV$. (b) Recalling that the saturation voltage $V_{DSsat} \cong \phi_t(\sqrt{1 + i_f} + 3)$, we find that $V_{DSsat} = 175\,mV$ for both M_1 and M_2, which gives $0.18\,V < V_O < 3.12\,V$. (c) The voltage gain is given by (7.1.18). The data of the problem give $A_{V0} = -60$. (d) The input-voltage variation such that the output voltage remains within the voltage range calculated in (b) is given by $\Delta V_{IN} = \Delta V_O / A_{V0} = 2950/60 \cong 49\,mV$. (e) Let us assume that the parasitic capacitances associated with the output node are negligible. In this case, the slew rate is determined by the load capacitance C_L. The maximum sourcing current is $I_B = 4.5\,\mu A$, which gives a maximum rising rate of change of the output equal to $2.25\,V/\mu s$. The maximum sinking current is $I_{D1}(V_{IN} = V_{DD}) - I_B$. The current that flows through M_1 can be calculated using the approximate formula for the UICM $V_P = (V_{DD} - V_{T0})/n = \phi_t[\sqrt{1 + i_{f1}} - 2 + \ln(\sqrt{1 + i_{f1}} - 1)]$ with $V_G = V_{DD}$. We then find $i_{f1} \cong 8250$ or $I_{D1} = I_{S1}i_{f1} \cong 2.48\,mA$, a value considerably higher than I_B. Therefore, the maximum falling rate of change of the output is around $1.2\,V/ns$. (f) The transconductance g_{m1} is calculated from the expression for the transconductance-to-current ratio in saturation, which gives $g_{m1} = 60\,\mu A/V$. Then, using expression (7.1.25) to calculate the input-referred thermal noise, we find that $\overline{e_n^2}/\Delta f = 8 \times 10^{-16}\,V^2/Hz$. (g) Using (7.1.26) we find $\overline{e_n^2}/\Delta f = 0.32 \times 10^{-10}(1/f)\,V^2/Hz$, which is the PSD of the input-referred flicker noise (h). On equating the results obtained in (f) and (g) we find that the corner frequency $f_C = 40\,kHz$.

7.1.5   The push–pull amplifier (static CMOS inverter)

The conventional static CMOS inverter, which is shown in Figure 7.8(a), can be used as a push-pull amplifier [4], [5]. The output characteristics of both n-channel (solid line) and

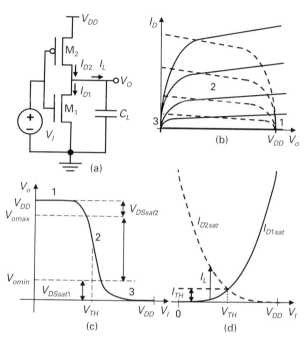

Fig. 7.8 (a) The CMOS static inverter as a push–pull amplifier. (b) The n-channel (solid lines) and p-channel (dashed lines) MOSFET output characteristics. (c) The voltage transfer characteristic. (d) The output current capability.

p-channel (dashed line) MOS transistors are given in Figure 7.8(b), while the voltage transfer curve is shown in Figure 7.8(c). In region 1, the input voltage is low and the current is also low; the n-channel device operates in saturation whereas the p-channel device operates in the triode region. In region 2, both transistors operate in saturation, the voltage gain reaches its maximum value, and the current that flows through them is equal to I_{TH}, the bias current for the operation of the CMOS inverter as an analog amplifier. In region 3, which is associated with high input voltages, M_1 enters the triode region whereas M_2 operates in saturation; the current becomes progressively lower for increasing input voltage. The CMOS inverter operates as an amplifier in region 2, where the gain can be relatively high. The class-AB drive capability of the amplifier is shown in Figure 7.8(d). The currents labeled I_{D1sat} and I_{D2sat} represent the maximum currents provided by M_1 and M_2, respectively, for a given input voltage. Note that the static current I_{TH} of the amplifier is much lower than the maximum sinking or sourcing currents.

V_{TH} (denoting the gate threshold voltage in the case of the CMOS static inverter), the approximate input voltage in region 2, can be calculated through the application of the UICM to both M_1 and M_2, yielding, for the case of strong inversion,

$$\frac{V_{TH} - V_{T0N}}{n_1 \phi_t} \cong \sqrt{i_{f1}}, \qquad \frac{V_{DD} - V_{TH} + V_{T0P}}{n_2 \phi_t} \cong \sqrt{i_{f2}}. \qquad (7.1.27)$$

Since both transistors operate in saturation, we have

$$\frac{i_{f2}}{i_{f1}} = \frac{I_D/I_{S2}}{I_D/I_{S1}} = \frac{I_{S1}}{I_{S2}}. \tag{7.1.28}$$

The combination of (7.1.27) and (7.1.28) leads to the following value for V_{TH}:

$$V_{TH} = \frac{V_{DD} + V_{T0P} + V_{T0N}\sqrt{\dfrac{\mu_n\, n_2\, (W/L)_1}{\mu_p\, n_1\, (W/L)_2}}}{1 + \sqrt{\dfrac{\mu_n\, n_2\, (W/L)_1}{\mu_p\, n_1\, (W/L)_2}}}. \tag{7.1.29}$$

On the other hand, if both transistors operate in weak inversion we write

$$\frac{V_{TH} - V_{T0N}}{n_1 \phi_t} + 1 \cong \ln\left(\frac{i_{f1}}{2}\right); \qquad \frac{V_{DD} - V_{TH} + V_{T0P}}{n_2 \phi_t} + 1 \cong \ln\left(\frac{i_{f2}}{2}\right). \tag{7.1.30}$$

For the sake of simplicity let us assume that $n_1 = n_2 = n$. Thus, from (7.1.30) we find

$$V_{TH} \cong \frac{V_{DD} + V_{T0P} + V_{T0N}}{2} - \frac{n\phi_t}{2} \ln\left(\frac{I_{S1}}{I_{S2}}\right). \tag{7.1.31}$$

Note that for operation of both M_1 and M_2 in weak inversion the value of V_{TH} must be within the range

$$V_{DD} + V_{T0P} < V_{TH} < V_{T0N}. \tag{7.1.32}$$

The rightmost and leftmost inequalities in (7.1.32) are required for operation of M_1 and M_2, respectively, in weak inversion. Consequently, the maximum power-supply voltage cannot exceed the sum of the absolute values of the threshold voltages. Also note that operation in saturation for both M_1 and M_2 requires that $V_{DD} > 200\,\mathrm{mV}$.

Once we have calculated the value of V_{TH} for either strong inversion or weak inversion, we can use either (7.1.27) or (7.1.30) for the calculation of the transistor inversion level and, as long as we know the value of the normalization current, we can find the bias current. The insertion of the value of V_{TH} obtained in (7.1.29) into any of the expressions shown in (7.1.27) yields

$$I_{TH} \cong \left(\frac{1}{n_1/\sqrt{I_{S1}} + n_2/\sqrt{I_{S2}}}\right)^2 \left(\frac{V_{DD} + V_{T0P} - V_{T0N}}{\phi_t}\right)^2. \tag{7.1.33}$$

The value of I_{TH} in (7.1.33) can be interpreted as the current of an n-MOS transistor in saturation for which the gate is driven by a voltage equal to $V_{DD} + V_{T0P}$ and for which the equivalent specific current I_{SN} is described by the following rule for the series association of M_1 and M_2:

$$\frac{n_N}{\sqrt{I_{SN}}} = \frac{n_1}{\sqrt{I_{S1}}} + \frac{n_2}{\sqrt{I_{S2}}}, \tag{7.1.34}$$

where n_N ($\cong n_1$) is the slope factor of an n-channel MOSFET measured at $V_G = V_{DD} + V_{T0P}$. A similar interpretation can be given for an equivalent p-MOS transistor. An important aspect of the conventional CMOS inverter amplifier is the strong dependence of the bias current on both the supply voltage and the technology.

The frequency response of the push–pull amplifier can be analyzed using the small-signal equivalent circuit in Figure 7.6 with some differences in parameter values. The input capacitance must also include the gate-to-source and gate-to-bulk capacitances of the p-channel transistor, while the Miller capacitor must also include the gate-to-drain overlap capacitance of M_2. The most important difference is that the amplifier transconductance is now the sum of both transconductances g_{m1} and g_{m2}. Thus, the low-frequency voltage gain of the inverting amplifier in Figure 7.8 is

$$A_{V0} = -\frac{g_{m1} + g_{m2}}{g_o}$$

$$= -\frac{1}{\left(\dfrac{\phi_t}{V_{A1}} + \dfrac{\phi_t}{V_{A2}}\right)}\left(\frac{2/n_1}{\sqrt{1 + \dfrac{I_{TH}}{I_{S1}}} + 1} + \frac{2/n_2}{\sqrt{1 + \dfrac{I_{TH}}{I_{S2}}} + 1}\right). \qquad (7.1.35)$$

The input-referred noise source e_n can be readily calculated using the expression

$$\frac{\overline{e_n^2}}{\Delta f} = \frac{1}{(g_{m1} + g_{m2})^2}\left(\frac{\overline{i_{n1}^2}}{\Delta f} + \frac{\overline{i_{n2}^2}}{\Delta f}\right) \qquad (7.1.36)$$

together with the PSD of the channel (thermal or flicker) noise current of a MOSFET in saturation.

7.2   Common-gate amplifiers

The scheme of the common-gate amplifier is shown in Figure 7.9(a). The input signal is applied to the source and the output signal is taken at the drain. The gate is connected to a dc voltage V_G that allows the appropriate biasing conditions for the amplifier. Figure 7.9(b) gives the output characteristics of both n-channel and p-channel devices. The increase in the input voltage tends to decrease the current flowing through the transistors and, thus, to increase the output voltage, as can be seen in the voltage transfer characteristic of Figure 7.9(c). Region 2 is the most appropriate region for the operation of the circuit in Figure 7.9 as an amplifier. Note that the common-gate amplifier is a non-inverting amplifier.

Let us calculate the approximate value of the input voltage V_{TH} at which both M_1 and M_2 operate in saturation. The drain current is approximately I_B, whereas the reverse current is negligible for both devices. Assuming that the bulk of M_1 is connected to ground, we can write the approximate UICM expression for M_1 as

$$\frac{V_P - V_{SB,n}}{\phi_t} = \frac{V_G - V_{T0,n}}{n_1\phi_t} - \frac{V_{TH}}{\phi_t} = \sqrt{1 + i_{f1}} - 2 + \ln\left(\sqrt{1 + i_{f1}} - 1\right), \qquad (7.2.1)$$

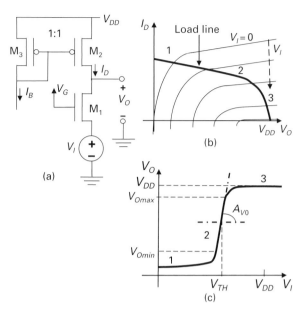

Fig. 7.9   (a) A common-gate amplifier. (b) The n-channel and p-channel MOSFET output characteristics. (c) The voltage transfer characteristic.

which gives

$$V_{TH} = \frac{V_G - V_{T0,n}}{n_1} - \phi_t\left[\sqrt{1 + i_{f1}} - 2 + \ln\left(\sqrt{1 + i_{f1}} - 1\right)\right], \qquad (7.2.2)$$

where $i_{f1} = I_B/I_{S1}$.

The maximum output voltage is limited by the saturation voltage of M_2 according to

$$V_{omax} = V_{DD} - |V_{DSsat2}|, \qquad (7.2.3)$$

whereas the output voltage must remain above a minimum value for M_1 to be in saturation, i.e.

$$V_{omin} = V_{DSsat1} + V_{TH} = \frac{V_G - V_{T0,n}}{n_1} + \phi_t\left[5 - \ln\left(\sqrt{1 + i_{f1}} - 1\right)\right]. \qquad (7.2.4)$$

The drain-to-source saturation voltages of M_1 and M_2 are given by

$$V_{DSsat1(2)} = (-)\phi_t\left(\sqrt{1 + i_{f1(2)}} + 3\right). \qquad (7.2.5)$$

The negative sign in (7.2.5) is required for the calculation of the drain-to-source saturation voltage of the p-channel device.

Figure 7.10 shows the common-gate amplifier together with its small-signal equivalent circuit and the asymptotic frequency-response magnitude of the voltage gain. C_X represents the sum of all capacitances connected between node X and the ac ground, namely the intrinsic capacitances C_{gs1} and C_{bs1} (=$(n-1)C_{gs1}$) and the extrinsic overlap (C_{gs1ov}) and junction (C_{jbs1}) capacitances. C_L is composed of both the load capacitance and other capacitances associated with the output node, namely C_{gd1ov}, C_{gd2ov}, C_{jbd1}, and C_{jbd2}.

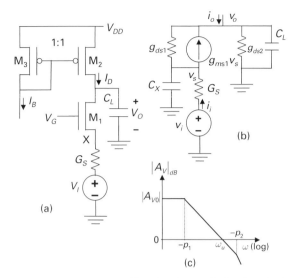

Fig. 7.10   (a) A common-gate amplifier. (b) The small-signal equivalent circuit. (c) The magnitude
response of the voltage gain.

Using the small-signal equivalent of Figure 7.10(b), circuit analysis leads to the
following results for the transadmittance Y_m,

$$Y_m(s) = -\frac{i_o}{v_i}\bigg|_{v_o=0} = \frac{(g_{ms1} + g_{ds1})G_S}{sC_X + g_{ms1} + g_{ds1} + G_S},\qquad (7.2.6)$$

and the output admittance Y_o,

$$Y_o(s) = -\frac{i_o}{v_o}\bigg|_{v_i=0}$$

$$= \frac{g_{ds2}(g_{ms1} + g_{ds1} + G_S) + g_{ds1}G_S}{sC_X + g_{ms1} + g_{ds1} + G_S}$$

$$+ \frac{s[C_L(g_{ms1} + g_{ds1} + G_S) + C_X g_{ds2}] + s^2 C_X C_L}{sC_X + g_{ms1} + g_{ds1} + G_S}.\qquad (7.2.7)$$

Assuming that the usual approximations $C_X g_{ds2} \ll C_L(g_{ms1} + G_S)$ and $g_{ms1} \gg g_{ds1}$,
and the dominant-pole approximation, are valid, we find that the voltage gain of the
common-gate amplifier is

$$A_V(s) = \frac{v_o}{v_i}\bigg|_{i_o=0} = \frac{Y_m}{Y_o} \cong \frac{g_{ms1}G_S}{C_X C_L(s - p_1)(s - p_2)},\qquad (7.2.8)$$

where

$$p_1 \cong -\frac{g_{ds2}(g_{ms1} + G_S) + g_{ds1}G_S}{C_L(g_{ms1} + G_S)},\qquad p_2 \cong -\frac{g_{ms1} + G_S}{C_X}.$$

The low-frequency gain and the unity-gain frequency are

$$A_{V0} = \frac{g_{ms1}}{g_{ds1} + g_{ds2}} \left(1 + \frac{g_{ms1}}{G_S} \frac{g_{ds2}}{g_{ds1} + g_{ds2}} \right)^{-1},$$

$$\omega_u = -A_{V0} p_1 = \frac{g_{ms1}}{C_L (1 + g_{ms1}/G_S)}. \tag{7.2.9}$$

Using the results of (7.2.8) and (7.2.9), we can readily find

$$-\frac{p_2}{\omega_u} = \frac{C_L}{C_X} \left(1 + \frac{g_{ms1}}{G_S} \right) \left(1 + \frac{G_S}{g_{ms1}} \right), \tag{7.2.10}$$

which shows that the frequency of the secondary pole is typically higher (by a factor of at least four times the ratio C_L/C_X) than the unity-gain frequency (see Figure 7.10(c)). The input admittance is

$$Y_i = \frac{i_i}{v_s} = \frac{g_{ms1} + g_{ds1}}{1 + g_{ds1}/g_{ds2}} \left(\frac{1 + sC_L/g_{ds2}}{1 + sC_L/(g_{ds1} + g_{ds2})} \right) + sC_X. \tag{7.2.11}$$

The input admittance under low-frequency operation is approximately $g_{ms1}/2$ for $g_{ds1} = g_{ds2}$.

The input-referred noise sources i_n and e_n can be determined in terms of the channel noise associated with each transistor and from the small-signal equivalent circuit. For low frequencies, the PSD of the input-referred noise current is negligible, whereas the PSD of the input-referred noise voltage is

$$\frac{\overline{e_n^2}}{\Delta f} = \frac{1}{(g_{ms1} + g_{ds1})^2} \left(\frac{\overline{i_{n1}^2}}{\Delta f} + \frac{\overline{i_{n2}^2}}{\Delta f} \right) \simeq \frac{1}{g_{ms1}^2} \left(\frac{\overline{i_{n1}^2}}{\Delta f} + \frac{\overline{i_{n2}^2}}{\Delta f} \right). \tag{7.2.12}$$

The use of the expressions for the thermal and flicker noise of a single transistor presented in Chapter 4 can be directly applied to (7.2.12) for the computation of the input-referred equivalent noise source.

7.3   Source followers

In source followers, such as the one represented in Figure 7.11, the input signal is applied to the gate and the output signal is taken at the source. The local substrate can be either connected to the source $(V_B = V_O)$[2] of the input transistor or to V_{DD} $(V_B = V_{DD})$.

The output characteristics of transistor M_1 are shown in Figure 7.11(b). Note that, if $V_O < V_{DD}$, the current is higher for $V_B = V_O$ than for $V_B = V_{DD}$. The voltage transfer characteristics in Figure 7.11(c) show that in the linear region the gain of the source follower with $V_B = V_O$ is very close to unity, whereas that with $V_B = V_{DD}$ is around $1/n$, where n is the slope factor. This dependence of the gain on n represents not only

---

[2] In the case of the p-channel input transistor of the source follower, a connection between bulk and source is allowed for n-well or triple-well technologies.

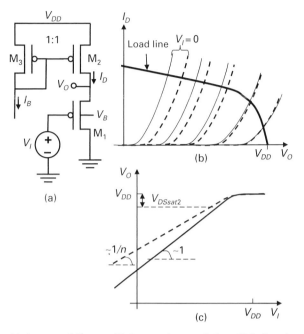

Fig. 7.11   (a) A source follower. (b) Output characteristics of M_1 for a local substrate connected either to the output (solid lines) or to the V_{DD} supply (dashed lines). (c) Voltage transfer characteristics for a local substrate connected either to the output (solid line) or to the V_{DD} supply (dashed line).

a gain loss but also distortion since n is a function, even though slight, of the input voltage.

We once again make use of the UICM to determine the voltage transfer characteristics of the source follower. For the configuration in which the substrate is connected to the source we have

$$V_O|_{V_B=V_O} \cong V_I - V_{T0} + n\phi_t \left[\sqrt{1 + i_{f1}} - 2 + \ln\left(\sqrt{1 + i_{f1}} - 1\right) \right], \qquad (7.3.1)$$

whereas for the case in which the substrate is connected to V_{DD} we have

$$V_O|_{V_B=V_{DD}} \cong V_{DD} + \frac{V_I - V_{DD} - V_{T0}}{n} + \phi_t \left[\sqrt{1 + i_{f1}} - 2 + \ln\left(\sqrt{1 + i_{f1}} - 1\right) \right], \qquad (7.3.2)$$

where $i_{f1} = I_B/I_{S1}$. Expressions (7.3.1) and (7.3.2) are approximations for calculating the dependence of the output voltage on the input voltage. If one assumes n and i_{f1} to be constant over the input-signal variation, the low-frequency gain of the follower with separate wells for the active and load transistors, which is calculated using (7.3.1), equals unity. On the other hand, the voltage gain of the follower with the active and load transistors in a common well equals $1/n$. The model used to derive both (7.3.1) and (7.3.2) assumes that the transistor current in saturation is independent of the drain voltage. Next, the application of the small-signal model shows how the voltage gain is affected by the output conductance of the MOS transistor.

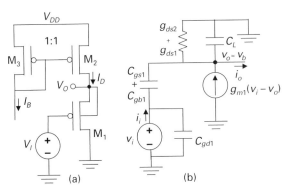

Fig. 7.12   (a) A source follower and (b) its small-signal equivalent circuit.

The ac parameters of the source follower can be calculated using the small-signal equivalent circuit drawn in Figure 7.12. C_L is the sum of all capacitances connected between the output node and the ac ground. The analysis of the circuit in Figure 7.12(b) gives

$$Y_m(s) = \frac{i_o}{v_i}\bigg|_{v_o=0} = g_{m1} + s(C_{gs1} + C_{gb1}), \tag{7.3.3}$$

$$Y_o(s) = -\frac{i_o}{v_o}\bigg|_{v_i=0} = s(C_L + C_{gs1} + C_{gb1}) + g_{ds2} + g_{m1} + g_{ds1}. \tag{7.3.4}$$

For low frequency, the output admittance in (7.3.4) reduces to the value of the transistor transconductance, which is a well-known result.

The voltage gain of the source follower is

$$A_V(s) = \frac{v_o}{v_i}\bigg|_{i_o=0} = \frac{Y_m}{Y_o} = \frac{1}{1 + \dfrac{g_{ds1} + g_{ds2}}{g_{m1}}} \cdot \frac{1 + \dfrac{s(C_{gs1} + C_{gb1})}{g_{m1}}}{1 + \dfrac{s(C_L + C_{gs1} + C_{gb1})}{g_{m1} + g_{ds1} + g_{ds2}}}. \tag{7.3.5}$$

The zero frequency in (7.3.5) is equal to the intrinsic transition (unity-gain) frequency. The voltage gain $A_{V0} \cong 1 - (g_{ds1} + g_{ds2})/g_{m1}$ is slightly below unity for frequencies up to the pole frequency, which is approximately equal to the intrinsic transition frequency when the output capacitance is lower than the gate capacitance. For load capacitances much higher than the transistor's gate capacitance, the pole frequency is approximately g_{m1}/C_L. The input admittance determined from the ac equivalent circuit in Figure 7.12 is

$$Y_i = \frac{I_I}{V_I} = sC_{gd1} + s(C_{gs1} + C_{gb1})(1 - A_V). \tag{7.3.6}$$

For frequencies below the pole frequency, the input admittance in (7.3.6) is

$$Y_i = sC_{gd1} + s(C_{gs1} + C_{gb1})\left(\frac{g_{ds1} + g_{ds2}}{g_{m1}}\right) \cong sC_{gd1}. \tag{7.3.7}$$

The result in (7.3.7) is not surprising since the output voltage follows the input voltage; thus, the resulting current flowing through capacitor $C_{gs1} + C_{gb1}$ is considerably lower than that through C_{gd1}.

The PSDs of the input-referred noise voltage and noise current are

$$\frac{\overline{e_n^2}}{\Delta f} = \frac{\overline{i_{n1}^2}/\Delta f + \overline{i_{n2}^2}/\Delta f}{g_{m1}^2 + \omega^2 \left(C_{gs1} + C_{gb1} \right)^2} = \frac{1/g_{m1}^2}{1 + \omega^2/\omega_u^2} \left(\frac{\overline{i_{n1}^2}}{\Delta f} + \frac{\overline{i_{n2}^2}}{\Delta f} \right), \tag{7.3.8}$$

$$\frac{\overline{i_n^2}}{\Delta f} = \omega^2 \left(C_{gs1} + C_{gb1} + C_{gd1} \right)^2 \frac{\overline{e_n^2}}{\Delta f}. \tag{7.3.9}$$

In (7.3.8) ω_u is the intrinsic unity-gain frequency. The PSD of the input-referred noise voltage follows the PSD of the channel currents of M_1 and M_2 up to frequencies approaching ω_u.

Example 7.4

Assume that the body of transistor M_1 in Figure 7.12 is connected to $V_{DD} = 5$ V. The bias current is $I_B = 8$ μA. For the sake of simplicity let us assume that the specific currents are not dependent on the gate voltage and that their values are $I_{S1} = I_{S2} = I_{S3} = 1$ μA. Also, assume that the Early voltage of all of the transistors is $V_A = 10$ V and that the slope factor of transistor M_1 is $n(V_G = 1$ V$) = 1.12$ and $n(V_G = 4$ V$) = 1.25$. (a) What is the approximate maximum value of the input voltage such that M_2 remains in saturation? (b) What is the small-signal voltage gain in terms of the small-signal parameters? (c) What is the range of the voltage gain for 1 V $< V_G < 4$ V? Assume $V_{T0} = -0.6$ V and $\phi_t = 25$ mV.

Answer

(a) For the data given we have $i_{f2} = i_{f1} = 8$. The source–drain saturation voltage of M_2 is

$$V_{SDsat2} \cong \phi_t \left(\sqrt{1 + i_{f2}} + 3 \right) = 6\phi_t.$$

Using the UICM for M_1, we find the maximum value at which M_2 remains in saturation:

$$-\left(\frac{V_{GB1} - V_{T0}}{n_1 \phi_t} - \frac{V_{SB1}}{\phi_t} \right) = -\left(\frac{V_{Gmax} - V_{DD} - V_{T0}}{n_1 \phi_t} + \frac{V_{SDsat2}}{\phi_t} \right)$$

$$= \sqrt{1 + i_{f1}} - 2 + \ln\left(\sqrt{1 + i_{f1}} - 1 \right)$$

$$= 1 + \ln 2.$$

Assuming $n \cong 1.12$ for V_{Gmax} and using the previous expression, we find

$$V_{Gmax} = V_{DD} + V_{T0} - n_1 \phi_t (6 + 1 + \ln 2) \cong 4.18 \text{ V}.$$

Note that the value obtained for V_{Gmax} is close to 4 V. Therefore, the assumption $n = 1.12$ is quite acceptable.

(b) Using the small-signal transistor model we find that

$$A_{V0} = \frac{V_O}{V_I}\bigg|_{I_O=0} = \frac{g_{m1}}{g_{ms1} + g_{md2}} = \frac{1}{n_1 + g_{md2}/g_{m1}}.$$

(c) The output conductance $g_{md2} = 8/10 = 0.8\,\mu\text{A/V}$. The transconductance g_{m1} is given by

$$g_{m1} = \frac{I_D}{n_1\phi_t}\frac{2}{\sqrt{1 + i_{f1}} + 1} = \frac{8 \times 10^3}{25n_1}\frac{2}{4} = \frac{160}{n_1}\,\mu\text{A/V}.$$

The voltage gains for $1\,\text{V} < V_G < 4\,\text{V}$ are in the range $0.888 > A_{V0} > 0.796$.

7.4   Cascode amplifiers

The MOS cascode amplifier is a common-source amplifier loaded by a common-gate amplifier [1], [4], [6], [7]. Cascode amplifiers can have quite large gains and reduce significantly the Miller effect, which gives them an improved frequency response compared with that of common-source amplifiers [6]. However, cascode topologies suffer from reduced output-voltage swing due to stacked transistors. This is particularly troublesome for more advanced CMOS technologies, which operate at low supply voltages.

7.4.1   Telescopic- and folded-cascode amplifiers

The basic topologies of the telescopic- and folded-cascode amplifiers are shown in Figure 7.13.

In either topology, the bias current I_{B3} and transistor M_3 can be properly designed to allow maximum output-voltage swing [8], [9]. To reach this goal, the bias generator

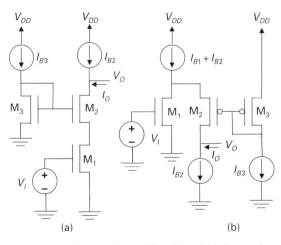

(a)   (b)

Fig. 7.13   (a) A telescopic-cascode amplifier. (b) A folded-cascode amplifier.

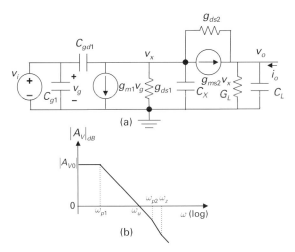

Fig. 7.14   (a) The small-signal equivalent circuit of the telescopic- and folded-cascode amplifiers of Figure 7.13. (b) The magnitude response of the voltage gain.

composed of M_3 and I_{B3} is designed in order to bias M_1 with a drain-to-source voltage slightly higher than the saturation voltage. The design equations for reaching such a goal are explained in Chapter 5, in the subsection on high-swing cascode current mirrors. Note that the power consumption of the folded-cascode amplifier is higher than that of the telescopic-cascode amplifier. Despite this drawback, the folded-cascode amplifier has an exceptional advantage over its telescopic counterpart; when used in a differential amplifier, the common-mode input range and the output-voltage swing are higher in folded-cascode amplifiers than in telescopic-cascode amplifiers.

To evaluate the frequency response of the cascode amplifiers in Figure 7.13 we use the small-signal equivalent circuit represented in Figure 7.14. The signal source is assumed to be ideal. As we will see later, this approximation is not of major consequence since the effect of the gate-to-drain capacitance C_{gd1} is not as high as in the common-source amplifier. Note that $C_{g1} = C_{gs1} + C_{gb1}$ and that G_L and C_L represent the load plus the parasitic capacitances connected to the output node. C_X is the total capacitance associated with node v_X. The short-circuit and output transadmittances of the cascode amplifier are, respectively,

$$Y_m(s) = -\frac{i_o}{v_i}\bigg|_{v_o=0} = -\frac{g_{m1}}{1 + \dfrac{g_{ds1}}{g_{ms2} + g_{ds2}}} \frac{1 - \dfrac{sC_{gd1}}{g_{m1}}}{1 + \dfrac{s(C_{gd1} + C_X)}{g_{ms2} + g_{ds2} + g_{ds1}}}, \tag{7.4.1}$$

$$Y_o(s) = \frac{i_o}{v_o}\bigg|_{v_i=0} = sC_L + G_L + \frac{g_{ds2}\left[s(C_{gd1} + C_X) + g_{ds1}\right]}{s(C_{gd1} + C_X) + g_{ms2} + g_{ds2} + g_{ds1}}. \tag{7.4.2}$$

Equation (7.4.1) shows that, as expected, the transconductance gain of the cascode amplifier is very close to g_{m1}, the transconductance of the common-source amplifier.

A small difference between the two transconductances is originated by a slight reduction in the drain voltage of M_1 due to a small (but not zero) resistance seen by the M_1 drain. The output conductance is the sum of G_L and the output conductance of the cascode device, being approximately equal to $g_{ds1}g_{ds2}/g_{ms2}$, which is interpreted very simply as the output conductance of M_1 attenuated by the voltage gain of the common-gate amplifier. Now, considering the approximation $g_{ms2} \gg g_{ds1}$, g_{ds2}, G_L and the output capacitance to be at least of the same order of magnitude as the internal capacitances, we find that

$$A_V(s) \cong - \frac{\dfrac{g_{m1}g_{ms2}}{G_L g_{ms2} + g_{ds1}g_{ds2}} \left(1 - \dfrac{sC_{gd1}}{g_{m1}}\right)}{\left(1 + \dfrac{sC_L g_{ms2}}{G_L g_{ms2} + g_{ds1}g_{ds2}}\right)\left(1 + \dfrac{s\left(C_{gd1} + C_X\right)}{g_{ms2}}\right)}. \tag{7.4.3}$$

Once again we have used the dominant-pole approximation to find the poles of the frequency response of the voltage gain. From (7.4.3) we note that the low-frequency gain and the pole and zero frequencies are

$$A_{V0} = \frac{g_{m1}g_{ms2}}{G_L g_{ms2} + g_{ds1}g_{ds2}}, \qquad \omega_{p_1} \cong \frac{G_L g_{ms2} + g_{ds1}g_{ds2}}{C_L g_{ms2}},$$

$$\omega_u = A_{V0}\omega_{p_1} = \frac{g_{m1}}{C_L}, \qquad \omega_{p_2} \cong \frac{g_{ms2}}{C_{gd1} + C_X}, \qquad \omega_z = \frac{g_{m1}}{C_{gd1}}. \tag{7.4.4}$$

We note that the low-frequency gain for the case of an ideal current source ($G_L = 0$) is equal to the product of the low-frequency gains of the common-source and common-gate amplifiers. Assuming that the two gains are approximately the same, the gain of the cascode amplifier is equal to the square of the intrinsic transistor gain g_m/g_{ds}. To obtain a significant increase in the gain through the use of a cascode topology, the source conductance G_L must be of the order of $g_{ds1}g_{ds2}/g_{ms2}$ or lower, which means that the output impedance of the current source must be high. When M_1 and M_2 are, as is common, transistors with the same dimensions, $g_{ms2} > g_{m1}$ because $g_{ms2} = n_2 g_{m2} \cong n_1 g_{m1}$. Since C_L is typically higher than $C_{gd1} + C_X$, the secondary pole and zero frequencies are higher than the unity-gain frequency, which is an important benefit for the stability and transient response of feedback amplifiers.

An important advantage of the cascode amplifier compared with the common-source amplifier is the reduction in the Miller effect. In the cascode amplifier, the effect of C_{gd1} on the input capacitance can be readily calculated by noting that the voltage gain at the intermediate node v_x is equal to $g_{m1}/g_{ms2} \cong 1/n$ for equally sized M_1 and M_2. Therefore, C_{gd1} is seen by the input source as a capacitance equal to

$$C_{gd1}\left(1 + g_{m1}/g_{ms2}\right) \cong C_{gd1}(n+1)/n < 2C_{gd1},$$

since the slope factor n is greater than unity, which clearly shows that the cascode stage contributes a much smaller Miller capacitance than does the common-source stage. This advantage is especially important in applications where the input source impedance is not low and can cause the Miller effect to be the dominant factor in determining the roll-off characteristics of amplifiers.

One might think of enhancing the gain by increasing the number of stacked transistors. For each additional cascode level, the output impedance increases by a factor equal to g_{ms}/g_{ds}. The gain of a triple-cascode amplifier, for example, is approximately $(g_m/g_{ds})^3$. However, increasing the number of levels of stacked transistors has two main disadvantages [1], [10]: (i) the addition of each cascode level reduces the output swing by at least (if proper bias is employed) one saturation voltage; and (ii) each transistor included in the signal path introduces an extra pole in the transfer function. To restore the phase margin, the load capacitance has to be increased, thus reducing the unity-gain frequency of the amplifier [10].

Example 7.5

For the circuit in Figure 7.13(a) $W_1 = W_2 = 40\,\mu m$, $W_3 = 2\,\mu m$, $L_1 = L_2 = L_3 = 1.0\,\mu m$, $V_{DD} = 3.3$ V, $I_{B2} = I_{B3} = 9\,\mu A$, $V_{TON} = 0.5$ V, $C'_{ox} = 5\,fF/\mu m^2$, $\mu_N = 400\,cm^2/V$ per s, and $C_L = 2\,pF$. For the sake of simplicity, assume that $n_1 = n_2 = 1.2$, $V_{A1} = V_{A2} = 9$ V, and $\phi_t = 25$ mV. Calculate (a) V_{DS1} assuming that M_2 operates in saturation. Is M_1 operating in saturation? Calculate (b) the minimum output voltage such that M_2 operates in saturation, (c) the transconductance of the cascode stage, (d) the output resistance, (e) the voltage gain, (f) the unity-gain frequency, and (g) the dominant-pole frequency.

Answer

(a) For the given data we have $I_{S1} = I_{S2} = 3\,\mu A$ and $I_{S3} = 0.15\,\mu A$. Thus, for saturated transistors we have $i_{f1} = i_{f2} = 3$ and $i_{f3} = 60$. Using the UICM for M_3, we find that $V_{P3} = 7.73\phi_t$. Since the gates of M_2 and M_3 are connected to each other, $V_{P2} = V_{P3}$. Now, on applying the UICM to M_2, we find that

$$V_{SB2} = V_{P2} - \phi_t\left[\sqrt{1 + i_{f2}} - 2 + \ln\left(\sqrt{1 + i_{f2}} - 1\right)\right] = V_{P2}.$$

Since the substrate of M_2 is connected to ground, $V_{DS1} = V_{SB2} = 7.73\phi_t$. Note that V_{DS1} is slightly greater than $V_{DSsat1} = \phi_t\left[\sqrt{1 + i_{f1}} + 3\right] = 5\phi_t$; therefore, M_1 is operating in saturation. Also note that $V_{DSsat2} = V_{DSsat1}$ because $i_{f1} = i_{f2}$. (b) Thus, the minimum output voltage such that M_2 operates in saturation is $V_{Omin} = 12.73\phi_t \cong 318$ mV. (c) The transconductance of the cascode stage is approximately equal to g_{m1}, which can be calculated from the transconductance-to-current ratio

$$g_{m1} = \frac{2I_{D1}}{n\phi_t(\sqrt{1 + i_{f1}} + 1)} = \frac{2 \times 9 \times 10^{-6}}{1.2 \times 25 \times 10^{-3} \times 3}\,\mu A/V = 0.2\,mA/V.$$

(d) The output conductance is, from (7.4.2),

$$\frac{g_{ds2}g_{ds1}}{g_{ms2} + g_{ds2} + g_{ds1}} \cong \frac{g_{ds2}g_{ds1}}{g_{ms2}} = \frac{10^{-6} \times 10^{-6}}{1.2 \times 0.2 \times 10^{-3}} = 4.17\,nA/V.$$

Thus, the output resistance $R_O = 240\,M\Omega$.

(e) $A_{V0} = -g_m R_O = -48\,000$.

(f) The unity-gain frequency $f_u = g_{m1}/(2\pi C_L) \cong 16$ MHz.

(g) The pole frequency $f_P = 16\,MHz/48\,000 \cong 330$ Hz.

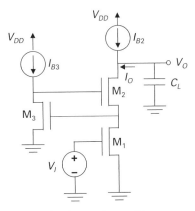

Fig. 7.15   The regulated cascode amplifier.

As the technology is scaled down, the transition frequency of transistors increases but the intrinsic gain decreases. Thus, achieving high voltage gains becomes tougher in more advanced technologies [10]. In some cases, the gain achievable with a two-level cascode stage is not sufficient for the application and an additional cascode level introduces the previously mentioned disadvantages. To achieve high voltage gains without introducing the drawbacks of triple (or quadruple) cascode stages, the gain-boost technique [10]–[13] introduced in the next section has been proposed.

7.4.2   The gain-boost technique

One of the circuits that has a greater output impedance than that of a cascode stage is the regulated cascode circuit shown in Figure 7.15 [12], [13]. Transistor M_3 implements a negative-feedback amplifier that reduces the drain voltage (and current) variations of M_1 in response to the changes in the output voltage V_O, thus increasing the output resistance.

The regulated cascode in Figure 7.15 was conceived to operate in strong inversion, even though the general concepts for its proper operation can be applied to other regions. In order to have the full benefit of the regulated cascode circuit, the output voltage must be

$$V_O > V_{DS1} + V_{DSsat2} \qquad (7.4.5)$$

for operation of M_2 in saturation. In order to maximize the output swing, V_{DS1} should be approximately equal to the drain-to-source saturation voltage of M_1. However, the value of $V_{DS1} = V_{GS3}$ is mainly determined by the feedback amplifier composed of M_3 and I_{B3}. If M_1 is designed to operate in weak/moderate inversion ($i_f < 100$), V_{GS3} must be within the approximate range 100–350 mV for an optimally biased circuit. Such values of the gate-to-source voltage are extremely hard to obtain, since a very low I_{B3} and/or a large aspect ratio for M_3 would expend too much silicon area.

The transconductance of the regulated cascode circuit is $G_m \cong g_{m1}$. The output conductance is

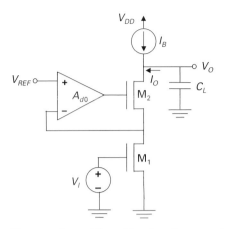

Fig. 7.16   The cascode amplifier with gain enhancement.

$$G_0 = \frac{g_{ds1}g_{ds2}}{g_{m2}A_3 + g_{ms2} + g_{ds1} + g_{ds2}} \cong \frac{g_{ds1}g_{ds2}g_{ds3}}{g_{m2}g_{m3}}, \qquad (7.4.6)$$

where A_3 is the voltage gain v_{d3}/v_{g3} (see Problem 7.6). The output impedance of the regulated cascode amplifier is higher than that of a cascode amplifier by a factor approximately equal to the voltage gain of the additional amplifier composed of M_3 and current source I_{B3}. Thus, the low-frequency voltage gain is greater than that of a simple cascode amplifier. To calculate the unity-gain frequency, we simply note that the amplifier voltage gain equals the product of the transconductance gain and the output impedance. At high frequencies, the output impedance is dominated by the output capacitance C_L. Therefore, the unity-gain frequency is $f_u = g_{m1}/(2\pi C_L)$.

Another circuit for the gain-boost principle, which is based on an operational amplifier, is shown in Figure 7.16 [10]. This implementation, also called the active cascode amplifier [1], allows a larger output swing and a higher low-frequency gain than those of the regulated cascode circuit in Figure 7.15.

For further details on the frequency response of the active cascode amplifier the reader is referred to [10]. A well-designed active cascode amplifier has the benefit of an improved low-frequency gain while keeping the same unity-gain frequency as that of a simple cascode amplifier [10].

The transconductance of the gain-boost cascode circuit is $G_m \cong g_{m1}$ and its output conductance is

$$G_0 = \frac{g_{ds1}g_{ds2}}{g_{m2}A_{d0} + g_{ms2} + g_{ds1} + g_{ds2}} \cong \frac{g_{ds1}g_{ds2}}{g_{m2}A_{d0}}, \qquad (7.4.7)$$

where A_{d0} is the low-frequency voltage gain of the operational amplifier. The overall dc gain of the cascode stage with the additional operational amplifier is

$$A_{V0} = -\frac{G_m}{G_0} \cong -\frac{g_{m1}}{g_{ds1}}\frac{g_{m2}}{g_{ds2}} A_{d0}, \qquad (7.4.8)$$

i.e. the dc gain of the active cascode stage is now equal to that of the simple cascode amplifier multiplied by the dc gain of the operational amplifier.

7.5   Differential amplifiers

The differential amplifier, one of the most widely used blocks of analog integrated circuits, is aimed at amplifying the voltage difference between two input signals and rejecting the signal common to both inputs [1], [6]. Differential amplifiers are important to detect weak signals contaminated by common-mode signals, which can be supply/ground noise, cross-talk, etc. An important property of differential amplifiers is that they can be coupled to one another without the need for either interstage coupling capacitors or level shifting [1], [6].

7.5.1   The source-coupled pair

The main block of differential amplifiers is the source-coupled pair shown in Figure 7.17. In its idealized version, it is composed of two perfectly matched transistors M_1 and M_2 and an ideal constant-current source. Transistors M_1 and M_2 operate in saturation and the currents through them are assumed to be independent of the drain voltages. The local substrate can be connected either to the source or to ground.

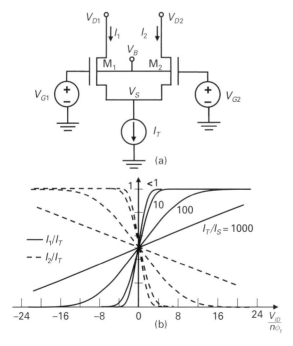

Fig. 7.17   (a) A schematic diagram of the source-coupled pair and (b) its dc transfer characteristics.

7.5.1.1 The dc transfer characteristics

For the sake of simplicity we will assume that the slope factor n and the specific currents I_S of both M_1 and M_2 are independent of the gate voltage. Recalling that

$$V_{P1} \cong (V_{GB1} - V_{T0})/n, \qquad V_{P2} \cong (V_{GB2} - V_{T0})/n \qquad (7.5.1)$$

and defining the differential output current

$$I_{OD} = I_1 - I_2, \qquad (7.5.2)$$

the application of the UICM to both M_1 and M_2 yields

$$\frac{V_{P1(2)} - V_{SB}}{\phi_t} = \sqrt{1 + \frac{I_{1(2)}}{I_S}} - 2 + \ln\left(\sqrt{1 + \frac{I_{1(2)}}{I_S}} - 1\right). \qquad (7.5.3)$$

The use of (7.5.3) together with (7.5.1) and (7.5.2) gives the differential input voltage in terms of the differential output current as

$$\frac{V_{ID}}{n\phi_t} = \frac{V_{G1} - V_{G2}}{n\phi_t} = \sqrt{1 + \frac{I_T + I_{OD}}{2I_S}} - \sqrt{1 + \frac{I_T - I_{OD}}{2I_S}}$$

$$+ \ln\left[\left(\sqrt{1 + \frac{I_T + I_{OD}}{2I_S}} - 1\right) \Big/ \left(\sqrt{1 + \frac{I_T - I_{OD}}{2I_S}} - 1\right)\right].$$

$$(7.5.4)$$

If both M_1 and M_2 operate in weak inversion, the relationship between the differential output current and the differential input voltage is

$$\frac{I_{OD}}{I_T} = \frac{I_1 - I_2}{I_T} = \tanh\left(\frac{V_{ID}}{2n\phi_t}\right). \qquad (7.5.5)$$

Formula (7.5.5) shows that the differential output current does not change significantly if the differential input voltage is greater than $4n\phi_t$, for any inversion level, as can be seen in Figure 7.17 for $I_T/I_S < 1$ and in Figure 7.18, where I_1, I_2, and the difference between them are plotted.

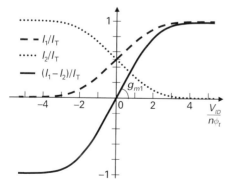

Fig. 7.18 The dc transfer characteristics of the differential pair in weak inversion.

On the other hand, the differential output current in strong inversion is

$$
\frac{I_{OD}}{I_T} =
\begin{cases}
\dfrac{V_{ID}}{n\phi_t\sqrt{I_T/I_S}}\sqrt{2 - \left(\dfrac{V_{ID}}{n\phi_t\sqrt{I_T/I_S}}\right)^2} & \text{for } \dfrac{|V_{ID}|}{n\phi_t\sqrt{I_T/I_S}} \le 1, \\[2ex]
1 & \text{for } \dfrac{V_{ID}}{n\phi_t\sqrt{I_T/I_S}} > 1, \\[2ex]
-1 & \text{for } \dfrac{V_{ID}}{n\phi_t\sqrt{I_T/I_S}} < -1.
\end{cases}
\tag{7.5.6}
$$

The output current saturates ($|I_{OD}| = I_T$) for $|V_{ID}| \ge n\phi_t\sqrt{I_T/I_S}$.

7.5.1.2   The common-mode input range

So far we have assumed the current source to be ideal and both M_1 and M_2 to operate in saturation, which, in reality, is not always the case. To calculate the range of input voltages for which the differential pair operates appropriately, let us assume that the drain voltages of both transistors are equal to V_D and that the two inputs are tied together and connected to a supply voltage equal to V_G. If V_G increases, the voltage V_S at the common-source node also increases and the drain-to-source voltage of both transistors decreases. M_1 and M_2 will remain in saturation as long as $V_{DS} > V_{DSsat}$. On the other hand, decreasing V_G will reduce V_S, which must be higher than a certain voltage for the proper operation of the current source. In the analysis that follows, we assume that $V_B = 0$. In this case, the use of the UICM, which is given by (7.5.3), allows us to write

$$
V_S = \frac{V_G - V_{T0}}{n} - \phi_t\left[\sqrt{1 + \frac{I_T}{2I_S}} - 2 + \ln\left(\sqrt{1 + \frac{I_T}{2I_S}} - 1\right)\right].
\tag{7.5.7}
$$

The current source operates appropriately if a minimum voltage difference is established between its terminals; in other words, one must have $V_S > V_{Smin}$. On the other hand, the drain-to-source voltage of both M_1 and M_2 must be greater than the transistor saturation voltage. These two conditions are written as

$$
V_{Smin} \le V_S \le V_D - V_{DSsat1}.
\tag{7.5.8}
$$

By substituting for the value of V_S in (7.5.8) that given in (7.5.7), we can readily find the common-mode input range for proper operation of the source-coupled pair.

Example 7.6

The source-coupled pair in Figure 7.17 is biased with $I_T = 140\,\text{nA}$ and $V_{D1} = V_{D2} = 3.3\,\text{V}$. The voltage difference between the terminals of the current source must be at least $100\,\text{mV}$ for correct operation. Assume that $V_{T0} = 0.5\,\text{V}$, $n = 1.2$, and $I_{SH} = 70\,\text{nA}$. Calculate the common-mode input voltage for aspect ratios of (a) $W/L = 100$ and (b) $W/L = 0.01$.

Answer

Using (7.5.7) and (7.5.8), we find that

$$
n\left[V_{Smin} + \phi_t F\left(\frac{I_T}{2I_S}\right)\right] \le V_G - V_{T0} \le n\left[V_D - V_{DSsat1} + \phi_t F\left(\frac{I_T}{2I_S}\right)\right]
\tag{E7.6}
$$

with

$$F\left(\frac{I_T}{2I_S}\right) = \sqrt{1 + \frac{I_T}{2I_S}} - 2 + \ln\left(\sqrt{1 + \frac{I_T}{2I_S}} - 1\right).$$

(a) For the data given, $I_T/(2I_S) = i_{f1} = i_{f2} = 0.01$, which gives $V_{DSsat1} = \phi_t[\sqrt{1 + i_{f1}} + 3] \cong 100$ mV. Using formula (E7.6), we find that 0.43 V $\leq V_G \leq 4.15$ V. (b) $I_T/(2I_S) = i_{f1} = i_{f2} = 100$, $V_{DSsat1} = 325$ mV, and 0.93 V $\leq V_G \leq 4.37$ V.

7.5.1.3   The input offset voltage

So far we have analyzed the differential pair assuming M_1 and M_2 to be matched. In this case, the currents flowing through M_1 and M_2 are equal if $V_{ID} = 0$. Now, let us evaluate the impact of mismatch on the dc performance of the differential pair. In the case of an MOS differential amplifier, the input offset voltage is a quantity that represents the effect of all the component mismatches within the amplifier on its dc performance [1]. In the case of the simple differential pair under analysis, the input offset voltage is the voltage difference required at the input in order to make the differential output current equal to zero.

A very simple model assumes that the mismatch between two transistors can be represented by an imbalance between the transistor threshold voltages and normalization currents. Using this simple model we write

$$V_{T01} = V_{T02} + \Delta V_T, \qquad I_{S1} = I_{S2} + \Delta I_S. \tag{7.5.9}$$

The drain current of a long-channel transistor can be written as

$$I_D = I_S[i_f(V_G - V_{T0}, V_S) - i_r(V_G - V_{T0}, V_D)]. \tag{7.5.10}$$

Since both M_1 and M_2 operate in saturation, the last term in (7.5.10) can be neglected; thus, for a small differential input signal we can write the difference between the currents flowing through M_1 and M_2 as

$$\frac{\Delta I_D}{I_D} \cong \frac{1}{I_D}\left(\frac{\partial I_D}{\partial I_S}\Delta I_S + \frac{\partial I_D}{\partial V_G}\Delta V_G + \frac{\partial I_D}{\partial V_{T0}}\Delta V_{T0}\right)$$

$$= \frac{\Delta I_S}{I_S} + \frac{g_m}{I_D}(\Delta V_G - \Delta V_{T0}), \tag{7.5.11}$$

where $\Delta I_D = I_1 - I_2$ and $\Delta V_G = V_{G1} - V_{G2}$. Note that, according to (7.5.10), the derivatives of the current with respect to the gate voltage or to the threshold voltage are equal but with opposite signs. In order to have $I_2 = I_1$, that is $\Delta I_D = 0$, (7.5.11) gives

$$\Delta V_G = V_{OS} = \Delta V_{T0} - \frac{I_D}{g_m}\frac{\Delta I_S}{I_S} = \Delta V_{T0} - n\phi_t\frac{1 + \sqrt{1 + I_T/(2I_S)}}{2}\frac{\Delta I_S}{I_S}. \tag{7.5.12}$$

According to the model in (7.5.12), the offset voltage is equal to the threshold-voltage mismatch plus an additional term associated with the specific-current mismatch that increases for increasing inversion levels. In order to have a low offset voltage, the input transistors of a differential pair should operate in the weak/moderate-inversion region to minimize the impact of the specific current mismatch on the offset voltage. If one assumes Pelgrom's model to be valid and the mismatch parameters to be uncorrelated, the variance of the input offset voltage calculated in (7.5.12) can be written as

$$\sigma^2(V_{OS}) = \sigma^2(V_T) + \left(\frac{I_T}{2g_m}\right)^2 \frac{\sigma^2(I_S)}{I_S^2}$$

$$= \frac{A_{VT}^2}{WL} + \left(n\phi_t \frac{\sqrt{1 + I_T/2I_S} + 1}{2}\right)^2 \frac{A_{IS}^2}{WL}. \tag{7.5.13}$$

Using the model derived in Chapter 4 for current mismatch, we can write

$$\sigma^2(V_{OS}) = \frac{\sigma_{I_D}^2}{g_m^2} = \left(\frac{I_T}{2g_m}\right)^2 \frac{1}{WL} \left[\frac{N_{oi}}{N^{*2}} \frac{1}{i_f} \ln(1 + i_f) + A_{IS}^2\right], \tag{7.5.14}$$

where $I_T/(2g_m)$ is given by the same expression as in (7.5.13).

Example 7.7

The nominal specific current of transistors M_1 and M_2 in Figure 7.17 is $I_S = 1\,\mu A$. The transistor dimensions are $W = 40\,\mu m$ and $L = 2.5\,\mu m$. Assume that $n = 1.2$, $\phi_t = 25\,mV$, $A_{VT} = 10\,mV\,\mu m$, and $A_{IS} = 2\%\,\mu m$. (a) Using (7.5.13), calculate the standard deviation of the offset voltage for inversion levels from 0.1 to 4000. At which current level is the contribution of the threshold-voltage mismatch to the offset voltage equal to the contribution of the specific-current mismatch? (b) Using (7.5.14), calculate the standard deviation of the offset voltage for inversion levels from 0.1 to 4000.

Answer

(a) Using (7.5.13) we have

$$\sigma^2(V_{OS}) = \frac{(10^{-8})^2}{10^{-10}} + \left(1.2 \times 25 \times 10^{-3} \frac{\sqrt{1 + i_f} + 1}{2}\right)^2 \frac{(2 \times 10^{-8})^2}{10^{-10}},$$

which gives a standard-deviation range of 1.0 mV to 2.17 mV for i_f in the range 0.1 to 4000. The contributions of V_T mismatch and of I_S mismatch to the offset voltage are equal when $i_f = 1045$. (b) Recalling that $N_{oi}/N^{*2} = [A_{VT}/(n\phi_t)]^2$, the application of (7.5.14) gives a standard-deviation range of 1.0 mV to 2.41 mV for i_f in the range 0.1 to 4000.

7.5.1.4   Small-signal analysis

Before starting with the ac analysis of the differential pair, let us recall some important definitions. For the differential pair in Figure 7.17(a), the differential (v_{id}) and common-mode (v_{icm}) input voltages are defined as

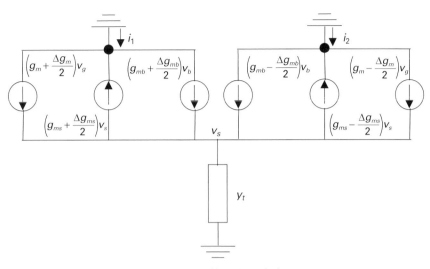

Fig. 7.19   The small-signal circuit equivalent to the differential pair for $v_{g1} = v_{g2} = v_g$.

$$v_{id} = v_{g1} - v_{g2} \tag{7.5.15}$$

and

$$v_{icm} = \frac{v_{g1} + v_{g2}}{2}, \tag{7.5.16}$$

respectively. Note that the input voltages can be written in terms of the differential and common-mode components as

$$v_{g1} = v_{icm} + \frac{v_{id}}{2}, \qquad v_{g2} = v_{icm} - \frac{v_{id}}{2}. \tag{7.5.17}$$

Ideally, the differential pair is not sensitive to the common-mode input voltage, i.e. the differential output current is dependent only on the differential input voltage.

We now define the small-signal differential transconductance of the differential pair as

$$g_{dm} = \left. \frac{dI_{OD}}{dV_{ID}} \right|_{V_{ID}=0} = g_{m1(2)} = \frac{2I_S}{n\phi_t} \left(\sqrt{1 + \frac{I_T}{2I_S}} - 1 \right). \tag{7.5.18}$$

The transconductance of the differential pair equals that of a single transistor of the pair. In effect,

$$\frac{dI_{OD}}{dV_{ID}} = \frac{dI_1}{dV_{G1}} \frac{dV_{G1}}{dV_{ID}} - \frac{dI_2}{dV_{G2}} \frac{dV_{G2}}{dV_{ID}} = g_{m1} \frac{1}{2} - g_{m2} \left(-\frac{1}{2} \right) = g_{m1}.$$

One effect of transistor mismatch on the differential pair is the non-zero common-mode transadmittance [14] $y_{cm-dm} = i_{od}/v_{icm}$ that contributes to degrading the common-mode rejection ratio (CMRR), an important parameter to be defined later. To calculate y_{cm-dm}, we show in Figure 7.19 the small-signal circuit equivalent to the scheme in Figure 7.17(a) for $v_{g1} = v_{g2} = v_g = v_{icm}$. Note that the small-signal model is composed of the superposition of differential and common-mode models. The common-mode small-signal parameters are $g_m = (g_{m1} + g_{m2})/2$, and so on, whereas the differential parameters are $\Delta g_m = g_{m1} - g_{m2}$, and so on.

Now, using nodal analysis, we obtain

$$v_s = \frac{g_m}{y_t/2 + g_{ms} - g_{mb}} v_g = \frac{g_m}{y_t/2 + g_m + g_{md}} v_g \qquad (7.5.19)$$

when $v_b = v_s$ (substrate connected to the source) and

$$v_s = \frac{g_m}{y_t/2 + g_{ms}} v_g \qquad (7.5.20)$$

when $v_b = 0$ (substrate connected to ground). The differential output current is readily calculated from the circuit in Figure 7.19 by eliminating the common-mode components of the voltage-controlled current sources. The remaining circuit gives

$$y_{cm-dm}|_{v_b=v_s} = \frac{i_{od}}{v_g}\bigg|_{v_b=v_s} = \frac{(y_t/2 + g_{md})\Delta g_m}{y_t/2 + g_m + g_{md}} \cong (y_t/2 + g_{md})\frac{\Delta g_m}{g_m} \qquad (7.5.21)$$

and

$$y_{cm-dm}|_{v_b=0} = \frac{i_{od}}{v_g}\bigg|_{v_b=0} = \frac{(y_t/2 + g_{ms})\Delta g_m - g_m \Delta g_{ms}}{y_t/2 + g_{ms}} \cong \frac{y_t}{2}\frac{\Delta g_m}{ng_m} - g_m\frac{\Delta n}{n}. \qquad (7.5.22)$$

In (7.5.21) we have assumed that the variations in g_{md} are negligibly small compared with other transconductance variations. Both approximations in (7.5.21) and (7.5.22) are quite acceptable since the transistor transconductance is usually much higher than the magnitude of the output admittance of the current source, except for very high frequencies. Recalling that the relationship between transconductance and current in a MOSFET operating in saturation is

$$g_m = \frac{2I_S}{n\phi_t}\left(\sqrt{1 + \frac{I_D}{I_S}} - 1\right), \qquad (7.5.23)$$

the normalized transconductance mismatch is

$$\frac{\Delta g_m}{g_m} = -\frac{\Delta n}{n} + \frac{1}{2}\frac{\frac{\Delta I_D}{I_D}\left(\sqrt{1 + \frac{I_D}{I_S}} + 1\right) + \frac{\Delta I_S}{I_S}\left(\sqrt{1 + \frac{I_D}{I_S}} - 1\right)}{\sqrt{1 + \frac{I_D}{I_S}}}. \qquad (7.5.24)$$

$\Delta g_m/g_m$ is dependent not only on the structural asymmetry of the differential pair, through Δn and ΔI_S, but also on the functional asymmetry, through ΔI_D [14], which in turn is dependent on the differential input voltage. The transconductance mismatch reduces to

$$\frac{\Delta g_m}{g_m} = -\frac{\Delta n}{n} + \frac{\Delta I_D}{I_D} \qquad (7.5.25)$$

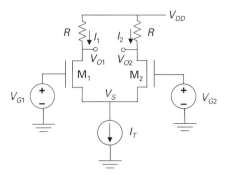

Fig. 7.20   A resistive-load differential amplifier.

in weak inversion and to

$$\frac{\Delta g_m}{g_m} = -\frac{\Delta n}{n} + \frac{1}{2}\left(\frac{\Delta I_D}{I_D} + \frac{\Delta I_S}{I_S}\right) \tag{7.5.26}$$

in strong inversion. When the input voltage equals the offset voltage, $\Delta I_D/I_D = 0$.

7.5.2   Resistive-load differential amplifiers

The output currents of the source-coupled pair can be converted into voltages by means of resistors, as shown in Figure 7.20. In the linear region of the transfer characteristic, the output currents and, consequently, the output voltages V_{O1} and V_{O2} are proportional to the differential input voltage and $180°$ out of phase.

   The use of resistors in amplifiers in CMOS technologies is, generally, not recommended because they expend too much silicon area and have very large parasitic capacitance. Furthermore, the voltage gain of a resistive load amplifier is limited by the power-supply voltage [1]. Consider, for instance, the differential amplifier in Figure 7.20. The differential voltage gain $A_d = (v_{o2} - v_{o1})/(v_{g1} - v_{g2}) = g_m R$. Note that output V_{O1} is in phase with input V_{G2} and $180°$ out of phase with input V_{G1}. The maximum value of g_m for a given drain current is $I_D/(n\phi_t)$, which is achieved in weak inversion. Thus, the maximum voltage gain is $A_{dmax} = RI_D/(n\phi_t)$. Clearly, the voltage drop RI_D cannot exceed the supply voltage V_{DD}. For the 0.35-μm CMOS technology, $V_{DD} = 3.3$ V is the maximum supply voltage and, therefore, assuming that $n\phi_t \cong 30$ mV, a voltage gain of the order of 100 would be unfeasible with the topology of Figure 7.20.

   To overcome the problem of voltage gain limited by the supply voltage in a resistive-load amplifier, transistors operating as current sources can be used as loads to benefit from their high (non-linear) output resistance. This type of amplifier, however, does not have a stable common-mode output voltage. This problem can be solved by the use of a common-mode feedback scheme, which will be discussed in the chapter on operational amplifiers.

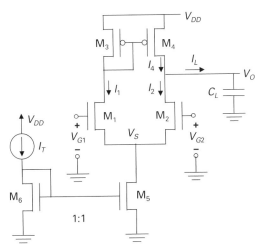

Fig. 7.21   A current-mirror-load differential amplifier.

7.5.3   Current-mirror-load differential amplifiers

A popular configuration of differential amplifiers uses a current mirror for the load, as shown in Figure 7.21.

We start by analyzing the differential amplifier as an operational transconductance amplifier (OTA), a device that converts a differential input voltage (V_{ID}) into an output current (I_L). Note that this asymmetrically loaded amplifier converts a differential input voltage into a single-ended output. For the purpose of a first-order analysis, let us assume that the pairs M_1–M_2, M_3–M_4, and M_5–M_6 are perfectly matched and all transistors operate in saturation. Also, let us assume that the output node (V_O) is connected to a constant-voltage source and that the Early voltage of all transistors is infinite. If $V_{ID} = V_{G1} - V_{G2} = 0$, then $I_1 = I_2$ and the (ideal) current mirror forces I_4 to be equal to I_3; thus, $I_L = 0$. If $V_{ID} > 0$, then $I_1 > I_2$ and, since $I_4 = I_1$ we have $I_L > 0$. The current transfer characteristic of the OTA resembles the solid curve in Figure 7.18. The gate of M_1 is the non-inverting input since an increase in V_{G1} in relation to V_{G2} forces an outgoing current at the amplifier output. When a resistive load is connected to the output, the output voltage will be in phase with the gate voltage of M_1. Similarly, the gate of M_2 is the inverting input since an increase in V_{G2} with respect to V_{G1} will give rise to an incoming current at the output.

7.5.3.1   Voltage transfer characteristics

The plot of the output voltage in terms of the input voltages is shown in Figure 7.22. For $V_{ID} = 0$, the currents in M_1 and M_2 are the same and the output voltage, for matched pairs of transistors, equals the gate voltage of the current-mirror transistors. If we increase V_{ID}, this will force I_1 to be greater than I_2 and, consequently, the gate voltage of M_3 will decrease, which tends to increase I_4. However, for static conditions, the currents I_2 and I_4 must be equal, thus, the output voltage has to increase in such a way that the equality

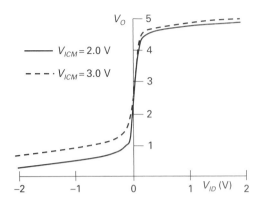

Fig. 7.22 Voltage transfer characteristics of the differential amplifier of Figure 7.21 for common-mode input voltages of 2 V and 3 V and $V_{DD} = 5$ V.

$I_2 = I_4$ holds. Note that the gate of M_1 is the non-inverting input since an increase in V_{G1} will force V_O to increase. The voltage transfer characteristic of the differential amplifier has a center region for which the gain is relatively high and two regions where the gain is low. In the center region both M_2 and M_4 operate in saturation, a condition that leads to the voltage gain having a high value. The segment of the curve to the right of the high-gain region corresponds to M_4 in the triode region while the segment to the left corresponds to the situation in which M_2 is in the triode region. Note that the range of the output voltage associated with the high-gain region is dependent on the common-mode input voltage. This drawback of the current-mirror-load amplifier inhibits its use in applications for which the variation in the common-mode input voltage is significant.

7.5.3.2 The common-mode input range

The proper operation of the differential amplifier requires the transistors that compose it to be in saturation in order to provide it with a high gain and to keep the tail current at an almost constant value. The minimum voltage required at node V_S to keep M_5 in saturation is

$$V_{Smin} = V_{DSsat5} = \phi_t\left(\sqrt{1 + I_T/I_{S5}} + 3\right). \tag{7.5.27}$$

The common-mode input voltage V_{ICM} required to keep M_5 in saturation must be greater than a minimum value V_{ICMmin}. Using the relationship between V_S and the common-mode input voltage V_G given by (7.5.7) together with the minimum value of V_S given by (7.5.27), we find that

$$\frac{V_{ICMmin} - V_{T0N}}{n_N} = \phi_t\left[\sqrt{1 + \frac{I_T}{I_{S5}}} + \sqrt{1 + \frac{I_T}{2I_{S1}}} + 1 + \ln\left(\sqrt{1 + \frac{I_T}{2I_{S1}}} - 1\right)\right]. \tag{7.5.28}$$

Now let us determine the maximum input voltage for which M_1 remains in saturation. When the differential amplifier operates in the high-gain region, the current provided by

M_5 is equally split between M_1 and M_2. The drain voltage $V_{D1} = V_{D3} = V_{G3}$ can be calculated from the application of the UICM to M_3, yielding

$$V_{D1} = V_{DD} + V_{T0P} - n_P \phi_t \left[\sqrt{1 + \frac{I_T}{2I_{S3}}} - 2 + \ln\left(\sqrt{1 + \frac{I_T}{2I_{S3}}} - 1 \right) \right]. \qquad (7.5.29)$$

If V_{ICM} increases, the voltage V_S at the common-source node also increases and the drain-to-source voltage of M_1 decreases, since V_{D1} is a constant given by (7.5.29). M_1 remains in saturation as long as $V_{DS1} > V_{DSsat1}$. By calculating V_{S1} from (7.5.7) and V_{D1} from (7.5.29), and using the expression for the saturation voltage in terms of the inversion level, we find that

$$\frac{V_{ICMmax} - V_{T0N}}{n_N} = V_{DD} + V_{T0P} + \phi_t \left[-5 + \ln\left(\sqrt{1 + \frac{I_T}{2I_{S1}}} - 1 \right) \right]$$

$$- n_P \phi_t \left[\sqrt{1 + \frac{I_T}{2I_{S3}}} - 2 + \ln\left(\sqrt{1 + \frac{I_T}{2I_{S3}}} - 1 \right) \right]. \qquad (7.5.30)$$

The appropriate operation of the differential amplifier is achieved when both M_2 and M_4 also operate in saturation. In the next section, we determine the range of the output voltage appropriate to obtain this condition.

7.5.3.3 The output voltage range

The maximum output voltage, which is limited by the drain-to-source saturation voltage of M_4, is

$$V_{omax} = V_{DD} - |V_{DSsat4}| = V_{DD} - \phi_t\left(\sqrt{1 + I_T/2I_{S4}} + 3 \right). \qquad (7.5.31)$$

The output voltage, which must be greater than a minimum value V_{omin} in order for M_2 to remain in saturation, is given by

$$V_{omin} = V_S + V_{DSsat2} = \frac{V_{ICM} - V_{T0N}}{n_N} - \phi_t \left[-5 + \ln\left(\sqrt{1 + \frac{I_T}{2I_{S2}}} - 1 \right) \right]. \qquad (7.5.32)$$

Equation (7.5.32) imposes a severe constraint on the use of a differential amplifier as a single-stage amplifier, since the minimum output voltage is dependent on the common-mode input range. This constraint can be seen in Figure 7.22, which shows that the minimum output voltage required in order for the amplifier to remain in the high-gain region is higher for $V_{ICM} = 3$ V than for $V_{ICM} = 2$ V. We will see in Chapter 8 that the inclusion of some transistors in the topology of Figure 7.21 allows an increase in the output voltage range of the differential amplifier, but at the expense of higher power dissipation.

7.5.3.4 The offset voltage

If matching is ideal in the differential amplifier of Figure 7.21, the output voltage equals V_{G3} and $I_L = 0$ when the differential input voltage is equal to zero. The input-referred offset voltage is the differential input voltage required to give $I_L = 0$.

Using once again the simple model that represents transistor mismatch by imbalances between the threshold voltages and between the normalization currents, we write

$$V_{T03} = V_{T04} + \Delta V_{TOP}, \qquad I_{S3} = I_{S4} + \Delta I_{SP},$$

$$V_{TOP} = \frac{V_{T03} + V_{T04}}{2}, \qquad I_{SP} = \frac{I_{S3} + I_{S4}}{2}. \qquad (7.5.33)$$

A mismatch between the Early voltages also affects the offset voltage but is not of major impact, owing to the much lower sensitivity of the current to the Early voltage than to the threshold voltage or to the specific current. The common-mode input voltage also affects the offset voltage [15] since the tail current is dependent on it, but we will ignore this effect due to its secondary importance to the offset voltage.

We have previously calculated the offset voltage owing to mismatch in the pair M_1–M_2 (see (7.5.12)). To calculate the overall offset voltage we must add the offset term V_{OS2} due to mismatch in the pair M_3–M_4.

The difference $(I_3 - I_4)$ is given by

$$\Delta I_D = I_3 - I_4 \cong \frac{\partial I_D}{\partial I_{SP}} \Delta I_{SP} + \frac{\partial I_D}{\partial V_{TOP}} \Delta V_{TOP}$$

$$= \frac{I_T \Delta I_{SP}}{2 \, I_{SP}} + g_{mp} \Delta V_{TOP}. \qquad (7.5.34)$$

If the pair M_1–M_2 is perfectly matched, we can compute V_{OS2} by noting that $I_1 - I_2 = I_3 - I_4$ is required in order to have $I_L = 0$. The difference $(I_1 - I_2)$ referred to the differential input becomes

$$V_{OS2} = \frac{I_1 - I_2}{g_{mn}} = \Delta V_{TOP} \frac{g_{mp}}{g_{mn}} + \frac{I_T/2}{g_{mn}} \frac{\Delta I_{SP}}{I_{SP}}. \qquad (7.5.35)$$

Finally, by adding the offset voltage calculated in (7.5.12) (with the inclusion of the subscript N to make it clear that we are referring to n-channel devices) to that calculated in (7.5.35), we find the overall offset voltage of the differential amplifier as

$$V_{OS} = \Delta V_{TON} - \frac{I_T}{2g_{mn}} \frac{\Delta I_{SN}}{I_{SN}} + \Delta V_{TOP} \frac{g_{mp}}{g_{mn}} + \frac{I_T}{2g_{mn}} \frac{\Delta I_{SP}}{I_{SP}}$$

$$= \Delta V_{TON} + \Delta V_{TOP} \frac{n_N \sqrt{1 + I_T/(2I_{SN})} + 1}{n_P \sqrt{1 + I_T/(2I_{SP})} + 1}$$

$$+ n_N \phi_t \frac{\sqrt{1 + I_T/(2I_{SN})} + 1}{2} \left(\frac{\Delta I_{SP}}{I_{SP}} - \frac{\Delta I_{SN}}{I_{SN}} \right). \qquad (7.5.36)$$

For the sake of simplicity, let us assume that the inversion levels of the input transistors and of the current mirror are about the same. In this case, the offset voltage is equally affected by the threshold-voltage mismatch of the input pair of transistors and that of the current-mirror load. The term of the offset voltage corresponding to the specific-current mismatch is proportional to the current-to-transconductance ratio, which increases for high inversion levels. As previously explained, the specific-current mismatch does not play an important role in the determination of mismatch in transistors except at high

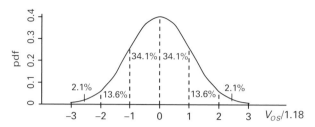

Fig. 7.23   The probability density function (pdf) of the offset voltage for the differential amplifier of Example 7.8.

inversion levels. Thus, as expected, the offset voltage of differential amplifiers is mostly affected by threshold-voltage mismatch.

Example 7.8

The nominal specific current of transistors M_1 through M_4 in Figure 7.21 is $I_{SN} = I_{SP} = 1\ \mu A$. The transistor dimensions are $W_{1,2} = 40\ \mu m$, $L_{1,2} = 2.5\ \mu m$ and $W_{3,4} = 100\ \mu m$, $L_{3,4} = 2.5\ \mu m$. Assume that $n_N = n_P = 1.2$, $\phi_t = 25\ mV$, $A_{VTN} = A_{VTP} = 10\ mV\ \mu m$, and $A_{ISN} = A_{ISP} = 2\%\ \mu m$. Determine the standard deviation of the offset voltage for $I_T = 0.2$ and $200\ \mu A$, assuming that the mismatch parameters are uncorrelated. Draw the graph of the probability density function (pdf) of the offset voltage for $I_T = 0.2\ \mu A$, assuming that it follows a normal distribution.

Answer

(a) Using

$$\sigma^2(V_{OS}) = \sigma^2(V_{T0N}) + \sigma^2(V_{T0P})$$

$$+ \left(n\phi_t \frac{\sqrt{1 + I_T/(2I_{SN})}}{2} \right)^2 [\sigma^2(I_{SN}) + \sigma^2(I_{SP})],$$

$$\sigma^2(V_{T0N(P)}) = \frac{A_{VTN(P)}^2}{W_{1(3)}L_{1(3)}} \quad \text{and} \quad \sigma^2(I_{SN(P)}) = \frac{I_{SN(P)}^2}{W_{1(3)}L_{1(3)}},$$

we find that

$$\sigma(V_{OS}) \cong \begin{cases} 1.18\ mV & \text{for } I_T = 0.2\ \mu A, \\ 1.25\ mV & \text{for } I_T = 200\ \mu A. \end{cases}$$

Now, assuming that the offset voltage follows a Gaussian distribution, we plot in Figure 7.23 the pdf of the offset voltage normalized to its standard deviation, which is 1.18 mV for $I_T = 0.2\ \mu A$. Note that, for the normal distribution, 68.2% of the samples are expected to have an offset voltage that lies within $\pm 1.18\ mV$.

7.5.3.5   Small-signal analysis – differential voltage gain

To analyze the small-signal behavior of the differential amplifier, we assume that it is operating in a closed-loop configuration, which implies that the differential input voltage is very small. In this case, the differential amplifier can be assumed to operate linearly and, consequently, the output can be calculated as the superposition of the effects of both the differential and the common-mode input voltage. We will first analyze the response of the amplifier to the differential input voltage.

A common simplification employed to ease the analysis of the response to the differential input voltage is to assume that the differential amplifier is perfectly balanced. This is the case, for example, when the differential pair shown in Figure 7.20, loaded with a pair of perfectly identical resistors, is perfectly matched. In the balanced case, the half-circuit concept [1], [6], [16] can be used to calculate the amplifier response. This concept consists of determining the overall ac performance of the amplifier by analyzing the behavior of one half of the circuit for a set of symmetric (common-mode) and antisymmetric (differential-mode) input signals [6].

To calculate the response to the differential input voltage, we first make $v_{icm}=0$; thus $v_{g1}=v_{i1}=v_{id}/2$ and $v_{g2}=v_{i2}=-v_{id}/2$. If the differential amplifier were fully balanced, the ac voltages and currents in each half of the circuit would vary antisymmetrically; thus, the voltage at node v_s would remain unchanged, i.e. $v_s=0$. However, the current-mirror-load differential amplifier is not balanced and, rigorously speaking, the assumption of ac ground at the sources of M_1–M_2 is not valid. However, as verified in [17], the response of the simplified circuit in Figure 7.21 to a differential input voltage is approximately equal to that obtained by assuming that node V_S is an ac ground.

We first make a simplified ac analysis of the circuit in Figure 7.21, assuming that the internal capacitances are equal to zero. As previously discussed, we also assume that V_S is an ac ground for differential signals. The differential transconductance of the amplifier is calculated by connecting the output node to a constant-voltage source and assuming that the current mirror has a unity gain. In this case, the amplifier transconductance G_m is given by

$$G_m = \left.\frac{i_l}{v_{id}}\right|_{V_O=0} \simeq \frac{g_{m1}v_{id}/2 - (-g_{m2}v_{id}/2)}{v_{id}} = g_{m1} = g_{m2}. \tag{7.5.37}$$

The admittance seen at the output node is $Y_O = sC_L + G_O$, where $G_O = g_{md2} + g_{md4}$. Therefore, the differential voltage gain is

$$A_d = \frac{v_o}{v_{id}} = \frac{g_{m1}}{G_o}\frac{1}{1+sC_L/G_o}. \tag{7.5.38}$$

In a first-order analysis, the differential voltage gain is a single-pole function with a low-frequency value of g_{m1}/G_O and unity-gain angular frequency, in rad/s, of g_{m1}/C_L. Note that, for low-frequency signals, the gate of M_1 is the non-inverting input since an increase in V_{G1} will cause an increase in the output voltage. On the other hand, the gate of M_2 is the inverting input since the output voltage is 180° out of phase in relation to V_{G2}.

The analysis of the differential gain in the case of mismatched transistor pairs will not be shown here, since mismatch does not affect significantly the differential voltage gain. For more details, the reader is referred to [1].

To obtain a more accurate result for the frequency response of the differential voltage gain, let us now include the frequency-dependent response of the current mirror, for

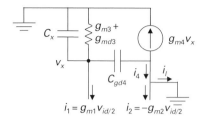

Fig. 7.24   The small-signal model of the current-mirror load for the calculation of the frequency response of the differential voltage gain.

which the small-signal model is given in Figure 7.24. Capacitance C_x is the sum of all capacitances connected between node v_x and the ac ground; C_x includes the gate-to-source and gate-to-bulk capacitances of M_3 and M_4 as well as C_{db3} and C_{db1}. Using the approximation $g_{m3} + g_{md3} \cong g_{m3}$, the amplifier transadmittance Y_d can be written as

$$Y_d = \frac{i_l}{v_{id}} \cong g_{m1} \frac{1 + s(2C_{gd4} + C_x)/(2g_{m3})}{1 + s(C_{gd4} + C_x)/g_{m3}}. \qquad (7.5.39)$$

As can be seen in (7.5.39), the action of the current mirror over just one half of the input voltage introduces a pole–zero doublet with frequency values very close to each other. The doublet does not affect significantly the frequency response of the amplifier, but can greatly modify the time response [15], [18], [19], especially when the doublet frequencies are much lower than the amplifier unity-gain frequency. When the doublet frequencies are of the order of the amplifier unity-gain frequency, their effect on the settling time can be inferred from their phase margin, a topic that will be introduced in Chapter 8.

Usually $C_x \gg C_{gd4}$; thus, (7.5.39) can be simplified to

$$Y_d = \frac{i_l}{v_{id}} \cong g_{m1} \frac{1 + sC_x/(2g_{m3})}{1 + sC_x/g_{m3}}. \qquad (7.5.40)$$

Note that the zero frequency is twice that of the pole frequency. Using the result in (7.5.40), the differential voltage gain is now written as

$$A_d = \frac{v_o}{v_{id}} = \frac{Y_d}{Y_o} = \frac{g_{m1}}{G_o} \frac{1}{1 + sC_L/G_o} \frac{1 + sC_x/(2g_{m3})}{1 + sC_x/g_{m3}}. \qquad (7.5.41)$$

A graph of the differential voltage-gain magnitude versus frequency for the differential amplifier in Figure 7.21 is shown in Figure 7.25. Two plots are shown, one assuming the current mirror to have a frequency-independent gain (dashed line), and the other (solid line) with the inclusion of a pole–zero doublet with frequencies lower than the amplifier unity-gain frequency.

7.5.3.6   Small-signal analysis – common-mode gain and CMRR

The common-mode gain is a measure of the sensitivity of the output voltage to the common-mode input voltage. For the single-ended amplifier of Figure 7.21 we define the common-mode voltage gain as

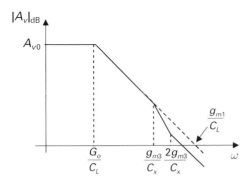

Fig. 7.25   The differential voltage-gain magnitude versus frequency for the differential amplifier with an ideal current mirror (dashed line) and a real current mirror (solid line).

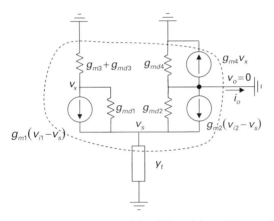

Fig. 7.26   The small-signal model of the differential amplifier used to calculate the common-mode transadmittance.

$$A_{cm} = \frac{v_o}{v_{icm}}. \tag{7.5.42}$$

The common-mode voltage gain can be determined from both the common-mode transadmittance and the output admittance using

$$A_{cm} = Y_{cm}/Y_o. \tag{7.5.43}$$

To calculate Y_{cm} we use the circuit in Figure 7.26. Note that we have assumed that the body of M_1 and M_2 is connected to the source. We have also eliminated the capacitances associated with node v_x and C_{db2} since these capacitances do not play a major role in the frequency response of the common-mode transadmittance.

Noting that $v_{i1} = v_{i2} = v_{icm}$, the KCL equations for nodes v_s and v_x are

$$\begin{pmatrix} y_t + g_{md1} + g_{md2} + g_{m1} + g_{m2} & -g_{md1} \\ -(g_{m1} + g_{md1}) & g_{md1} + g_{md3} + g_{m3} \end{pmatrix} \begin{pmatrix} v_s \\ v_x \end{pmatrix}$$
$$= \begin{pmatrix} g_{m1} + g_{m2} \\ -g_{m1} \end{pmatrix} v_{icm}. \tag{7.5.44}$$

By applying the KCL equations to the closed surface in Figure 7.26 we find that

$$i_o + y_t v_s + (g_{m3} + g_{md3} + g_{m4})v_x = 0. \tag{7.5.45}$$

To make explicit the effects of mismatch on the common-mode transconductance, we write below the transconductances and conductances as a combination of common-mode components and differential-mode components:

$$
\begin{aligned}
g_{m1} &= g_{mi} + \Delta g_{mi}/2, & g_{m2} &= g_{mi} - \Delta g_{mi}/2 \\
g_{m3} &= g_{ml} + \Delta g_{ml}/2, & g_{m4} &= g_{ml} - \Delta g_{ml}/2 \\
g_{md1} &= g_{mdi} + \Delta g_{mdi}/2, & g_{md2} &= g_{mdi} - \Delta g_{mdi}/2 \\
g_{md3} &= g_{mdl} + \Delta g_{mdl}/2, & g_{md4} &= g_{mdl} - \Delta g_{mdl}/2
\end{aligned}
\tag{7.5.46}
$$

Finally, using expressions (7.5.44)–(7.5.46), we find that

$$Y_{cm} = \frac{i_o}{v_{icm}} = Y_{cms} + Y_{cmr}, \tag{7.5.47}$$

where

$$Y_{cms} = -y_t \frac{g_{mi}(g_{mdi} + g_{mdl})}{g_{ml}(y_t + 2g_{mi})} \tag{7.5.48}$$

is the systematic component of Y_{cm}, whereas

$$Y_{cmr} = \frac{g_{mi}}{y_t + 2g_{mi}} \left[y_t \left(\frac{\Delta g_{mi}}{g_{mi}} - \frac{\Delta g_{ml}}{g_{ml}} \right) + 2g_{mdi} \left(\frac{\Delta g_{mi}}{g_{mi}} - \frac{\Delta g_{mdi}}{g_{mdi}} \right) \right] \tag{7.5.49}$$

is the random component of Y_{cm}.

The sum of the systematic and random components of the common-mode transadmittances gives the total common-mode transadmittance as

$$Y_{cm} = \frac{g_{mi}}{y_t + 2g_{mi}} (y_t \alpha_1 + 2g_{mdi}\alpha_2), \tag{7.5.50}$$

where

$$\alpha_1 = \frac{\Delta g_{mi}}{g_{mi}} - \frac{\Delta g_{ml}}{g_{ml}} - \frac{g_{mdi} + g_{mdl}}{g_{ml}}, \qquad \alpha_2 = \frac{\Delta g_{mi}}{g_{mi}} - \frac{\Delta g_{mdi}}{g_{mdi}}.$$

Note that α_1 has both systematic and random components, whereas α_2 has a random component only. In a well-designed amplifier, both factors α_1, $\alpha_2 \ll 1$. We now assume that the current-source admittance y_t can be represented as a parallel association of a resistor and a capacitor, i.e. $y_t = G_t + sC_t$. For practical differential amplifiers we have $g_{mi} \gg G_t/2$; thus, using (7.5.50), Y_{cm} becomes

$$Y_{cm} \cong \frac{G_t \alpha_1 + 2g_{mdi}\alpha_2}{2} \frac{1 + sC_t \alpha_1/(G_t \alpha_1 + 2g_{mdi}\alpha_2)}{1 + sC_t/(2g_{mi})}. \tag{7.5.51}$$

The common-mode transadmittance is a first-order rational function for which the low-frequency value is composed of both the systematic component G_{cms} and the random component G_{cmr} given below:

$$G_{cms} = -\frac{G_t}{2} \frac{g_{mdi} + g_{mdl}}{g_{ml}},$$

$$G_{cmr} = \frac{G_t}{2} \left(\frac{\Delta g_{mi}}{g_{mi}} - \frac{\Delta g_{ml}}{g_{ml}} \right) + g_{mdi} \left(\frac{\Delta g_{mi}}{g_{mi}} - \frac{\Delta g_{mdi}}{g_{mdi}} \right).$$

(7.5.52)

The systematic common-mode transconductance equals half the conductance of the current source multiplied by the error due to the non-zero output conductances of M_2 and M_3. G_{cms} is negative since an increase in the common-mode input voltage causes the load current to decrease. To understand why, suppose that v_{icm} increases; this causes the voltage at the common-source node to increase, which, in turn, increases the tail current due to the non-zero output conductance of the current source. The currents I_1 flowing through both M_1 and M_3, and I_2 through M_2, also increase, but the increase in I_1 is slightly smaller than that in I_2 (note that the drain of M_1 is connected to a diode-connected transistor whereas the drain of M_2 is connected to a constant-voltage source). Also, the increase in the current flowing through M_4 is smaller than the increase in the current flowing through M_3 (note that the current-mirror gain is slightly below unity due to the non-zero output conductance of M_3). Consequently, the increase in the load current is negative due to the larger increase in the current through M_2 compared with that through M_4. The factor $(g_{mdi} + g_{mdl})/g_{ml}$ corresponds to the relative error of the currents flowing through M_2 and M_4. The common-mode current that flows through M_1 and M_2 is equal to $(G_t/2)v_{icm}$, as can be observed by drawing the scheme of the common-mode half circuit; this is the reason for the presence of the factor $G_t/2$ in (7.5.52).

The random component of the common-mode transconductance is dependent on the mismatch of the devices as given in (7.5.52) as well as on the output conductances of both the current source and the input transistor. Note that a high-impedance current source is not sufficient for a low common-mode transconductance; input transistors with low output conductance are also required in order to achieve low G_{cm}.

The output voltage (or current) of the differential amplifier, written as the super-position of the effects of the differential and common-mode input voltages, is

$$v_o = A_d v_{id} + A_{cm} v_{icm}$$

(7.5.53)

or

$$i_o = Y_d v_{id} + Y_{cm} v_{icm}.$$

(7.5.54)

An important parameter of differential amplifiers is the common-mode rejection ratio (CMRR), which is the ratio of the differential to common-mode gain or of the differential to common-mode admitttance [1]:

$$\text{CMRR} = \left| \frac{A_d}{A_{cm}} \right| = \left| \frac{Y_d}{Y_{cm}} \right|.$$

(7.5.55)

Usually, the CMRR is expressed in dB. Since $Y_d \cong g_{mi}$ up to frequencies close to the unity-gain frequency and Y_{cm} is given by (7.5.51), we can directly plot the dependence of the CMRR on frequency, as shown in Figure 7.27. Note that, in general, both the

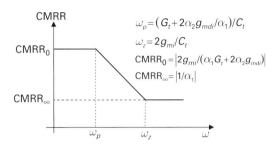

CMRR

$$\omega_p = (G_t + 2\alpha_2 g_{mdi}/\alpha_1)/C_t$$
$$\omega_z = 2g_{mi}/C_t$$
$$\text{CMRR}_0 = |2g_{mi}/(\alpha_1 G_t + 2\alpha_2 g_{mdi})|$$
$$\text{CMRR}_\infty = |1/\alpha_1|$$

Fig. 7.27   The common-mode rejection ratio versus frequency for the differential amplifier loaded with a current mirror.

frequency location of the pole and the value of the CMRR at low frequencies are dependent on mismatch.

The common-mode rejection ratio is the sum of the systematic and random components, as written below

$$\frac{1}{\text{CMRR}} = \left| \frac{Y_{cmr}}{Y_d} + \frac{Y_{cms}}{Y_d} \right| = \frac{1}{\text{CMRR}_r} + \frac{1}{\text{CMRR}_s}. \tag{7.5.56}$$

Example 7.9

The nominal specific current of transistors M_1 through M_4 in Figure 7.21 is $I_{SN} = I_{SP} = 1\,\mu\text{A}$. Assume that $n_N \phi_t = n_P \phi_t = 40\,\text{mV}$, $I_T = 16\,\mu\text{A}$, and the Early voltage of all transistors is $V_A = 8\,\text{V}$. (a) Calculate the differential transconductance and differential voltage gain. (b) Calculate the systematic common-mode transconductance and common-mode voltage gain. (c) What is the systematic CMRR? (d) Calculate the standard deviation of the random component of the CMRR assuming that the relative standard deviations of the transconductances and output conductances of the input and load transistors are all equal to 1% and uncorrelated.

Answer

(a) Refer to Figure 7.21. We first calculate the inversion levels of the input and load transistors. The current through M_1 to M_4 is $8\,\mu\text{A}$; thus, their inversion levels are $i_f = 8\,\mu\text{A}/1\,\mu\text{A} = 8$. The transconductances of the input and load transistors are calculated from the universal transconductance-to-current ratio:

$$g_{mi} = g_{ml} = \frac{2I_S}{n\phi_t}\left(\sqrt{1 + i_f} - 1\right) = \frac{2 \times 10^{-6}}{40 \times 10^{-3}}2 = 100\,\mu\text{A/V}.$$

The output conductances are

$$g_{mdi} = g_{mdl} = (I_T/2)/V_A = 8 \times 10^{-6}/8 = 1\,\mu\text{A/V},$$
$$G_t = I_T/V_A = 16 \times 10^{-6}/8 = 2\,\mu\text{A/V}.$$

The differential transconductance is $g_{mi} = 100\,\mu\text{A/V}$, whereas the differential voltage gain is

$$A_{d0} = \frac{g_{mi}}{g_{mdi} + g_{mdl}} = \frac{100 \times 10^{-6}}{(1+1) \times 10^{-6}} = 50.$$

(b) The systematic common-mode transconductance and voltage gain are

$$G_{cms} = -\frac{G_t}{2}\frac{g_{mdi} + g_{mdl}}{g_{ml}} = -\frac{2 \times 10^{-6}}{2}\frac{(1+1) \times 10^{-6}}{100 \times 10^{-6}} = -20 \text{ nA/V},$$

$$A_{cms0} = \frac{G_{cms}}{g_{mdi} + g_{mdl}} = \frac{-20 \times 10^{-9}}{(1+1) \times 10^{-6}} = -10^{-2}.$$

(c) The systematic CMRR at low frequency is

$$CMRR_0 = \left| \frac{G_d}{G_{cms}} \right| = \frac{100 \times 10^{-6}}{20 \times 10^{-9}} = 5 \times 10^3,$$

$$CMRR_0|_{dB} = 20 \log CMRR_0 = 74 \text{ dB}.$$

(d) The random component of the CMRR is

$$CMRR_{r0} = \left| \frac{Y_{d0}}{Y_{cmr0}} \right| = \frac{G_d}{G_{cmr}}.$$

Using (7.5.52) and noting that the standard deviations of the (trans)conductances are uncorrelated, we can write

$$\sigma^2(G_{cmr}) = \left(\frac{G_t}{2}\right)^2 \left(\sigma^2\left(\frac{\Delta g_{mi}}{g_{mi}}\right) + \sigma^2\left(\frac{\Delta g_{ml}}{g_{ml}}\right)\right)$$
$$+ g_{mdi}^2 \left(\sigma^2\left(\frac{\Delta g_{mi}}{g_{mi}}\right) + \sigma^2\left(\frac{\Delta g_{mdi}}{g_{mdi}}\right)\right),$$

which gives $\sigma(G_{cmr}) = 20\text{nA/V}$.

7.5.3.7 Small-signal analysis – power-supply rejection ratio

In integrated circuits the power-supply buses are susceptible to noise and interference, which contribute as an added disturbance at the output of analog blocks. In modern integrated circuits, analog and digital circuits usually coexist on the same chip. Even though separate analog and digital supply lines are often employed in integrated circuits, there is always some coupling of the digital noise to analog supply lines [1], [20]. This digital noise will inevitably be added to the output of the differential amplifier, as illustrated in Figure 7.28, where v_{dd} and v_{ss} are noise signals superimposed on the positive and negative supply lines. Assuming that the common-mode input voltage is zero, the output of the block shown in Figure 7.28 is

$$v_o = A_d v_{id} + A_{vdd} v_{dd} + A_{vss} v_{ss}, \tag{7.5.57}$$

where A_{vdd} and A_{vss} are the small-signal voltage gains from the positive and negative power supplies to the output, respectively. Ideally, both A_{vdd} and A_{vss} equal zero.

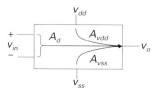

Fig. 7.28　An illustration of the coupling of noise signals from the supply lines to the output of an analog block.

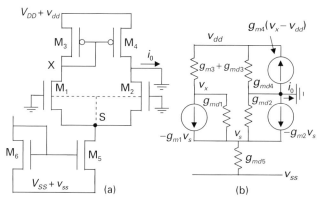

Fig. 7.29　(a) A differential amplifier scheme to determine the small-signal transconductances from the positive and negative power supplies to the output. (b) The small-signal low-frequency equivalent circuit.

The power-supply rejection ratio (PSRR) of a differential amplifier is a measure of the relative influences of the differential input and the supply-voltage variation on the output. Simply stated, the PSRRs relating to each power supply are defined as

$$\mathrm{PSRR}_{vdd} = \frac{A_d}{A_{vdd}} = \frac{G_d}{G_{vdd}}, \qquad \mathrm{PSRR}_{vss} = \frac{A_d}{A_{vss}} = \frac{G_d}{G_{vss}}, \tag{7.5.58}$$

where G_{vdd} and G_{vss} are the small-signal transconductances from the positive and negative power supplies to the output, respectively.

Figure 7.29 will be used to determine the PSRR for the positive and negative power supplies [21]. For the calculation of the small-signal transconductance G_{vdd} from the positive power supply to the output we make $v_{ss}=0$. The (almost) symmetric circuit allows us to conclude that $v_s \ll v_{dd}$ (you can prove this inequality using circuit analysis and noting that $g_m \gg g_{md}$). Also, for a practical differential amplifier, $g_{m3} \gg g_{md1}$. Under these conditions, the voltage at node X is $v_x \cong v_{dd}$. Therefore, a current equal to $g_{md1}v_{dd}$ flows through g_{md1}. Since $v_s \ll v_{dd}$, we have $g_{md1}v_{dd} \cong (g_{m1}+g_{m2})v_s$, i.e. one half of $g_{md1}v_{dd}$ flows into the output node via the source of M_2 and the other half through current mirror M_3–M_4 [21]. An additional current component equal to $g_{md4}v_{dd}$ also flows into the output node, thus yielding

$$G_{vdd} = \frac{i_o}{v_{dd}} = g_{md1} + g_{md4}. \tag{7.5.59}$$

The small-signal transadmittance from the positive power supply to the output [21] is

$$Y_{vdd} = \frac{i_o}{v_{dd}} \cong g_{md1} + g_{md4} + s(C_{p1} + C_{p4}), \tag{7.5.60}$$

where C_{p1} and C_{p4} are the parasitic drain-to-bulk capacitances in parallel with g_{md1} and g_{md4}. In (7.5.60) we have assumed that the remaining capacitances do not affect appreciably the transadmittance, which is true when the frequency is not very high. Using (7.5.60), the positive power-supply rejection ratio becomes

$$PSRR_{vdd} = \frac{Y_d}{Y_{vdd}} \cong \frac{g_{m1}}{|g_{md1} + g_{md4} + j\omega(C_{p1} + C_{p4})|}. \tag{7.5.61}$$

For low frequencies, $PSRR_{vdd}$ equals the differential voltage gain, whereas for higher frequencies the rejection of spurious signals at the positive power supply becomes poorer owing to the parasitic capacitances.

To compute the small-signal transadmittance from the negative power supply to the output, we make $v_{dd} = 0$. The small-signal source v_{ss} injects a current equal to $v_{ss}(g_{md5} + sC_{p5})$ into node S (note that $v_s \ll v_{ss}$), which is split into two components: the current that flows through M_1 is proportional to $g_{m1}/(g_{m1} + g_{m2})$, whereas the one that flows through M_2 is proportional to $g_{m2}/(g_{m1} + g_{m2})$. The current in M_1 is equal to that of M_3, which, in turn, is replicated at M_4 with a current gain equal to $g_{m4}/(g_{m3} + g_{md3})$. Thus, the transadmittance from the negative power supply to the output is

$$Y_{vss} = \frac{i_o}{v_{ss}} \cong (g_{md5} + sC_{p5})\left(\frac{-\Delta g_{mi}}{2g_{mi}} + \frac{\Delta g_{ml} - g_{md3}}{2g_{ml}}\right), \tag{7.5.62}$$

where the subscripts i and l refer to the input and load devices, respectively (refer to (7.5.46) for the meaning of the symbols). The negative-power-supply rejection ratio

$$PSRR_{vss} = \frac{G_d}{G_{vss}} \cong \frac{g_{m1}}{\left|(g_{md5} + j\omega C_{p5})\left(\dfrac{-\Delta g_{mi}}{2g_{mi}} + \dfrac{\Delta g_{ml} - g_{md3}}{2g_{ml}}\right)\right|} \tag{7.5.63}$$

is dependent on the output impedance of the current source and on the mismatch of both input and load transistor pairs. Note that the $PSRR_{vss}$ of an n-channel-input differential amplifier is typically much greater than the $PSRR_{vdd}$.

7.5.3.8 Noise

The differential amplifier is the dominant source of noise in operational amplifiers and, consequently, in many analog integrated circuits. To analyze the noise performance of the differential amplifier loaded with a current mirror, the reader is referred to Figure 7.30, where the current sources i_{n1} through i_{n4} represent the thermal and flicker noise of each corresponding transistor. The current source will add practically no noise to the output since, in a first approximation, its noise current splits equally between M_1 and M_2. Owing to the action of the current mirror, the current in M_4, which is an inverted replica of the current in M_1, will add to that in M_2, resulting in a zero current at the output. Assuming

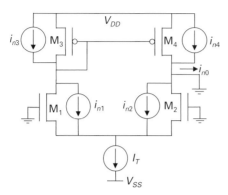

Fig. 7.30   A differential amplifier scheme to determine the equivalent input noise representation.

that the frequency response of the current mirror is flat over the frequency range of analysis, and since the noise currents are uncorrelated, we can write

$$\overline{i_{no}^2} = \overline{i_{n1}^2} + \overline{i_{n2}^2} + \overline{i_{n3}^2} + \overline{i_{n4}^2} = 2\left(\overline{i_{n1}^2} + \overline{i_{n3}^2}\right). \tag{7.5.64}$$

Now, introducing the simplified formulas for the thermal and flicker noise presented in Chapter 4, we find that the output current is

$$\frac{\overline{i_{no}^2}}{\Delta f} = 8kT(g_{m1} + g_{m3}) + \frac{2g_{m1}^2}{C_{ox}'}\left(\frac{K_{F,n}}{W_1 L_1} + \frac{K_{F,p}}{W_3 L_3}\frac{g_{m3}^2}{g_{m1}^2}\right)\frac{1}{f}. \tag{7.5.65}$$

To refer the output noise current to the input, we divide (7.5.65) by g_{m1}^2, thus obtaining

$$\frac{\overline{e_{ni}^2}}{\Delta f} = \frac{8kT}{g_{m1}}\left(1 + \frac{g_{m3}}{g_{m1}}\right) + \frac{2K_{F,n}}{W_1 L_1 C_{ox}'}\left(1 + \frac{K_{F,p}W_3 L_3 g_{m3}^2}{K_{F,n}W_1 L_1 g_{m1}^2}\right)\frac{1}{f}, \tag{7.5.66}$$

where $K_{F,n}$ and $K_{F,p}$ are the flicker-noise coefficients for n-MOS and p-MOS transistors, respectively. Note that the thermal noise is a function of the input and load transconductances, which are independent of transistor area. In general, the transconductance of the input device is set by the required unity-gain frequency of the differential amplifier. Thermal and flicker noise can be reduced by choosing low g_{m3}, but this approach is generally not recommended since a low g_{m3} can lead to a poor frequency response of the current mirror. This, in turn, will degrade the overall performance of the differential amplifier for high frequencies and even adversely affect the settling response. Note that the flicker noise (and also mismatch) is inversely proportional to the device area; thus, sometimes, the designer can play with the dimensions in order to reduce the effect of the load transistors on flicker noise.

For a CMOS amplifier, the effect of the input-referred noise current is usually negligible compared with that of the input-referred noise voltage, particularly for low source-impedance levels. Thus, for most practical situations, assuming that the operating

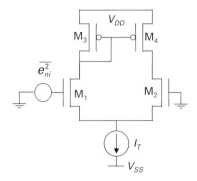

Fig. 7.31   An approximate noise model of the CMOS differential amplifier. The mean-square value of the noise voltage is given by (7.5.66).

frequency is not very high, the noise model for the CMOS differential amplifier in Figure 7.31 is quite acceptable.

7.5.3.9   The slew rate and settling response

The differential amplifier can supply only a maximum amount of current equal to the tail current I_T to charge/discharge the parasitic and load capacitances. As a result, the large-signal response of the amplifier is degraded compared with the small-signal response. To understand why, consider that the amplifier in Figure 7.21 is initially set at $V_{G1} = V_{G2}$ and that the input voltage V_{G1} changes abruptly to a value such that $V_{G1} \gg V_{G2}$. This makes the current flowing through M_1 equal to I_T and that through M_2 equal to zero. As a result, the load current reaches a value equal to the tail current I_T, and the rate of change of the output voltage becomes

$$dV_O/dt = I_T/C_L. \tag{7.5.67}$$

Since the load current cannot be greater than I_T, (7.5.67) gives the maximum rate of change of the output voltage. The output slew rate (SR), defined as the maximum rate of change of the output voltage, is I_T/C_L for the differential amplifier under analysis. Note that for the differential amplifier in Figure 7.21 the maximum charging and discharging currents are both equal to I_T; thus, the positive and negative slew rates are equal. In order to obtain a handy model of the differential amplifier to ease the transient analysis, we convert the non-linear voltage-to-current transfer characteristic of the differential trans-conductor into the piecewise linear characteristic shown in Figure 7.32. According to this simplified model, the transconductor behaves as a linear transconductor when the differential input voltage is less than I_T/g_{m1}; otherwise, the output current is limited to $\pm I_T$.

Just to give an example of how the slew rate affects the transient response of the differential amplifier, let us consider it connected in the unity-gain configuration, as in Figure 7.33. We assume that the differential amplifier can be modeled as the association of the piecewise linear transfer characteristic of Figure 7.32 and the load capacitor

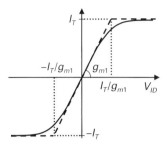

Fig. 7.32   Non-linear (solid line) and piecewise linear (dashed line) transfer characteristics of the differential amplifier.

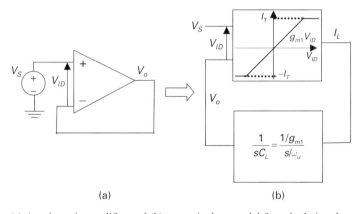

(a)   (b)

Fig. 7.33   (a) A unity-gain amplifier and (b) an equivalent model for calculating the transient response.

represented by the integrator in the block diagram [22]. Note that in this simplified model the output conductance is assumed to be zero. This assumption is quite acceptable for the purpose of the first-order transient analysis to be shown next. Recall that $\omega_u = g_{m1}/C_L$ is the unity-gain frequency of the open-loop voltage gain.

To calculate the settling response of the unity-gain buffer, let us first assume that the input voltage is a step with an amplitude that does not exceed I_T/g_{m1}. In this case, the closed-loop circuit behaves linearly and the relationship between output voltage and input voltage is given by the following differential equation:

$$\frac{dV_o}{\omega_u \, dt} + V_o = V_S. \tag{7.5.68}$$

Assuming that $V_S = V_M u(t)$, where $u(t)$ is the unit step, the output voltage is

$$V_o = V_M[1 - \exp(-\omega_u t)] \tag{7.5.69}$$

when $V_M < I_T/g_{m1}$.

Let us now assume that the magnitude of the step input $V_M > I_T/g_{m1}$. For this condition, the initial differential voltage V_{ID} applied to the input causes the output current to be equal to I_T and the output voltage to increase linearly with time, as shown in Figure 7.34. In this case we have

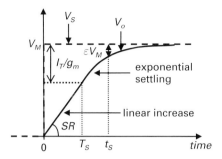

Fig. 7.34   The transient response (solid line) of the unity-gain buffer to a step input (dashed line) with a magnitude that exceeds the linearity range of the differential amplifier [22].

$$v_o = \mathrm{SR} \cdot t = \frac{I_T}{C_L} t. \tag{7.5.70}$$

Equation (7.5.70) holds for $t < T_S$, where T_S is given by

$$T_S = \frac{V_M}{\mathrm{SR}} - \frac{1}{\omega_u}. \tag{7.5.71}$$

For $t > T_S$ the differential equation in (7.5.68) describes the transient response of the unity-gain amplifier. In this situation, the initial condition is $V_o(t = T_S) = V_M - I_T/g_{m1}$; thus, the solution to the differential equation becomes

$$V_o = V_M - \frac{I_T}{g_{m1}} \exp[-\omega_u(t - T_S)]. \tag{7.5.72}$$

The relative dynamic error is the difference between the input and the output voltages normalized with respect to the input voltage,

$$\frac{V_M - V_o}{V_M} = \frac{I_T}{g_{m1} V_M} \exp[-\omega_u(t - T_S)]. \tag{7.5.73}$$

Settling within a relative error equal to ε occurs in a time t_s given [22] by

$$t_s = T_S + \frac{1}{\omega_u} \ln\left(\frac{I_T}{\varepsilon g_{m1} V_M}\right) = \frac{V_M}{\mathrm{SR}} + \frac{1}{\omega_u}\left[\ln\left(\frac{\mathrm{SR}}{\varepsilon \omega_u V_M}\right) - 1\right] \tag{7.5.74}$$

when $V_M > I_T/g_{m1}$, otherwise

$$t_s = \frac{1}{\omega_u} \ln\left(\frac{1}{\varepsilon}\right). \tag{7.5.75}$$

Example 7.10

The nominal specific current of transistors M_1 to M_4 in Figure 7.21 is $I_{SN} = I_{SP} = 1\,\mu\mathrm{A}$. Assume that $n\phi_t = 40\,\mathrm{mV}$, $I_T = 160\,\mu\mathrm{A}$, and $C_L = 4\,\mathrm{pF}$. (a) Calculate the unity-gain frequency of the differential amplifier's voltage gain assuming that the pole frequency introduced by the current mirror is much greater than the amplifier's unity-gain frequency. (b) Calculate the slew rate. (c) Calculate the 0.1% settling time for the unity-gain configuration for V_M within the range 10 mV to 2 V.

Answer

(a) Refer to Figure 7.21. We first calculate the inversion level of the input transistors. The current through both M_1 and M_2 is 80 μA. The inversion level of both M_1 and M_2 is $i_f = 80 \, \mu A/1 \, \mu A = 80$. The transconductance of the input transistor is calculated from the universal transconductance-to-current ratio:

$$g_m = \frac{I_D}{n\phi_t} \frac{2}{\sqrt{1 + i_f} + 1} = \frac{80 \, \mu A}{40 \, mV} \frac{2}{\sqrt{81} + 1} = 0.4 \, mA/V.$$

The unity-gain frequency of the open-loop amplifier is

$$\omega_u = \frac{g_m}{C_L} = \frac{0.4 \, mA/V}{4 \, pF} = 100 \, Mrad/s.$$

(b) The positive (and negative) slew rate is

$$SR = \frac{I_T}{C_L} = \frac{160 \, \mu A}{4 \, pF} = 40 \, V/\mu s.$$

(c) The 0.1% settling time is

$$t_s = \frac{1}{\omega_u} \ln\left(\frac{1}{\varepsilon}\right) = \frac{1}{100 \, Mrad/s} \ln(1000) = 69 \, ns$$

for $V_M < I_T/g_m$, i.e. $V_M < 0.4 \, V$ and the settling time in nanoseconds for $V_M > 0.4 \, V$ is

$$t_s = \frac{V_M}{SR} + \frac{1}{\omega_u} \left[\ln\left(\frac{SR}{\varepsilon \omega_u V_M}\right) - 1 \right] = 25 V_M - 10 + 10 \ln\left(\frac{400}{V_M}\right).$$

7.6   Sizing and biasing of MOS transistors for amplifier design

The transistor used as the active component of an amplifier is biased in saturation, where the reverse current is negligible compared with the forward current. Thus, the following set of equations can be used for the design of MOSFETs operating in saturation:

$$\frac{\phi_t n g_m}{I_D} = \frac{2}{1 + \sqrt{1 + i_f}}, \tag{7.6.1}$$

$$i_f = \frac{I_D}{I_S}, \tag{7.6.2}$$

$$I_S = \mu C'_{ox} n \frac{\phi_t^2}{2} \frac{W}{L}, \tag{7.6.3}$$

$$f_T \cong \frac{\mu \phi_t}{2\pi L^2} 2\left(\sqrt{1 + i_f} - 1\right), \tag{7.6.4}$$

$$\frac{V_{DSsat}}{\phi_t} \cong \sqrt{1 + i_f} + 3. \tag{7.6.5}$$

It is of interest to compare here MOSFET design with bipolar design [23]–[25]. The transconductance-to-current ratio of MOSFETs is given by (7.6.1), whereas, due to the exponential relationship between output current and input voltage, $\phi_t g_m / I_C = 1$ (I_C is the collector current) for bipolar transistors. If g_m is defined in a bipolar design, so is I_C. However, in a MOSFET design, the specification of g_m allows the designer to choose from a range of currents, according to (7.6.1). Equation (7.6.4) gives an approximation for the intrinsic cut-off frequency f_T in terms of the inversion level i_f. Expressions (7.6.1) through (7.6.5) constitute a set of fundamental expressions with which to design MOS amplifiers for any inversion level.

7.6.1 Sizing and biasing of a common-source amplifier

Most of the concepts used to design the intrinsic gain stage shown in Figure 7.3 can be extended to more elaborate blocks such as differential amplifiers or class-AB stages. Using the common-source amplifier as a demonstration tool, we will provide closed expressions for the bias current, the aspect ratio, the saturation voltage, and the intrinsic cut-off frequency. Note that the only technology-dependent parameter used in our methodology is I_{SH}, the sheet-specific current.

In the ideal common-source amplifier in Figure 7.3, the transconductance required to achieve a gain–bandwidth product (GB), in hertz, for a load capacitance equal to C_L is $g_m = 2\pi GB C_L$. Using this equation together with (7.6.1) allows us to write the bias current and the aspect ratio as

$$I_D = I_B = 2\pi GB C_L n \phi_t \frac{1 + \sqrt{1 + i_f}}{2}, \tag{7.6.6}$$

$$\frac{W}{L} = \frac{2\pi GB C_L}{\mu C'_{ox} \phi_t} \left(\frac{1}{\sqrt{1 + i_f} - 1}\right). \tag{7.6.7}$$

Equations (7.6.6) and (7.6.7) show that an infinite set of solutions is available to satisfy the required gain–bandwidth product. Figure 7.35 shows plots of the bias current in (7.6.6) and the aspect ratio in (7.6.7) as functions of i_f. Both curves have been normalized with respect to the corresponding values at $i_f = 8$. A tradeoff between area and power consumption can be reached by an appropriate choice of i_f. It is important to note that the power consumption is low but the aspect ratio is high for low inversion levels. Moreover, the lowest current to meet the specified GB is obtained for weak inversion ($i_f < 1$).

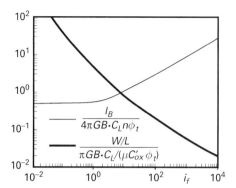

Fig. 7.35   The bias current and aspect ratio versus the inversion coefficient. Both curves are normalized with respect to the values of I_B and W/L at $i_f = 8$.

Inversion levels i_f close to unity can be a good choice when low power is required. On the other hand, a normalized drain current much less than unity leads to a prohibitively high aspect ratio for a negligible reduction in power consumption as compared with the case $i_f \cong 1$.

Assuming that the Early voltage V_A in a first-order approximation is proportional to the transistor channel length ($V_A = V_E L$), the following equation is used to determine the absolute value of the dc voltage gain A_{V0}:

$$A_{V0} = \frac{g_m}{g_{md}} = \frac{g_m}{I_D} V_E L = \frac{V_E L}{n\phi_t} \frac{2}{\sqrt{1 + i_f} + 1}. \tag{7.6.8}$$

7.6.2   The design procedure for a common-source amplifier

Assuming that the gain–bandwidth product and dc gain are the specifications to be met for a given load capacitance, the following design procedure for a common-source amplifier is suggested.

1. For the specified C_L and GB values, use expressions (7.6.6) and (7.6.7) to determine I_D and W/L as functions of the inversion level. Plots of both the bias current and the aspect ratio required to fulfill the gain–bandwidth-product specification are shown in Figure 7.35. Choose a value for i_f resulting in a pair (I_D, W/L) that satisfies the design requirements of power consumption and area.
2. Select dimensions, W and L, of the input transistor to satisfy the aspect ratio determined in step 1. Verify whether the unity-gain frequency of the input transistor is consistent with the gain–bandwidth product specified. In general, the value of f_T must be at least four times higher than GB in order to avoid the parasitic diffusion and overlap capacitances of the MOSFET being of the same order of magnitude as the load capacitance. From (7.6.4) and (7.6.7), we can write

$$W \cong 2\frac{C_L}{LC'_{ox}}\frac{GB}{f_T}. \tag{7.6.9}$$

The diffusion and overlap capacitances are both proportional to W; therefore, the parasitic capacitance of the MOSFET drain is proportional to GB/f_T. Thus, a GB/f_T ratio close to or greater than unity should be avoided since the ratio of the parasitic capacitance to the load capacitance could become very large.

3. Verify whether the gain specification is satisfied. If not, an increase in channel length and/or a reduction in i_f can be employed to increase the gain.

4. Design the current source. For a low-voltage design, the saturation voltage of the current source should be as small as possible. Therefore, inversion coefficients close to unity can be used in order to achieve a good compromise between area and saturation voltage. Note that a low inversion level can lead to a wide transistor for the current source, which, in turn, can add significant parasitic capacitance.

7.6.3   MOSVIEW: a graphical interface for MOS transistor design

Circuit topology and transistor bias current and dimensions are the basic factors that determine the performance of an analog circuit. Frequency response, voltage gain, output voltage swing, power dissipation, and silicon area are some of the performance metrics of amplifiers. The transistor-level design problem consists of determining bias currents and channel widths and lengths such that the amplifier (or other analog circuit) specifications are fulfilled.

Simulation programs are important tools to demonstrate the influence of various parameters on the overall amplifier performance. An interesting design tool for sizing and biasing MOS transistors for use in analog building blocks was presented in [26]. MOSVIEW [27], [28], a transistor-level design tool, was developed to assist integrated-circuit designers with their tasks. MOSVIEW is based on the MOSFET model presented in Chapters 2 and 4 and can be applied to CMOS processes. A valuable characteristic of MOSVIEW is its graphical interface, described by a plane that shows the channel length as the abscissa and the intrinsic low-frequency gain as the ordinate. Families of curves of constant performance, such as transition frequency and current, can be plotted and only the region denominated ***design space*** meets the input specifications. With MOSVIEW we can easily analyze design tradeoffs and develop an insight into transistor sizing for analog circuits. In addition, MOSVIEW can show the user how the transistor parameters influence the amplifier performance and, thus, avoid the need for time-consuming SPICE-like simulations.

Figure 7.36 shows the list of technological input parameters for MOSVIEW along with the design requirements. MOSVIEW handles four cases of common-source amplifiers, namely amplifier I, amplifier II, transconductor, and current mirror, classified according to the input specifications.

Some comments are in order regarding the list of technological parameters. (i) The slope factor n is, in fact, a parameter that is dependent on the gate voltage; however, it is here assumed to be a constant for the sake of simplicity. (ii) The Early voltage per unit length V_E is also a rough simplification for the calculation of the MOSFET output conductance and can be estimated from experimental data on test transistors or from simulations run on SPICE-like simulators. (iii) N_{ot} is a parameter usually not provided by

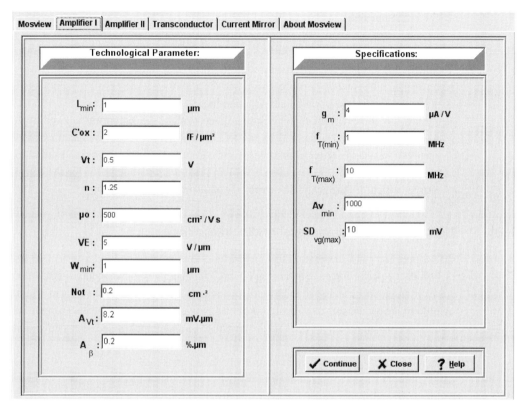

Fig. 7.36   A MOSVIEW window of technological parameters and specifications for the common-source amplifier.

foundries, but it can be determined from the flicker-noise parameter K_F, as shown in the section on flicker noise in Chapter 4.

The specifications to be satisfied by the common-source amplifier exemplified here are a transconductance $g_m = 4\,\mu A/V$, an intrinsic cut-off frequency in the range $f_{Tmin} = 1\,MHz$, $f_{Tmax} = 10\,MHz$, and a minimum value for the voltage gain $A_{Vmin} = 1{,}000$.

Figure 7.37 gives the low-frequency gain against the transistor channel length. The ascending straight line represents the maximum allowable gain ($i_f \to 0$ in (7.6.8)) for the specified technological parameters. The bottommost curve shows the variation in the gain in terms of the channel length for $f_T = f_{Tmax}$, while the uppermost parabola corresponds to $f_T = f_{Tmin}$. The white region represents the design space, the region for which the specifications are met. The specifications shown on the right of Figure 7.37 correspond to the mark selected inside the design space. The bias current is about 0.58 μA and the transistor dimensions are $W = 8.26\,\mu m$ and $L = 36.9\,\mu m$. The list also contains the power-spectral density of the thermal noise and the corner frequency as well as the standard deviation of the gate voltage if the transistor is in the diode configuration.

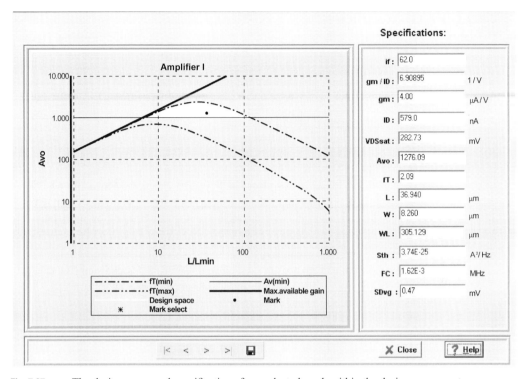

Fig. 7.37.   The design space and specifications for a selected mark within the design-space contour.

7.7   Reuse of MOS analog design

The scaling down of physical dimensions of transistors has resulted in a continuous reduction in the power–delay product, an important measure of the performance of digital circuits. On the other hand, analog circuits do not have a clear benefit when technology is scaled down [29]. In fact, simply reducing transistor dimensions can adversely affect some properties or functions of analog circuits. The migration of an analog design to a new technology is not straightforward and usually involves a complete redesign [30].

The MOSFET model employed throughout this textbook is particularly suited for design reuse because it can be used for all the MOSFET operating regions and employs a small set of parameters that are associated with device physics. The fundamental equations presented in Chapter 2 are rewritten below, since they contain the basic information required to define scaling rules for analog design. For a transistor operating in saturation we have

$$ng_m\phi_t/I_D = 2/\left(1 + \sqrt{1 + i_f}\right), \tag{7.7.1}$$

$$i_f = I_D/I_S, \tag{7.7.2}$$

$$I_S = \mu C'_{ox} n \frac{\phi_t^2}{2} \frac{W}{L}, \tag{7.7.3}$$

$$g_{md} = I_D/V_A = I_D/(V_E L), \tag{7.7.4}$$

$$f_T \cong \frac{g_m}{2\pi[(1/2)C'_{ox}WL]} = \frac{\mu\phi_t}{\pi L^2}\left(\sqrt{1+i_f} - 1\right). \tag{7.7.5}$$

7.7.1   Effects of scaling on analog circuits

For some applications, the basic specifications for the gain–bandwidth product (GB) and the dynamic range (DR), defined in [30] as the maximum signal-to-noise ratio (SNR), should be maintained for a voltage amplifier designed in a scaled-down (or scaled-up) technology [31]. The methodology for design reuse developed next is based on keeping constant values for both SNR and GB, but other criteria could be used for migration of the design to a scaled-down technology.

For the sake of simplicity, we will apply the rules for design reuse to a common-source amplifier, composed of a transistor, an ideal current source, and a load capacitance C_L. In this amplifier, the low-frequency voltage gain and GB are, respectively,

$$A_{V0} = -g_m/g_{md} = -(g_m/I_D)V_E L, \tag{7.7.6}$$

$$GB = g_m/(2\pi C_L). \tag{7.7.7}$$

In practice, one should limit the frequency of operation of the MOSFET, and thus the value of GB/f_T, to a fraction of f_T, e.g. 1/4 or less. This limitation has two main objectives [25]. The first is to prevent the non-quasi-static effects (related to the channel length) from affecting the frequency response of the amplifier for frequencies approaching GB (in moderate and strong inversion the lumped model we have used so far is approximately valid for frequencies up to the order of $f_T/4$). The second is to prevent both the input capacitance and the parasitic output capacitance from becoming a significant fraction of the load capacitance.

When a sinusoidal signal voltage with a peak-to-peak value equal to V_{pp} is applied across C_L, the signal power is $V_{pp}^2/8$. To keep distortion levels within acceptable values, V_{pp} must be less than the total supply voltage V_{DD}. To simplify matters, we assume here that the maximum value of V_{pp} can be equal to V_{DD}. On the other hand, the minimum power N associated with thermal noise is limited by C_L according to the expression $N = kT/C_L$ [30] (see Problem 4.4), where $k = 1.38 \times 10^{-23}$ J/K is Boltzmann's constant and T is the absolute temperature. In this case, the dynamic range DR is

$$\text{DR} = V_{DD}^2 C_L/(8kT). \tag{7.7.8}$$

We conclude from Equations (7.7.7) and (7.7.8) that, to keep the same DR and GB values, both C_L and g_m must increase by the square of the supply-voltage scaling factor (K_V^2) as long as the voltage supply is scaled down by a factor equal to K_V.

In the following we present the procedure for reusing the design of analog circuits [29]. A set of very simple expressions allows the calculation of the transistor dimensions and bias of a given circuit in a new-generation technology, starting from a previous design of the same circuit in an earlier technology.

Table 7.1. MOS transistor resizing rules using generalized scaling $V_{DD} \rightarrow K_V^{-1} V_{DD}$, $C'_{ox} \rightarrow K_{ox} C'_{ox}$, $L \rightarrow K_L^{-1} L$, and $V_E \rightarrow K_E V_E$

Quantity	Constant inversion-level scaling	Channel-length scaling	
L	1	K_L^{-1}	
W	$K_V^2 K_{ox}^{-1}$	$K_V^2 K_{ox}^{-1} K_L$	
		Weak inversion	Strong inversion
I_D	K_V^2	K_V^2	$K_V^2 K_L^{-2}$
i_f	1	K_L^{-2}	K_L^{-4}
A_{V0}	K_E	$K_E K_L^{-1}$	$K_E K_L$

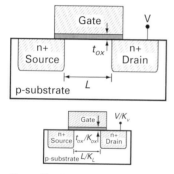

Fig. 7.38   Generalized scaling.

7.7.2   Analog resizing rules

To generalize the resizing rules for MOS analog-design reuse, we now define different scaling factors for the voltage supply (K_V), oxide thickness (K_{ox}), and channel length (K_L), as Figure 7.38 shows. We also include the scaling factor K_E associated with the Early voltage per unit length, which is expected to increase as technology is scaled down due to the increase in substrate doping concentration.

In designing transistors for analog circuits, designers usually consider three independent parameters: channel width W, channel length L, and bias current I_D [15]. For the following resizing rules we maintain the original dynamic-range and gain–bandwidth-product specifications. As long as the amplifier frequency specification (GB) does not change in the new technology, we also maintain the transistor's original transition frequency. Constant-inversion-level and channel-length scaling approaches are the strategies presented next for reusing analog cells. Table 7.1 lists the transistor-resizing rules.

7.7.2.1   Constant-inversion-level scaling

As Equation (7.7.1) shows, g_m/I_D is a function of the inversion level i_f. Because n is almost insensitive to technology, constant-inversion-level scaling keeps g_m/I_D constant. The scaling factor for g_m is K_V^2, so the current must also be multiplied by K_V^2. Assuming that the carrier mobility is constant (the reader is referred to [29] for considerations on

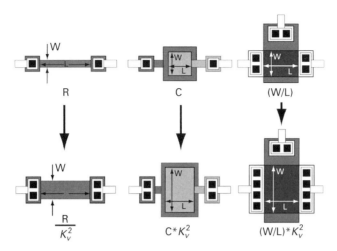

Fig. 7.39   Component redesigning in the same technology for a lower supply voltage ($K_{ox}=1$) (after [29]).

non-constant-mobility scaling), it follows from (7.7.5) that L must remain the same in order for the transition frequency to remain constant. We deduce the scaling factor for W from Equations (7.7.2) and (7.7.3). In constant-inversion-level scaling, static power consumption increases by K_V and the active area is scaled up by $K_V^2 K_{ox}^{-1}$. Finally, Equation (7.7.6) shows that the voltage gain is proportional to V_E; that is, the voltage gain will increase slightly.

An interesting case is the redesign of an analog cell for a lower supply voltage in the same technology ($K_{ox}=1$, $K_E=1$). Table 7.1 demonstrates that the new analog circuit can be viewed as a width-scaled replica of the original circuit [32]. In effect, as the channel width scales up by K_V^2, transistor lengths and current densities remain the same as in the original design. Figure 7.39 illustrates the redesign of an analog circuit, showing how to scale the three main components of a CMOS integrated circuit to comply with the specifications when the supply voltage is scaled down by K_V.

7.7.2.2   Channel-length scaling: $L \rightarrow L/K_L$

Channel-length scaling ($L \rightarrow L/K_L$) is a natural choice to take advantage of the smaller dimensions of a new-generation technology. For the same set of analog specifications, channel-length scaling implies a reduction of the inversion level, which can be advantageous in analog circuits.

We first scale down the channel length by the factor K_L. As follows from Equation (7.7.5), for the transition frequency to remain the same as in the original design, the scaling factor for W must be $K_V^2 K_{ox}^{-1} K_L$. Using the scaling factors for L and W and Equation (7.7.3), we note that the normalization current I_S is scaled by a factor of $K_V^2 K_L^2$. We calculate the inversion-level reduction associated with channel-length scaling from Equation (7.7.1). Finally, we derive the scaling factor for I_D from (7.7.2).

We now consider the important case in which the transistors operate in strong inversion and the channel shortens by a factor of K_L. Using the strong-inversion ($i_f > 100$)

Table 7.2. Simulated performance of a common-source amplifier in 1.2-μm and 0.35-μm CMOS technologies ($K_V = 2.5$, $K_{ox} = 3$, and $K_L = 3.4$)

	i_f	W (μm)	L (μm)	I_D (μA)	$\frac{g_m}{I_D}$ (V^{-1})	A_{V0} (dB)	GB (MHz)	Noise (μV)
Original design[a]	24	160	10	6.0	10.7	67.3	1.03	11.2
Constant-i_f scaling[b]	24	323	10	37.5	8.1	64.3	0.80	4.9
Channel-length scaling[b]	0.8	1109	2.8	14.6	22.1	63.3	0.86	4.9

[a] 1.2-μm Technology.
[b] 0.35-μm Technology.

approximation for Equation (7.7.5), we derive the inversion-level scaling factor K_L^{-4}. A scaling factor of $K_V^2 K_L^{-2}$ for I_D follows from (7.7.2). In channel-length scaling in strong inversion, the inversion level decreases and the g_m/I_D ratio is multiplied by K_L^2. Compared with constant inversion-level scaling, static power consumption decreases by a factor of K_L^2. Finally, as Equation (7.7.6) shows, the voltage gain increases by $K_E K_L$.

The results for weak inversion in Table 7.1 are easily derived from the weak-inversion ($i_f < 1$) approximation of Equation (7.7.5). Reference [29] gives a list of redesign equations for the general case of arbitrary inversion level.

The choice between constant-inversion-level and channel-length scaling depends on several factors. Strong-inversion operation is not power-efficient, but it is unavoidable in some high-frequency applications. On the other hand, moderate inversion achieves the best compromise between consumption and speed [2]. Consequently, designers can more conveniently reuse a design for transistors operating in strong inversion by using channel-length scaling to reduce the inversion level and thus decrease power consumption, rather than using constant-i_f scaling. For transistors operating in moderate inversion, designers can use either channel-length or constant-i_f scaling, depending on the specific requirements of the design.

7.7.2.3 A design-reuse example

As an example, we here design a common-source amplifier with a p-channel driver biased by an ideal current source and with $C_L = 10$ pF as the load, for a 1.2-μm CMOS technology and 5-V power supply. Using both the constant-i_f and the channel-length scaling rule, we recalculate the amplifier load capacitance ($C_L = 62.5$ pF), transistor dimensions, and bias current for a 0.35-μm technology and 2-V power supply. Our noise calculation assumes the amplifier operates with a feedback factor of 1.

Table 7.2 shows the results of the simulation to verify the suitability of the scaling rules applied in this design. For constant-i_f scaling, power consumption increases by K_V and the specifications are almost the same as in the original design. Channel-length scaling leads to lower power consumption than does constant-i_f scaling. The results show that the variation in low-frequency gain (A_{V0}) is not large. However, designers should not rely completely on simulation results to assess voltage gain because the output conductance is generally not well modeled in circuit simulators. Finally, due to variations in the mobility and slope factor, some fine tuning may be required after the scaling rules have been applied.

Problems

7.1 Assume that in Figure 7.3 M_1 operates in saturation and the output voltage is kept constant by replacing capacitor C_L with a constant voltage source. (a) Using the unified current-control model (UICM) for the long-channel transistor, determine the first-, second-, and third-order derivatives of the output current, in terms of the input voltage and of the inversion level. Assume that the slope factor n is constant. (b) If the input voltage is a sine wave $V_I = V_G + V_M \sin(\omega_o t)$, where V_G is a dc voltage for which M_1 remains in saturation, calculate the second- and third-order harmonic distortion in terms of the inversion level using the result of (a). Determine the relative values of the second- and third-order harmonic distortion for inversion levels of 0.1, 1, 10, 100, and 1000.

7.2 Analyze the frequency response of the voltage gain of the common-source amplifier in Figure 7.6(a) for $R_S \neq 0$. Determine the poles, assuming that the dominant-pole approximation is valid. What condition needs to be imposed on the small-signal parameters to obtain a non-dominant pole greater than the unity-gain frequency?

7.3 For the CMOS inverter circuit in Figure 7.8, $W_1 = 10\,\mu m$, $W_2 = 25\,\mu m$, $L_1 = L_2 = 2.5\,\mu m$, $V_{DD} = 3.3\,V$, $V_{TON} = 0.5\,V$, $V_{TOP} = -0.7\,V$, $C'_{ox} = 5\,fF/\mu m^2$, $\mu_N = 400\,cm^2/V$ per s, $\mu_P = 200\,cm^2/V$ per s, and $C_L = 2\,pF$. For the sake of simplicity, assume that $n_1 = n_2 = 1.2$, $\phi_t = 25\,mV$, $V_{A1} = V_{A2} = 9\,V$. (a) Calculate the voltage at which $V_O = V_I$. (b) Calculate the output voltage range at which M_1 and M_2 remain in saturation. (c) What is the voltage gain? (d) What is the input voltage range at which the output voltage remains within the voltage range calculated in (b)? (e) What are the approximate maximum and minimum rates of change of the output? (f) What is the PSD of the input-referred thermal noise? (g) What is the input-referred flicker noise when $N_{ot,p} = N_{ot,n} = 3 \times 10^7\,cm^{-2}$? (h) What is the corner frequency?

7.4 Calculate the dc voltage gain, pole, and unity-gain frequency of the common-gate amplifier when $R_S = 0$.

7.5 The bias current of the voltage follower (body of M_1 tied to source) in Figure 7.12 is $I_B = 99\,\mu A$. For the sake of simplicity, let us assume that the specific currents are not dependent on the gate voltage and that their values are $I_{S1} = I_{S2} = I_{S3} = 1\,\mu A$. Also, assume that the Early voltage of all transistors is $V_A = 10\,V$ and that the slope factor of transistor M_1 is $n = 1.2$ (in this voltage-follower configuration the gate-to-bulk voltage is almost constant over the input voltage range; thus, the value of n is not dependent on the input voltage). (a) What is the approximate maximum value of the input voltage at which M_2 remains in saturation? (b) What is the small-signal voltage gain in terms of the small-signal parameters? (c) What is the numerical value of the voltage gain? Assume that $V_{TOP} = -0.6\,V$ and $\phi_t = 25\,mV$.

7.6 Regulated cascode. For the circuit shown in Figure 7.15, do the following. (a) Derive a formula to calculate the inversion level of M_3 in terms of the inversion level of M_1, to bias M_1 on the edge of saturation. (b) Determine the range of i_{f3} for $0.1 < i_{f1} < 1000$ using the result of (a) and the corresponding range of V_{G1} for $0.1 < i_{f1} < 1000$. (c) Find combinations of $(W/L)_3$ and I_{B3} for the lowest value found for i_{f3} and comment on the

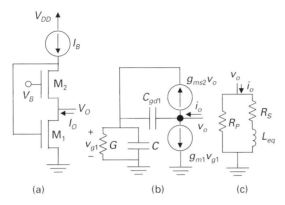

Fig P7.7. (a) A CMOS active inductor [33]. (b) The small-signal equivalent circuit excluding the output conductance and the output capacitance of M_1. (c) The approximate equivalent impedance.

results. (d) Verify the correctness of expression (7.4.6). Calculate the dc voltage gain for $W/L = 10\,\mu\text{m}/1\,\mu\text{m}$ for all transistors, $I_{B3} = 2\,\mu\text{A}$, $I_{B2} = 200\,\mu\text{A}$. Use $V_{TO} = 0.35\,\text{V}$, $I_{SH} = 200\,\text{nA}$, $n = 1.1$, $\phi_t = 26\,\text{mV}$, and $V_E = 10\,\text{V}/\mu\text{m}$.

7.7 An active circuit that emulates an inductive component is shown in Figure P7.7 [33]. In fact, the circuit shown is equivalent to an LC tank, but for the sake of clarity we have omitted both the output conductance and the output capacitance from the small-signal equivalent circuit shown in Figure P7.7(b). (a) Demonstrate that the impedance seen by node v_o is equivalent to the circuit shown in Figure P7.7(c). (b) Calculate the approximate values of R_S, R_P and L_{eq} in terms of the small-signal parameters. Assume that $R_P \gg R_S$. (c) Assuming that all the extrinsic capacitances are equal to zero, determine the self-resonance frequency ω_0 of the LC tank and the intrinsic quality factor Q_0 at the resonance frequency. Show that $\omega_0 \simeq \sqrt{\omega_{T1}\omega_{T2}}$; $Q_0 \simeq \sqrt{\omega_{T1}/\omega_{T2}}$ [33], where ω_T is the transistor's intrinsic unity-gain frequency.

7.8 (a) Determine the minimum common-mode input voltage of the source-coupled pair of Figure 7.17 when the bulk is connected to the source. (b) Determine the maximum common-mode input voltage of a source-coupled pair of p-MOS transistors for cases of both bulk connected to source and bulk connected to V_{DD}.

7.9 Redraw the circuit of Figure 7.29(b) for the case in which the body of the input transistors is tied to V_{SS} and recalculate the low-frequency negative power-supply rejection ratio.

7.10 Calculate the transient response of the unity-gain buffer in Figure 7.33, assuming that the differential amplifier can be represented as in Figure 7.33 but with the integrator replaced with $(1/g_{m1})[1/(s/\omega_u + 1/A_0)]$, where A_0 is equal to $g_{m1}/g_o \gg 1$ and g_o is the output conductance. Calculate the transient response for both the cases of an input signal step less than and greater than I_T/g_{m1}. What is the steady-state difference between the output and input voltages?

7.11 Using the complete low-frequency small-signal MOSFET model with transconductances g_{ms}, g_{mg}, g_{mb}, and g_{md}, verify that the low-frequency circuit associated with Figure 7.12(b) and the circuit in Figure 7.29(b) are appropriate for low-frequency analysis of the source follower and the differential amplifier shown in the respective figures.

References

[1] P. R. Gray, P. J. Hurst, S. H. Lewis, and R. G. Meyer, *Analysis and Design of Analog Integrated Circuits*, 4th edn., New York: John Wiley & Sons, 2001.

[2] F. Silveira, D. Flandre, and P. G. A. Jespers, "A g_m/I_D based methodology for the design of CMOS analog circuits and its application to the synthesis of a silicon-on-insulator micropower OTA," *IEEE Journal of Solid-State Circuits*, vol. **31**, no. 9, pp. 1314–1319, Sep. 1996.

[3] K. Bult and R. Blauschild, "CMOS op-amp design," *Advanced Engineering Course on Transistor-Level Analog IC Design*, Lausanne, EPFL, Oct. 2002.

[4] P. Horowitz and W. Hill, *The Art of Electronics*, 2nd edn., Cambridge: Cambridge University Press, 1989.

[5] E. Vittoz, "Micropower techniques," in *Design of Analog-Digital VLSI Circuits for Telecommunications and Signal Processing*, 2nd edn., ed. J. E. Franca and Y. Tsividis, Englewood Cliffs, NJ: Prentice-Hall, 1994, pp. 53–96.

[6] A. B. Grebene, *Bipolar and MOS Analog Integrated Circuit Design*, Hoboken, NJ: Wiley Interscience, 2003.

[7] D. A. Johns and K. Martin, *Analog Integrated Circuit Design*, New York: Wiley, 1997.

[8] V. C. Vincence, C. Galup-Montoro, and M. C. Schneider, "A high-swing MOS cascode bias circuit," *IEEE Transactions on Circuits and Systems II*, vol. **47**, no. 11, pp. 1325–1328, Nov. 2000.

[9] P. Aguirre and F. Silveira, "Bias circuit design for low-voltage cascode transistors," *Proceedings of SBCCI 2006*, pp. 94–98, Sep. 2006.

[10] K. Bult and G. J. G. M. Geelen, "A fast-settling CMOS op amp for SC circuits with 90-dB DC gain," *IEEE Journal of Solid-State Circuits*, vol. **25**, no. 6, pp. 1379–1384, Dec. 1990.

[11] B. J. Hosticka, "Improvement of the gain of MOS amplifiers," *IEEE Journal of Solid-State Circuits*, vol. **14**, no. 6, pp. 1111–1114, Dec. 1979.

[12] E. Säckinger and W. Guggenbühl, "A high-swing, high-impedance MOS cascode circuit," *IEEE Journal of Solid-State Circuits*, vol. **25**, no. 1, pp. 289–298, Feb. 1990.

[13] H. C. Yang and D. J. Allstot, "An active-feedback cascode current source," *IEEE Transactions on Circuits and Systems.*, vol. **37**, no. 5, pp. 644–646, May 1990.

[14] E. Vittoz, "Elementary building blocks," in *Intensive Summer Course on CMOS VLSI Design: Analog & Digital*, Lausanne, 1989.

[15] K. R. Laker and W. M. C. Sansen, *Design of Analog Integrated Circuits and Systems*, New York: McGraw-Hill, 1994.

[16] P. E. Allen and Douglas R. Holberg, *CMOS Analog Circuit Design*, 2nd edn., New York: Oxford University Press, 2002.

[17] B. Razavi, *Design of Analog CMOS Integrated Circuits*, Boston, MA: McGraw-Hill, 2001.

[18] B. Y. Kamath, R. G. Meyer, and P. R. Gray, "Relationship between frequency response and settling time of operational amplifiers," *IEEE Journal of Solid-State Circuits*, vol. **9**, no. 6, pp. 347–352, Dec. 1974.

[19] G. Palmisano and G. Palumbo, "Analysis and compensation of two-pole amplifiers with a pole-zero doublet," *IEEE Transactions on Circuits and Systems I*, vol. **46**, no. 7, pp. 864–868, July 1999.

[20] P. R. Gray and R. G. Meyer, "MOS operational amplifier design – a tutorial overview," *IEEE Journal of Solid-State Circuits*, vol. **17**, no. 6, pp. 969–982, Dec. 1982.

[21] M. S. J. Steyaert and W. Sansen, "Power supply rejection ratio in operational transconductance amplifiers," *IEEE Transactions on Circuits and Systems*, vol. **37**, no. 9, pp. 1077–1084, Sep. 1990.

[22] J. Dostal, *Operational Amplifiers*, 2nd edn., Boston, MA: Butterworth-Heinemann, 1993.

[23] A. I. A. Cunha, M. C. Schneider, and C. Galup-Montoro, "An MOS transistor model for analog circuit design," *IEEE Journal of Solid-State Circuits*, vol. **33**, no. 10, pp. 1510–1519, Oct. 1998.

[24] R. L. Oliveira Pinto, A. I. A. Cunha, M. C. Schneider, and C. Galup-Montoro, "An amplifier design methodology derived from a MOSFET current-based model," *Proceedings of IEEE ISCAS 1998*, Monterey, CA, USA, pp. I-301–I-304, June 1998.

[25] R. L. O. Pinto, M. C. Schneider, and C. Galup-Montoro, "Sizing of MOS transistors for amplifier design," *Proceedings of of IEEE ISCAS 2000*, Geneva, vol. **4**, pp. 185–188, May 2000.

[26] D. M. Binkley, C. E. Hopper, S. D. Tucker, *et al.*, "A CAD methodology for optimizing transistor current and sizing in analog CMOS design," *IEEE Transactions on Computer-Aided Design*, vol. **22**, no. 2, pp. 225–237, Feb. 2003.

[27] P. Giacomelli, M. C. Schneider, and C. Galup-Montoro, "MOSVIEW: a graphical tool for MOS analog design," *2003 International Conference on Microelectronic Systems Education*, pp. 43–44, May 2003.

[28] P. Giacomelli, C. R. Machado, M. C. Schneider, and C. Galup-Montoro, MOSVIEW: a graphical tool for MOS analog design, http://www.eel.ufsc.br/lci/mosview.

[29] C. Galup-Montoro, M. C. Schneider, and R. M. Coitinho, "Resizing rules for MOS analog-design reuse," *IEEE Design & Test of Computers*, vol. **19**, no. 2, pp. 50–58, Mar.–Apr. 2002.

[30] E. A. Vittoz, "Future of analog in the VLSI environment", *Proceedings of IEEE ISCAS*, pp. 1372–1375, 1990.

[31] H.-T. Ng, R. M. Ziazadeh, and D. J. Allstot, "A multistage amplifier technique with embedded frequency compensation", *IEEE Journal of Solid-State Circuits*, vol. **34**, no. 3, pp. 339–347, Mar. 1999.

[32] B. Nauta, "Analog CMOS low-power design considerations," *Digest of Technical Papers of "Low power-low voltage workshop" at ESSCIRC' 96*, 1996.

[33] Y. Wu, W. Ding, M. Ismail, and H. Olsson, "RF bandpass filter design based on CMOS active inductors," *IEEE Transactions on Circuits and Systems II*, vol. **50**, no. 12, pp. 942–949, Dec. 2003.

8   Operational amplifiers

One of the most useful building blocks for analog integrated circuits is the operational amplifier, or op amp for short. The fundamentals for the op amp were established by Harry Black in 1934 [1], [2]. Black's basic idea to make an amplifier stable over temperature changes and power-supply variations was to first build an amplifier that had more gain (say 40 dB) than the application required and then include negative feedback around this amplifier to stabilize the gain. The use of the term operational amplifier and the systematic presentation of its applications came much later in a 1947 paper [3]. Op amps became popular circuit-building blocks only in the late 1960s with the development of the bipolar analog integrated-circuit technology [4]. Since then, the op amp has been used in a number of applications such as instrumentation amplifiers, continuous-time and switched-capacitor filters, D/A and A/D converters, non-linear analog operators, signal generators, and voltage regulators. This chapter deals with the analysis and design of CMOS operational amplifiers. The design of op amps is an important and sometimes challenging task owing to the specifications to be met and to the boundary conditions such as process, supply voltage, and temperature [5]. We start the chapter with an introduction to definitions, applications, and performance parameters. We then present classical topologies of single-ended op amps together with the analysis of large-signal and small-signal characteristics. Next, we give some introductory concepts on rail-to-rail amplifiers and class-AB output stages. Finally, we introduce fully-differential operational amplifiers and the associated common-mode feedback circuits, which have become progressively more important as CMOS technologies migrate to lower supply voltages.

8.1   Applications and performance parameters

8.1.1   The ideal operational amplifier

The operational amplifier in simplified form is represented in Figure 8.1. It is a five-terminal device, two of them for the power-supply lines, an input pair of terminals, and an output terminal. The ideal op amp is classically represented in textbooks as the differential amplifier in Figure 8.1(a) with a voltage gain that tends toward infinity. The ideal op amp has infinite input impedance and zero output impedance. Also, it has a voltage gain equal to zero for common-mode signals and the bandwidth is infinite. Although the

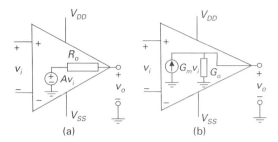

Fig. 8.1   (a) A macromodel of the ideal operational amplifier. (b) An alternative macromodel representation of the operational amplifier.

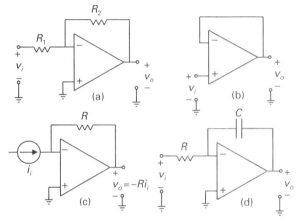

Fig. 8.2   (a) An inverting amplifier. (b) A voltage follower. (c) A transimpedance amplifier. (d) A voltage integrator.

model in Figure 8.1(a) is widely employed in textbooks, here we will use more often the equivalent op-amp model in Figure 8.1(b). The latter has more physical meaning since the description of the main components of op amps, be they bipolar or MOS transistors, as voltage-controlled current sources is much closer to the physics than is considering them as voltage-controlled voltage sources.

When the op amp is represented by the equivalent model of Figure 8.1(b), it is usually called an operational transconductance amplifier or OTA for short. Even though there is no difference between these two types of amplifiers, we will reserve the denomination OTA for the op amps which provide an output current proportional to the differential input voltage. Unless noted otherwise, for the remaining part of this chapter we will simply use the term operational amplifier, in spite of the widespread use of the acronym OTA for operational amplifiers with a voltage at the output.

8.1.2   Basic applications of operational amplifiers

The relevant specifications of an op amp depend on the particular application of the device. In Figure 8.2 we represent some important configurations of op amps for

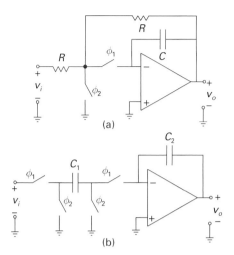

Fig. 8.3   (a) An inverting sample–hold amplifier. (b) A switched-capacitor integrator.

continuous-time circuits. Assuming that the op amps in the topologies shown in Figure 8.2 are ideal, the inverting amplifier has a gain of $-R_2/R_1$, and the voltage follower has a gain of unity. The transimpedance amplifier converts the input current into an output voltage with a transimpedance gain of $-R$. In the voltage integrator, the input resistor converts the input voltage into a current, which is subsequently integrated into the feedback capacitor, giving an output voltage that is the negative integral of the input voltage scaled by the factor $1/(RC)$. The gain–bandwidth product of the op amp, together with the values of the components connected to it, defines the range of frequencies at which the topologies show a behavior close to that of the ideal op amp. Note that the common-mode input range of the buffer amplifier in Figure 8.2(b) is equal to the input range, whereas the common-mode input range of the other topologies is zero. The input-referred offset voltages of the op amps are transferred to the output nodes with gains equal to $(1 + R_2/R_1)$, 1, and 1, for the structures in Figures 8.2(a), (b), and (c), respectively. The offset voltage, as well as the input-referred noise voltage of the op amp in Figure 8.2 (d), is also integrated at the output. The slew rate can be an important specification depending on the output-voltage swing and frequency specifications. In the particular case of the integrator, when the magnitude of the input signal is evenly distributed over its frequency span, the slew rate is not an important specification since the magnitude of the output voltage decreases with the inverse of the frequency. Other specifications such as the CMRR and PSRR are equally important for all topologies, but their minimum values are strongly dependent on the surroundings of the op amps to be designed. The op amps of applications in Figures 8.2(a) and (c) must drive, in general, not only feedback resistors but also capacitive loads (not shown in the figure). On the other hand, the voltage follower and the integrator in Figures 8.2(b) and (d), in some cases, drive capacitive loads only.

Figure 8.3 shows two applications for switched-capacitor circuits, namely the sample–hold (S/H) amplifier and the integrator. Switches ϕ_1 and ϕ_2 cannot close simultaneously, i.e. when one of them is closed the other is open. When the S/H amplifier goes from the hold mode to the sample mode, switch ϕ_1 closes while switch ϕ_2 opens. After a short time, determined by both the RC time constant and the settling time of the op amp, the output becomes an inverted copy of the input. In the hold mode, ϕ_1 is open while ϕ_2 is closed; thus, the charge stored on the capacitor remains constant and, consequently, so does the output voltage.

In sampled-data circuits such as the two shown in Figure 8.3, an important specification is the settling time. When the circuits switch to the sample mode, the output voltage takes some time to change from the previous value to the new value. If the switching frequency is, as in many situations such as in oversampled converters, much higher than the signal frequency, the small-signal settling behavior is sufficient to characterize the transient response of the circuit. However, when the switching frequency is close to the Nyquist frequency (equal to twice the maximum input frequency) the output-voltage swing between two consecutive samples can be high, in which case the slew rate becomes an important factor limiting the settling response. In most sampled-data circuits, the settling behavior of the output is not affected by the slew rate since either the signal is oversampled or the components of the signal at higher frequencies are of small magnitude. It is worth noting that the circuits in Figure 8.3 are subject to coupling of the digital signal (clock signal) that turns the switches on and off to the outputs via power supply lines, via sensitive input nodes of the op amps, or through the substrate.

Example 8.1

Suppose that, for low frequencies, the op amp of the inverting amplifier in Figure 8.2(a) is modeled as shown in Figure E8.1. (a) Determine the output voltage in terms of the input and offset voltages. (b) What is the output voltage for $g_m \rightarrow \infty$? (c) What is the relative gain error compared with the idealized voltage gain $-R_2/R_1$ when $g_m = 10\,\text{mA/V}$, $g_o = 10\,\mu\text{A/V}$, $R_1 = 10\,\text{k}\Omega$, and $R_2 = 100\,\text{k}\Omega$?

Answer
(a) Kirchoff's current law (KCL) for the op-amp inverting input gives

$$\frac{V_I - V_X}{R_1} + \frac{V_o - V_X}{R_2} = 0,$$

where $V_X = -V_{ID}$ is the voltage at the inverting input. The application of KCL to the output node gives

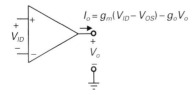

Fig. E8.1   A macromodel of the op amp at low frequencies.

$$g_m(V_X + V_{OS}) + g_o V_o + \frac{V_O - V_X}{R_2} = 0.$$

Using these two KCL equations we find, after some algebra,

$$V_O = -\frac{\dfrac{R_2}{R_1}\left(1 - \dfrac{1}{g_m R_2}\right)V_I + \left(1 + \dfrac{R_2}{R_1}\right)V_{OS}}{1 + \dfrac{1}{g_m}\left[g_o\left(1 + \dfrac{R_2}{R_1}\right) + \dfrac{1}{R_1}\right]}.$$

(b) When $g_m \to \infty$, the expression for the output voltage becomes

$$V_O = -\frac{R_2}{R_1}V_I - \left(1 + \frac{R_2}{R_1}\right)V_{OS},$$

which is the classical result obtained when the open-loop dc gain of the op amp is infinite.

(c) From the expression for the output voltage we find, for small gain error ε, that

$$\varepsilon = \frac{\text{actual gain} - \text{ideal gain}}{\text{ideal gain}} \cong -\frac{1}{g_m}\left[\frac{1}{R_2} + g_o\left(1 + \frac{R_2}{R_1}\right) + \frac{1}{R_1}\right],$$

which gives $\varepsilon \cong -2.2\%$.

Example 8.2

Suppose that, for low frequencies, the op amp of the inverting integrator in Figure 8.2(d) is modeled as shown in Figure E8.1. (a) Determine the differential equation for the output voltage in terms of the input voltage and of the circuit parameters. (b) Find the solution of the differential equation for constant input voltage and initial condition $V_O(t=0)=0$. (c) What is the output voltage for $g_m \to \infty$? (d) What is the error in the output voltage compared with the ideal case of a ramp signal at the output?

Answer

(a) The application of KCL to the op-amp inverting input and output gives the following differential equation relating the output voltage to the input voltage:

$$RC(1 + \varepsilon_1 + \varepsilon_2)\frac{d[V_O + (V_{OS} - V_I/(g_m R))/(1 + \varepsilon_1 + \varepsilon_2)]}{dt} + \varepsilon_1 V_O$$
$$= -(V_I + V_{OS}),$$

where $\varepsilon_1 = g_o/g_m$ and $\varepsilon_2 = 1/(g_m R)$.

(b) Since V_{OS} is, in general, a slowly varying signal and, in this example, V_I is constant, we can write

$$RC(1 + \varepsilon_1 + \varepsilon_2)\frac{dV_O}{dt} + \varepsilon_1 V_O = -(V_I + V_{OS}),$$

which gives the solution

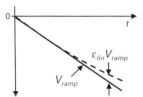

Fig. E8.2   Error in the linearity of the ramp generator.

$$V_O(t) = -\frac{V_I + V_{OS}}{\varepsilon_1}\left[1 - \exp\left(-\frac{\varepsilon_1 t}{(1 + \varepsilon_1 + \varepsilon_2)RC}\right)\right]. \qquad\text{(E8.1.1)}$$

(c) When $g_m \to \infty$, both ε_1 and $\varepsilon_2 \to 0$. Using the approximation $\exp(x) \cong 1 + x$ for $|x| \ll 1$, we find the following result for $V_O(t)$:

$$V_O(t) = -(V_I + V_{OS})\left(\frac{t}{RC}\right).$$

In the ideal case ($V_{OS} = 0$), the output voltage is a ramp the value of which is the negative of the input voltage after a time slot equal to RC, the integrator time constant.

(d) To calculate the error due to the non-infinite op-amp transconductance, we use the approximation $\exp(x) \cong 1 + x + x^2/2$ for $|x| \ll 1$ in (E8.1.1). The output voltage then becomes

$$V_O(t) \cong -(V_I + V_{OS})\left[\frac{t}{(1 + \varepsilon_1 + \varepsilon_2)RC} - \frac{\varepsilon_1}{2}\left(\frac{t}{(1 + \varepsilon_1 + \varepsilon_2)RC}\right)^2\right].$$

For a small value of ε_1, the output voltage is composed of both a linear and a quadratic term in t. The time constant of the linear term is now modified by the errors $\varepsilon_1 = g_o/g_m$ and $\varepsilon_2 = 1/(g_m R)$, which must be small. The quadratic term gives rise to a relative error ε_{lin} (see Figure E8.2) in the linearity of the ramp generator given by

$$\varepsilon_{lin} \cong \frac{\varepsilon_1}{2}\frac{t}{1 + \varepsilon_1 + \varepsilon_2}.$$

8.1.3   Performance parameters

In this subsection we present a set of parameters that represent appropriately the characteristics of the op amp for most of its applications. Most of them have already been described in the section on differential amplifiers of the previous chapter, but for the sake of completeness and review, we present here the performance parameters of differential (or operational) amplifiers, namely the input offset voltage, common-mode input range, output-voltage swing, differential voltage gain, gain–bandwidth product, input impedance, output impedance, noise, slew rate, settling time, common-mode rejection ratio, power-supply rejection ratio, distortion, current consumption, and silicon area. Herein, we

Table 8.1 Measured dc and ac characteristics of the buffered op amp of reference [6]

Supply voltage	5 V
DC gain (163 Ω – 33 pF load)	67.2 dB
Gain–bandwidth product	11.4 MHz
SR^+ (50 Ω – 33 pF load)	20.4 V/μs
SR^- (50 Ω – 33 pF load)	18.8 V/μs
Offset voltage	1 mV
V_{out+}	4.12 V
V_{out-}	0.88 V
Power dissipation	10 mW

Table 8.2 Experimental performance of the operational amplifiers of reference [7](V_{supply} = 1 V, 1.2-μm CMOS technology, C = 15 pF)

Parameter	Amplifier I	Amplifier II
Active die area	0.81 mm^2	0.26 mm^2
I_{DD} (supply current)	410 μA	208 μA
DC gain	87 dB	70.5 dB
Unity-gain frequency	1.9 MHz	2.1 MHz
Phase margin	61°	73°
SR^+	0.8 V/μs	0.9 V/μs
SR^-	1 V/μs	1.7 V/μs
THD (0.5 V_{pp} at 1 kHz)	−54dB	−77dB
THD (0.5 V_{pp} at 40 kHz)	−32dB	−57dB
v_{ni} (at 1 kHz)	267 nV/Hz$^{1/2}$	359 nV/Hz$^{1/2}$
v_{ni} (at 10 kHz)	91 nV/Hz$^{1/2}$	171 nV/Hz$^{1/2}$
v_{ni} (at 1 MHz)	74 nV/Hz$^{1/2}$	82 nV/Hz$^{1/2}$
CMRR	62 dB	58 dB
$PSRR^+$	54.4 dB	56.7 dB
$PSRR^-$	52.1 dB	51.5 dB

consider in all cases that the gate current is negligible; consequently, the input bias current is zero. Just to give the reader an idea of common specifications of op amps, we reproduce here two sets of op-amp specifications provided in the technical literature.

Table 8.1 describes some performance parameters of the buffered operational amplifier of [6], designed to drive an RC load. The op amp was fabricated through the MOSIS service in a 2-μm, n-well CMOS, double-polysilicon, double-metal technology. Most of the dc characteristics were tested with the op amp connected in the voltage-follower configuration. The ac characteristics were measured with a negative-feedback network consisting of a 3-MΩ resistor between the output and the inverting input and a 2.2-nF capacitor between ground and the inverting input. Details on the op-amp topology can be found in [6].

Table 8.2 describes the experimental performance of two operational amplifiers presented in [7]. These two op amps were designed to drive a purely capacitive load. Note that the harmonic distortion is specified for two signal frequencies. The power-spectral density (PSD) of the equivalent input noise voltage is measured at three frequencies. The measurement of the PSD at several frequencies is generally required for op amps since they will mostly be employed in applications for which the noise comprises not only thermal noise (flat spectrum) but also flicker ($1/f$) noise.

8.2   The differential amplifier as an operational amplifier

Some topologies of differential amplifiers studied in Chapter 7, usually referred to as single-stage operational amplifiers, are reviewed next.

8.2.1   The simple-stage differential amplifier

The differential amplifier in its simplest version is shown in Figure 8.4. We recall here that the differential voltage gain is given by $g_{m1}/(g_{d2}+g_{d4})$. Let us assume for now that, owing to the circuit specifications, we must keep both the amplifier transconductance and the bias current constant. Since the g_m/I_D ratio should remain constant, so should the inversion levels of the transistors. If we use minimum channel lengths for all transistors, the voltage gain will be typically between 10 and 100. This range of gains is, in general, not acceptable for common applications. One way to increase the voltage gain is to increase the transistor channel length (recall that the transistor Early voltage is approximately proportional to the channel length). However, increasing the channel length has two adverse effects: (i) for a constant inversion level, which is the case under analysis, the transistor's unity-gain frequency decreases with the square of the channel length, which means that we cannot increase the channel length above a limit value given by frequency constraints; and (ii) the increase in channel length leads to an increase in channel width when both the inversion level and the current are constant, which, in turn, results in increases in silicon area (the gate area is proportional to L^2) and in leakage through the reverse-biased p–n junctions. Therefore, for relatively high-frequency applications, the

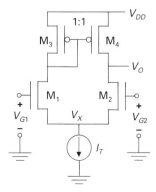

Fig. 8.4   A current-mirror-load differential amplifier.

voltage gain of the topology in Figure 8.4 is limited to a value that is, in general, lower than 100. Another limitation of the topology in Figure 8.4 is the reduced output-voltage swing. For a constant common-mode input range, the output voltage must be within limits that ensure that both M_2 and M_4 remain in saturation. Example 8.3 shows how to calculate the common-mode input range that leads to the optimum output-voltage swing.

Example 8.3

Determine the value of the common-mode input voltage of the differential amplifier shown in Figure 8.4 that maximizes the output-voltage swing. The nominal specific current of transistors M_1 to M_4 in Figure 8.4 is $I_{SN}=I_{SP}=1\,\mu A$, $V_{T0N}=0.5\,V$, $V_{T0P}=-0.7\,V$, $V_{DD}=3.3\,V$, and $I_T=160\,\mu A$. Assume that $n_N=n_P=1.2$ and $\phi_t=25\,mV$. Assume that the current source is implemented with a simple transistor with a specific current equal to twice that of M_1.

Answer

To keep the nominal performance of the differential amplifier, the output voltage must be within limits that ensure that all transistors operate in saturation. The maximum output voltage must be such that

$$V_{omax} = V_{DD} - |V_{DSsat4}|$$

to keep M_4 in saturation. The dc current through M_1–M_4 is $80\,\mu A$. Since the specific current of transistors M_1 to M_4 is $1\,\mu A$, their inversion level is equal to 80. The maximum output voltage is thus

$$V_{omax} = V_{DD} - \phi_t\left(\sqrt{1 + i_{f4}} + 3\right) = 3.0\ \text{V}.$$

On the other hand, the minimum output voltage required for M_2 to remain in saturation is

$$V_{omin} = V_{DSsat2} + V_X.$$

The goal is to minimize V_{omin}. V_{DSsat2} is given by

$$V_{DSsat2} = \phi_t\left(\sqrt{1 + i_{f2}} + 3\right) = 0.3\ \text{V}.$$

V_{Xmin} is the minimum value of V_X required in order to keep the current source I_T in saturation, i.e.

$$V_{Xmin} = V_{DSsatT} = 0.3\ \text{V}.$$

Note that the minimum output voltage required for the current source to remain in saturation is the same as that required for M_1–M_2 because the inversion levels are the same.

To calculate the common-mode input voltage required to set the current source on the verge of saturation, we use the UICM. We assume here that the substrate of the n-channel transistors is connected to ground. In this case, the UICM is written as

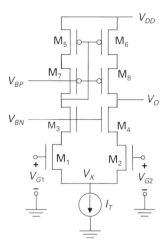

Fig. 8.5   A telescopic-cascode differential amplifier.

$$\frac{V_{ICMmin} - V_{TON}}{n_N} - V_{Xmin} = \phi_t\left[\sqrt{1 + i_{f1}} - 2 + \ln\left(\sqrt{1 + i_{f1}} - 1\right)\right] = 0.227 \text{ V}.$$

Thus, $V_{ICMmin} = 1.13$ V.

The common-mode input voltage can be generated using the bias circuits described in [8], [9] and given in Chapter 5.

Since the common-mode input voltage was optimized to impose a voltage at node V_X equal to the saturation voltage of the tail current source, the output voltage swing for $V_{ICM} = 1.13$ V is, thus, equal to 0.6 V $\leq V_O \leq$ 3.0 V.

8.2.2   The telescopic-cascode differential amplifier

For applications in which the amplifier is required to operate at high frequencies, increasing the channel length to achieve a high gain is not an appropriate choice due to the dependence of the transition frequency on the reciprocal of the square of the channel length. Another way to increase the gain of an amplifier is through the use of cascode devices. A typical topology of a differential amplifier is shown in Figure 8.5, where load devices and input devices are stacked in a specific sequence. This telescopic-cascode differential amplifier has a voltage gain that can be considerably higher, typically by a factor of 10–50, than that of the simple differential amplifier configuration. The price to be paid, however, is a reduced output swing due to the stacked transistors. The circuits for generating V_{BN} and V_{BP} must be designed in order to bias M_1–M_2 and M_5–M_6, respectively, on the edge of saturation. Also, to allow for maximum output swing, the common-mode input voltage must be set to a value that will bias the tail-current source on the verge of saturation (see Example 8.3). If we assume that the telescopic-cascode amplifier is optimally biased and that the saturation voltages are all equal to V_{DSsat}, the maximum output swing will be $V_{DD} - 5V_{DSsat}$. The data of Example 8.3 give a maximum output swing of $3.3 - (5 \times 0.3) = 1.8$ V, which is acceptable. However, for more advanced technologies, $V_{DD} = 1.8$ V or even less. In this case, the telescopic-cascode amplifier imposes serious constraints on the signal swing and it is rarely employed, except for situations where transistors are biased at low inversion levels, for which the saturation voltage of a transistor is around 100 mV.

A single-stage differential amplifier with a topology similar to the telescopic-cascode is the folded-cascode amplifier. This will be discussed extensively in this chapter due to its undeniable importance in analog integrated circuits. Before moving to the section on folded-cascode amplifiers, we present the very simple topology of the symmetric operational amplifier.

8.3   The symmetric operational amplifier

The inclusion of current mirrors in the topology of Figure 8.4, as shown in Figure 8.6, can be used to increase the output-voltage range of the differential amplifier at the expense of higher power dissipation. The topology in Figure 8.6 is referred to as either symmetric op amp [10], due to the symmetry of the input stage, or current-mirror op amp [11]. There are two important advantages of this topology compared with the simple topology in Figure 8.4: (i) there is no high-impedance node associated with the input pair M_1–M_2, which translates into a negligible Miller effect; and (ii) the output node is not connected to the input stage; thus, the output-voltage swing depends only on the topology and bias of the output stage, being insensitive to the common-mode input voltage. In order to increase the voltage gain, one can use high-output impedance current mirrors. However, to focus on the first-order analysis of the symmetric op amp, we use the simple topology shown in Figure 8.6.

8.3.1   DC characteristics

We first analyze the op amp in Figure 8.6 as a device that converts the differential input voltage ($V_{ID} = V_{G1} - V_{G2}$) into an output current (I_L). For the analysis, let us assume that

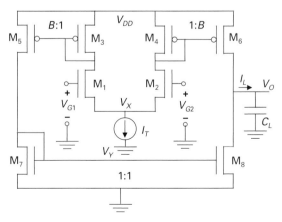

Fig. 8.6   A symmetric operational amplifier.

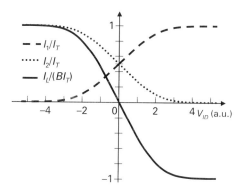

Fig. 8.7 The dc transfer characteristic (I_L against V_{ID}) of the symmetric op amp. The differential input voltage is expressed in arbitrary units.

all transistors operate in saturation. Also, let us assume that the output node (V_O) is connected to a constant-voltage source and that the Early voltage of all transistors is infinite. If $V_{ID} = V_{G1} - V_{G2} = 0$, then $I_1 = I_2$ and the (ideal) current mirrors force $I_L = 0$. If $V_{ID} > 0$, then $I_1 > I_2$ and, since $I_L = B(I_2 - I_1)$ we have $I_L < 0$. The current transfer characteristic of the OTA is shown by the solid-line curve of Figure 8.7. The gate of M_1 is the inverting input since an increase in V_{G1} relative to V_{G2} forces an incoming current at the amplifier output; if a resistive load is connected to the output, the output voltage will be $180°$ out of phase with the gate voltage of M_1. Similarly, the gate of M_2 is the non-inverting input since an increase in V_{G2} with respect to V_{G1} will give rise to an outgoing current at the amplifier output.

The shape of the voltage transfer characteristic of the symmetric op amp is very similar to that shown in Figure 8.7. At sufficiently high negative values of V_{ID}, the output voltage saturates at V_{DD}, whereas at sufficiently high positive values of V_{ID}, the output voltage saturates at the ground level.

The common-mode input range is equal to that calculated in Chapter 7 for the differential pair with current-mirror load. In fact, the common-mode input voltage must be greater than a certain value for the current source to remain in saturation and less than a certain value for the input transistor to remain in saturation. The output-voltage swing is limited by the need to keep the output transistors in saturation.

The offset voltage caused by the mismatch component can be calculated using the result previously obtained for the current-mirror-load differential amplifier (see Section 7.5.3.4 of Chapter 7) added to the contributions from the mismatch of transistor pairs M_5-M_6 and M_7-M_8. In the analysis that follows we define the offset voltage $V_{OS} = V_{G1} - V_{G2}$ as the differential input voltage required for the output current to be equal to zero when M_6 and M_8 operate in saturation. We are interested in determining the major contributors of the offset voltage; therefore, in this analysis we assume that all the Early voltages are infinite. The offset voltage is split into four components, $V_{OSi,i+1}$ with $i = 1, 3, 5,$ and 7, each being associated with a specific pair of transistors. Using Pelgrom's model for the mismatch (see Section 4.6.1), the components of the offset voltage can be written as

$$V_{OS1-2} = V_{T01} - V_{T02} - \frac{I_T}{g_{m1}} \frac{I_{S1} - I_{S2}}{I_{S1} + I_{S2}},$$

$$V_{OS3-4} = \frac{g_{m3}}{g_{m1}} (V_{T03} - V_{T04}) + \frac{I_T}{g_{m1}} \frac{I_{S3} - I_{S4}}{I_{S3} + I_{S4}},$$

$$V_{OS5-6} = -\frac{g_{m3}}{g_{m1}} (V_{T05} - V_{T06}) - \frac{I_T}{g_{m1}} \frac{I_{S5} - I_{S6}}{I_{S5} + I_{S6}}, \qquad (8.3.1)$$

$$V_{OS7-8} = -\frac{g_{m7}}{B g_{m1}} (V_{T07} - V_{T08}) + \frac{I_T}{g_{m1}} \frac{I_{S7} - I_{S8}}{I_{S7} + I_{S8}}$$

$$= -(V_{T07} - V_{T08}) + \frac{I_T}{g_{m1}} \frac{I_{S7} - I_{S8}}{I_{S7} + I_{S8}}.$$

Note that, to simplify the last term in (8.3.1), we assume that the aspect ratio of transistor M_7 is B times higher than that of transistor M_1; consequently, M_1 and M_7 operate at the same inversion level. The ratio I_T/g_m can be calculated using once again the universal current-to-transconductance ratio, which, for first-order analysis, is dependent on the inversion level and temperature only.

As previously explained for differential amplifiers, the specific-current mismatch does not play an important role in the determination of mismatch in transistors except for high inversion levels. Thus, as expected, the offset voltage of differential amplifiers is mostly affected by threshold-voltage mismatch or, equivalently, by fluctuations in the number of doping atoms below the MOSFET channel [12].

8.3.2   Small-signal characteristics and noise

To calculate the response to the differential input voltage we make $v_{g1} = v_{id}/2$ and $v_{g2} = -v_{id}/2$. We first perform a simplified ac analysis of the circuit in Figure 8.6 assuming that the internal capacitances are equal to zero. The differential transconductance is calculated with the output node connected to a constant-voltage source.

The op-amp transconductance (G_m), which is equal to the transconductance (g_{m1}) of input transistor M_1 or M_2 multiplied by the current-mirror gain B, i.e. $G_m = -B g_{m1}$, can be summarily explained as follows. For a small variation of the input voltage v_{id}, the ac currents that flow through M_3 and M_4 are $g_{m1} v_{id}/2$ and $-g_{m1} v_{id}/2$. Both currents are amplified by factor B through the action of the current mirrors on the upper part of the scheme. The current that flows through M_5 is inverted and added at the output to the current that flows through M_6, giving an ac load current equal to $-B g_{m1} v_{id}$. Mathematically, we have

$$G_m = \left. \frac{i_l}{v_{id}} \right|_{V_O} \cong B \frac{-g_{m1} v_{id}/2 - g_{m1} v_{id}/2}{v_{id}} = -B g_{m1}. \qquad (8.3.2)$$

The admittance seen at the output node is $Y_o = sC_L + G_o$, where $G_o = g_{d6} + g_{d8}$. Hence, the differential voltage gain is

$$A_d = \frac{v_o}{v_{id}} = -B \frac{g_{m1}}{G_o} \frac{1}{1 + sC_L/G_o}. \qquad (8.3.3)$$

In a first-order analysis, the differential voltage gain is a single-pole function with a low-frequency value equal to Bg_{m1}/G_o and a unity-gain angular frequency of Bg_{m1}/C_L. A more accurate result for the frequency response of the differential voltage gain must include the transfer function of the current mirrors. Note that current mirrors M_3–M_5 and M_4–M_6 introduce equal delays in the current paths; as a consequence, they do not introduce any asymmetry into the frequency response. On the other hand, current mirror M_7–M_8 acts on the current that flows on the left side of the circuit, thus giving rise to a pole–zero doublet, as studied in Section 7.5.3.5. Summarizing, the inclusion of the frequency response of the current mirrors leads to the following expression for the voltage gain:

$$A_d = \frac{-Bg_{m1}/G_o}{1+sC_L/G_o}\frac{1}{1+s/\omega_{mir}}\frac{1+s(2C_{gd8}+C_y)/(2g_{m8})}{1+s(C_{gd8}+C_y)/g_{m8}}, \tag{8.3.4}$$

where $\omega_{mir}=g_{m4}/C_4$, with C_4 equal to the sum of the capacitances between the gates of M_4 and M_6 and the ac ground, and C_y is the sum of all capacitances between node V_Y and the ac ground. In order to avoid the frequency response of the current mirror adversely affecting the op-amp response, it is generally advisable to set $B=1$. In some low-frequency applications, however, the current-mirror gain B can be much smaller than unity in order to provide a very low transconductance [13].

If the reader wishes to calculate the common-mode gain and the voltage gains relative to the supply rails, they can proceed as previously shown in Chapter 7 for the differential amplifier.

The noise current at the output of the op amp is calculated by adding the uncorrelated channel currents at the ac grounded output. A simple ac analysis for the symmetric op amp in Figure 8.6 gives the following result for the PSD of the output noise current:

$$\frac{\overline{i_{no}^2}}{\Delta f} = 2\left[B^2\left(\frac{\overline{i_{n1}^2}+\overline{i_{n3}^2}}{\Delta f}\right) + \frac{\overline{i_{n5}^2}+\overline{i_{n7}^2}}{\Delta f}\right]. \tag{8.3.5}$$

On referring the output noise current to the input, we obtain

$$\frac{\overline{e_{ni}^2}}{\Delta f} = \frac{2}{g_{m1}^2}\left(\frac{\overline{i_{n1}^2}+\overline{i_{n3}^2}}{\Delta f} + \frac{\overline{i_{n5}^2}+\overline{i_{n7}^2}}{B^2\Delta f}\right). \tag{8.3.6}$$

The channel noise currents can be replaced with their expressions associated with both thermal and flicker noise, as given in Chapter 4.

8.3.3 Slew rate

The maximum rising and falling rates of the output voltage for the topology of Figure 8.6 are given by

$$\left|\frac{dV_O}{dt}\right|_{max} = \frac{BI_T}{C_L}, \tag{8.3.7}$$

since the maximum sink or source current at the output is BI_T.

8.4   The folded-cascode operational amplifiers

One important problem with symmetric op amps is that the input transistors see a diode-connected load, which contributes to reducing the common-mode input-voltage range. On the other hand, the main drawback of the telescopic-cascode amplifier is the stacking of five transistors between the supply lines, which reduces the allowable output-voltage swing. Thus, for many applications, the folded-cascode amplifier shown in Figure 8.8 is the amplifier of choice. It is composed of a p-channel differential input pair (M_1, M_2), followed by common-gate stages (M_3, M_4) and current sources (M_9, M_{10}), and a self-biased cascode current mirror (M_5–M_8) that inverts the current signal. To allow for greater output-voltage swing, (i) the self-biased cascode current mirror can be replaced with a high-swing cascode current mirror, and (ii) V_{bias2} can be generated by a bias circuit such that the current sources (M_9, M_{10}) operate on the edge of saturation [8], [9]. The disadvantage of the folded-cascode amplifier compared with the telescopic-cascode amplifier is a higher current consumption.

8.4.1   DC characteristics

According to the notation in Figure 8.8, the direct current flowing through M_3 and M_4 is $I_B - I_T/2$. Therefore, I_B must be higher than $I_T/2$. If I_B is slightly higher than $I_T/2$, a mismatch between current sources I_B and I_T can lead to very small (or even zero) currents through M_3 and M_4. A value of I_B much higher than $I_T/2$ leads to high current consumption with no benefit to dynamic parameters such as the gain–bandwidth product and slew rate. In most cases, the bias current I_B is made equal to the tail current I_T.

Let us now analyze the circuit in Figure 8.8 as an OTA, assuming that all transistors operate in saturation. The goal is initially to calculate the output current for constant output voltage. When the differential input voltage $V_{ID} = V_{G1} - V_{G2} = 0$, then $I_{D1} = I_{D2} = I_T/2$; and when $V_{ID} > 0$, we have $I_{D1} < I_{D2}$ or, equivalently, $\Delta i > 0$. The output

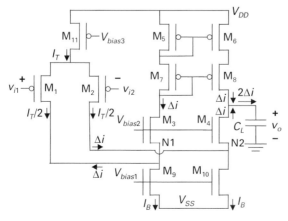

Fig. 8.8   A folded-cascode operational amplifier.

current, which equals $2\,\Delta i$, flows from the amplifier output into the load; thus, the gate of M_1 is the non-inverting input of the op amp.

In order to calculate the common-mode input range we note that the proper operation of the input stage requires transistors M_1, M_2, and M_{11} to operate in saturation. The maximum common-mode input voltage is limited by the saturation voltage of M_{11}. To keep M_{11} in saturation we must ensure that the value of V_{S1} is below a maximum value given by

$$V_{S1max} = V_{DD} - |V_{DSsat11}| = V_{DD} - \phi_t\left(\sqrt{1 + I_T/I_{S11}} + 3\right). \qquad (8.4.1)$$

The common-mode input voltage V_{ICM} at which M_{11} is in saturation must be smaller than a maximum value V_{ICMmax}. Application of the UICM, which gives the relationship between the inversion level i_f and the bias voltages, to the input transistor leads to

$$V_{ICMmax} = V_{DD} - |V_{DSsat11}| + V_{T0P} - n_P\phi_t F[I_T/(2I_{S1})], \qquad (8.4.2)$$

where

$$F[I_T/(2I_{S1})] = \sqrt{1 + I_T/(2I_{S1})} - 2 + \ln\left(\sqrt{1 + I_T/(2I_{S1})} - 1\right),$$

for the case in which the substrate (n-well) is connected to the source (see Problem 8.3). When the common-mode input voltage decreases, the voltage at the common-source node of the input transistors also decreases. V_{D1}, the drain voltage of M_1, is determined by both V_{bias2} and the forward inversion level of M_3, being independent of the input voltage. Let us now assume that V_{bias2} has been designed so that M_9 operates on the edge of saturation. In this case we have

$$V_{D1} = V_{SS} + V_{DSsat9}. \qquad (8.4.3)$$

Since V_{D1} is constant, a reduction in the common-mode input voltage causes the source-to-drain voltage of M_1 to decrease. The minimum value of the common-mode input voltage at which M_1 remains in saturation is thus determined by the saturation voltage of M_1. Using once again the UICM (see Problem 8.3), we find that

$$V_{ICMmin} = V_{SS} + V_{DSsat9} + |V_{DSsat1}| + V_{T0P} - n_P\phi_t F[I_T/(2I_{S1})]. \qquad (8.4.4)$$

The minimum output voltage is limited by the drain-to-source saturation voltages of M_4 and M_{10} if, as previously mentioned, V_{bias2} has been designed so that M_{10} operates on the edge of saturation. In this case we have

$$V_{omin} = V_{SS} + V_{DSsat4} + V_{DSsat10}. \qquad (8.4.5)$$

The output voltage must be lower than a maximum value V_{omax} in order for M_8 to be in saturation. For the topology of the current mirror in Figure 8.8, the maximum output voltage is

$$V_{omax} = V_{DD} - V_{SG6} - |V_{DSsat8}|. \qquad (8.4.6)$$

Note that in (8.4.6) we have used $V_{SD6} \cong V_{SG6}$. The offset voltage is caused by mismatch in pairs (M_1, M_2), (M_5, M_6), and (M_9, M_{10}). The effects of mismatch in pairs (M$_3$, M$_4$), (M$_7$, M$_8$) can be neglected for the calculation of the offset voltage since, as before, their mismatch affects the drain voltages of the remaining transistors and, again, we assume that the Early voltages play a minor role in the determination of the offset voltage. In the analysis that follows we define the offset voltage $V_{OS} = V_{G1} - V_{G2}$ as the differential input voltage required in order for the output current to be equal to zero. As in the analysis of the symmetric op amp, we split the offset voltage into three components, $V_{OSi-i+1}$ with $i = 1$, 5, and 9. The components of the offset voltage can thus be written as

$$
\begin{aligned}
V_{OS1-2} &= V_{T01} - V_{T02} + \frac{I_T}{g_{m1}} \frac{I_{S1} - I_{S2}}{I_{S1} + I_{S2}}, \\
V_{OS5-6} &= \frac{g_{m5}}{g_{m1}} (V_{T05} - V_{T06}) + \frac{2(I_B - I_T/2)}{g_{m1}} \frac{I_{S5} - I_{S6}}{I_{S5} + I_{S6}}, \\
V_{OS9-10} &= \frac{g_{m9}}{g_{m1}} (V_{T09} - V_{T010}) - \frac{2I_B}{g_{m1}} \frac{I_{S9} - I_{S10}}{I_{S9} + I_{S10}}.
\end{aligned}
\tag{8.4.7}
$$

As previously noted for differential pairs, the offset voltage is mainly affected by threshold-voltage mismatch or, equivalently, by the fluctuations in the number of doping atoms below the MOSFET channel [12], except for high inversion levels, at which the specific-current mismatch has an important effect on the offset voltage.

8.4.2   Small-signal characteristics and noise

The transconductance G_m of the folded-cascode op amp is equal to the transconductance (g_{m1}) of the input transistors M$_1$ or M$_2$. For a small variation of the input voltage $v_{id} = v_{i1} - v_{i2}$, two equal and opposite ac currents $\Delta i = g_{m1} v_{id}/2$ flow through M$_1$ and M$_2$. The current that flows through M$_5$ is inverted and added at the output to the current that flows through M$_4$, giving an ac load current equal to $g_{m1} v_{id}$. Therefore, the op-amp transconductance is equal to the input transistor transconductance, i.e.

$$
G_m = \frac{i_o}{v_{id}} \bigg|_{v_o = 0} = g_{m1}.
\tag{8.4.8}
$$

Assuming that the currents flowing through the input transistors are transferred to the output without any attenuation or delay, the differential voltage gain is given by

$$
A_d = \frac{v_o}{v_{id}} \cong \frac{g_{m1}}{G_o} \frac{1}{1 + sC_L/G_o},
\tag{8.4.9}
$$

where G_o is the op-amp output conductance. In this simplified model, the output impedance of the amplifier is represented by a parallel association of a resistance and a capacitance C_L. The unity-gain frequency is g_{m1}/C_L rad/s.

A more accurate result for the frequency response of the differential voltage gain must include the low-impedance nodes N1 and N2 as well as the frequency response of the current mirror. Note that nodes N1 and N2 introduce equal delays in the current paths. On the other hand, the current mirror M$_5$–M$_8$ affects the transference to the output of the

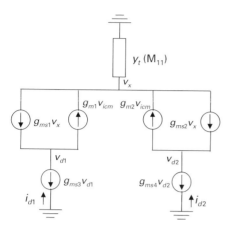

Fig. 8.9 A simplified small-signal model of the folded-cascode operational amplifier (current mirror not included) for determining the common-mode transadmittance.

current that flows through M_1. Replacing M_5–M_8 with a simple current mirror would give rise to a pole–zero doublet, as previously determined for the differential amplifier. For the self-biased current mirror shown in Figure 8.8 or another example, such as the high-swing current mirror, let us assume that its current transfer function is given by $H(s)$, where $H(s)=1$ for frequencies much lower than the transition frequency of M_5. The inclusion of both the influence of nodes N1 and N2 and the frequency response of the current mirrors leads to the following expression for the voltage gain:

$$A_d = \frac{g_{m1}/G_o}{1 + sC_L/G_o} \frac{1}{1 + sC_{N1}/g_{ms3}} \left(\frac{1 + H(s)}{2} \right), \qquad (8.4.10)$$

where C_{N1} is the sum of all capacitances between node N1 and the ac ground.

To calculate the CMRR we determine the common-mode transadmittance of the op amp, which is dependent not only on the impedance of the tail-current source but also on transistor mismatch. For a first-order approximation in the common-mode transadmittance and using a simplified notation, which gives more insight into the main parameters that affect the common-mode transadmittance, we note that (i) the short-circuit output current is $i_o = i_{d1}(1 - \varepsilon_{mir}) - i_{d2}$, where $1 - \varepsilon_{mir} \cong g_{m6}/g_{m5}$ is the current-mirror gain, ε_{mir} being the error gain due to mismatch between M_5 and M_6; (ii) since the drains of M_1, M_2, M_3, M_4,[1] M_9, and M_{10} are connected to low-impedance nodes, the output conductances of these transistors are not relevant for the determination of the common-mode gain; and (iii) for the purpose of obtaining an easily interpretable result we assume that the slope factors of the input pair of transistors are equal, i.e. $g_{ms1} = ng_{m1}$ and $g_{ms2} = ng_{m2}$.

For the small-signal circuit of the folded-cascode amplifier in Figure 8.9 we assume that the previous assumptions hold and that the local substrate of the input transistors is connected to V_{DD}. Using these simplifications,

---

[1] To calculate i_o the output is ac short-circuited to ground.

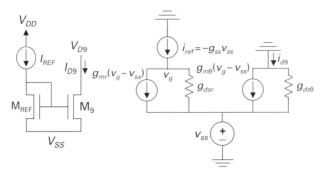

Fig. 8.10   A simplified small-signal model of the folded-cascode operational amplifier (current mirror not included) used for the determination of the transadmittance relative to the negative supply.

$$\frac{i_o}{v_{icm}} = \frac{i_{d1}(1 - \varepsilon_{mir}) - i_{d2}}{v_{icm}} \cong y_t \frac{g_{mi}}{y_t + 2ng_{mi}} \left(\varepsilon_{mir} - \frac{\Delta g_{mi}}{g_{mi}} \right),$$ (8.4.11)

where g_{mi} is the average transconductance of the input transistors. The common-mode transconductance is given by

$$G_{cm} = \left. \frac{i_o}{v_{icm}} \right|_{\substack{low \\ frequency}} \cong \frac{G_t}{2n} \left(\varepsilon_{mir} - \frac{\Delta g_{mi}}{g_{mi}} \right).$$ (8.4.12)

As in the case of the differential amplifier, the common-mode transadmittance is dependent on the mismatch both of the input transconductances and of the current mirror, as well as on the output impedance of the tail-current source. Finally, the CMRR can be calculated by recalling that $i_o/v_{id} = g_{mi}$ and using (8.4.11), which yields

$$\text{CMRR} = \frac{i_o/v_{id}}{i_o/v_{icm}} \cong \frac{1 + 2ng_{mi}/y_t}{\varepsilon_{mir} - \Delta g_{mi}/g_{mi}}.$$ (8.4.13)

The transconductance from the negative supply to the output is calculated by taking into account the influence of V_{SS} on the tail current I_T and on the bias currents flowing through M_9 and M_{10}.

The effect of V_{SS} on the currents in M_9 and M_{10} is calculated using the equivalent circuit shown in Figure 8.10. We assume that the bias currents are derived from a reference current I_{REF}, with sensitivity with respect to the negative supply voltage equal to g_{ss}, as shown in the small-signal model. Note that the drain of M_9 is ac grounded, which is satisfactory for our calculations since it is connected to a low-impedance node. The derivation that we use to calculate i_{d9} can also be applied to M_{10}. On applying KCL to the small-signal circuit in Figure 8.10, we find that

$$i_{d9} = -(g_{ss} + g_{ds9})v_{ss}.$$ (8.4.14)

Similarly, we have for M_{10}

$$i_{d10} = -(g_{ss} + g_{ds10})v_{ss}.$$ (8.4.15)

In a practical circuit, both the tail current I_T and the bias current I_B are obtained from the same reference current. Also, in general we have $I_T = I_B$. Therefore, the dependence of the tail current on the negative supply voltage is

$$i_t = -g_{ss} v_{ss}. \tag{8.4.16}$$

Using small-signal analysis for the input stage, we find that

$$i_{d1} = \frac{g_{ms1}}{g_{ms1} + g_{ms2}} g_{ss} v_{ss} \tag{8.4.17}$$

and

$$i_{d2} = \frac{g_{ms2}}{g_{ms1} + g_{ms2}} g_{ss} v_{ss}. \tag{8.4.18}$$

The transconductance G_{vss} relative to the negative supply is, from (8.4.14), (8.4.15), (8.4.17), and (8.4.18),

$$
\begin{aligned}
G_{vss} &= \frac{i_o}{v_{ss}} = \frac{(i_{d1} + i_{d9})(1 - \varepsilon_{mir}) - (i_{d2} + i_{d10})}{v_{ss}} \\
&\cong \frac{g_{ms1} - g_{ms2}}{g_{ms1} + g_{ms2}} g_{ss} + g_{ds10} - g_{ds9} + \varepsilon_{mir} \left(\frac{g_{ss}}{2} + g_{ds9} \right).
\end{aligned} \tag{8.4.19}
$$

The reciprocal of the negative power-supply rejection ratio is written as

$$
\begin{aligned}
\frac{1}{\text{PSRR}_{vss}} &= \frac{G_{vss}}{g_{m1}} \\
&= \frac{g_{ms1} - g_{ms2}}{g_{ms1} + g_{ms2}} \frac{g_{ss}}{g_{m1}} + \frac{g_{ds10} - g_{ds9}}{g_{m1}} + \varepsilon_{mir} \left(\frac{g_{ss}}{2g_{m1}} + \frac{g_{ds9}}{g_{m1}} \right).
\end{aligned} \tag{8.4.20}
$$

The first and second terms in (8.4.20) are associated with the mismatch (random error) of the input pair M_1–M_2 and the bias current pair M_9–M_{10}, respectively. The third term is associated with the current-mirror error, which is dependent on the mismatch of pair M_5–M_6.

An interesting result is obtained when the reference current is independent of the negative power supply, that is $g_{ss} = 0$. In this case, the input stage, as expected, does not contribute to PSRR_{vss}.

Using a procedure similar to that used to calculate G_{vss}, one can also determine the value of G_{vdd}, the transconductance relative to the positive power supply (see Problem 8.6). The reader must note, however, that, in comparison with the determination of G_{vss}, an additional component of the positive power supply is coupled to the output directly through the stacked transistors M_6 and M_8.

To calculate the equivalent noise voltage at the input, we first calculate the noise current at the ac-grounded output by adding the contributions of the uncorrelated channel currents. Small-signal analysis for the folded-cascode op amp in Figure 8.8 gives

$$\frac{\overline{i_{no}^2}}{\Delta f} \cong 2 \left(\frac{\overline{i_{n1}^2} + \overline{i_{n5}^2} + \overline{i_{n9}^2}}{\Delta f} \right) \tag{8.4.21}$$

for the PSD of the output noise current. To arrive at (8.4.21) we assumed that the output conductances of the MOS transistors are considerably lower than the source transconductances. With this simplification, we note that the magnitudes of the current transfer functions of the channel noise currents associated with M_1, M_2, M_5, M_6, M_9, and M_{10} to the output are slightly smaller than, but very close to, unity. On the other hand, the magnitudes of the current transfer functions of the channel noise currents associated with M_3, M_4, M_7, and M_8 are much smaller than unity (readers are invited to verify the validity of the approximations used to derive (8.4.21)). Note that transistor M_{11} contributes negligibly to the output noise current, since it generates a common-mode current. Once again, we assume that the noise at the input can be represented by the equivalent input noise voltage, for which the PSD is given by

$$\frac{\overline{e_{ni}^2}}{\Delta f} \cong \frac{2}{g_{m1}^2} \left(\frac{\overline{i_{n1}^2} + \overline{i_{n5}^2} + \overline{i_{n9}^2}}{\Delta f} \right). \tag{8.4.22}$$

The channel noise currents can be replaced with their expressions associated with both thermal and flicker noise, which were given in Chapter 4. Note, however, that in this case the noise of the reference current, which is transferred to both M_9 and M_{10}, must also be taken into account.

8.4.3   Slew rate

For the folded-cascode op amp in Figure 8.8, the maximum sink or source current at the output is I_T, as long as $I_T \leq I_B$. The currents that flow through M_3 and M_4 during slewing are equal to $I_B - I_T$ and I_B or vice versa, thus resulting in a maximum sink or source current equal to I_T. Note that, when $I_T > I_B$, for a sufficiently high differential input voltage $V_{G1} - V_{G2}$ all the current provided by M_{11} flows through M_2. Since KCL applied to the source of M_4 is $I_{D2} + I_{D4} = I_{D10} = I_B$, the node voltage at the source of M_4 increases in order to ensure that $I_{D2} + I_{D4} = I_B$. This voltage rise at the source of M_4 turns M_4 off and also decreases the current I_{D2} to a value equal to I_B. At the same time, the voltage at the drain of M_{11} increases to force the current flowing through it to be equal to I_B. Thus, when $I_T > I_B$, the slew rate will be determined by I_B. Summarizing, the maximum rising and falling rates of the output voltage are given by

$$\left| \frac{dV_O}{dt} \right|_{max} = \frac{\min(I_T, I_B)}{C_L}. \tag{8.4.23}$$

Example 8.4

Figure E8.4 shows a folded-cascode amplifier and its bias circuit. Consider the following approximate nominal values of a 0.5-μm CMOS technology: $I_{SHN} = 40$ nA, $I_{SHP} = 16$ nA, slope factors $n_N = n_P = 1.2$, $V_{EN} = V_{EP} = 10$ V/μm, $\phi_t = 25$ mV, $V_{TON} = 0.7$ V, $V_{TOP} = -0.9$ V, $C'_{ox} = 2.5$ fF/μm^2, $C_L = 1$ pF, $V_{DD} = 5$ V, and $V_{SS} = 0$. The reference current is $I_{REF} = 0.6$ μA. Transistor dimensions are given in Table E8.4

Table E8.4 Channel widths and lengths of transistors of the folded-cascode op amp

	W (μm)	L (μm)		W (μm)	L (μm)
M_1, M_2, M_5-M_8	12.5	1	$M_{12}-M_{14}$	10	4
M_3, M_4	5	1	M_{15}	6	16
M_9, M_{10}	10	1	M_{17}, M_{18}	4	4
M_{11}	25	1	M_{16}	7.5	50

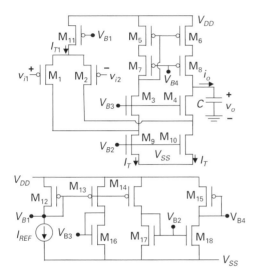

Fig. E8.4   A folded-cascode op amp and bias circuit.

(a) Give the drain currents and forward inversion levels of all transistors. (b) Determine the low-frequency differential voltage gain and the extrapolated unity-gain frequency of the voltage gain. (c) Give a rough estimate of the pole associated with the source node of M_3. (d) Determine the slew rate. (e) Estimate the standard deviation of the offset voltage assuming that $A_{VT}=10$ mV μm and that the errors in the specific currents are negligible. (f) Estimate the standard deviation of 1/CMRR assuming that the relative standard deviations of the input transconductances and current mirror are both equal to 1% and uncorrelated.

Answer

(a)

Transistor	I_D (μA)	i_f	Transistor	I_D (μA)	i_f
M_1, M_2, M_5-M_8	3	15	$M_{12}-M_{14}$	0.6	15
M_3, M_4	3	15	M_{15}	0.6	100
M_9, M_{10}	6	15	M_{17}, M_{18}	0.6	15
M_{11}	6	15	M_{16}	0.6	100

(b) The low-frequency voltage gain is $A_{d0} = g_{m1}/G_o$. The transconductance g_{m1} of the input transistor is calculated from the transconductance-to-current ratio of a transistor in saturation

$$g_{m1} = \frac{2I_{SHP}(W/L)_1}{n\phi_t}\left(\sqrt{1 + i_{f1}} - 1\right) = 40 \ \mu A/V.$$

The output conductance of the op amp is given by

$$G_o \cong \frac{g_{md10} + g_{md2}}{g_{ms4}/g_{md4}} + \frac{g_{md6}}{g_{ms8}/g_{md8}} = \frac{0.6 + 0.3}{48/0.3} + \frac{0.3}{48/0.3}$$
$$\cong 7.5 \ nA/V.$$

The low-frequency voltage gain is $A_{d0} = g_{m1}/G_o \cong 5300$ V/V. The gain–bandwidth product is

$$GB = g_{m1}/C_L = 40 \ Mrad/s.$$

(c) As a rough approximation, let us consider that the major capacitance for calculating the pole associated with the source node of M_3 is $C_{gs3} + C_{gb3}$. The values of these capacitances for a transistor in saturation are given by

$$C_{gs} = \frac{2}{3}WLC'_{ox}\frac{1 + 2\alpha}{(1+\alpha)^2}\frac{\sqrt{1 + i_f} - 1}{\sqrt{1 + i_f}}, \qquad C_{gb} = \frac{n-1}{n}\left(WLC'_{ox} - C_{gs}\right),$$

where

$$\alpha = \frac{1}{\sqrt{1 + i_f}}.$$

The resulting values of the capacitances are $C_{gs} = 6$ fF and $C_{gb} = 1.1$ fF. The source transconductance g_{ms3} is calculated from

$$g_{ms3} = \frac{2I_{SHN}(W/L)_3}{\phi_t}\left(\sqrt{1 + i_{f3}} - 1\right) = 48 \ \mu A/V.$$

The pole frequency is

$$\omega_{p3} = \frac{g_{ms3}}{C_{gs3} + C_{gb3}} \cong 6.8 \ Grad/s \gg GB.$$

The secondary pole exceeds the unity-gain frequency by more than two decades. Hence, the secondary pole affects neither the time response nor the frequency response over the amplifier frequency range of application.

(d) The slew rate is $SR = I_T/C_L = 6$ V/μs.

(e) Using (8.4.7), we find that

$$\sigma^2(V_{OS}) \cong \sigma^2(V_{T01}) + \left(\frac{g_{m5}}{g_{m1}}\right)^2\sigma^2(V_{T05}) + \left(\frac{g_{m9}}{g_{m1}}\right)^2\sigma^2(V_{T09}).$$

Given that $\sigma^2(V_{T0}) = A_{VT}^2/(WL)$, we find $\sigma^2(V_{T01}) = 8$ mV^2, $\sigma^2(V_{T05}) = 8$ mV^2, and $\sigma^2(V_{T09}) = 10$ mV^2. Note that $g_{m5}/g_{m1} = 1$ and $g_{m9}/g_{m1} = 2$. Using these values in the formula given above, we find that $\sigma(V_{OS}) \cong 7.5$ mV.

(f) The reciprocal of the CMRR is, from (8.4.13),

$$\frac{1}{\text{CMRR}} \cong \left(\varepsilon_{mir} - \frac{\Delta g_{mi}}{g_{mi}} \right) \Big/ (1 + 2ng_{mi}/g_t).$$

The conductance of M$_{11}$ is $g_t = I_T/(V_E L) = 6 \times 10^{-6}/(10 \times 1) = 0.6$ μA/V. The transconductance of the input stage is $g_{mi} = 40$ μA/V. From the preceding formula we have

$$\sigma^2(1/\text{CMRR}) \cong (\sigma^2(\varepsilon_{mir}) + \sigma^2(\Delta g_{mi}/g_{mi}))/(1 + 2ng_{mi}/g_t)^2$$
$$= 2 \times 10^{-4}/(1 + 2 \times 1.2 \times 40/0.6)^2,$$

thus

$$\sigma(1/\text{CMRR}) = 8.8 \times 10^{-5}.$$

Example 8.5 Folded-cascode design

Consider the approximate MOSFET parameters of Table E8.5.1 for a 0.35-μm CMOS technology.

Design a folded-cascode amplifier, with p-channel input as in Figure E8.4, to meet the following specifications: $C_L = 1$ pF, power-supply voltage 3.3 V, gain–bandwidth product 100 MHz, low-frequency voltage gain greater than 80 dB, slew rate greater than 50 V/μs, common-mode input voltage 1.65 V, and output-voltage swing greater than 2 V. After completing the design, estimate the standard deviation of the offset voltage and the input-referred noise.

Answer

We first calculate the parameters shown in Table E8.5.2, with $\phi_t = 25.9$ mV and $n_i = 10^{10}$ cm^{-3}. For the calculation of the slope factor n we assume that the transistors

Table E8.5.1 MOSFET model parameters for a 0.35-μm CMOS technology

Parameter	n-Channel	p-Channel
V_{T0} (V)	0.60	−0.75
γ (V$^{1/2}$)	0.63	0.33
$K'(\mu_0 C'_{ox}/2)$ (μA/V^2)	80	30
NCH (cm^{-3})	2.4×10^{17}	8.5×10^{16}
TOX (nm)	7.7	7.7
V_E (V/μm)	20	20
A_{VT} (mV μm)	10	10
A_{ISH} (% μm)	2	2
N_{ot} (cm^{-2})	2.5×10^7	2.5×10^7

Table E8.5.2 Calculated model parameters (n and I_{SH}) of a 0.35-μm CMOS technology

Parameter	n-Channel	p-Channel
$2\phi_F = 2\phi_t \ln(\text{NCH}/n_i)$ (mV)	880	826
$n = 1 + \gamma/\left(2\sqrt{2\phi_F + V_P}\right)$		
$\cong 1 + \gamma/\left(2\sqrt{2\phi_F}\right)$	1.34	1.18
$I_{SH} = \mu_0 C'_{ox} n\phi_t^2/2$ (nA)	72	23.75
$C'_{ox} = \varepsilon_{ox}/\text{TOX}$ (fF/μm^2)	4.5	4.5
$\mu_0 = 2K'/C'_{ox}$ (cm^2/V per s)	356	133

operate with $V_P \cong 0$. Note that the value of n is not very sensitive to V_P. For first-order calculations, we assume that both the slope factor and the sheet-specific current are independent of the gate voltage.

The transconductance of the folded-cascode stage, which is equal to the transconductance of the input transistors, is given by

$$g_{m1} = 2\pi \cdot GB \cdot C_L = 6.28 \times 100 \times 10^6 \times 10^{-12} = 628 \ \mu\text{A/V}.$$

The universal transconductance-to-current ratio of the MOSFET is

$$g_{m1}n\phi_t/I_D = 2/(\sqrt{1 + i_f} + 1).$$

Therefore, since the tail current $I_T = 2I_{D1}$, from the transconductance-to-current ratio we find that the current required for obtaining g_{m1} is

$$I_T = g_{m1}n\phi_t(\sqrt{1 + i_f} + 1) = 628 \times 1.18 \times 25.9 \times 10^{-3}(\sqrt{1 + i_f} + 1)$$
$$I_T = 19.2(\sqrt{1 + i_f} + 1) \ \mu\text{A}.$$

If we impose that the input transistors operate deep in weak inversion ($i_f \ll 1$), the tail current for the required transconductance is $I_T = 38.4 \ \mu\text{A}$. On the other hand, the slew rate imposes a minimum current given by

$$I_T = \text{SR} \cdot C_L \geq 50 \frac{\text{V}}{\mu\text{s}} \cdot 1 \ \text{pF}; \qquad I_T \geq 50 \ \mu\text{A}.$$

In order to meet the requirements of the slew rate, the input transistors must operate at an inversion level higher than unity. However, the inversion level should not be high; otherwise, power consumption increases considerably. For this design, we decided to set the inversion level of all transistors (except those in the bias network) equal to 8. Therefore, the tail current becomes

$$I_T = 19.2(\sqrt{1 + i_f} + 1) = 76.8 \ \mu\text{A}.$$

The aspect ratio of each transistor can thus be calculated as

Table E8.5.3 Drain current and transistor dimensions for the folded-cascode amplifier

Transistor	i_f	I_D (μA)	S	W (μm)	L (μm)
M_1, M_2, M_5–M_8	8	38.4	202	202	1
M_{11}	8	76.8	404	404	1
M_3, M_4	8	38.4	66.7	66.7	1
M_9, M_{10}	8	76.8	133.3	133.3	1

Table E8.5.4 Transistor saturation and gate-to-source voltages for the folded-cascode amplifier in Figure E8.4.

Transistor	i_f	V_{DSsat} (mV)	V_{GS} (mV)	V_{DS} (mV)
M_1, M_2[a]	8	−155	−927	−2395
M_{11}	8	−155	−802	−723
M_5, M_6	8	−155	−802	−181
M_7, M_8[b]	8	−155	−834	−621[d]
M_3, M_4[c]	8	155	720	2317[d]
M_9, M_{10}	8	155	659	181

[a] V_{GS} calculated for $V_G = 1.65$ V.
[b] V_{GS} calculated assuming that $V_{DS5} = V_{DSsat5} - \phi_t = -7\phi_t$
[c] V_{GS} calculated assuming that $V_{S3} = V_{DSsat9} + \phi_t = 7\phi_t$
[d] The drain-to-source voltages of M_4 and M_8 were calculated assuming that $V_{D4} = V_{D3}$.

$$S = W/L = I_D/(I_{SH}i_f).$$

Table E8.5.3 gives the values for the current, inversion level, and aspect ratio of transistors M_1–M_{11}. We have chosen a channel length of 1 μm to satisfy the voltage-gain requirement and to enable the use of a lumped model for the transistors up to the unity-gain frequency of the op amp, as will now be shown.

Table E8.5.4 shows the values for the saturation voltages given by

$$V_{DSsat} = (-)\phi_t(\sqrt{1 + i_f} + 3)$$

together with the values for the gate-to-source voltages, which are calculated from the UICM as

$$V_{GS} = V_{T0} + V_{SB}(n - 1) + (-)n\phi_t[\sqrt{1 + i_f} - 2 + \ln(\sqrt{1 + i_f} - 1)].$$

The minus sign in parentheses in the two preceding formulas must be used for p-channel devices.

To calculate the low-frequency voltage gain we have to compute the op amp's output conductance, which is given by

$$G_o \cong \frac{g_{md10} + g_{md2}}{g_{ms4}/g_{md4}} + \frac{g_{md6}}{g_{ms8}/g_{md8}}.$$

The output conductance and the source transconductance are given by

$$g_{md} = \frac{I_D}{V_E L}, \qquad g_{ms} = \frac{1}{\phi_t} \frac{2I_D}{\sqrt{1 + i_f} + 1}$$

and the resulting output conductances by

$$g_{md2} = g_{md4} = g_{md6} = g_{md8} = \frac{38.4 \times 10^{-6}}{20 \times 1} = 1.92 \ \mu\text{A/V},$$

$$g_{md10} = \frac{76.8 \times 10^{-6}}{20 \times 1} = 3.84 \ \mu\text{A/V},$$

whereas the source transconductances are

$$g_{ms4} = g_{ms8} = \frac{1}{25.9 \times 10^{-3}} \frac{2 \times 38.4 \times 10^{-6}}{\sqrt{1 + 8} + 1} = 741 \ \mu\text{A/V}.$$

The op amp's output conductance is

$$G_o \cong \frac{g_{md10} + g_{md2}}{g_{ms4}/g_{md4}} + \frac{g_{md6}}{g_{ms8}/g_{md8}} = \frac{(3.84 + 1.92) \times 10^{-6}}{741/1.92} + \frac{1.92 \times 10^{-6}}{741/1.92} = 19.9 \ \text{nA/V}.$$

Finally, the resulting low-frequency voltage gain is

$$A_{d0} = \frac{g_{m1}}{G_0} = \frac{628 \ \mu\text{A/V}}{19.9 \ \text{nA/V}} = 31\,558,$$

$$A_{d0}|_{dB} \cong 90 \ \text{dB}.$$

Note that we select $L = 1 \ \mu\text{m}$ for all transistors, which allows the voltage gain to be of the order of 90 dB. Also note that the lumped model becomes increasingly less accurate as $\omega\tau_1 \to 1$, where τ_1 is the time constant given by

$$\tau_1 = \frac{4}{15} \frac{L^2}{\mu_0 \phi_t} \frac{1}{\sqrt{1 + i_f}} \frac{1 + 3\alpha + \alpha^2}{(1 + \alpha)^3}, \qquad \alpha = \frac{\sqrt{1 + i_r}}{\sqrt{1 + i_f}}.$$

In this design, $i_f = 8$ and $i_r \cong 0$ (transistor in saturation), which gives $\alpha = 1/3$. The value of $\tau_1 = 182$ ps for $\mu_0 = 170 \ \text{cm}^2/\text{V}$ per s, $L = 1 \ \mu\text{m}$, and $i_f = 8$. Since the gain–bandwidth product is 628 Mrad/s, at this frequency $\omega\tau_1 = 0.114 \ll 1$ and the lumped model of the MOSFET is still very accurate. The reader should note here that, when the lumped model of the transistors loses accuracy, a simulation using a segmented transistor could be used to obtain a more accurate model at high frequencies, albeit at the expense of increased simulation time.

To calculate the output-voltage swing we note that the minimum and maximum output voltages must ensure that transistors M_4 and M_8 remain in saturation, i.e.

$$V_{DS4} = V_o - V_{DS10} \geq V_{DSsat4}, \qquad V_o \geq (155 + 181) \times 10^{-3} = 336 \ \text{mV};$$

$$V_{DS8} = -V_{DD} + V_o - V_{DS6} \leq V_{DSsat8}, \qquad V_o \leq (3300 - 181 - 155) \times 10^{-3}$$
$$= 2964 \ \text{mV}.$$

The output-voltage swing is around 2.63 V.

To calculate the random-offset voltage, we write, from (8.4.7),

$$\sigma^2(V_{OS}) \cong \sigma^2(V_{T01}) + \left(\frac{g_{m5}}{g_{m1}}\right)^2 \sigma^2(V_{T05}) + \left(\frac{g_{m9}}{g_{m1}}\right)^2 \sigma^2(V_{T09})$$

with the assumption that the contributions due to the mismatch in the specific currents are negligible in this case. Using $\sigma^2(V_{T0}) = A_{VT}^2/(WL)$ we find $\sigma^2(V_{T01}) = 0.5 \text{ mV}^2$, $\sigma^2(V_{T05}) = 0.5 \text{ mV}^2$, and $\sigma^2(V_{T09}) = 0.75 \text{ mV}^2$. Note that $g_{m5}/g_{m1} = 1$ and $g_{m9}/g_{m1} = (n_P/n_N)(I_{D9}/I_{D1}) = (1.18/1.34)(76.8/38.4) = 1.76$. Using these values in the formula for the offset voltage, we find that $\sigma(V_{OS}) \cong 1.82 \text{ mV}$.

The input-referred noise can be calculated from (8.4.22), which is given again below:

$$\frac{\overline{e_{ni}^2}}{\Delta f} \cong \frac{2}{g_{m1}^2} \left(\frac{\overline{i_{n1}^2} + \overline{i_{n5}^2} + \overline{i_{n9}^2}}{\Delta f}\right).$$

To simplify the calculations, we assume that the thermal noise can be approximated by

$$\left.\frac{\overline{i_n^2}}{\Delta f}\right|_{th} \cong 4kT\frac{g_{ms}}{2},$$

which is an acceptable approximation since the inversion level is relatively low. Because the inversion levels are the same for M_1 to M_{11}, we have

$$g_{ms1} = g_{ms5} = g_{ms9}/2$$
$$= I_{D1}/(2\phi_t) = 741 \text{ μA/V}.$$

The input-referred thermal noise thus becomes

$$\frac{\overline{e_{ni}^2}}{\Delta f} \cong \frac{4kT}{g_{m1}/(4n_P)} = \frac{4 \times 41.4 \times 10^{-22}}{628 \times 10^{-6}/(4 \times 1.18)} \cong 1.245 \times 10^{-16} \frac{\text{V}^2}{\text{Hz}}.$$

Roughly speaking, the input-referred noise resistance of the folded-cascode op amp is equal to $4n_P/g_{m1}$.

On the other hand, the PSD of the flicker noise is

$$\left.\frac{\overline{i_n^2}}{\Delta f}\right|_{fl} = \frac{q^2 N_{ot}\mu I_D}{nC_{ox}'L^2 f} \ln\left(\frac{1+i_f}{1+i_r}\right),$$

which gives

$$\left.\frac{\overline{i_{n1}^2}}{\Delta f}\right|_{fl} = \frac{(1.6 \times 10^{-19})^2 \times 2.5 \times 10^7 \times 133 \times 38.4 \times 10^{-6}}{1.18 \times 4.5 \times 10^{-7} \times 10^{-8} \times f} \ln\left(\frac{1+8}{1}\right) \frac{\text{A}^2}{\text{Hz}},$$

$$\left.\frac{\overline{i_{n5}^2}}{\Delta f}\right|_{fl} = \left.\frac{\overline{i_{n1}^2}}{\Delta f}\right|_{fl} = \frac{1.35 \times 10^{-18}}{f} \frac{\text{A}^2}{\text{Hz}}, \qquad \left.\frac{\overline{i_{n9}^2}}{\Delta f}\right|_{fl} = \frac{6.36 \times 10^{-18}}{f} \frac{\text{A}^2}{\text{Hz}}.$$

Table E8.5.5 Dimensions and currents of the bias network in Figure E8.4.

Transistor	V_{GS} (mV)	i_f	I_D (µA)	W (µm)	L (µm)
M_{12}–M_{14}	−802	8	7.7	40.4	1
M_{15}	−1015	73	7.7	44.4	10
M_{16}	901	73	7.7	14.6	10
M_{17}, M_{18}	659	8	7.7	13.3	1

The flicker noise referred to the op-amp input is

$$\frac{\overline{e_{ni}^2}}{\Delta f} \simeq \frac{2}{g_{m1}^2} \frac{(1.35 + 1.35 + 6.36) \times 10^{-18}}{f} = \frac{46 \times 10^{-12}}{f} \frac{V^2}{Hz}.$$

By equating the thermal noise and the flicker noise we find that the corner frequency f_C is

$$f_C = \frac{46 \times 10^{-12}}{1.245 \times 10^{-16}} \cong 370 \text{ kHz}.$$

To complete the design, we show in Table E8.5.5 a possible solution for the transistor dimensions of the bias network of the folded-cascode amplifier in Figure E8.4 and the corresponding drain currents. The bias current I_{REF} was set at 1/10 of the tail current I_T of the differential amplifier. The inversion level and dimensions of M_{15} and M_{16} were calculated using the UICM.

8.5   Two-stage operational amplifiers

8.5.1   Cascade versus cascode amplifiers

We have seen that one way to increase the voltage gain of a single-stage amplifier is to resort to cascoded stages to increase the output impedance, as in the telescopic- and folded-cascode amplifiers. However, the stacking of transistors has the consequence of reducing the output swing for a given supply voltage. This problem becomes more severe for more advanced CMOS technologies, which require lower supply voltages. To circumvent the reduction in voltage swing inherent to cascode amplifiers, we can instead use cascaded stages, with which the low-frequency gain is the product of the voltage gains of the individual stages. The price to be paid for the use of multiple-stage amplifiers is the challenging task of frequency compensation, which is required in order to ensure an acceptable transient response. This is a clear disadvantage of multistage amplifiers compared with cascode amplifiers, since in the latter secondary poles are usually at frequencies much higher than the unity-gain frequency, which leads to an acceptable time response.

Figure 8.11 shows a two-stage CMOS amplifier composed of a single-ended differential amplifier followed by a common-source amplifier, where M_6 is the driver transistor and M_7 acts as the current-source load. The current that flows through the second gain

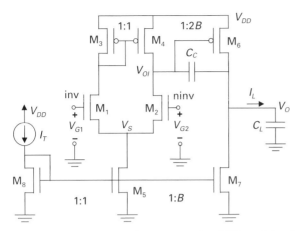

Fig. 8.11   A two-stage CMOS amplifier as a cascade of differential and common-source amplifiers.

stage is B times higher than the tail current I_T. Labels inv and ninv are associated with the inverting and non-inverting inputs, respectively. In fact, when the voltage V_{G1} applied to the gate of M_1 increases, the voltage V_{OI} at the output of the differential amplifier also increases, which, in turn, forces the output V_O of the op amp to decrease through the action of the common-source stage M_6–M_7. Thus, the gate of transistor M_1 is the op-amp inverting input since the output signal is $180°$ out of phase with respect to the input V_{G1}. Capacitor C_C connected between the outputs of the differential and common-source stages is used for frequency compensation, as will be explained later.

8.5.2   DC characteristics of the two-stage amplifier

For small differential input signals, the current flowing through M_1 and M_2 is approximately $I_T/2$. Since the op amp is, in general, connected in a feedback configuration, it will operate in the high-gain region, where transistors M_6 and M_7 of the output stage operate in saturation. In this case, the current that flows through M_6 and M_7 is BI_T.

The common-mode input range, which is defined by the input stage, was calculated in the section on differential amplifiers in Chapter 7. We note again that the maximum common-mode input voltage is limited by the saturation voltage of M_1. In fact, when both input voltages increase, the voltage V_S at the common-source node of the input transistors also increases. M_1 operates in saturation as long as V_S is lower than the drain voltage of M_1 minus its drain-to-source saturation voltage. On the other hand, the minimum common-mode input voltage is that required for transistor M_5 to remain in saturation.

The output-voltage swing is limited by the drain-to-source saturation voltages of output transistors M_6 and M_7, i.e.

$$V_{DSsat7} \leq V_O \leq V_{DD} - |V_{DSsat6}|. \tag{8.5.1}$$

Next, we calculate the offset voltage, which can be split into systematic and random components. We start with the systematic offset voltage, which is calculated by assuming that the pairs (M_1, M_2) and (M_3, M_4) are perfectly matched.

Let us initially establish the condition for the aspect ratios of M_6 and M_7 such that the systematic offset voltage of the op amp is close to zero. We assume that the differential input voltage is zero. Since the input devices and the current-mirror load are assumed to be perfectly matched, the value of the output voltage of the differential amplifier is $V_{OI} = V_{D3}$. Under this condition, transistors M_1 and M_2 operate with the same set of bias voltages, and the same applies to M_3 and M_4. The aspect ratios of M_3, M_5, M_6, and M_7 are referred to as S_3, S_5, S_6, and S_7, respectively. When both output transistors operate in saturation, the currents through M_6 and M_7 are relatively insensitive to the output voltage and we write

$$I_{D6} = (S_6/S_4)I_{D4} = (S_6/S_4)I_{D5}/2 \qquad (8.5.2)$$

and

$$I_{D7} = (S_7/S_5)I_{D5}. \qquad (8.5.3)$$

Finally, taking $S_7/S_5 = B$, as in the scheme of Figure 8.11, and noting that $I_{D6} = I_{D7}$ we find that

$$S_6/S_4 = 2(S_7/S_5) \qquad (8.5.4)$$

or, equivalently, S_6/S_4 must equal $2B$ in order to ensure a systematic offset voltage of approximately zero.

The random offset voltage is caused by mismatch in the pairs (M_1, M_2), (M_5, M_7), and in the triplet (M_3, M_4, M_6). Since the gain of the first stage is relatively high and the transistors that compose the second stage are usually of larger area than those of the first stage, we can neglect the contribution of the second stage to the random offset voltage. Thus, the random offset voltage becomes equal to that previously calculated for the differential amplifier, as repeated below:

$$
\begin{aligned}
V_{OS} &= \Delta V_{T0N} - \frac{I_T}{2g_{mn}}\frac{\Delta I_{SN}}{I_{SN}} + \Delta V_{T0P}\frac{g_{mp}}{g_{mn}} + \frac{I_T}{2g_{mn}}\frac{\Delta I_{SP}}{I_{SP}} \\
&= \Delta V_{T0N} + \Delta V_{T0P}\frac{n_N}{n_P}\frac{\sqrt{1 + I_T/(2I_{SN})} + 1}{\sqrt{1 + I_T/(2I_{SP})} + 1} \\
&\quad + n_N\phi_t\frac{\sqrt{1 + I_T/(2I_{SN})} + 1}{2}\left(\frac{\Delta I_{SP}}{I_{SP}} - \frac{\Delta I_{SN}}{I_{SN}}\right),
\end{aligned} \qquad (8.5.5)
$$

where subscript N refers to input devices M_1 and M_2, while subscript P refers to load devices M_3 and M_4.

8.5.3   Small-signal characteristics of the two-stage Miller-compensated op amp

To derive a first-order approximation of the small-signal characteristics of the two-stage op amp shown in Figure 8.11 we make use of the simplified scheme in Figure 8.12 with the following correspondence between parameters: $g_{mI} = g_{m1}$, $g_{mII} = g_{m6}$, $g_{oI} = g_{ds2} + g_{ds4}$, and $g_{oII} = g_{ds6} + g_{ds7}$. C_{oI} and C_L are the capacitances between the corresponding nodes and ac ground.

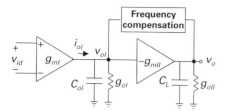

Fig. 8.12 A two-stage operational amplifier represented as an association of two transconductors. The frequency-compensation network is included for stability.

In Figure 8.11, the frequency-compensation network is composed of the feedback capacitor C_C only. Other compensation networks will be described later in this chapter. Possible implementations of C_C are through a double-poly capacitor, a metal–metal capacitor, or even a non-linear gate capacitor [14].

Frequency compensation is generally required in a two-stage op amp in order to avoid the op amp in a feedback configuration being unstable or having an unacceptably underdamped oscillatory time response. In fact, a necessary condition for the stability of the feedback amplifier is that the phase lag introduced by the loop gain (op-amp open-loop gain combined with feedback network) at the unity-gain frequency does not exceed 180°. In order to obtain a time response without excessive overshoot, a phase margin of 45° or more is required [5].

For low-frequency analysis, the frequency-compensation network can be removed. The differential stage, represented by transconductor g_{mI}, converts the differential input voltage into a current of $g_{mI}V_{id}$, which, in turn, is converted back into a voltage equal to $g_{mI}V_{id}/g_{oI}$. The output voltage of the first stage is converted into a current equal to $-g_{mII}g_{mI}V_{id}/g_{oI}$, which, through the action of the second stage, is converted into the output voltage $v_o = -g_{mII}g_{mI}V_{id}/(g_{oI}g_{oII})$. Thus, the differential low-frequency voltage gain is $g_{mII}g_{mI}/(g_{oI}g_{oII})$.

Let us now calculate the frequency response of the amplifier in Figure 8.11, which is represented by the small-signal equivalent circuit in Figure 8.12. We first recall that, in order to include the effect of the pole–zero doublet in the frequency response, the designer simply needs to write the transconductance g_{mI} of the first stage as

$$g_{mI} = g_{m1} \frac{1 + sC_x/(2g_{m3})}{1 + sC_x/g_{m3}}, \tag{8.5.6}$$

where C_x is the capacitance between the gate of M_3 and the ac ground. Now, assuming that for the range of frequencies under analysis $g_{mI} \cong g_{m1}$, nodal analysis shows that

$$\begin{pmatrix} s(C_{oI} + C_C) + g_{oI} & -sC_C \\ g_{mII} - sC_C & s(C_L + C_C) + g_{oII} \end{pmatrix} \begin{pmatrix} V_{oI} \\ V_o \end{pmatrix} = \begin{pmatrix} g_{mI} \\ 0 \end{pmatrix} V_{ID}. \tag{8.5.7}$$

The frequency response of the voltage gain calculated from (8.5.7) is

$$\frac{V_o}{V_{ID}} = \frac{-g_{mI}(g_{mII} - sC_C)}{D(s)}, \tag{8.5.8}$$

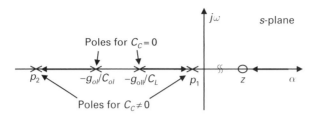

Fig. 8.13   Locus of poles and zeros of the operational amplifier scheme in Figure 8.12. The arrows on the horizontal axis indicate the displacement of poles and zero after the inclusion of the compensation capacitance C_C. The zero for $C_C = 0$ is at infinity.

where

$$D(s) = s^2[C_{oI}(C_L + C_C) + C_L C_C]$$
$$+ s[g_{oI}(C_L + C_C) + g_{oII}(C_{oI} + C_C) + g_{mII}C_C] + g_{oI}g_{oII}.$$

Let us start the analysis of the voltage gain assuming that the compensation capacitance is zero.[2] In this case, the voltage gain becomes

$$\left.\frac{V_o}{V_{ID}}\right|_{C_C=0} = \frac{-g_{mI}g_{mII}/(g_{oI}g_{oII})}{(1 + sC_{oI}/g_{oI})(1 + sC_L/g_{oII})}. \tag{8.5.9}$$

Typically, the pole frequencies g_{oI}/C_{oI} and g_{oII}/C_L are located relatively close to each other, as shown in Figure 8.13. When the compensation capacitance is included, the poles move to frequencies considerably farther apart, given by the roots of $D(s)$ in (8.5.8). This technique of placing the poles widely distant from each other, called pole splitting [15], is widely employed for frequency compensation of amplifiers. Next, we calculate the position of the poles after the inclusion of C_C, assuming that the dominant-pole approximation [15] is valid, i.e. the frequency of the secondary pole p_2 is much higher than that of the dominant pole p_1. Using such an approximation, we find that

$$p_1 = \frac{-g_{oI}g_{oII}}{g_{oI}(C_L + C_C) + g_{oII}(C_{oI} + C_C) + g_{mII}C_C} \cong \frac{-g_{oI}g_{oII}}{g_{mII}C_C}, \tag{8.5.10}$$

$$p_2 = -\frac{g_{oI}(C_L + C_C) + g_{oII}(C_{oI} + C_C) + g_{mII}C_C}{C_{oI}(C_L + C_C) + C_L C_C}$$
$$\cong -\frac{g_{mII}C_C}{C_{oI}(C_L + C_C) + C_L C_C}. \tag{8.5.11}$$

The approximation applied to both (8.5.10) and (8.5.11) is generally valid since transconductances are typically much higher than output conductances. The value of the dominant pole in (8.5.10) can be interpreted with the help of the Miller effect described in Chapter 7. The output of the first stage in Figure 8.12 sees a capacitor equivalent to the parallel association of C_{oI} and the Miller capacitance with a value equal

---

[2] Note that, in the scheme of Figure 8.11, the capacitance between the input and output of the second stage is never equal to zero due to the drain overlap capacitance of M_6.

to $C_C(1 + |A_{V0II}|)$, where $|A_{V0II}| = g_{mII}/g_{oII}$ is the low-frequency gain of the second stage. Thus, the pole frequency given in (8.5.10) is simply determined by the output conductance g_{oI} of the first stage and the Miller capacitance, which is approximately equal to $C_C g_{mII}/g_{oII}$.

The overall gain of the op amp can be written as the product of the voltage gains of the two stages, i.e.

$$\frac{V_o}{V_{ID}} = A_{VI}A_{VII}, \qquad A_{VI} = \frac{V_{oI}}{V_{ID}}, \qquad A_{VII} = \frac{V_o}{V_{oI}}. \qquad (8.5.12)$$

Using (8.5.7), we find that

$$A_V \cong -\frac{g_{mI}g_{mII}}{g_{oI}g_{oII}} \frac{1 - sC_C/g_{mII}}{(1 - s/p_1)(1 - s/p_2)},$$

$$A_{VI} \cong \frac{g_{mI}}{g_{oI}} \frac{1 + s(C_L + C_C)/g_{oII}}{(1 - s/p_1)(1 - s/p_2)}, \qquad (8.5.13)$$

$$A_{VII} = -\frac{g_{mII}}{g_{oII}} \frac{1 - sC_C/g_{mII}}{1 + s(C_L + C_C)/g_{oII}},$$

where the values of p_1 and p_2 are given by (8.5.10) and (8.5.11).

Note that there is a zero in the transfer function of the voltage gain given by

$$z = \frac{g_{mII}}{C_C}, \qquad (8.5.14)$$

which can be understood with the aid of Figure 8.12. Two current paths, one through the compensation capacitance and another through the second transconductor, inject current into the output impedance. There is a complex frequency, given by (8.5.14), for which the sum of the two currents is zero. This zero on the positive real axis of the s-plane has a deleterious effect on the phase margin of the amplifier, as we will see next. Figure 8.13 shows the loci of poles and zeros for the two-stage op amp with and without compensation capacitance.

For C_{oI} much less than C_L and C_C, which is the typical situation, $p_2 \cong -g_{mII}/C_L$. Now, considering these cases, the secondary pole p_2 can be given a simple interpretation [5] with the aid of Figure 8.12. In a well-designed op amp, the secondary pole frequency is higher than ω_u. At frequencies exceeding ω_u, the reactance of the compensation capacitance is lower than $1/(\omega_u C_C) = 1/g_{mI}$. Since this reactance is much lower than that associated with C_{oI}, the voltages at the input and output nodes of the second stage are approximately equal. As a result, the output current of the second transconductor is $-g_{mII}V_o$, which yields an output impedance equivalent to the parallel association of the output capacitance and the second-stage transconductance.

Since the op amp is generally used in a negative-feedback configuration, its design should prevent the feedback circuit from being prone to instability or having a poor transient response. For stability analysis, we use the magnitude and phase plots of the voltage gain shown in Figure 8.14. The plots on the left represent the frequency response of the uncompensated amplifier. At low frequencies the voltage gain is $g_{mI}g_{mII}/(g_{oI}g_{oII})$. In the asymptotic magnitude plot, the gain drops at a rate of 20 dB/dec between the

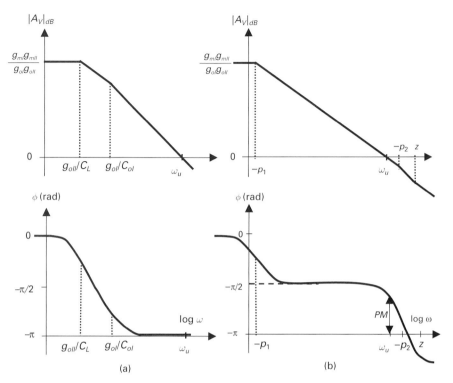

Fig. 8.14   Asymptotic magnitude and phase plots of the op-amp transfer function (a) without and (b) with compensation capacitance.

frequencies of the two poles and, for frequencies higher than g_{ol}/C_{ol}, the drop rate of the gain is 40 dB/dec. The phase plot in the lower part of Figure 8.14(a) shows that, at the unity-gain frequency ω_u, the output signal is around 180° out of phase with respect to the input signal. Thus, the uncompensated amplifier has a phase margin close to zero as a result of the close proximity of the two poles. The phase margin can even be negative, owing to the presence of other singularities in the transfer function that exist in a real amplifier and have not been accounted for in this simplified analysis. Therefore, some kind of modification of the frequency response of the uncompensated two-stage amplifier is required in order to give the feedback circuit stability and an acceptable time response.

Using C_C as the frequency-compensation network leads to the magnitude and phase plots shown in Figure 8.14(b). We note that, for the case illustrated, the frequencies of the two poles have moved farther apart. Also, the secondary-pole and zero frequencies are higher than the extrapolated unity-gain frequency ω_u, a condition that is generally desirable since it leads to acceptable phase margins. Calculating ω_u as the gain–bandwidth product yields

$$\omega_u = A_{V0}(-p_1) = \frac{g_{mI}g_{mII}}{g_{oI}g_{oII}}\frac{g_{oI}g_{oII}}{g_{mII}C_C} = \frac{g_{mI}}{C_C}. \tag{8.5.15}$$

The extrapolated unity-gain frequency (or gain–bandwidth product) is determined by the transconductance of the first stage and the compensation capacitance. From now on, we will use the value calculated in (8.5.15) for the unity-gain frequency of a two-stage amplifier.

Since we know the values of the secondary-pole, zero, and unity-gain frequencies, we can now determine the phase margin (PM), as defined in Figure 8.14(b). We first note that the dominant-pole frequency is considerably lower than the unity-gain frequency; there-fore, the contribution of the dominant pole to the phase of the voltage transfer function is $-\pi/2$ rad at frequency ω_u. Including the contributions of the secondary pole and zero to the phase margin, we find that

$$
\begin{aligned}
\text{PM} &= \frac{\pi}{2} - \tan^{-1}\left(\frac{\omega_u}{-p_2}\right) - \tan^{-1}\left(\frac{\omega_u}{z}\right) \\
&= \frac{\pi}{2} - \tan^{-1}\left[\frac{g_{mI}}{g_{mII}}\frac{C_L}{C_C}\left(1 + \frac{C_{oI}}{C_C} + \frac{C_{oI}}{C_L}\right)\right] - \tan^{-1}\left(\frac{g_{mI}}{g_{mII}}\right). \quad (8.5.16)
\end{aligned}
$$

We note that the zero on the positive real axis has a detrimental effect on the phase margin given by the last term in (8.5.16).

It is of interest to rewrite the expression for the phase margin as

$$
\text{PM} = \frac{\pi}{2} - \tan^{-1}\left[\frac{\omega_u C_L}{g_{mII}}\left(1 + \frac{C_{oI}}{C_C} + \frac{C_{oI}}{C_L}\right)\right] - \tan^{-1}\left(\frac{\omega_u C_L}{g_{mII}}\frac{C_C}{C_L}\right) \quad (8.5.17)
$$

since, for most practical cases, one of the specifications to be achieved is the unity-gain bandwidth ω_u for a given load capacitance C_L. We will now discuss the usefulness of writing the phase margin as in (8.5.17) for a particular design example.

Example 8.6

Design the two-stage operational amplifier shown in Figure 8.11 for the following specifications: $V_{DD}=5$ V, $C_L=1$ pF, $\omega_u=200$ Mrad/s, $A_{V0}\geq1000$, peak-to-peak output voltage swing ≥4 V, common-mode input voltage 2.5 V, PM$\geq45°$ in 0.5-μm CMOS technology. Assume that the technology parameters are $\mu_N=427\,\text{cm}^2/\text{V}$ per s, $\mu_P=171\,\text{cm}^2/\text{V}$ per s, slope factors $n_N=n_P=1.2$, $V_{EN}=V_{EP}=10$ V/μm, $\phi_t=25$ mV, $V_{TON}=0.7$ V, $V_{TOP}=-0.9$ V, and $C'_{ox}=2.5\,\text{fF}/\mu\text{m}^2$.

Answer

For the design that follows, all transistors operate at the same inversion level; however, the procedure developed next can be applied to more general cases.

We first calculate the constraints imposed by the common-mode input voltage and the output-voltage swing on the inversion level.

The pinch-off voltage of the input transistors is

$$
V_{P1} = \frac{V_{GB1} - V_{TON}}{n_N} = \frac{2.5 - 0.7}{1.2} = 1.5 \text{ V}.
$$

Application of the UICM to M_1 yields

$$\frac{V_{P1} - V_{SB1}}{\phi_t} = F(i_{f1}); \qquad F(i_f) = \sqrt{1 + i_f} - 2 + \ln\left(\sqrt{1 + i_f} - 1\right).$$

Since

$$V_{SB1} = V_{DS5} \geq V_{DSsat5} = \phi_t\left(\sqrt{1 + i_{f5}} + 3\right),$$

by combining this restriction with the UICM and noting that $i_{f1} = i_{f5}$ we find that

$$F(i_{f1}) + \sqrt{1 + i_{f1}} + 3 \leq V_{P1}/\phi_t = 60 \rightarrow i_{f1} \leq 780.$$

On the other hand, the output-voltage swing is constrained by the saturation voltages of M_6 and M_7, which are the same since the inversion levels of M_6 and M_7 are equal. In this case, we have

$$V_{DSsat6} = \phi_t\left(\sqrt{1 + i_{f6}} + 3\right) \leq 0.5 \text{ V} \rightarrow i_{f6} \leq 288.$$

Examination of Equation (8.5.17) for the phase margin can be used to give an idea of the values required for both the transconductance g_{mII} of the second stage and the compensation capacitance. C_C must be of the order of C_L or lower; otherwise, the value of the transconductance g_{mII} will need to be very high in order to achieve the appropriate phase margin. Also, C_C cannot be lower than the output capacitance C_{oI} of the first stage because the compensation would be ineffective. Since the value of C_{oI} is dominated by the input capacitance of transistor M_6, which drives the load capacitance, C_{oI} increases with C_L. Therefore, the value of C_C should be at least a significant fraction of C_L. As a rule of thumb, a value of C_C between 20% and 100% of C_L is suitable for two-stage amplifiers. For this design, we use $C_C = 0.5C_L$, but the reader is invited to test other solutions for different values of C_C. Now, using the value $C_C = 0.5C_L$, and assuming initially that C_{oI} is negligible, (8.5.17) becomes

$$\text{PM} = \frac{\pi}{2} - \tan^{-1}\left(\frac{\omega_u C_L}{g_{mII}}\right) - \tan^{-1}\left(\frac{\omega_u C_L}{2g_{mII}}\right) \geq 45°,$$

which gives $\omega_u C_L/g_{mII} \leq 0.56$. Since we have neglected the degradation of the phase margin due to C_{oI}, we attempt to compensate for this by using $\omega_u C_L/g_{mII} = 0.50$, i.e. $g_{mII} = 2\omega_u C_L = 400 \ \mu\text{A/V}$. The value of g_{mI} is then calculated from $g_{mI} = \omega_u C_C = 100 \ \mu\text{A/V}$.

In the following, we show two design alternatives provided by inversion levels in the moderate-inversion region, the first equal to 8 and the second to 80.

For the calculation of the transistor aspect ratios and direct currents from the inversion level, we must determine the sheet-specific currents. Using the technology parameters, we find $I_{SHN} \cong 40 \ \text{nA}$ and $I_{SHP} \cong 16 \ \text{nA}$.

Now, recalling that in saturation

$$I_D = g_m n \phi_t\left(\sqrt{1 + i_f} + 1\right)/2; \qquad S = W/L = I_D/\left(I_{SH} i_f\right),$$

we can find the values required for the drain current and aspect ratio of each transistor. Tables E8.6.1 and E8.6.2 show the values found for I_D and S for each transistor, when the inversion levels are 8 and 80, respectively.

Table E8.6.1 Drain currents and aspect ratios for the design with $i_f = 8^a$

Transistor	I_D (µA)	S
M_1, M_2	6	19
M_3, M_4	6	47
M_5, M_8	12	38
M_6	24	188
M_7	24	76

[a] Since the values of S were taken as integers close to the calculated values using an inversion level of either 8 or 80, the real values of the inversion level are slightly different from those used for the calculations.

Table E8.6.2 Drain currents and aspect ratios for the design with $i_f = 80^a$

Transistor	I_D (µA)	S
M_1, M_2	15	5
M_3, M_4	15	12
M_5, M_8	30	10
M_6	60	48
M_7	60	20

[a] Since the values of S were taken as integers close to the calculated values using an inversion level of either 8 or 80, the real values of the inversion level are slightly different from those used for the calculations.

A comparison between these two tables shows that, for the inversion level of 80, the current consumption is 2.5 times higher, but, assuming that the channel lengths are the same, the active area is about 4 times smaller.

For both designs, the channel length of all transistors is 1 µm. Thus, the value of the channel width W, in µm, is equal to S. As seen in Table E8.6.1, transistor M_6 has a gate area of 188 µm². The estimate for C_{gs6}, using the formula for its intrinsic value, is around 200 fF. In addition to C_{gs6}, other capacitances contribute to increasing the value of C_{ol}, which for the design with $i_f = 8$ can be around 50% of C_{gs6} or more. In this case, a simulation can be very useful to indicate whether the phase-margin requirement is still being met. On the other hand, for the design with $i_f = 80$, the value of C_{ol} is about four times lower than that with $i_f = 8$ and is not of major concern.

Finally, we calculate the dc voltage gain for both designs. The dc gains of the first and second stages are

$$A_{VOI} = \frac{g_{m1}}{g_{d2} + g_{d4}} = \frac{g_{m1}}{I_{D2}(1/V_{A2} + 1/V_{A4})},$$

$$A_{VOII} = \frac{g_{m6}}{g_{d6} + g_{d7}} = \frac{g_{m6}}{I_{D6}(1/V_{A6} + 1/V_{A7})}.$$

Since all transistors have the same channel length $L = 1$ µm, $V_A = 10$ V. The low-frequency voltage gains are $A_{VOI} = 250/3$, $A_{VOII} = 250/3$, and total op-amp gain $A_{VO} = 62\,500/9$ for $i_f = 8$, whereas $A_{VOI} = 100/3$, $A_{VOII} = 100/3$, and total op-amp gain $A_{VO} = 10^4/9$ for $i_f = 80$.

The common-mode rejection ratio of the two-stage op amp is the same as that of the differential amplifier calculated in Chapter 7, since the second stage is a single-input stage and, thus, amplifies any incoming signal by the same factor, whatever its origin.

The power-supply rejection ratio for either positive or negative supply can be calculated using the result obtained for the PSRR of the differential amplifier and including the contribution of the common-source stage.

First, let us calculate the PSRR for the negative (or ground) supply V_{SS}. For the two-stage op amp in Figure 8.11, we can write the output current i_{o1} of the first stage as a combination of the effects of the differential input voltage v_{id} and negative supply variation v_{ssI} according to

$$i_{o1} = Y_{mI}v_{id} + Y_{mvssI}v_{ssI}. \qquad (8.5.18)$$

Here, Y_{mI} and Y_{mvssI} are the small-signal transconductances from the differential input and negative supply, respectively, to the output of the first stage. v_{ssI} is the negative supply variation that is coupled to the output of the first stage through transistors M_8 and M_5, and also through current source I_T, as described in Chapter 7.

The output current i_o, obtained by superposing the effects of the output current of the first stage and those of V_{SS} coupled to the output via the second stage, is written as

$$i_o = A_{iII}i_{o1}\big|_{v_{ssII}=0} + Y_{vssII}v_{ssII}\big|_{v_{ssI}=0}. \qquad (8.5.19)$$

Here, v_{ssII} is the negative supply variation that is coupled to the output of the second gain stage through transistor M_7 and current source I_T. A_{iII} is the short-circuit current gain of the second stage. Now, substituting the expression of i_{o1} from (8.5.18) into (8.5.19) results in

$$i_o = A_{iII}(Y_{mI}v_{id} + Y_{mvssI}v_{ssI}) + Y_{mvssII}v_{ssII}\big|_{v_{ssI}=0}. \qquad (8.5.20)$$

In (8.5.19) and (8.5.20) we used subscripts I and II for v_{ss} to emphasize the conditions under which the gains are measured, but, of course, $v_{ssI} = v_{ssII}$. Y_{mvssII} is the transconductance of the output stage in response to a signal at the negative supply. By applying the definition of the PSRR to (8.5.20), we find that

$$\frac{1}{\text{PSRR}_{vss}} = \frac{1}{\text{PSRR}_{vssI}} + \frac{Y_{mvssII}}{A_{iII}Y_{mI}} = \frac{1}{\text{PSRR}_{vssI}} + \frac{Y_{oI}}{Y_{mI}\text{PSRR}_{vssII}}, \qquad (8.5.21)$$

where PSRR_{vssI} and PSRR_{vssII} are the power-supply rejection ratios for the negative supply of the first and second stage, respectively, and $A_{iII} = Y_{mII}/Y_{oI}$, with Y_{mII} equal to the transconductance of the second stage and Y_{oI} equal to the admittance at the output of the first stage, with the op amp's output connected to ground. Note that the ratio Y_{mI}/Y_{oI}, the voltage gain of the first stage with the output short-circuited to the ac ground, is given by

$$\frac{Y_{mI}}{Y_{oI}} = \frac{g_{mI}}{g_{oI} + s(C_C + C_{oI})} \cong \frac{g_{mI}}{g_{oI} + sC_C}. \qquad (8.5.22)$$

For the calculation of PSRR_{vssII} we use the equivalent small-signal circuit shown in Figure 8.15 with the output short-circuited. We assume that the variation in the reference

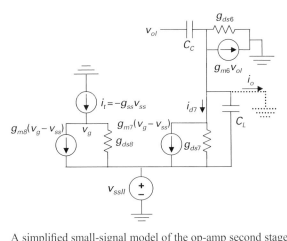

Fig. 8.15   A simplified small-signal model of the op-amp second stage for determining the power-supply rejection ratio.

current I_T with respect to the negative supply is g_{ss}. The drain current flowing through M_7 due to v_{ss} is

$$i_o = -i_{d7} = (Bg_{ss} + g_{ds7} + sC_L)v_{ss}. \tag{8.5.23}$$

The output current of the second stage for a signal equal to v_{ol} applied to its input is

$$i_o = Y_{mII}v_{ol} = -(g_{mII} - sC_C)v_{ol}, \tag{8.5.24}$$

where the transconductance of the second stage is equal to that of M_6, i.e. $g_{mII} = g_{m6}$. Thus, the PSRR of the second stage is

$$\text{PSRR}_{vssII} = \frac{Y_{mII}}{Y_{mvssII}} \cong \frac{-g_{mII}}{Bg_{ss} + g_{ds7} + sC_L}. \tag{8.5.25}$$

In (8.5.25) we have neglected the effect of the zero of the transconductance of the second stage on the PSRR. The value of g_{ss} can be written as $g_{ss}/I_T = -(\Delta I_T/I_T)/\Delta V_{SS}$, where g_{ss}/I_T is the relative sensitivity (regulation) of the current source with respect to V_{SS}. Since $g_{ds7} = BI_T/V_A$, (8.5.25) can be rewritten as

$$\text{PSRR}_{vssII} \cong \frac{-g_{mII}}{BI_T\left(\dfrac{1}{V_A} + \dfrac{\Delta I_T/I_T}{\Delta V_{SS}}\right) + sC_L} \cong \frac{-g_{mII}}{\dfrac{BI_T}{V_A} + sC_L}, \tag{8.5.26}$$

where the last approximation holds for well-regulated current sources.

The calculation of the positive PSRR is the task in Problem 8.9. The reader can note that the value of PSRR_{vddII} is $g_{m6}/g_{ms6} = 1/n_6$, which is slightly smaller than unity. The PSRR of the second stage referred to the input of the first stage must be multiplied by the first-stage gain.

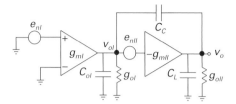

Fig. 8.16   A simplified small-signal model of the two-stage op amp for determining the input-referred noise.

To calculate the equivalent noise voltage at the input, we assume that the noise model can be represented by two noise-voltage sources connected at the inputs of the two amplifying stages, as shown in Figure 8.16. To determine the input-referred noise source $e_{nII,i}$, equivalent to e_{nII}, we first calculate the current $i_{o,nII}$ that flows through the grounded output due to e_{nII}. Circuit analysis shows that

$$i_{o,nII} = -g_{mII}e_{nII}. \qquad (8.5.27)$$

Therefore, the value of $e_{nII,i}$ is

$$e_{nII,i} = \frac{i_{o,nII}}{Y_m} = g_{mII}e_{nII}\frac{g_{oI} + s(C_C + C_{oI})}{g_{mI}(g_{mII} - sC_C)}, \qquad (8.5.28)$$

where Y_m is the op-amp transadmittance. Neglecting the effect of the zero of Y_m, the frequency of which is usually higher than the unity-gain bandwidth, and assuming that $C_C \gg C_{oI}$, a condition that is generally fulfilled, the input-referred PSD of $e_{nII,i}$ becomes

$$\frac{\overline{e_{nII,i}^2}}{\Delta f} = \frac{\overline{e_{nII}^2}}{\Delta f}\left[\left(\frac{g_{oI}}{g_{mI}}\right)^2 + \left(\frac{\omega}{\omega_u}\right)^2\right] = \frac{\overline{e_{nII}^2}}{\Delta f}\left[\frac{1}{A_{VOI}^2} + \left(\frac{\omega}{\omega_u}\right)^2\right]. \qquad (8.5.29)$$

For low frequencies, the input-referred PSD is equal to the PSD of $e_{nII,i}$ divided by the square of the first-stage voltage gain. For frequencies such that $\omega > \omega_u/A_{VOI}$, the term in square brackets in (8.5.29) increases proportionately to the square of the frequency. Note that for $\omega = \omega_u$ the term in square brackets becomes very close to unity.

Example 8.7

Assume that the small-signal parameters of the equivalent circuit of the op amp shown in Figure 8.16 are $g_{mI} = 50\ \mu\text{A/V}$, $g_{mII} = 500\ \mu\text{A/V}$, $g_{oI} = 1\ \mu\text{A/V}$, $g_{oII} = 10\ \mu\text{A/V}$, $C_C = 1\ \text{pF}$, $C_L = 2\ \text{pF}$, and $C_{oI} \ll C_C$. Assume that the noise sources are white and that their PSDs are

$$\overline{e_{nI}^2}/\Delta f = 8 \times 10^{-16}\ \text{V}^2/\text{Hz}; \qquad \overline{e_{nII}^2}/\Delta f = 0.5 \times 10^{-16}\ \text{V}^2/\text{Hz}.$$

Determine (a) the open-loop voltage gain and phase margin; (b) the op-amp transconductance; and (c) the PSD of e_{nII} referred to the input. Assume that the op amp is connected in the unity-gain configuration and that the effect of the noise associated

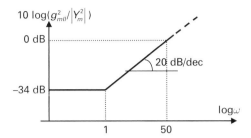

10 log($g_{mll}^2/|Y_m^2|$)

0 dB

20 dB/dec

−34 dB

logω

1   50

Fig. E8.7.1 The magnitude response of the transfer function required to refer the noise of the second stage to the op-amp input.

with the second gain stage can be neglected, and (d) calculate the noise power at the output both for the open-loop op amp and for the unity-gain buffer.

Answer

(a) Using (8.5.10), (8.5.11), and (8.5.13), we find

$$A_V \cong -\frac{g_{ml}g_{mll}}{g_{ol}g_{oll}}\frac{1-sC_C/g_{mll}}{(1-s/p_1)(1-s/p_2)} = -2500\frac{1-s/500}{(1+s/0.02)(1+s/250)}.$$

The unit of frequency in the formula above is Mrad/s. The unity-gain frequency of the operational amplifier is $\omega_u = 50$ Mrad/s. The phase margin is

$$\text{PM} = 90° - \tan^{-1}\left(\frac{\omega_u}{-p_2}\right) - \tan^{-1}\left(\frac{\omega_u}{z}\right) = 90° - \tan^{-1}(0.2) - \tan^{-1}(0.1) \cong 73°.$$

(b) The op-amp transconductance is

$$Y_m = \frac{g_{ml}g_{mll}}{g_{ol}}\frac{1-sC_C/g_{mll}}{1+sC_C/g_{ol}} = -25\,000\frac{1-s/500}{1+s/1}\ \mu\text{A/V}.$$

(c) From (8.5.29) we have

$$\frac{\overline{e_{nll,i}^2}}{\Delta f} = \frac{\overline{e_{nll}^2}}{\Delta f}\left[(g_{ol}/g_{ml})^2+(\omega/\omega_u)^2\right] = \frac{\overline{e_{nll}^2}}{\Delta f}\left[(1/50)^2+(\omega/50)^2\right].$$

The noise of the second stage is referred to the input stage by shaping its PSD with function $g_{mll}^2/|Y_m^2|$, the magnitude of which is dependent on the frequency, as shown in the plot of Figure E8.7.1

(d) The noise power at the output for the open-loop op amp is

$$\overline{e_{no}^2} = \int_0^\infty \frac{\overline{e_{nl}^2}}{\Delta f}|A_V|^2\,df \cong \frac{\overline{e_{nl}^2}}{\Delta f}\int_0^\infty \frac{A_{V0}^2}{1+(\omega A_{V0}/\omega_u)^2}\,df = \frac{\overline{e_{nl}^2}}{\Delta f}\frac{A_{V0}\omega_u}{2\pi}\tan^{-1}x\Big|_0^\infty.$$

Note that we have approximated the voltage gain of the op amp using a single-pole transfer function. Since $A_{V0} = 2500$ and $\omega_u = 50$ Mrad/s, the noise power at the output is

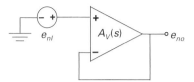

Fig. E8.7.2   A unity-gain buffer with equivalent noise source at the input.

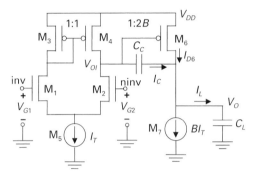

Fig. 8.17   A simplified schematic diagram of the two-stage operational amplifier to calculate the slew rate.

$$\overline{e_{no}^2} = \frac{\overline{e_{nI}^2}}{\Delta f}\frac{A_{V0}\omega_u}{2\pi}\tan^{-1}x\Big|_0^\infty = 8 \times 10^{-16}\frac{2500 \times 50 \times 10^6}{2\pi}\frac{\pi}{2} = 25 \times 10^{-6}\ \text{V}^2.$$

The rms value of the output noise voltage is 5 mV.

For the unity-gain buffer shown in Figure E8.7.2, the closed-loop transfer function A_{VCL} is

$$A_{VCL}(s) = \frac{A_V(s)}{1 + A_V(s)} \cong \frac{1}{1 + j\omega/\omega_u}.$$

Thus, the mean-square value of the output noise voltage of the unity-gain buffer is

$$\overline{e_{no}^2} = \int_0^\infty \frac{\overline{e_{nI}^2}}{\Delta f}|A_{VCL}|^2 df = \frac{\overline{e_{nI}^2}}{\Delta f}\frac{\omega_u}{2\pi}\tan^{-1}x\Big|_0^\infty = 8 \times 10^{-16}\frac{50 \times 10^6}{2\pi}\frac{\pi}{2} = 10^{-8}\ \text{V}^2.$$

The rms value of the output noise voltage for the closed-loop amplifier is 0.1 mV.

8.5.4   Slew rate

A simplified schematic diagram of the two-stage amplifier of Figure 8.11 is shown in Figure 8.17. In the following, we analyze the behavior of the operational amplifier when the differential input voltage is such that either M_1 or M_2 is cut off, thus making the differential amplifier behave as either a current source or a sink. In such cases, the rate of change of the output is limited by the maximum currents that can flow through either the compensation capacitance or the load capacitance.

For the analysis of the rising slew rate, assume first that the op amp is operating with $v_{id} = 0$. In this case, the currents flowing through M_1 and M_2 are equal to $I_T/2$ and the current in M_6 is BI_T. Now, assume that a voltage difference is applied to the input such that M_1 turns off or, equivalently, the bias current (I_T) of the differential stage flows entirely through M_2. In this case, the current through M_2 pulls down the gate voltage of M_6, which, in turn, increases the current flowing through M_6. After a very short transient, the gate voltage of M_6 becomes constant and the current I_T supplied by the differential stage flows through the compensation capacitance, resulting in a rate of change of the output voltage such that

$$SR^+ = \frac{I_T}{C_C}. \tag{8.5.30}$$

Note that for this rate of change of the output voltage, the gate voltage of M_6 is such that the drain current through M_6 is $I_{D6} = (B + 1 + C_L/C_C)I_T$. In general, transistor M_6 can provide such a current in a well-designed op amp.

To analyze the falling slew rate let us assume that a step voltage applied to the amplifier input turns off M_2. Thus, the tail current I_T flows through M_1 and the action of current mirror M_3–M_4 sets a current equal to I_T in M_4 which, in turn, pushes up the gate voltage of M_6. After a quick transient required to charge the gate of M_6, the current in M_4 flows entirely through C_C, thus generating a voltage drop across C_C with a rate of change equal to

$$\frac{d(V_o - V_{ol})}{dt} \cong \frac{dV_o}{dt} = -\frac{I_T}{C_C}. \tag{8.5.31}$$

The approximation in (8.5.31) is valid as long as V_{ol} has reached the steady state or its variation is much smaller than that of V_o. In this case, the current that flows through M_6 is

$$I_{D6} = BI_T + (C_C + C_L)\frac{dV_o}{dt} = \left[B - \left(1 + \frac{C_L}{C_C}\right)\right]I_T. \tag{8.5.32}$$

Since $I_{D6} > 0$, (8.5.31) holds only for $B > 1 + C_L/C_C$. In other words, the descending slew rate given by the approximation in (8.5.31) is valid only when the current source BI_T is sufficient to discharge load capacitance C_L with a rate given by (8.5.31). However, when $B < 1 + C_L/C_C$, transistor M_6 shuts off and, as a result, the maximum falling rate of the output voltage [10] becomes

$$I_T = C_L\frac{dV_o}{dt} + BI_T, \quad \frac{dV_o}{dt} = -\frac{B-1}{C_L}I_T. \tag{8.5.33}$$

Therefore, the descending slew rate is

$$SR^- = \min\left(\frac{(B-1)I_T}{C_L}, \frac{I_T}{C_C}\right). \tag{8.5.34}$$

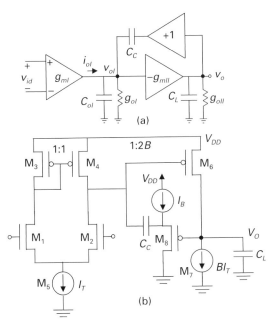

Fig. 8.18   (a) The small-signal equivalent circuit and (b) a schematic diagram of the two-stage op amp with elimination of the feedforward path using a unity-gain buffer.

8.5.5   Alternative forms of compensation of the two-stage op amp

8.5.5.1   Elimination of the feedforward effect of C_C using a buffer

A simple way to eliminate the right-half-plane (RHP) zero introduced by the feedforward path through the compensation capacitance is by placing a unity-gain buffer in series with the compensation capacitance, as shown in Figure 8.18 [16]. To calculate the open-loop transfer function of the op amp, we assume that the unity-gain buffer is ideal. Applying KCL to nodes v_{oI} and v_o in Figure 8.18 and using the dominant-pole approximation, we find that

$$\frac{v_o}{v_{id}} = \frac{-g_{mI}g_{mII}/(g_{oI}g_{oII})}{(1 - s/p_1)(1 - s/p_2)} \tag{8.5.35}$$

with

$$p_1 = \frac{-g_{oI}g_{oII}}{g_{oI}C_L + g_{oII}(C_{oI} + C_C) + g_{mII}C_C} \cong \frac{-g_{oI}g_{oII}}{g_{mII}C_C}, \tag{8.5.36}$$

$$p_2 = -\frac{g_{oI}C_L + g_{oII}(C_{oI} + C_C) + g_{mII}C_C}{(C_{oI} + C_C)C_L} \cong -\frac{g_{mII}}{C_L}\frac{1}{1 + C_{oI}/C_C}. \tag{8.5.37}$$

Note that, in comparison with the case in which a simple compensation capacitance is used, the zero in the RHP has been removed and the secondary pole has been moved to a slightly lower frequency, whereas the dominant pole is the same.

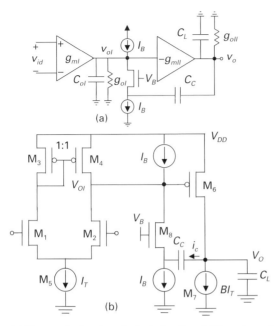

Fig. 8.19 (a) The small-signal equivalent circuit and (b) a schematic diagram of the two-stage op amp with elimination of the feedforward path using a common-gate amplifier.

The previous analysis is valid when the buffer is ideal. Designating the output resistance of the buffer as $1/g_{mo}$ and assuming that it is much lower than both the output resistances of the first and second stages results in an additional pole and a zero [5] given by

$$p_3 \cong \frac{-g_{mo}}{C_C}\left(1 + \frac{C_C}{C_{ol}}\right), \tag{8.5.38}$$

$$z \cong -\frac{g_{mo}}{C_C}. \tag{8.5.39}$$

This technique works well, but at the cost of more silicon area and power required by the unity-gain buffer [5], [17].

8.5.5.2 Elimination of the feedforward effect of C_C using a common-gate amplifier

Another simple way to remove the RHP zero introduced by the feedforward path through the compensation capacitance is by including a common-gate amplifier in the feedback path of the second stage, as shown in Figure 8.19 [15], [18]. Ideally, the source of the common-gate transistor M_8 acts as an ac ground. The current flowing through the compensation capacitor is given by $i_c = C_C d(v_o - v_{S8})/dt \cong C_C \, dv_o/dt$. This current flows through transistor M_8 and is fed back to the input of the second stage of the op amp. Using this simplified model of an ideal current-controlled current source for the common-gate transistor, the application of KCL to nodes v_{ol} and v_o in Figure 8.19 leads to

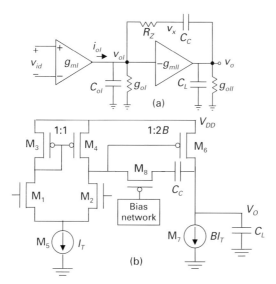

Fig. 8.20   (a) The small-signal equivalent circuit and (b) a schematic diagram of the two-stage op amp with insertion of a nulling resistor in series with the compensation capacitance.

$$\frac{v_o}{v_{id}} = \frac{-g_{mI}g_{mII}/(g_{oI}g_{oII})}{(1 - s/p_1)(1 - s/p_2)} \tag{8.5.40}$$

with

$$p_1 \cong \frac{-g_{oI}g_{oII}}{g_{mII}C_C}, \tag{8.5.41}$$

$$p_2 \cong -\frac{g_{mII}}{C_L + C_C}\frac{C_C}{C_{oI}}. \tag{8.5.42}$$

The extrapolated unity-gain frequency is, as before, given by $\omega_u = g_{mI}/C_C$.

8.5.5.3   A nulling resistor

Another approach to eliminating the RHP zero introduced by the compensation capacitor is to insert a resistor in series with the compensation capacitor, as illustrated in Figure 8.20. The application of KCL to nodes v_{oI}, v_x, and v_o in Figure 8.20 leads to the following voltage gain [5]:

$$\frac{v_o}{v_{id}} = \frac{-[g_{mI}g_{mII}/(g_{oI}g_{oII})](1 - s/z)}{(1 - s/p_1)(1 - s/p_2)(1 - s/p_3)} \tag{8.5.43}$$

with

$$z = \frac{1}{C_C(1/g_{mII} - R_z)}, \tag{8.5.44}$$

$$p_1 \cong \frac{-g_{oI}g_{oII}}{g_{mII}C_C},$$ (8.5.45)

$$p_2 \cong -\frac{g_{mII}}{C_L},$$ (8.5.46)

$$p_3 \cong -\frac{1}{R_zC_{oI}}.$$ (8.5.47)

The frequencies given by (8.5.45)–(8.5.47) have been calculated assuming that the poles are widely spaced [5]. Once again, the unity-gain frequency is g_{mI}/C_C. Note that, as expected, for $R_z = 0$ the zero is situated in the RHP and thus contributes to degrading the phase margin. When $R_z = 1/g_{mII}$, the zero is at infinity. Any increase in R_z places the zero in the left half-plane and, with further increases in R_Z, the zero can be positioned close to the secondary pole p_2 to improve the phase margin. For the complete cancellation of the secondary pole p_2 and z we equate (8.5.44) and (8.5.46), which yields

$$R_z = \frac{1}{g_{mII}}(1 + C_L/C_C).$$ (8.5.48)

Resistor R_z is commonly implemented as a transistor, which operates in the triode region [11], [19] since no direct current flows through it due to the series capacitor. The value of R_z is determined by the aspect ratio of the transistor and the dc voltages at transistor gate and source.

8.6 Three-stage operational amplifiers

Sometimes the cascade of two simple gain stages is not sufficient and a three-stage amplifier may be required in order to achieve the high voltage gain needed for the application. In this case, the frequency compensation becomes more complicated since the op amp will have three poles. Nested Miller compensation (NMC), schematically represented in Figure 8.21 for a three-stage amplifier, is a common method for frequency compensation of operational amplifiers implemented with more than two gain stages. This technique uses Miller capacitors to split poles, as presented previously for the two-stage op amp. The reader is referred to [15], [20]–[22] for more detailed analysis of the NMC op amp. The transfer function of an NMC amplifier is complicated by the zeros associated with pole splitting. An alternative compensation method, similar to the NMC technique, is also shown in Figure 8.21, where extra feedforward stages g_{mfI} and g_{mfII}, represented by the dashed lines, have been added to the original NMC topology. If transconductances g_{mfI} and g_{mfII} are equal to g_{mI} and g_{mII}, respectively, the zeros of the original amplifier are cancelled out. Further details on the design of nested transconductance–capacitance compensation (NGCC) op amps can be found in [15], [20], [23], [24].

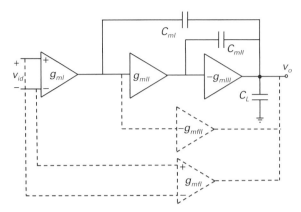

Fig. 8.21   A three-stage op amp with nested Miller compensation (NMC). The nested transconductance–capacitance compensation (NGCC) amplifier is obtained through the addition of two transconductors (dashed lines) to the op amp with NMC.

8.7   Rail-to-rail input stages

The common-mode (CM) input voltage for the circuits shown in Figure 8.2 is constant (and equal to zero), except for the voltage follower in Figure 8.2(b), for which the CM input equals v_i. In typical applications, the voltage applied to the voltage follower swings between the two supply rails. Therefore, the differential input stage of the op amp must be able to respond correctly to input signals between V_{SS} and V_{DD}. Another important application in which the input pair must be able to process properly an input signal that swings between the supply rails is the common-mode feedback amplifier required in fully differential amplifiers [20], to be analyzed in the next section.

The main limitation of using the conventional differential pair as the op-amp input stage is its reduced CM input range. To analyze the limitation of the conventional differential stage in processing rail-to-rail input voltages, we redraw in Figure 8.22 the input differential amplifier of the folded cascode of Figure 8.8.

The maximum CM input voltage must ensure that the current source M_{11} remains in saturation, i.e.

$$V_{SD11} = V_{DD} - (V_{G1} + V_{SG1}) \geq V_{SDsat11}. \tag{8.7.1}$$

Since the CM input voltage V_{ICM} equals V_{G1}, we have

$$V_{ICM} \leq V_{DD} - V_{SG1} - V_{SDsat11}. \tag{8.7.2}$$

The common-mode input voltage of the differential pair with p-channel input must be below V_{DD}, typically by several hundred mV. This is the main limitation of the p-channel differential pair. On the other hand, the CM input voltage must be such that M_1 and M_2 remain in saturation. To calculate the minimum CM input voltage for the saturation of M_1 (M_2) we write

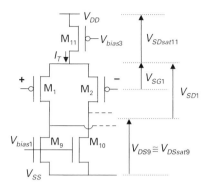

Fig. 8.22   Limitations of the CM input voltage of a differential amplifier with p-channel input. The drain-to-source voltage V_{DS9} is set by the coupled stage (see Figure 8.8).

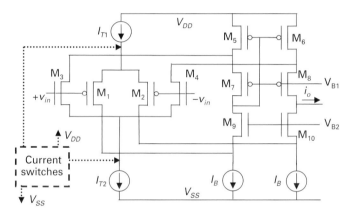

Fig. 8.23   A rail-to-rail input differential amplifier.

$$V_{SD1} = V_{S1} - V_{D1} \cong (V_{ICM} + V_{SG1}) - (V_{SS} + V_{DSsat9}) \geq V_{SDsat1}. \qquad (8.7.3)$$

The approximation in (8.7.3) is valid for an optimally biased folded cascode where the gate voltages of M_3 and M_4 (see Figure 8.8) are such that M_9 and M_{10} are approximately biased on the edge of saturation. Finally, we rewrite (8.7.3) as

$$V_{ICM} \geq V_{SS} + V_{DSsat9} + V_{SDsat1} - V_{SG1}. \qquad (8.7.4)$$

Typically, the source-to-gate voltage of the input transistor is higher than the sum of the saturation voltages of transistors M_9 and M_{11}. Therefore, the p-channel differential pair is able to operate with input voltages down to the negative supply but does not operate appropriately when the input voltage approaches V_{DD}. In contrast with the p-channel input pairs, the CM input voltage for the n-channel differential pairs can go up to V_{DD} but cannot be close to V_{SS}. A solution to circumvent the reduced CM input voltage of a conventional differential pair is to connect it to its complementary differential pair as shown in Figure 8.23 [25]. Note that, for a CM input voltage close to V_{DD}, the tail current I_{T1} tends to zero, thus turning off the p-channel input pair, whereas, for input voltages approaching the negative supply, I_{T2} becomes zero and, thus, the n-channel input devices are off. For an input voltage close to the mid-range power supply, the two tail currents provide the nominal bias currents to the corresponding differential pairs. The output currents of the differential pairs are added by connecting them at the high-swing cascode current mirror. Even though the topology shown in Figure 8.23 is able to deal with a rail-to-rail input range, it has several drawbacks. First, the offset voltage is not constant over the input voltage range. For low CM input, the offset voltage is strongly dependent on the p-channel input pair of transistors, whereas for a CM input approaching V_{DD} the offset voltage is strongly dependent on the n-channel input pair of transistors. In the inter-mediate range of the CM input, the offset voltage is dependent on a weighted sum of the mismatch of the two input pairs. An offset voltage that is dependent on the CM input increases the distortion. Mismatch also degrades the common-mode rejection ratio [26]. Another drawback of the topology in Figure 8.23 is the dependence of the amplifier's transconductance on the CM input voltage, which translates into variable low-frequency voltage gain, gain–bandwidth product, and noise. Several schemes have been proposed in the technical literature to provide a constant-transconductance input stage in order to avoid the aforementioned problems. The common approach to many of these schemes is to connect a set of current switches, as shown by the dotted/dashed lines in Figure 8.23. The purpose of the current switches is to control the bias currents of the differential pairs in such a way that the overall amplifier transconductance remains approximately constant over the CM input range. For input voltages close to midway between the positive and negative supplies, the direct currents for the p- and n-channel input pairs are I_{T1} and I_{T2}, respectively. For input voltages close to V_{DD}, the p-channel input pair turns off and I_{T1} is injected into the current-switch circuit, which, in turn, provides a scaled copy of I_{T1} that is added to I_{T2} to increase the transconductance of the n-channel input pair. This increase in the bias current of the n-channel input pair compensates for the loss of transconductance due to the turning-off of the p-channel input pair. A similar analysis holds for input voltages close to V_{SS}. For information on the schemes for the current switches the reader is referred to [21], [27]–[31].

8.8   Class-AB output stages for operational amplifiers

In some applications the operational amplifier must drive a large capacitance and/or a small resistance with acceptably low levels of signal distortion [32]. This is the case when the op amp has to drive off-chip loads [33], such as headphones for portable electronic devices [34]. Two desirable properties of output stages of operational amplifiers are large output current capability and voltage swing. Also, they must have low output impedance and low standby power [32], and, in addition, should not degrade the frequency response of the amplifier [15].

Class-A amplifiers have a poor efficiency and high power dissipation even when the output signal is zero. In order to reduce power consumption, more efficient output stages are required. One way to improve the power efficiency of operational amplifiers is the employment of class-AB output stages, which are introduced in this section.

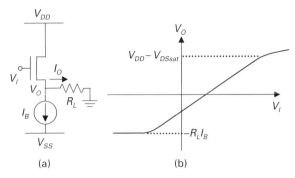

Fig. 8.24 (a) A CMOS class-A source follower and (b) its voltage transfer characteristic.

Before showing two commonly used CMOS class-AB output stages, we review the class-A source follower, which is shown in Figure 8.24. The small-signal properties of the unloaded source follower have been discussed previously, in Chapter 7. For the case of the resistively loaded voltage follower, when the substrate is connected to V_{SS}, the small-signal voltage gain is

$$A_{V0} = \frac{v_o}{v_i} = \frac{1}{n + 1/(g_m R_L)}, \tag{8.8.1}$$

where n is the transistor slope factor, g_m the gate transconductance, and R_L the load resistance. The output voltage is an attenuated copy of the input voltage. The gain is lower than unity and is approximately $1/n$ when $g_m R_L \gg 1$, but this inequality requires a high g_m value and, as a consequence, too much power. Also, the voltage gain is slightly dependent on the input voltage through n and g_m.

The voltage transfer characteristic of the source follower is shown in Figure 8.24(b). When the source follower is employed as the output stage of the op amp, distortion and variation in the voltage gain of the source follower are not of major concern. In common applications of op amps the negative feedback makes the overall gain less sensitive to variations in the op-amp gain and reduces the distortion effects of the output stage.

For the sake of simplicity, let us assume that in Figure 8.24 we have $V_{SS} = -V_{DD}$. The highest value of the current I_O that can be sourced into the resistive load is dependent on both the drive capability of the MOS transistor and the highest voltage that can be applied to the gate, which is usually close to V_{DD}. On the other hand, the highest current that can be sunk from the load is equal to I_B. Thus, to provide the amplifier with a current capability equal to I_B, a quiescent current equal to I_B must be supplied by the current source even when no signal power is delivered to the load [32]. Another limitation of this class-A amplifier is the highest output voltage that can be reached. Since the input voltage V_I is an internal voltage of the op amp, its highest value is slightly lower than V_{DD}. Thus, the highest value of the output voltage is around $V_{DD} - V_{GS}$. Since the gate-to-source voltage is typically several hundred mV, the highest value of the output voltage can be very low, especially for a low supply voltage, which in some cases is around 1 V. A class-A source follower has the advantages of simple topology and low output impedance, and

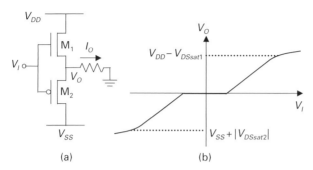

Fig. 8.25   (a) A CMOS class-B source follower and (b) its voltage transfer characteristic.

the main disadvantages are high quiescent power, asymmetric current-drive capability, and significant reduction in voltage swing for low power-supply voltages.

An alternative topology that can be used to avoid the excessive standby power consumption and limited current-sink capability of the class-A amplifier [15], [32] is the class-B source follower shown in Figure 8.25. For positive input voltages, the conductive n-channel transistor pushes current into the load whereas the p-channel device is off. For negative input voltages the p-transistor pulls current from the load whereas the n-transistor is off. The conduction of each transistor for alternate half cycles is the origin of the name push–pull [15].

One of the drawbacks of the class-B amplifier is the dead band in the voltage transfer characteristic around the origin, as shown in Figure 8.25(b). In effect, for small input voltages, the current through both M_1 and M_2 is very small and, as a result, the output current is negligible. Thus, the output voltage does not respond to input voltages lying within the dead zone. The effect of the dead band on the transfer function of the feedback op amp can be reduced with a large loop gain [21], but a non-linear static transfer characteristic, although not so prominent as the one shown in Figure 8.25, also appears. Also, since the stage previous to the class-B amplifier has a limited slew rate and, consequently, cannot impose a fast variation on V_I, a crossover distorted output occurs in response to quickly varying input signals [21].

In the class-AB amplifier in Figure 8.26(a), for which the transfer characteristic is shown in Figure 8.26(b), two diode-connected transistors, M_3 and M_4, are used to avoid the dead band of the transfer characteristic of class-B amplifiers. When the MOS transistors are replaced with bipolar transistors, the closed loop formed by the base–emitter junctions is denominated a *translinear* loop. Class-AB output stages are just one example of a class of circuits denominated *translinear* circuits [35]. The term translinear, coined by B. Gilbert in 1975 [36], is associated with the fact that the transconductance of a bipolar transistor is linearly dependent on the collector current. The translinear concept can also be applied to MOS transistors operating in weak inversion [37]. An extension of the translinear principle to MOS transistors operating in strong inversion was presented in [38].

On applying KVL to the closed loop that comprises the gate-to-source voltages of transistors M_1 to M_4 we find that

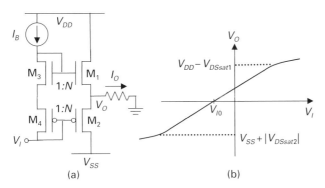

Fig. 8.26   (a) A CMOS class-AB source follower and (b) its voltage transfer characteristic.

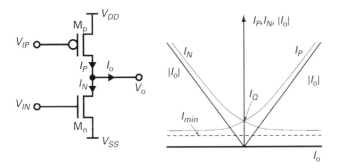

Fig. 8.27   A CMOS common-source output stage and push (I_P) and pull (I_N) currents versus output current.

$$V_{GS1} + V_{SG2} = V_{GS3} + V_{SG4}. \qquad (8.8.2)$$

To simplify the analysis of the translinear loop, let us assume that the voltage drop across the diode-connected transistors does not change with variations in input voltage, i.e. $V_{GS1} + V_{SG2}$ is constant (in fact, the drops in diode voltage change slightly with the input variations when the sources are not connected to their respective substrates). The quiescent current, which is the direct current when the signal current $I_O = 0$, is NI_B, where $N = S_1/S_3 = S_2/S_4$. The input voltage at which the output current is equal to zero is V_{I0}. When $V_I > V_{I0}$, the current flowing through M_1 is greater than that flowing through M_2, since $V_{GS1} + V_{SG2}$ is constant. For increasingly higher values of the input voltage, the output current can become several times greater than the quiescent current. When $V_I < V_{I0}$, the current in M_2 is greater than that in M_1 and, thus, the amplifier sinks current from the load.

As mentioned previously, one of the main drawbacks of the use of a source follower as the output stage of an op amp is that the output voltage cannot swing close to the supply rails. This is especially troublesome for deep-submicron technologies, where supply voltages are around 2 V or less. To achieve an output voltage range as large as possible, the output transistors are generally connected in the common-source configuration [39], as shown in the scheme of Figure 8.27, where v_{IP} and v_{IN} are in-phase signals, since, when I_P

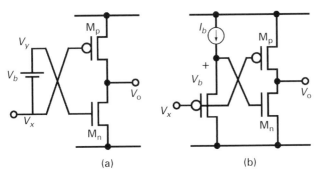

Fig. 8.28   A class-AB output stage with (a) floating source voltage and (b) an implementation of the floating source using a common-drain amplifier.

increases, I_N must decrease and vice versa. As the graph of the push–pull characteristics shows, the output (push or pull) current can be considerably higher than the quiescent current I_Q. Note that a desirable characteristic of these class-AB amplifiers is that the current through the p- or the n-channel transistor does not fall below a certain minimum level I_{min}. This prevents the turn-on delay of the non-active output transistor from being significant, which in turn reduces the crossover distortion [39], [40]. The output voltage of this stage is almost rail-to-rail since the output can reach either one of the supply voltages minus one drain-to-source saturation voltage [39] and still maintain the performance. When operating as an op-amp output stage, the transistors of the common-source amplifier can operate in the linear region (with V_{DS} equal to some tens of mV) and the op-amp gain can still be relatively high. The common-source amplifier is said to be rail-to-rail because it is capable of reaching the supply rails within some tens of mV.

The obvious choice for the push–pull output stage of Figure 8.27 is the static CMOS inverter, where the input voltages are tied together. The CMOS inverter has the push–pull characteristics with shapes close to those shown in Figure 8.27, but has two fundamental shortcomings: (i) the quiescent current is dependent on both the supply voltage and technological parameters; and (ii) the minimum current I_{min} is zero.

Another way to bias a class-AB output stage is to include a floating voltage source between the gates of the driver transistors, as Figure 8.28 shows. Probably, the simplest way to realize the floating source is by means of the common-drain amplifier, as shown in Figure 8.28(b). The lowest current in this topology, however, also tends to zero.

Figure 8.29 gives the simplified scheme of a class-AB amplifier [39], [41] that has current characteristics similar to those shown in Figure 8.27. For the dc analysis we assume that the currents provided by the top and bottom current sources are equal and that the aspect ratios obey [39] the following relationships:

$$\frac{S_3}{S_1} = \frac{S_4}{S_2} = \frac{S_5}{S_6} = \frac{S_P}{S_N}. \qquad (8.8.3)$$

One can see that the gate-to-source voltages of M_P and M_N are controlled by two translinear loops, namely M_1, M_2, M_6, M_N and M_3, M_4, M_5, M_P, respectively. The gate

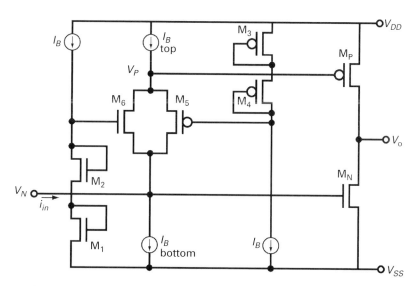

Fig. 8.29 A rail-to-rail output stage and bias circuit for class-AB operation (adapted from [41]).

voltage of M_5 (M_6), which is set by the stacked pair of diode-connected transistors M_3 and M_4 (M_1 and M_2), is independent of the input signal; thus, the currents flowing through the complementary devices M_P and M_N, which are set by their corresponding gate-to-source voltages, are determined in response to the complementary common-gate level shifters M_5 and M_6.

To gain an insight into the class-AB operation of the circuit in Figure 8.29, let us assume that v_N increases. The increase in v_N reduces the current flowing through M_6, which, in turn, forces the current in M_5 to increase and, thus, so does the voltage v_P. Therefore, v_P and v_N are in-phase voltages, as required in class-AB amplifiers.

For the determination of the quiescent current I_Q for $I_P = I_N$ we assume that, in addition to the relationships given in (8.8.3), we also have $I_{B,top} = I_{B,bottom} = 2I_B$. Thus, the currents that flow through M_5 and M_6 are equal to I_B. Since $S_5 = S_4$,[3] the inversion levels of M_4 and M_5 (both assumed to be in saturation) are equal and, consequently, their gate-to-source voltages are also equal. As a result, the gate-to-source voltage of M_P is equal to that of M_3. The same conclusion holds for the other translinear loop. The quiescent current is then given by

$$I_Q = \frac{S_P}{S_3} I_B = \frac{S_N}{S_1} I_B. \tag{8.8.4}$$

In the scheme of Figure 8.29, the current in the inactive output transistor cannot be lower than a certain fraction of the quiescent current I_Q. For decreasing input currents, the node voltage v_P can decrease to such an extent that M_5 shuts off. In this case, the bias current $2I_B$ flows completely through M_6; therefore, the voltage applied to the gate of M_N is clamped at a value equal to that at the gate of M_6 minus the gate-to-source voltage

---

[3] For the sake of simplicity, the aspect ratios and currents of our bias scheme have been modified with respect to the original paper [41].

across M_6. Thus, a minimum current, typically a fraction within the range 30%–60% of the quiescent current, is established in the inactive output transistor. A similar reasoning can be used to explain why the current through M_P is always greater than a minimum.

In order to analyze the small-signal behavior of the class-AB bias stage, we assign the symbols g_{bb} and g_{bt} to the output conductances of the bottom and top current sources, respectively. From KCL we can readily derive the following relationships:

$$v_p = v_n \frac{g_{ms6} + g_{md5}}{g_{ms5} + g_{md6} + g_{bt}}, \tag{8.8.5}$$

$$i_{in} = g_{bb}v_n + g_{bt}v_p = v_n \left(g_{bb} + g_{bt} \frac{g_{ms6} + g_{md5}}{g_{ms5} + g_{md6} + g_{bt}} \right). \tag{8.8.6}$$

For balancing reasons, M_5 and M_6 are usually designed such that their transconductances are equal in the quiescent state [42]. This is simply obtained by making the specific currents I_{S5} and I_{S6} of the level-shift transistors equal. When the input current increases, the node voltage v_n also increases and the voltage at node v_p follows that at node v_n. Thus, for operation of the class-AB amplifier close to the quiescent state, the association M_5–M_6 operates as a floating-voltage source. As previously explained, for high dc values at the input node, M_6 turns off and, consequently, g_{ms6} can be considered zero, which causes the voltage at node v_p to become almost unaffected by that at node v_n. This "disconnection" of the nodes that control the currents ensures that, when one of the output transistors is driving the load, the other is still conducting some current. It can be observed that the input conductance of the class-AB stage is usually low; under the quiescent condition it is approximately equal to the sum of the output conductances of the top and bottom current sources, when $g_{ms5} = g_{ms6}$.

Example 8.8

For the class-AB output stage in Figure 8.29, assume that all transistors operate in weak inversion. Assume that $n_N = n_P = 1.2$, $I_{Si} = 10\,\mu A$, for $i = 1,...,6$, $I_{SP} = I_{SN} = 1$ mA, and $I_B = 0.1\,\mu A$. (a) Derive an equation for the relationship between the output currents I_{Dp} and I_{Dn}. (b) Calculate the quiescent current. (c) What is the current flowing through M_P and M_N when the load current is $30I_Q$? The technology is n-well. The n-wells are connected to V_{DD}. Assume that (8.8.3) holds and $I_{Btop} = I_{Bbottom} = 2I_B$.

Answer

(a) We first rewrite the UICM for transistors operating in weak-inversion saturation,

$$\frac{V_{GB} - V_{T0}}{n\phi_t} - \frac{V_{SB}}{\phi_t} = (-)\ln\left(\frac{I_D}{I_S}\right),$$

where the sign (−) must be used for p-channel devices.

Using the UICM for the translinear loops M_1–M_2–M_6–M_N and M_3–M_4–M_5–M_P and considering that we have $V_{SB2} = V_{GS1}$ and $V_{GSn} = V_{SB6}$ and the analogous relation for the p-loop, we find that

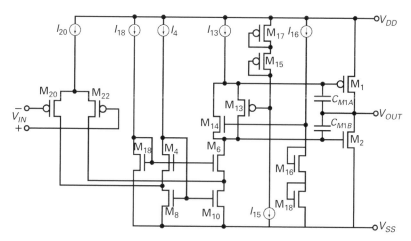

Fig. 8.30   A two-stage op amp with rail-to-rail class-AB output stage [42].

$$n_N \ln\left(\frac{I_B}{I_{S1}}\right) + \ln\left(\frac{I_B}{I_{S2}}\right) = n_N \ln\left(\frac{I_{DN}}{I_{SN}}\right) + \ln\left(\frac{I_{D6}}{I_{S6}}\right),$$

$$n_P \ln\left(\frac{I_B}{I_{S3}}\right) + \ln\left(\frac{I_B}{I_{S4}}\right) = n_P \ln\left(\frac{I_{DP}}{I_{SP}}\right) + \ln\left(\frac{I_{D5}}{I_{S5}}\right)$$

for the n- and p-loops, respectively, which can be rewritten as

$$\left(\frac{I_B}{I_{S1}}\right)^{n_N}\frac{I_B}{I_{S2}} = \left(\frac{I_{DN}}{I_{SN}}\right)^{n_N}\frac{I_{D6}}{I_{S6}} \quad \text{and} \quad \left(\frac{I_B}{I_{S3}}\right)^{n_P}\frac{I_B}{I_{S4}} = \left(\frac{I_{DP}}{I_{SP}}\right)^{n_P}\frac{I_{D5}}{I_{S5}}.$$

The substitution of $I_{D5} + I_{D6} = 2I_B$ into the two previous equations gives, after some algebra,

$$\left(I_Q/I_{DN}\right)^{n_N} + \left(I_Q/I_{DP}\right)^{n_P} = 2,$$

where $I_Q = (S_N/S_1)I_B$. Note that the specific currents of the n-channel devices are equal to those of the corresponding p-channel devices and also that M_2 (M_4) and M_6 (M_5) are equally sized. It is interesting to note that, if one of the output currents tends toward infinity, then the other tends toward $I_Q/2^{(1/n)}$. For typical slope factors, the minimum current is around $0.6I_Q$.

(b) The quiescent current $I_Q = (S_N/S_1)I_B = 10\ \mu A$.

(c) Using

$$\left(I_Q/30I_Q\right)^{1.2} + \left(I_Q/I_{DP}\right)^{1.2} = 2$$

we find that $I_{DP} \cong 0.565I_Q$, which is very close to $I_{DPmin} = 0.561I_Q$.

Figure 8.30 shows the scheme of a two-stage op amp with a rail-to-rail class-AB output stage. The first stage is a conventional folded-cascode amplifier and the second stage is the class-AB amplifier described previously. The two-stage topology is compensated using Miller capacitors C_{M1A} and C_{M1B}. One of the disadvantages of the circuit in Figure 8.30 is that it cannot be used for very low power-supply voltages [42] since the minimum supply voltage required for its proper operation is equal to two gate-to-source voltages plus one saturation voltage (see, for example, the mesh formed by I_{15}, transistors M_{15} and M_{17}, and power supply rails). Class-AB output stages for very low supply voltages are presented in [39], [42]–[44].

8.9   Fully-differential operational amplifiers

In this section we introduce the fully-differential (FD) op amp, which in an ideally balanced configuration has two inverted outputs (v_{o1} and v_{o2}) with the same dc voltage level, V_{ocm}, as shown in Figure 8.31. The magnitude of the differential output voltage (v_{od}), which is twice that of a single output, allows an improvement of the signal-to-noise ratio (SNR) by a factor of 3 dB [15] over that of a single-ended amplifier. The higher output-voltage swing is particularly important for operation at low power-supply voltage. Other advantages of FD circuits include the cancellation of common-mode signals and noise, as well as the power-supply noise [45]. The proper use of FD amplifiers can help reduce the effects of charge injection and clock feedthrough, which degrade the performance of switched-capacitor circuits. Overall, the use of FD circuits can improve the SNR by over one order of magnitude compared with their single-ended counterparts [46]. The absence of even-order non-linearities is another benefit of balanced FD amplifiers [15], [45], [46]. The disadvantage of FD op amps is the need for both two matched feedback networks and a common-mode feedback (CMFB) circuit, to be discussed later. The latter is required in order to set the value of the common-mode output voltage (V_{ocm} in Figure 8.31), which is usually around midway between the power-supply voltages to allow for maximum signal swing.

    In order to explain the need for the CMFB circuit,[4] we start by showing in Figure 8.32 the scheme of an idealized FD amplifier together with the bias circuit to the right of the dotted line. The FD amplifier is composed of matched transistor pairs (M_1, M_2),

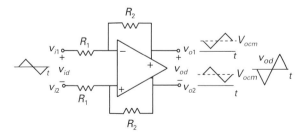

Fig. 8.31   A fully-differential op amp and representation of input and output waveforms.

---

[4] Even though FD amplifiers without CMFB have been discussed in the technical literature [47], we restrict our analysis to those that employ CMFB, which is widespread in FD amplifiers currently in use.

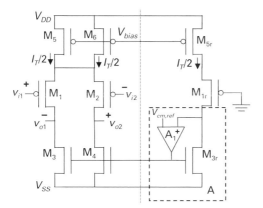

Fig. 8.32 An example of an idealized FD amplifier, where transistors are assumed to be matched, and the bias network employed to set the common-mode output voltage to $V_{cm,ref}$.

(M_3, M_4), and (M_5, M_6). The bias circuit is composed of "one half" of the main amplifier, in which transistors M_{1r}, M_{3r}, and M_{5r} are matched copies of M_1, M_3, and M_5, and an additional amplifier A_1. Amplifier A, the cascade of amplifiers A_1 and the common-source amplifier, composed of the drive transistor M_{3r} loaded by stacked transistors M_{1r} and M_{5r}, is configured as a voltage follower. For the sake of simplicity we assume that the common-mode (CM) input voltage is equal to zero (ground potential) and that amplifier A has a very high gain. Under such conditions, we can see that the bias circuit, except for amplifier A_1, is an exact replica of one of the branches of the main amplifier. Amplifier A_1 drives the gate of M_{3r} in such a way that the current through it is equal to $I_T/2$ while its drain voltage is approximately $V_{cm,ref}$. Since the dc voltages at the inputs of the FD amplifier are assumed to be zero and the bias circuit is matched to the FD amplifier, the dc voltages at both v_{o1} and v_{o2} equal the drain voltage of M_{3r}. Thus, for the idealized FD amplifier in Figure 8.32, the CM output voltage is approximately $V_{cm,ref}$. However, since mismatch is unavoidable, the dc voltage at the FD amplifier output nodes can be anywhere between the supply voltages.

To explain the effect of mismatch on the CM output voltage, let us assume that the op amp is connected in the unity-gain configuration, that is, $v_{i1} = v_{o1}$ and $v_{i2} = v_{o2}$. Of course, the currents that flow through M_1 and M_3 are the same, as are the currents that flow through M_2 and M_4. The mismatch between the pairs (M_1, M_2) and (M_3, M_4) is compensated for by means of a small voltage difference between the inputs (input-referred offset voltage). Thus, the feedback stabilizes the differential-mode output voltage, but what can be said about the CM output voltage? To see what happens with the CM output voltage, let us assume that the pairs (M_1, M_2), (M_3, M_4), and (M_5, M_6) are matched, which is acceptable for the analysis of the CM voltage since the mismatch between components is compensated for by the offset voltage. In this case, a p-type current source in a cascode configuration $(M_5, M_6$ and $M_1, M_2)$ must be balanced by an n-type current source (M_3, M_4). Note that, in the unity-gain configuration, a replica of the output voltage is transferred to the drains of M_5, M_6 through a voltage follower. Since the pairs (M_1, M_2) and (M_5, M_6) operate in saturation, a large variation of v_{o1} (or v_{o2}) can be

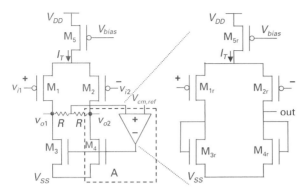

Fig. 8.33   A simplified scheme of an FD amplifier employed to illustrate how to set the common-mode output voltage to $V_{cm,ref}$. On the right is the scheme of the low-gain amplifier, a replica of the main amplifier but with diode-connected loads.

necessary in order to accommodate differences between the CM currents of the p-type and n-type current sources. Thus, the bias circuit shown in Figure 8.32 is unfeasible for controlling the CM output voltage.

A practical scheme that employs feedback to control the CM output voltage of the FD amplifier in Figure 8.32 is illustrated in Figure 8.33. When the FD amplifier operates properly, the output voltages are inverted signals and the resistive divider provides a sample of the CM output voltage $((v_{o1} + v_{o2})/2)$ to the non-inverting input of the high-gain operational amplifier A, which is composed of two stages. The first stage is the low-gain differential amplifier M_{1r}–M_{5r} which, except for the connections of the load transistors as diodes, is a replica of the main amplifier. The second stage is a double common-source stage, composed of M_3 and M_4 loaded by the input pair (M_1, M_2). The gate voltage of M_3/M_4 sets the CM output voltage at $V_{cm,ref}$. Note that amplifier A in the CM signal path is connected as a voltage follower.

The open-loop low-frequency gain of amplifier A is given by

$$A_{V0.CMFB} = \frac{v_{o,cm}}{v_{+r} - v_{-r}}$$

$$= \frac{-g_{m1,r}}{2(g_{m3,r} + g_{ds1,r} + g_{ds3,r})} \frac{g_{m3}}{g_{ds1} + g_{ds3}} \cong -\frac{1}{2} \frac{g_{m1,r}}{g_{ds1} + g_{ds3}}, \qquad (8.9.1)$$

where v_{+r} and v_{-r} are the input voltages of amplifier A. Note that the resistive network used to determine the CM output voltage does not affect the open-loop gain but affects the differential voltage gain of the FD amplifier. If we assume that the value of R is considerably higher than the parallel association of the output resistances of M_1 and M_3, then the differential amplifier gain, measured at one of the outputs, is given by (8.9.1). Even though the resistive CM detector is highly linear, its main drawback is its effect on the differential voltage gain for practical values of resistors integrated in CMOS technologies [46].

At this point we recall that the FD op amp operates in a closed-loop configuration; therefore, under no ac input signal, a differential input voltage equal to the offset voltage of the op amp forces the two outputs to have the same dc voltage, which, in turn, due to the action of the CMFB circuit, equals $V_{cm,ref}$.

As noted in [46], [48], a CMFB network must satisfy the following set of requirements: it must be possible (i) to set the CM voltage at a given value $V_{cm,ref}$ to allow for maximum signal swing and/or maximum differential voltage gain; (ii) to process the CM signal with speed and accuracy similar to those of the differential-mode (DM) component processed by the FD op amp; and (iii) to avoid the influence of the CMFB circuit on the DM output voltage.

In the example of Figure 8.33, we have seen that requirement (i) is achieved once we set $V_{cm,ref}$ at the proper value. Condition (ii) is also fulfilled since the scheme devised for the CMFB is identical to that of the FD amplifier. Strictly speaking, however, the CM signal path in the example of Figure 8.33 is slightly slower than the FD signal path. This is due to the extra delay generated in the CM signal path by the conversion of the current flowing through M_{2r} into a voltage across load M_{4r}, which is then converted back into currents that flow through M_3 and M_4. Requirement (iii) can be satisfied if the resistors of the CM detector do not affect too much the differential gain or if buffers are included at the FD op-amp outputs to avoid their being loaded by the resistive components. In general, the CM detection using the resistive network requires large silicon area and/or additional power overheads.

Since the CMFB circuit operates in a negative-feedback configuration, stability must be accounted for. As mentioned in [45], when the CM and DM signal paths at the very front end of the amplifier are merged and the remaining parts are identical, stability and accuracy of the CM loop are achieved automatically by the design of the DM path. The introduction of additional non-dominant poles in the CM path with respect to the DM path can, however, decrease the phase margin significantly. Thus, in this case, to obtain a phase margin of the CM loop close to or even better than that of the DM loop, one must introduce some kind of modification to the CM loop, as shown next.

Let us take a closer look at the frequency response of the FD amplifier in Figure 8.33. The extrapolated unity-gain frequency of the voltage gain, in rad/s, is, for a single output,

$$\omega_u = \frac{g_{m1}/2}{C_L} = \frac{I_{SH,P}(W/L)_1}{C_L n_P \phi_t} \left[\sqrt{1 + \frac{I_T}{2I_{SH,P}(W/L)_1}} - 1 \right], \qquad (8.9.2)$$

where g_{m1} is the transconductance of the input transistor of the FD amplifier and C_L the capacitance of the output node. In a typical op amp, the phase margin, degraded by secondary poles and right-half-plane zeros, is usually about 45° or more. On the other hand, as regards the open-loop gain associated with the CMFB loop in Figure 8.33, the extrapolated unity-gain frequency is also given by (8.9.2). However, the phase margin of the CMFB loop is degraded by an additional secondary pole. In the example of Figure 8.33, the extra path in the CM feedback loop relative to the DM gain creates an additional pole that reduces the phase margin of the CMFB loop gain. To avoid this degradation being significant or even causing instability of the feedback amplifier, a possible solution to increase the phase margin of the CMFB loop is to reduce the unity-gain frequency of the op amp [15].

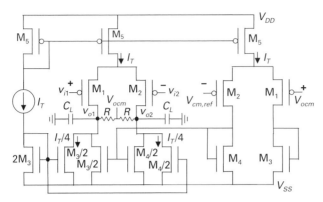

Fig. 8.34   An FD amplifier with resistive CM detector and the error amplifier on the right employed to set the common-mode output voltage to $V_{cm,ref}$.

In the FD amplifier shown in Figure 8.34,[5] which is a modified version of the op amp in Figure 8.33, the unity-gain frequency of the CMFB loop is about half that of the DM amplifier. Note that this decrease is obtained through a reduction in the effective transconductance of the differential amplifier on the right, since just half of the current through M_4 is now transferred to transistor $M_4/2$. This modification not only decreases the unity-gain frequency of the CM control amplifier with respect to that in Figure 8.33 but also increases slightly the frequency of the secondary pole introduced by the diode-connected transistor M_4 due to the lower capacitance seen by the drain of M_4. A disadvantage of this approach to reducing the unity-gain frequency of the CM control amplifier is the simultaneous reduction in the dc voltage gain of the CMFB loop. In most cases, however, a smaller dc gain is not of major impact on the accuracy of the (closed-loop) voltage follower employed to control the CM output voltage.

Some other comments about the bandwidth required for the CMFB loop are in order. As described in [45], the bandwidth of the CM loop has to be at least as wide as the highest frequency at which output balancing is required. In integrated circuits there are several sources of CM signals, which can be coupled to the FD amplifier outputs. The CM input signal and power-supply noise are coupled to the amplifier outputs due to finite CMRR and PSRR, respectively. Substrate-induced noise can also appear at the amplifier outputs via transistors or coupling capacitances. Readers are referred to [15] for further details on the requirements regarding the bandwidth of the CMFB loop. In the examples that follow we assume that, unless stated otherwise, the unity-gain frequency of the CMFB loop has been designed to be approximately equal to the unity-gain frequency of the DM amplifier. This assumption is, in general, not needed in practice, but it will be used here just for the sake of exemplification.

Figure 8.35 shows a modified version of the FD amplifier of Figure 8.34. Note that, to simplify the schematic drawing, we do not show the load capacitors. Two major

---

[5] In some of the schemes, we have assigned the symbol αM_i to transistors. Here α is a number equal to the aspect ratio of the transistor under consideration relative to that of transistor M_i. For example, $M_3/2$ is a transistor whose aspect ratio is half the aspect ratio of M_3.

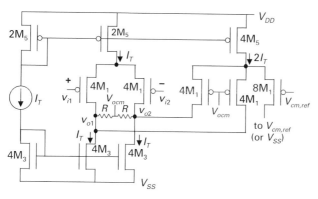

Fig. 8.35   An FD amplifier where the two CM transconductors are replicas of the DM transconductor.

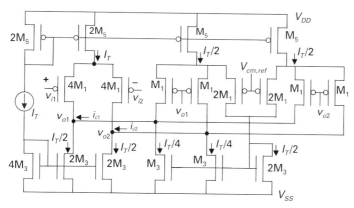

Fig. 8.36   An FD amplifier employing CM detection based on differential pairs that are replicas of the DM transconductor.

modifications relative to Figure 8.34 are (i) the output currents of the CM transconductors (the two rightmost transistors $4M_1$) are directly injected into the DM amplifier output nodes, and (ii) the CM transconductor is biased with a current that is twice that of the DM transconductor. Thus, the input differential pairs of the CM and DM amplifiers both operate at the same inversion level and their transconductances are the same. Consequently, except for mismatching, the frequency response of the two amplifiers is the same since they share a common load. Note that the transistor labels in Figure 8.35 represent parallel associations of transistors; this representation is employed for the sake of comparison with the circuit shown in Figure 8.36.

The topology of Figure 8.35 has two drawbacks: it makes use of resistors and the current that flows through the rightmost transistor $8M_1$ is wasted.

An often employed alternative to the use of resistors for a CM detector is based on two identical differential pairs. For one of them the inputs are v_{o1} and $V_{cm,ref}$, whereas for the other the inputs are v_{o2} and $V_{cm,ref}$. The drain currents of the transistors, the gates of which are connected to v_{o1} and v_{o2}, are approximately proportional to the differences

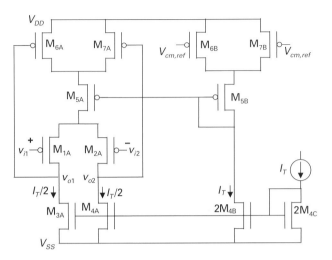

Fig. 8.37   An FD amplifier using transistors in the triode region to set the common-mode output voltage to $V_{cm,ref}$ [15], [49].

$v_{o1} - V_{cm,ref}$ and $v_{o2} - V_{cm,ref}$, respectively. If these two drain currents are added, a current that increases in proportion with the CM signal results. This is the principle used in the FD amplifier of Figure 8.36 to detect the CM signal. Another advantage of the circuit in Figure 8.36 over that in Figure 8.35 is the power saving due to the advantageous use of the current flowing through the transistors with the gates connected to $V_{cm,ref}$. Note that the transconductances of the CM transconductors in Figure 8.35 and 8.36 are the same even though the total current in the core cell of the FD amplifier of Figure 8.35 is 50% higher than that in Figure 8.36.

The action of the CMFB amplifier in Figure 8.36 can be explained as follows. Assuming that the CM output voltage $(v_{o1} + v_{o2})/2 = V_{cm,ref}$, the output currents i_{c1} and i_{c2} injected at nodes v_{o1} and v_{o2} equal zero. Now, assume that a disturbance causes $(v_{o1} + v_{o2})/2 > V_{cm,ref}$. In this case, the currents i_{c1} and i_{c2} are negative, i.e. they flow out of nodes v_{o1} and v_{o2}. Assume for a while that the currents that flow through the input transistors $4M_1$ and the load transistors $2M_3$ of the main differential amplifier do not change with the CM output voltage. In this case, i_{c1} and i_{c2} produce the discharge of the capacitances at nodes v_{o1} and v_{o2} down to the CM reference voltage, reinstating $(v_{o1} + v_{o2})/2 = V_{cm,ref}$ and $i_{c1} = i_{c2} = 0$. Thus, the CMFB circuit provides a negative feedback, a condition required for stability. Note that the transconductances relating to the CM and DM signals are about the same.

The main disadvantage of the circuit in Figure 8.36 compared with that in Figure 8.35 is the non-linearity of the CM detector. This non-linearity makes the detector sensitive not only to CM signals, which is a desirable characteristic, but also to DM signals, which is undesirable.

A simple principle used to stabilize the CM voltage is shown in Figure 8.37, where transistors M_{6A}, M_{7A}, M_{6B}, and M_{7B} are biased in the triode region [15], [49]. The subscript A refers to the main amplifier, whereas B refers to the bias network. Transistors

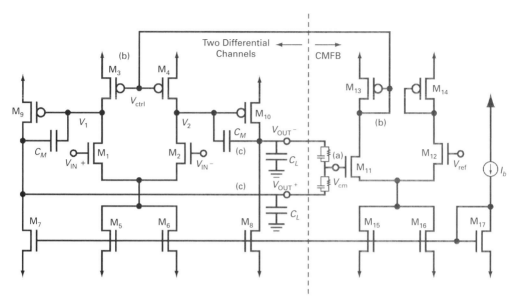

Fig. 8.38   A complete scheme of an FD two-stage op amp including the CM control circuit [50].

indicated by the same subscript numbers are matched. For an initial analysis of the CMFB loop, let us assume that the currents flowing through the triode-biased transistors M_{6A} and M_{7A} respond linearly to their gate voltages. With this simplifying assumption, the current flowing through M_{5A} is proportional to the sum of the gate voltages applied to M_{6A} and M_{7A}. Assuming the perfect matching of networks A and B and that the effect of the drain voltage of M_{5A} on its current is negligible, the sum of the gate voltages of M_{6A} and M_{7A} is forced to be equal to $2V_{cm,ref}$. In this way, the control of the CM output voltage is achieved.

Transistors M_{6A} and M_{7A} in the topology of Figure 8.37 can also operate in saturation but the dependence of the drain current on the gate voltage becomes more non-linear, thus introducing a CM detection that has a greater dependence on the DM output voltage. It is also worth mentioning that this kind of CM detection is not appropriate for rail-to-rail output stages since transistors M_{6A} and M_{7A} turn off for gate voltages approaching V_{DD} [15]. In this topology the CM loop gain is relatively low due to the employment of transistors M_{6A} and M_{7A} in the triode region, which contributes to reducing their transconductances in comparison with their operation in saturation. The same principle as shown in Figure 8.37 for a simple differential amplifier has also been applied to a folded-cascode amplifier [49].

An FD two-stage Miller amplifier that employs resistive CM detection is shown in Figure 8.38. The integrated resistors are, in fact, RC lines distributed between their terminals and the substrate/well. Therefore, the high-frequency components of the CM signal are delayed at node V_{cm} and can severely degrade the phase margin of the CMFB loop. The purpose of the capacitors being added in parallel with the resistors is to provide a high-frequency bypass of the distributed capacitance between the integrated resistors and the substrate to guarantee the stability of the CMFB loop [45]. Note that the CM

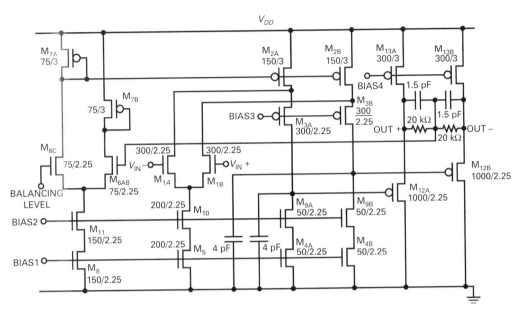

Fig. 8.39   A complete scheme of an FD folded-cascode op amp fabricated in 1.75-μm CMOS technology [45].

transconductor (M_{11}, M_{12}, M_{15}, M_{16}) is a replica of the DM transconductor (M_1, M_2, M_5, M_6), but, as previously mentioned, the phase margin of the CM loop is smaller than that of the DM loop due to the two-output current mirror (M_{13}, M_3–M_4). Since the DM and CM paths are identical, except for slightly different phase margins, capacitors C_M provide the compensation required for both DM and CM loops.

    An example of an FD folded-cascode amplifier designed for a MOSFET-C continuous-time filter application in a 1.75-μm CMOS technology is shown in Figure 8.39 [45]. MOSFET-C filters will be introduced in the next chapter. The differential pair (M_{1A}, M_{1B}), biased by the cascode current source (M_5, M_{10}), converts the differential input voltage into a differential current, which, in turn, is converted into a differential output voltage at the drain nodes of M_{9A} and M_{9B}. Two source followers, M_{12A}–M_{13A} and M_{12B}–M_{13B}, "sense" the amplifier output voltages and provide two outputs, OUT^+ and OUT^-, which are dc-shifted copies of the amplifier output voltages. The resistive network averages the values of OUT^+ and OUT^-. The CM voltage at the middle node of the resistive network is then compared with the balancing level at the CM input stage. The balancing level, which is a dc-shifted copy of the CM reference level $V_{cm,ref}$, is 3.5 V. The CM error is injected into the amplifier output through M_{7A}. In order to obtain a low offset voltage, the authors of [45] used no direct injection of the CM signal into the output stage even though this was at the expense of a lower CM loop gain.

    Some other schemes for detecting and controlling the CM output voltage of FD amplifiers exist. For a description of the most commonly employed CM detectors in CMOS technology together with such important properties such as gain or transconductance, sensitivity to mismatch in devices, and the (undesirable) dependence of CM detection on DM signals, the reader is referred to [46].

Appendix

A8.1 Systematic offset of a two-stage op amp

Refer to Figure 8.11 of a two-stage CMOS amplifier. In the following we estimate the value of the systematic offset voltage due to modulation of the current by the drain voltage. Since devices M_5 and M_7 are fully matched, except for a scaling factor, the current flowing through M_7 can be written as

$$I_{D7} = BI_{D5} + g_{ds7}(V_O - V_S), \tag{A8.1.1}$$

where the last term in (A8.1.1) has been included to account for the difference between the drain voltages of M_5 and M_7. In the following, we calculate the current I_{D6} that flows through M_6 as a result of an incremental variation v_{ol}, around the dc value V_{G6}, at the output node of the differential amplifier. Recalling that the current through M_6 is a replica of the current through M_4 (or M_3) with slight deviations due to differences in their gate and drain voltages, we can write the current through M_6 as

$$I_{D6} = 2B\frac{I_{D5}}{2} + g_{m6}(V_{G3} - v_{ol} - V_{G3}) + g_{ds6}(V_{G3} - V_O). \tag{A8.1.2}$$

Note that I_{D6} is a replica of the current that flows through M_3 (or M_4) for differential input voltage equal to zero plus two additional components that account for differences between both the gate and the drain voltages of M_6 and M_4. V_{G3} is the dc gate voltage of M_3 for $V_{G1} = V_{G2}$ and v_{ol} is the small variation required in the output voltage of the differential amplifier such that the op-amp output voltage equals V_O. The value of v_{ol} can be calculated in terms of V_{OS} as

$$v_{ol} = \frac{g_{m1}}{g_{ds2} + g_{ds4}} V_{OS}, \tag{A8.1.3}$$

since V_{OS} is the input voltage required in order to obtain a voltage equal to V_O at the op amp output. Now, equating (A8.1.1) and (A8.1.2), and using (A8.1.3), we find that the systematic offset voltage $V_{OS,s}$ is

$$V_{OS} = \frac{1}{A_{V0}} \left[\frac{g_{ds6}}{g_{ds6} + g_{ds7}} (V_{G3} - V_O) + \frac{g_{ds7}}{g_{ds6} + g_{ds7}} (V_S - V_O) \right], \tag{A8.1.4}$$

where A_{V0} is the low-frequency voltage gain of the two-stage op amp:

$$A_{V0} = \frac{g_{m1}g_{m6}}{(g_{ds2} + g_{ds4})(g_{ds6} + g_{ds7})}.$$

We note that the value of the systematic offset voltage is dependent on the value of the output voltage. Suppose, for example, that the offset voltage is defined as the differential input voltage such that the output voltage equals $V_{DD}/2$. In this case, we have

$$V_{OS}\big|_{V_O = V_{DD}/2} = \frac{1}{A_{V0}} \left(\frac{g_{ds6}V_{G3} + g_{ds7}V_S}{g_{ds6} + g_{ds7}} - \frac{V_{DD}}{2} \right). \tag{A8.1.5}$$

Also, note in the above equation that the systematic offset voltage is dependent on the CM input range through V_S. It is interesting to note that, if we select $V_O = V_{G3}$ to define the systematic offset, the influence of V_{G3} on the offset voltage disappears, as can be readily observed from (A8.1.4).

Problems

8.1 The output current of the OTA employed for the integrator of Figure P8.1 is given by $I_o = g_m V_X + g_o V_o$. Plot the magnitude and phase response of the voltage transfer function of the non-ideal integrator. What are the dc gain and the unity-gain frequency of the integrator?

8.2 Derive Equation (8.3.1) for the offset voltage of the symmetric op amp in Figure 8.6. Examine and interpret carefully the sign of each term. Calculate the CM input range for an n-well technology.

8.3 Calculate the CM input range of the folded-cascode amplifier in Figure 8.8 when the n-well of the input devices is connected to the positive supply. Also, derive the equation of the offset voltage and interpret carefully the sign of each term.

8.4 Refer to Example 8.4 and the data therein to solve this problem. (a) Calculate the CM input range. (b) Verify whether M_5 and M_9 operate in saturation. (c) Determine the maximum output-voltage swing. (d) Verify whether the expression for the output conductance is correct. (e) Estimate the standard deviation of $1/PSRR_{VSS}$ assuming that the relative standard deviations of the transconductances, conductances, and current-mirror gain are 1% and that the sensitivity of I_{REF} to the negative supply is 1%/V.

8.5 Use the data given in Example 8.4 for a 0.5-μm CMOS technology to design a folded-cascode amplifier with a unity-gain frequency of 1000 Mrad/s for a load capacitance of 1 pF and $V_{DD} = 5$ V. Calculate the drain currents and dimensions of all transistors, including the bias network. The dc voltage gain must be higher than 500. Calculate the standard deviation of the offset voltage for $A_{VT} = 10$ mV μm and $A_{ISH} = 2\%$ μm. What is the slew rate? What is the intrinsic unity-gain frequency of the input transistors and of the transistors in the current mirror? What is the input-referred thermal noise? How would you change your design in order to save power?

8.6 Determine $PSRR_{vdd}$ in terms of the ac and mismatch parameters for the folded-cascode amplifier shown in Figure 8.8.

8.7 Check whether the values in Table E8.5.4 are appropriate for the design of Example E8.5.

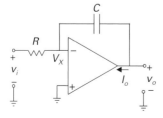

Fig. P8.1   An active RC integrator with non-ideal OTA.

Table P8.12 Dimensions of transistors in Figure 8.33

Transistor	W (µm)	L (µm)
M_1, M_2, M_{1r}, M_{2r}	125	1
M_5, M_{5r}	250	1
M_3, M_4, M_{3r}, M_{4r}	312.5	1

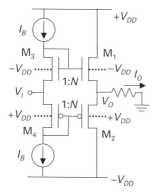

Fig. P8.11   A CMOS class-AB source follower.

8.8 Use the approximate data given in Example 8.4 for a 0.5-µm CMOS technology to design the two-stage Miller-compensated amplifier in Figure 8.11 with a unity-gain frequency of 1000 Mrad/s, load capacitance of 1 pF, and $V_{DD}=5$ V. Calculate the drain currents and dimensions of all transistors. The dc voltage gain must be higher than 500. Calculate the standard deviation of the offset voltage for $A_{VT}=10$ mV µm and $A_{ISH}=2\%$ µm. What is the slew rate?

8.9 Determine PSRR$_{vdd}$ in terms of the ac and mismatch parameters for the two-stage Miller-compensated amplifier shown in Figure 8.11.

8.10 Show that the small-signal voltage gain of the amplifier in Figure 8.24(a) is

$$A_{V0} = \frac{v_o}{v_i} = \frac{1}{1 + 1/(g_m R_L) + g_{md}/g_m}$$

when the bulk is connected to the source.

8.11 Analyze the class-AB amplifier in Figure P8.11 in either weak inversion or strong inversion, assuming that $n_P=n_N$, $I_{S1}=I_{S2}$, and $V_{TOP}=-V_{TON}$. Write the output current in terms of the input voltage for both cases.

8.12 Assume that the dimensions of the transistors in Figure 8.33 are those given in Table P8.12. The tail current $I_T=150$ µA and $R=1$ MΩ. The load capacitance at each output is 10 pF. Use the data given in Example 8.4 for a 0.5-µm CMOS technology to calculate (a) the DM and CM open-loop gains and (b) the unity-gain frequency of the differential mode gain.

8.13 Verify whether the scheme of the amplifier in Figure P8.13, which employs both direct current injection and current injection via current mirrors, has approximately the same transconductances for the DM input and CM output signals. This can be

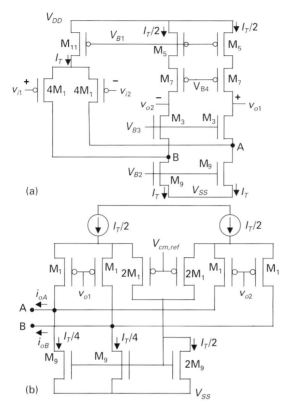

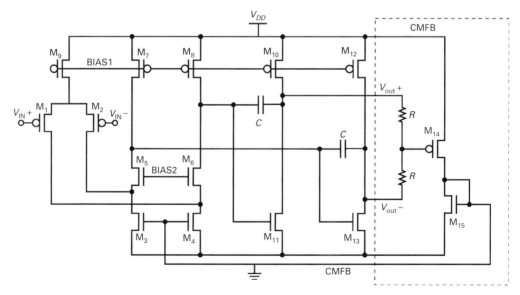

Fig. P8.13   Simplified schematic diagrams of (a) a folded-cascode op amp and (b) a common-mode feedback circuit.

Fig. P8.14   The fully differential amplifier of [51] for low-voltage applications.

carried out by comparing current i_{oA}, due to a variation in the CM output signal, with that flowing through $4M_1$, due to a variation in the DM input signal. For the sake of simplicity, the load capacitors and the bias circuit that generates V_{B1}–V_{B4} are not shown.

8.14 The circuit shown in Figure P8.14 is an FD amplifier for low-voltage applications [51]. (a) Explain the behavior of the DM amplifier and why the compensation capacitors are needed. (b) Assuming that the transconductance of the folded-cascode stage is equal to that of the common-source stage, determine the relationship between the aspect ratios of M_{11} and M_1 for equal drain currents flowing through them and $I_{SHN}=2I_{SHP}$. (c) Give the dc currents flowing through the transistors of the DM circuit. (d) What is the CM reference voltage of this circuit?

References

[1] H. S. Black, "Stabilized feed-back amplifiers," *Proceedings of the IEEE*, vol. **87**, no. 2, pp. 379–385, Feb. 1999. Reprinted from *Electrical Engineering*, vol. **53**, no. 1, pp. 114–120, Jan. 1934.

[2] R. Mancini, editor, *Op Amps for Everyone*, Dallas, TX: Texas Instruments, 2002.

[3] J. R. Ragazzini, R. H. Randall, and F. A. Russell, "Analysis of problems in dynamics by electronic circuits," *Proceedings of the IRE*, vol. **35**, pp. 444–452, May 1947.

[4] T. H. Lee, "Tales of the continuum: a subsampled history of analog circuits," Solid-State Circuit Society Newsletter, http://www.ieee.org/portal/site/sscs/menuitem.f07ee9e3b2a01d06bb9305765 bac26c8/index.jsp?&pName=sscs_level1_article&TheCat=2171&path=sscs/07Fall&file=Lee.xml.

[5] P. E. Allen and D. R. Holberg, *CMOS Analog Circuit Design*, 2nd edn., New York: Oxford University Press, 2002.

[6] L. Moldovan and H. H. Li, "A rail-to-rail, constant gain, buffered op-amp for real time video applications," *IEEE Journal of Solid-State Circuits*, vol. **32**, no. 2, pp. 169–173, Feb. 1997.

[7] J. F. Duque-Carrillo, J. L. Ausín, G. Torelli, J. M. Valverde, and A. Dominguez, "1-V rail-to-rail operational amplifiers in standard CMOS technology," *IEEE Journal of Solid-State Circuits*, vol. **35**, no. 1, pp. 33–44, Jan. 2000.

[8] V. C. Vincence, C. Galup-Montoro, and M. C. Schneider, "A high-swing MOS cascode bias circuit," *IEEE Transactions on Circuits and Systems II*, vol. **47**, no. 11, pp. 1325–1328, Nov. 2000.

[9] P. Aguirre and F. Silveira, "Bias circuit design for low-voltage cascode transistors," *Proceedings of SBCCI* 2006, pp. 94–98, Sep. 2006.

[10] K. R. Laker and W. M. C. Sansen, *Design of Analog Integrated Circuits and Systems*, New York: McGraw-Hill, 1994.

[11] D. A. Johns and K. Martin, *Analog Integrated Circuit Design*, New York: Wiley, 1997.

[12] C. Galup-Montoro, M. C. Schneider, H. Klimach, and A. Arnaud, "A compact model of MOSFET mismatch for circuit design," *IEEE Journal of Solid-State Circuits*, vol. **40**, no. 8, pp. 1649–1657, Aug. 2005.

[13] A. Arnaud, R. Fiorelli, and C. Galup-Montoro, "Nanowatt, sub-nS OTAs, with sub-10-mV input offset using series-parallel current mirrors," *IEEE Journal of Solid-State Circuits*, vol. **41**, no. 9, pp. 2009–2018, Sep. 2006.

[14] A. T. Behr, M. C. Schneider, S. Noceti Filho, and C. G. Montoro, "Harmonic distortion caused by capacitors implemented with MOSFET gates," *IEEE Journal of Solid-State Circuits*, vol. **27**, no. 10, pp. 1470–1475, Oct. 1992.

[15] P. R. Gray, P. J. Hurst, S. H. Lewis, and R. G. Meyer, *Analysis and Design of Analog Integrated Circuits*, 4th edn., New York: John Wiley & Sons, 2001.

[16] Y. P. Tsividis and P. R. Gray, "An integrated NMOS operational amplifier with internal compensation," *IEEE Journal of Solid-State Circuits*, vol. **11**, no. 6, pp. 748–753, Dec. 1976.

[17] P. R. Gray and R. G. Meyer, "MOS operational amplifier design – a tutorial overview," *IEEE Journal of Solid-State Circuits*, vol. **17**, no. 6, pp. 969–982, Dec. 1982.

[18] B. Ahuja, "An improved frequency compensation technique for CMOS operational amplifiers," *IEEE Journal of Solid-State Circuits*, vol. **18**, no. 6, pp. 629–633, Dec. 1983.

[19] S. Nicolson and K. Phang, "Improvements in biasing and compensation CMOS opamps," *Proceedings of IEEE ISCAS* 2004, pp. I-665–I-668.

[20] W. Sansen, *Analog Design Essentials*, Dordrecht: Springer, 2006.

[21] J. H. Huijsing, *Operational Amplifiers Theory and Design*, Boston, MA: Kluwer, 2001.

[22] R. G. H. Eschauzier and J. H. Huijsing, *Frequency Compensation Techniques for Low-Power Operational Amplifiers*, Boston, MA: Kluwer, 1995.

[23] X. Xie, M. C. Schneider, E. Sánchez-Sinencio, and S. H. K. Embabi, "Sound design of nested transconductance–capacitance compensation amplifiers," *Electronics Letters*, vol. **35**, no. 12, pp. 956–958, June 10, 1999.

[24] F. You, S. H. K. Embabi, and E. Sánchez-Sinencio, "Multistage amplifier topologies with nested G_m–C compensation," *IEEE Journal of Solid-State Circuits*, vol. **32**, no. 12, pp. 2000–2011, Dec. 1997.

[25] J. A. Fisher and R. Koch, "A highly linear CMOS buffer amplifier," *IEEE Journal of Solid-State Circuits*, vol. **22**, no. 3, pp. 330–334, June 1987.

[26] F. You, S. H. K. Embabi, and E. Sánchez-Sinencio, "On the common mode rejection ratio in low voltage operational amplifiers with complementary N–P input pairs," *IEEE Transactions on Circuits and Systems II*, vol. **44**, no. 8, pp. 678–683, Aug. 1997.

[27] J. H. Huijsing and D. Linebarger, "Low-voltage operational amplifier with rail-to-rail input and output ranges," *IEEE Journal of Solid-State Circuits*, vol. **20**, no. 6, pp. 1144–1150, Dec. 1985.

[28] J. N. Babanehzad, "A rail-to-rail CMOS op amp," *IEEE Journal of Solid-State Circuits*, vol. **23**, no. 6, pp. 1414–1417, Dec. 1988.

[29] W.-C. S. Wu, W. J. Helms, J. A. Kuhn, and B. E. Byrkett, "Digital-compatible high-performance operational amplifier with rail-to-rail input and output ranges," *IEEE Journal of Solid-State Circuits*, vol. **29**, no. 1, pp. 63–66, Jan. 1994.

[30] R. Hogervorst, J. P. Tero, R. G. H. Eschauzier, and J. H. Huijsing, "A compact power-efficient 3 V CMOS rail-to-rail input/output operational amplifier for VLSI cell libraries," *IEEE Journal of Solid-State Circuits*, vol. **29**, no. 12, pp. 1505–1513, Dec. 1994.

[31] G. Ferri and W. Sansen, "A rail-to-rail constant-g_m low-voltage CMOS operational transconductance amplifier," *IEEE Journal of Solid-State Circuits*, vol. **32**, no. 10, pp. 1563–1567, Oct. 1997.

[32] A. B. Grebene, *Bipolar and MOS Analog Integrated Circuit Design*, Hoboken, NJ: Wiley Interscience, 2003.

[33] R. Castello, "CMOS buffer amplifier," in *Analog Circuit Design, Operational Amplifiers, Analog to Digital Convertors, Analog Computer Aided Design*, ed. J. Huijsing, R. van de Plassche, and W. Sansen, Dordrecht: Kluwer, 1993, pp. 113–138.

[34] V. Dhanasekaran, J. Silva-Martínez, and E. Sánchez-Sinencio, "A 1.2 mW 1.6 V_{pp}-swing class-AB 16 Ω headphone driver capable of handling load capacitance up to 22 nF," *International Solid-State Circuits Conference Digest of Technical Papers*, 2008, pp. 434–435.

[35] B. Gilbert, "Current-mode circuits from a translinear viewpoint: a tutorial," in *Analogue IC Design: The Current-Mode Approach*, ed. C. Toumazou, F. J. Lidgey, and D. G. Haigh, London: Peter Peregrinus Ltd., 1990, pp. 11–91.

[36] B. Gilbert, "Translinear circuits: a proposed classification," *IEE Electronics Letters*, vol. **11**, no. 1, pp. 14–16, Jan. 9, 1975.

[37] E. Vittoz and J. Fellrath, "CMOS analog integrated circuits based on weak inversion operation," *IEEE Journal of Solid-State Circuits*, vol. **12**, no. 3, pp. 224–231, June 1977.

[38] E. Seevinck and R. J. Wiegerink, "Generalized translinear circuit principle," *IEEE Journal of Solid-State Circuits*, vol. **26**, no. 8, pp. 1098–1102, Aug. 1991.

[39] R. Hogervorst and J. H. Huijsing, *Design of Low-Voltage, Low-Power Operational Amplifier Cells*, Boston, MA: Kluwer, 1996.

[40] E. Seevinck, W. de Jager, and P. Buitendik, "A low-distortion output stage with improved stability for monolithic power amplifiers," *IEEE Journal of Solid-State Circuits*, vol. **23**, no. 3, pp. 794–801, June 1988.

[41] D. M. Monticelli, "A quad CMOS single-supply op amp with rail-to-rail output swing," *IEEE Journal of Solid-State Circuits*, vol. **21**, no. 6, pp. 1026–1034, Dec. 1986.

[42] K.-J. de Langen and J. H. Huijsing, "Compact low-voltage power-efficient operational amplifier cells," *IEEE Journal of Solid-State Circuits*, vol. **33**, no. 10, pp. 1482–1496, Oct. 1998.

[43] V. C. Vincence, *Amplificador operacional Classe-AB para baixa tensão de alimentação*, Ph.D. Thesis, Universidade Federal de Santa Catarina, Feb. 2004, http://eel.ufsc.br/~lci/pdf/Tese_Volney.pdf.

[44] G. Palmisano, G. Palumbo, and R. Salerno, "CMOS output stages for low-voltage power supplies," *IEEE Transactions on Circuits and Systems II*, vol. **47**, no. 2, pp. 96–104, Feb. 2000.

[45] M. Banu, J. M. Khoury, and Y. Tsividis, "Fully differential operational amplifiers with accurate output balancing," *IEEE Journal of Solid-State Circuits*, vol. **23**, no. 6, pp. 1410–1414, Dec. 1988.

[46] J. F. Duque-Carrillo, "Control of the common-mode component in CMOS continuous-time fully differential signal processing," *Journal of Analog Integrated Circuits and Signal Processing*, vol. **4**, pp. 131–140, Sep. 1993.

[47] P. D. Walker and M. M. Green, "An approach to fully differential circuit design without common-mode feedback," *IEEE Transactions on Circuits and Systems II*, vol. **43**, no. 11, pp. 752–762, Nov. 1996.

[48] E. Sánchez-Sinencio, J. Silva-Martínez, and J. F. Duque-Carrillo, "Advanced common-mode control techniques for low voltage analog signal processors," http://www.techonline.com/zarticle/pdf/showPDFinIE.jhtml?id=2024038761.

[49] T. C. Choi, R. T. Kaneshiro, R. W. Brodersen *et al.*, "High-frequency CMOS switched-capacitor filters for communications application," *IEEE Journal of Solid-State Circuits*, vol. **18**, no. 6, pp. 652–664, Dec. 1983.

[50] G. Xu and S. H. K. Embabi, "A systematic approach in constructing fully differential amplifiers," *IEEE Transactions on Circuits and Systems II*, vol. **47**, no. 122, pp. 1547–1550, Dec. 2000.

[51] S. Xiao, J. Silva, U.-K. Moon, and G. Temes, "A tunable duty-cycle-controlled switched-RMOSFET-C CMOS filter for low-voltage and high-linearity applications," *Proceedings of IEEE ISCAS* 2004, pp. I-433–I-436.

9 Fundamentals of integrated continuous-time filters

Analog filters, essential parts in many different electronic systems, can be implemented in CMOS using switched-capacitor (SC), switched-current (SI), active-*RC*, MOSFET-C, or OTA-C techniques. Owing to their considerable importance in analog CMOS circuits, Chapter 10 will focus on SC circuits. In this chapter, we study the MOSFET-C and OTA-C techniques, which have been the most important ones for the realization of continuous-time (CT) integrated filters in CMOS technologies. Applications of CT filters include anti-aliasing and reconstruction filters, read channels of disk drives, data-communication circuits, and high-speed data links. A disadvantage of CT filters is that the filter coefficients are sensitive to process and temperature variations and aging. Therefore, tuning of the components that determine the frequency response is required. In this chapter, we will concentrate on the fundamental aspects of filter design at the component level and on the basic building blocks for each technique. We will describe some of the main limitations of the MOSFET-C and OTA-C techniques and the concepts of some tuning schemes used for keeping the filter transfer function close to the nominal specifications. The synthesis of high-order filters, which can be found in more specialized texts such as [1]–[5], is not the subject of this chapter.

9.1 Basics of MOSFET-C filters

The MOSFET-C (MOSFET-capacitor) is a mature technique [6]–[10] for the integration of CT filters in CMOS technologies. MOSFET-C circuits are similar to the active-*RC* topologies composed of operational amplifiers, resistors, and capacitors.

MOSFET-C filters make use of floating capacitors, which can be implemented by, e.g., poly-to-poly capacitors. The implementation of floating capacitors can pose difficult challenges in technologies where linear capacitors are unavailable and the gate capacitors are used for floating capacitors. We will not discuss in this chapter the implementation of capacitors for use in either MOSFET-C or OTA-C filters. The reader is referred to Chapter 3 for material on capacitors in CMOS technologies or to [9] for the use of capacitors in CT filters.

MOSFET-C filters use MOS transistors in the ohmic region to replace the resistors of their active-*RC* counterparts. MOS transistors, however, are inherently non-linear resistors that limit drastically the dynamic range [6]. The use of fully balanced networks significantly reduces the distortion introduced by the triode-biased MOS transistors.

Apart from the harmonic distortion due to the MOSFET non-linear resistance, time constants of MOSFET-C filters suffer from high variability owing to temperature variations, process deviations, and aging. Therefore, some kind of tuning is required in order to keep the frequency response close to its nominal specification. Usually, MOSFET-C filters are tuned by controlling the gate voltage of MOS transistors, as will be seen next. Alternatives to tune the frequency response of MOSFET-C filters involve digital programmability of either MOSFET-only current dividers or banks of capacitors.

9.1.1   The MOSFET as a tunable resistor

The output characteristics of a MOSFET for two different values of the gate voltage are shown in Figure 9.1. We can readily note that the MOSFET in the triode region acts as a voltage controlled resistor, i.e. the (non-linear) resistance between source and drain is controlled by the gate voltage. The value of the ac resistance R can be calculated for $V_{DS}=0$ as

$$\left.\frac{1}{R}\right|_{V_{DS}=0} = \left.\frac{dI_D}{dV_D}\right|_{V_{DS}=0} = -\left.\frac{dI_D}{dV_S}\right|_{V_{DS}=0} = g_{ms} = \frac{2I_S}{\phi_t}\left(\sqrt{1+i_f}-1\right). \tag{9.1.1}$$

The value of the inversion level i_f is dependent on the values of the source and gate voltages according to the unified current-control model (UICM). By controlling the gate voltage, we can obtain a variable resistance, the linearity of which will be examined in Appendix A9.1.

Assuming that the MOSFET operates in strong inversion, (9.1.1) can be rewritten as

$$\left.\frac{1}{R}\right|_{V_{DS}=0} = g_{ms(d)} = \mu C'_{ox}\frac{W}{L}\left(V_G - V_{T0} - nV_Q\right), \tag{9.1.2}$$

where V_Q is the dc source (or drain) voltage. Thus, the MOSFET conductance in strong inversion is a linear function of the gate voltage. It is also dependent on technological parameters, namely mobility, oxide capacitance, and threshold voltage. As the MOSFET enters the weak-inversion regime, the conductance becomes an exponential function of the gate voltage (see Example 9.2).

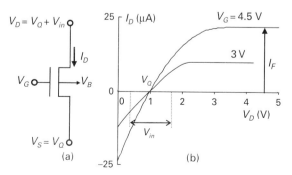

Fig. 9.1   (a) MOSFET symbol and applied voltages. (b) MOSFET output characteristics for $V_G=4.5$ V and 3 V, with $V_Q=1$ V and $V_B=0$ V.

Example 9.1

Consider that, for the MOSFET in Figure 9.1, $\mu = 500\ \text{cm}^2/\text{V}$ per s, $n = 1.2$, $V_{T0} = 0.8$ V, $C'_{ox} = 2\ \text{fF}/\mu\text{m}^2$, and $W/L = 4\ \mu\text{m}/40\ \mu\text{m}$. Determine the transistor conductance for $V_{DS} = 0$ V, $V_G = 4.5$ V, and $V_G = 3$ V.

Answer

Using (9.1.2) we find that

$$g_{ms} = 500 \times 2 \times 10^{-7} \times 1.2 \times \frac{4}{40}\left(\frac{4.5 - 0.8}{1.2} - 1\right) = 25\ \mu\text{A}/\text{V for } V_G = 4.5\ \text{V}$$

and

$$g_{ms} = 500 \times 2 \times 10^{-7} \times 1.2 \times \frac{4}{40}\left(\frac{3 - 0.8}{1.2} - 1\right) = 10\ \mu\text{A}/\text{V for } V_G = 3\ \text{V}.$$

Example 9.2

Assume that the MOSFET in Figure 9.1(a) is biased in weak inversion. What is the expression for the transistor conductance for $V_{DS} = 0$?

Answer

The application of (9.1.1) to operation in weak inversion gives

$$g_{ms} \cong \frac{2I_S}{\phi_t}\left(\frac{1}{2}i_f\right) = \frac{2I_S}{\phi_t}\exp\left(\frac{V_G - V_{T0} - nV_Q + n\phi_t}{n\phi_t}\right).$$

The distortion in the I–V characteristic of the MOSFET shown in Figure 9.1 is analyzed in Appendix A9.1 [11]. As expected, the second-order harmonic is the main component of distortion. Theoretically, the even-order terms of the harmonic distortion can be canceled out in fully-differential topologies, as we will see in the next section.

9.1.2   Balanced transconductors for MOSFET-C filters

A balanced transconductor, which can be used to reduce harmonic distortion since it cancels out the even-order harmonics, is shown in Figure 9.2. The device terminals of the right-hand side, as will be shown later, are at the same potential due to the action of a feedback op amp. To see how the even harmonics are eliminated, we write for the currents I_{D1} and I_{D2} flowing through the matched transistors M_1 and M_2

$$I_{D1} = I_S f(V_Q + V_{in}, V_Q, V_G); \qquad I_{D2} = I_S f(V_Q - V_{in}, V_Q, V_G). \qquad (9.1.3)$$

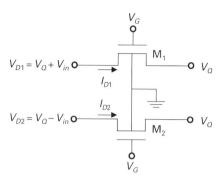

Fig. 9.2   A fully-differential MOSFET transconductor.

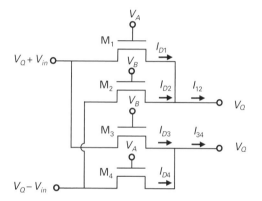

Fig. 9.3   A fully-differential MOSFET transconductor that can cancel out even and odd harmonics. The substrate is connected to V_{SS}.

As we will see later, the output of the basic building blocks of a MOSFET-C filter is proportional to the differential output current $I_{OD} = I_{D1} - I_{D2}$. Since I_{OD} is an odd function of V_{in}, i.e.

$$I_{OD}(V_{in}) = -I_{OD}(-V_{in}), \qquad (9.1.4)$$

the even-order terms of its power series are zero. Note that the complete canceling out of the even harmonics requires the transconductor to be fully balanced. In practical cases, however, the even-order harmonic distortion can be higher than the odd-order harmonic distortion owing to op-amp offsets, transistor mismatching, and unbalanced input signals [10].

There are various techniques, summarized in [7], that can ideally eliminate the transconductor non-linearities. One of them [7], [12] is shown in Figure 9.3. This circuit can, in principle, cancel out both even- and odd-order harmonics when biased in strong inversion (see Problem 9.1). The voltage-to-current conversion factor is dependent on the difference between the gate control voltages. In practice, however, the topology in Figure 9.3 does not cancel out the harmonics completely due to the non-idealities mentioned previously, in addition to differences in slope factor and mobility of the

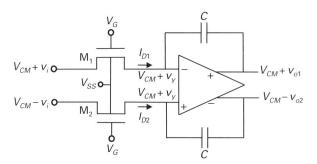

Fig. 9.4	A MOSFET-C integrator. The time constant of the integrator is tuned by the gate voltage V_G.

transistors owing to the different gate voltages. As pointed out in [7], either the scheme in Figure 9.2 or that in Figure 9.3 may be preferable, depending on practical constraints in a given application.

### 9.1.3	MOSFET-C integrators

In this section, we analyze the integrator, a basic building block for MOSFET-C filters. Among the different schemes for the transconductor, we have chosen the one in Figure 9.2 as the input of the MOSFET-C integrator, which is shown in Figure 9.4. The time constant RC of the integrator is adjusted by means of the gate voltage V_G. Some possible implementations of the fully-differential op amp have been described in Chapter 8 and will not be addressed in this section. The effects of some non-ideal op-amp characteristics will be examined later. The floating capacitors can be either linear capacitors from analog CMOS technologies or non-linear gate capacitors [13] with appropriate bias schemes [9] in any CMOS technology.

In the analysis that follows, we assume that the op-amp voltage gain is infinite and the offset voltage is zero. Also, the dc levels at the integrator inputs and outputs are both equal to V_{CM}. As seen in Chapter 8, the common-mode output voltage can be set to a given V_{CM} through the action of a common-mode feedback network. For an ideal op amp we can write

$$v_{o1}(t) = v_y(t) - \frac{1}{C}\int I_{D1}(\tau)d\tau,$$
$$-v_{o2}(t) = v_y(t) - \frac{1}{C}\int I_{D2}(\tau)d\tau. \tag{9.1.5}$$

The combination of the two equations in (9.1.5) gives, for the differential output voltage,

$$v_o = v_{o1} - v_{o2} = -\frac{1}{C}\int (I_{D1} - I_{D2})dt. \tag{9.1.6}$$

Note that v_y appears as a common-mode component at the output. Since $I_{D1} - I_{D2}$ is an odd function of v_i, so is the differential output voltage v_o. Neglecting the odd-order harmonics, we can finally write for v_o

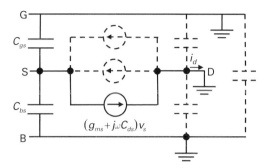

Fig. 9.5   The equivalent ac circuit for the MOSFET in Figure 9.4.

$$v_o = v_{o1} - v_{o2} = -\frac{2}{RC}\int v_i\, dt, \qquad (9.1.7)$$

where $R = 1/g_{ms}$ is the linearized MOSFET resistance calculated for $V_{DS} = 0$. In a practical MOSFET-C filter, some kind of tuning is required in order to set the time constants of the integrators to accurate values. This can be done by adjusting the value of g_{ms} by means of the gate voltage.

As regards non-idealities, the effects of component mismatch, unbalanced input signals, and op-amp offset voltage and finite gain on integrator distortion have been analyzed elsewhere [8]. We will restrict our analysis to the effects of the parasitics and unity-gain frequency of the op amp on the integrator performance. Later, we will also describe the effects of the non-ideal integrator frequency response on the transfer function of a biquadratic section.

To analyze the effects of transistor capacitances on the performance of the integrator, we represent in Figure 9.5 the small-signal equivalent to M_1 (see Figure 2.18 in Chapter 2) of Figure 9.4. To draw the equivalent ac circuit of Figure 9.5, we make the following assumptions: (i) the gate and the bulk are connected to ideal dc voltage sources; (ii) the rightmost terminal voltage of M_1 is constant due to the action of the feedback op amp; and (iii) the lumped model is an acceptable equivalent ac circuit of the distributed MOSFET, at least for the input frequency range of the problem under analysis.

The MOSFET transadmittance y_{ds} in Figure 9.5 is given by

$$y_{ds} = \frac{i_d}{v_s} = g_{ms} + j\omega C_{ds} = g_{ms}\left(1 - \frac{j\omega L^2}{6\mu\phi_t}\frac{1}{\sqrt{1+i_f}}\right). \qquad (9.1.8)$$

To arrive at the result in (9.1.8) we calculate the value of the drain-to-source capacitance for $V_{DS} = 0$ or, equivalently, for $\alpha = 1$

$$C_{ds}\big|_{\alpha=1} = -\frac{4}{15}nC'_{ox}WL\frac{1+3\alpha+\alpha^2}{(1+\alpha)^3}\bigg|_{\alpha=1}\frac{q'_{IS}}{q'_{IS}+1} = -\frac{nC'_{ox}WL}{6}\frac{q'_{IS}}{q'_{IS}+1}$$

and

$$g_{ms} = \mu\frac{W}{L}nC'_{ox}\phi_t q'_{IS}.$$

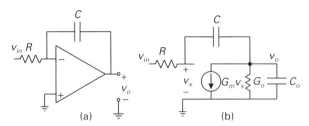

Fig. 9.6   (a) A simplified scheme of the half-circuit of the MOSFET-C integrator where the input transistor has been replaced with a resistor. (b) The corresponding small-signal circuit with the op amp represented as a transconductance. G_o is the op amp's output conductance in parallel with the input conductance of the following stage, while C_o includes all capacitances associated with the output node.

The quasi-static model in Figure 9.5 is valid as long as

$$g_{ms} \gg -\omega C_{ds} \qquad \text{or} \qquad \frac{\omega L^2}{6\mu\phi_t}\frac{1}{\sqrt{1+i_f}} \ll 1.$$

It is interesting to note here that the lumped model of Figure 9.5 gives the same result for the transadmittance as the distributed MOSFET model (see Problem 9.2), provided that the signal frequency $\omega \ll -g_{ms}/C_{ds}$. Expression (9.1.8) shows that the current i_d that flows into the capacitor is delayed with respect to the input voltage. This delay affects the integrator response and, as a consequence, it affects the performance of biquadratic filters based on integrators, as we will see later.

In the previous analysis we have assumed that the integrator input is a low-impedance node; in this case, neither the intrinsic nor the extrinsic capacitances of the transistor associated with the input node affect the conversion of the input voltage into a current. Additionally, the inverting input of the op amp has been assumed to be an ac ground.

After having analyzed how the distributed capacitance of the MOSFET affects the voltage-to-current conversion, we now proceed to determine the deviation in the ideal integration due to both the op-amp finite gain and the integrator output capacitance. For a first-order analysis we will use the ac model of the half-circuit of the integrator represented in Figure 9.6. The capacitance between the inverting input of the op amp and ground is assumed to be a very small fraction of C that does not have a significant influence on the voltage transfer function. Nodal analysis in Figure 9.6(b) gives the following transfer function

$$\frac{V_O}{V_{IN}}(s) = \frac{-(G_m - sC)}{s^2 CC_oR + [C(G_m + G_o)R + (C + C_o)]s + G_o}. \qquad (9.1.9)$$

Since the integrator is designed to have a unity-gain frequency (equal to $1/(RC)$) much higher than the low-frequency pole of the integrator, the dominant-pole approximation can be applied to the transfer function in (9.1.9). With such an approximation, we have, for the dominant pole p_1,

$$p_1 \cong -\frac{G_o/C}{(G_m + G_o)R + (1 + C_o/C)} \cong -\frac{G_o}{G_m}\frac{1}{RC} = -\frac{1}{A_{V0}RC}, \qquad (9.1.10)$$

where A_{V0} is the low-frequency voltage gain of the op amp. To simplify (9.1.10) we have assumed that $G_m R \gg 1$, i.e. the op-amp transconductance is high enough to allow the op amp to have a high voltage gain even when it is loaded with a load resistance equal to R. On the other hand, the non-dominant pole is

$$p_2 \cong -\left(\frac{G_m + G_o}{C_o} + \frac{C + C_o}{RCC_o}\right) \cong -\frac{G_m}{C_o}. \tag{9.1.11}$$

The integrator transfer function in (9.1.9) can thus be written as

$$\frac{V_O}{V_{IN}}(s) \cong -\frac{1 - sC/G_m}{RC[s + 1/(A_{V0}RC)](1 + sC_o/G_m)} \tag{9.1.12}$$

The ideal result $-1/(RC)$ is obtained for negligible C_o, infinite A_{V0}, and infinite unity-gain frequency G_m/C of the op amp. Finally, on substituting the value of y_{ds} in (9.1.8) for $1/R$ in (9.1.12) we find that

$$\frac{V_O}{V_{IN}}(s) \cong -\frac{g_{ms}(1 - s\tau_1)(1 - sC/G_m)}{C[s + g_{ms}/(A_{V0}C)](1 + sC_o/G_m)}, \tag{9.1.13}$$

where $\tau_1 = L^2/(6\mu\phi_t\sqrt{1 + i_f})$.

Since the zero and secondary pole frequencies must be much higher than the integrator unity-gain frequency, (9.1.13) can be simplified to

$$\frac{V_O}{V_{IN}}(s) \cong -\frac{g_{ms}}{C[s + g_{ms}/(A_{V0}C)](1 + \tau_{eq}s)}, \tag{9.1.14}$$

where

$$\tau_{eq} = \tau_1 + (C + C_o)/G_m,$$

for frequencies $\omega \leq g_{ms}/C$. To derive (9.1.14) from (9.1.13) we make the approximation $1 - x \cong 1/(1 + x)$ for $|x| \ll 1$ and neglect second-order terms.

As shown in Figure 9.7, the magnitude and phase plots of the ideal and non-ideal integrators are almost coincident, except for low frequencies and for frequencies exceeding the unity-gain frequency g_{ms}/C. An important factor to be considered in the implementation of filters is the phase error at the unity-gain frequency, which can profoundly affect the filter performance, especially filters of high selectivity, as we will see next.

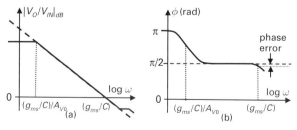

Fig. 9.7   (a) Magnitude and (b) phase plots for the ideal (dashed line) and non-ideal (solid line) MOSFET-C integrators.

The quality factor, or simply Q-factor, is an important parameter that indicates the deviation of the integrator from ideality. In the ideal integrator, the transfer function contains a purely imaginary term in the denominator (the ideal transfer function of the integrator in (9.1.14) would be $-g_{ms}/(sC)$), the deviation from ideality being the real part of the denominator. The ideal integrator would have a phase shift of $-\pi/2$ rad. In an actual MOSFET-C integrator, non-ideal effects such as the finite unity-gain frequency of the op amp, the distributed model of the MOS transistor, and the finite dc gain of the op amp cause phase deviations relative to the ideal case. These can give rise to significant variations in relation to the desired filter transfer function. Next, we analyze the quality factor of the integrator, which corresponds to the reciprocal of the phase deviation.

In the case of the integrator, the quality factor is the ratio between the imaginary and real parts of the denominator of (9.1.14), i.e.

$$Q = \omega\,\frac{1 + \tau_{eq}g_{ms}/(A_{V0}C)}{g_{ms}/(A_{V0}C) - \tau_{eq}\omega^2}. \tag{9.1.15}$$

In the ideal case, $A_{V0} \to \infty$ and $\tau_{eq} \to 0$; thus, $Q \to \infty$. The quality factor Q_I[1] at the integrator unity-gain frequency (g_{ms}/C), an important limitation for the implementation of biquad and higher-order filters, is given by

$$\frac{1}{Q_I} \cong \frac{1}{A_{V0}} - \frac{\tau_{eq}g_{ms}}{C}. \tag{9.1.16}$$

To calculate Q_I, we assume that $\tau_{eq}g_{ms}/(A_{V0}C) \ll 1$ in (9.1.15), a quite acceptable approximation since the phase deviation $\tau_{eq}g_{ms}/C$ must be small at the integrator unity-gain frequency. The reciprocal of Q_I is the phase deviation of the integrator in relation to the phase $\pi/2$ rad of the ideal integrator at the integrator unity-gain frequency, as indicated in Figure 9.7. In an actual MOSFET-C integrator, both the finite unity-gain frequency of the op amp and the distributed effect of the MOS transistor generate phase lag, whereas the finite dc gain of the op amp causes phase lead. The phase deviation relative to the ideal integrator at the unity-gain frequency, which is generally dominated by phase lag, is an important factor that can degrade significantly the performance of filters and must be compensated for in several cases, especially in highly selective and/or high-frequency filters.

Example 9.3

Assume that the MOSFET-C integrator in Figure 9.4 is implemented in a 0.35-μm technology for which $\mu = 356\ \text{cm}^2/\text{V}$ per s, $C'_{ox} = 4.5\ \text{fF}/\mu\text{m}^2$, $n \cong 1.34$, and $V_{T0} = 0.60\ \text{V}$. Assume that the op amp has a negligible output capacitance and that $G_m = 0.1\ \text{mA/V}$ and $G_o = 0.1\ \mu\text{A/V}$. Discuss the implementation of the MOSFET-C integrator for a unity-gain frequency equal to 1 Mrad/s with integrating capacitances of (a) 1 pF and (b) 10 pF. Consider that the parasitic capacitance at the output node of the integrator is 25% of the integrating capacitor and that the minimum channel width of

---

[1] In this chapter, the notation Q_I, which has been used to represent the integrator Q-factor at the unity-gain frequency, should not be mistakenly taken as the transistor inversion charge.

this technology is 1 μm. The values of the gate, substrate, and drain-to-source dc voltages are $V_G = 4.6$ V, $V_B = 0$ V, and $V_Q = 1.5$ V, respectively.

Answer

Expression (9.1.2), which can be written as

$$\frac{L}{W} = \mu C'_{ox} R(V_G - V_{T0} - nV_Q), \tag{E9.3.1}$$

will be used to calculate the value of the transistor aspect ratio.

(a) For $C = 1$ pF, the required resistance is $R = 1$ MΩ. From (E9.3.1) we find that

$$\frac{L}{W} = \mu C'_{ox} R(V_G - V_{T0} - nV_Q)$$

$$= 356 \times 4.5 \times 10^{-7} \times 10^6 [4.6 - 0.6 - (1.34 \times 1.5)] \cong 320.$$

Choosing the minimum value of W so as to avoid an excessively long-channel device, we have $L = 320$ μm. This channel length results in

$$\tau_1 = \frac{L^2}{6\mu(V_P - V_{SB})} = \frac{(320 \times 10^{-4})^2}{6 \times 356[(4.6 - 0.6)/1.34 - 1.5]} = 0.32 \text{ μs.}$$

The equivalent time constant is

$$\tau_{eq} = \tau_1 + (C + C_o)/G_m = 0.32 + (1 + 0.25)/100 \cong 0.333 \text{ μs,}$$

which is dominated by the distributed capacitance of the MOSFET. Using (9.1.16) we find that, for the integrator Q-factor,

$$\frac{1}{Q_I} = \frac{G_o}{G_m} - \frac{\tau_{eq}}{RC} = \frac{1}{1000} - \frac{0.333}{1} \to Q_I \cong -3,$$

a value that is not acceptable for most applications.

(b) For $C = 10$ pF, the required resistance is $R = 0.1$ MΩ and $L/W = 32$, as calculated through (E9.3.1). If we choose the minimum value of W once again, we have $L = 32$ μm. This channel length results in $\tau_1 = 3.2$ ns, which is 100 times smaller than the previous one. In this case, the equivalent time constant is

$$\tau_{eq} = \tau_1 + (C + C_o)/G_m = 3.2 + (10 + 2.5)/0.1 \cong 128 \text{ ns.}$$

Note that the value of τ_{eq} is now dominated by the unity-gain frequency of the op amp. Using (9.1.16), we find for the integrator Q-factor

$$\frac{1}{Q_I} = \frac{G_o}{G_m} - \frac{\tau_{eq}}{RC} = \frac{1}{1000} - \frac{0.128}{1} \to Q_I \cong -8.$$

A value of the integrating capacitor between 1 and 10 pF can be chosen in order to minimize the phase deviation at the unity-gain frequency (see Problem 9.3).

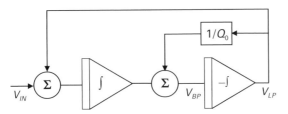

Fig. 9.8   A two-integrator loop for implementing a biquad.

We now proceed to analyzing the effect of the non-infinite quality factor of the integrator on the transfer function of a biquad filter. Specifically, we will exemplify the effect of the finite Q-factor of the integrator on the two-integrator loop shown in Figure 9.8. For the sake of simplicity, we assume here that the integrators are identical. Ideally, the integrator transfer function is ω_0/s, where ω_0 is the unity-gain frequency. Two transfer functions are achievable, associated with the low-pass (V_{LP}) and band-pass (V_{BP}) outputs. Using the ideal integrator transfer function, we obtain the following transfer function for the band-pass output:

$$\frac{V_{BP}}{V_{IN}}(s) = \frac{\omega_0 s}{s^2 + (\omega_0/Q_0)s + \omega_0^2} = \frac{\omega_0 s}{D(s)}. \tag{9.1.17}$$

Now, let us assume that the integrator transfer function is given by (9.1.14). After some algebraic manipulations and neglecting second-order terms [9], we find that

$$D(s) = s^2 + (\omega_N/Q_N)s + \omega_N^2; \quad \omega_N^2 \cong \omega_0^2, \quad \frac{1}{Q_N} = \frac{1}{Q_0} + \frac{2}{Q_I}. \tag{9.1.18}$$

The pole frequency ω_N of the filter implemented with non-ideal integrators is close to that obtained assuming the integrators to be ideal; however, the pole Q-factor Q_N can be strongly affected by the integrator Q-factor Q_I. As an example, assume the design of a two-integrator loop filter with $Q_0 = 20$, but, due to non-idealities of the op amp and distributed effects of the input transistor, the value of the integrator Q-factor is $Q_I = -100$; in this case, $Q_N \cong 33$, which is considerably different from the nominal value. One might think of pre-distorting the nominal quality factor Q_0; indeed, if we choose $Q_0 = 100/7$ in this example, the resulting filter will have $Q_N \cong 20$. This pre-distortion strategy, however, is not appropriate since the value of the integrator quality factor is seldom known with accuracy. Compensation of the phase error of the integrators can be achieved using (automatically tuned) small resistors in series with the integrating capacitors or small capacitors in parallel with the MOSFETs [14].

9.1.4   Filter examples

The simplest form of a MOSFET-C filter is the first-order low-pass filter (or lossy integrator) shown in Figure 9.9. The ideal transfer function of the filter is

$$\frac{v_o}{v_i}(s) = -\frac{R_2/R_1}{1 + sCR_2}, \tag{9.1.19}$$

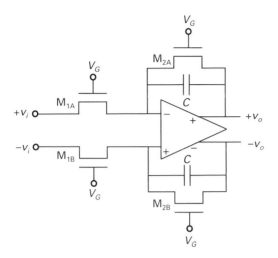

Fig. 9.9 A first-order low-pass MOSFET-C filter.

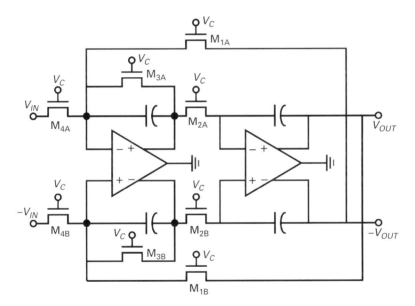

Fig. 9.10 A second-order MOSFET-C filter.

where R_2 and R_1 are the small-signal resistances for $V_{DS}=0$. The -3 dB frequency can be controlled by means of the transistor gate voltage. The dc gain of the filter, $-R_2/R_1$, however, is not affected by the gate voltage since the resistances of the input and feedback transistors have the same dependence on the gate control voltage; the ratio $-R_2/R_1$ is, in principle, equal to the ratio $(L/W)_2/(L/W)_1$.

In Figure 9.10, a second-order Tow–Thomas biquad composed of the two-integrator loop filter of Figure 9.8 is shown. The output of the op amp on the left is band-pass whereas the output of the op amp on the right is low-pass. This biquad can be orthogonally tuned [2] by the following procedure: (i) adjusting simultaneously the resistances of the pairs M_{1A}, M_{1B} and M_{2A}, M_{2B} to a specified value of the center frequency ω_0 of the filter; (ii) M_{3A}, M_{3B} can then be adjusted to give the specified value of the quality factor Q_0 without changing ω_0; and, finally, (iii) M_{4A}, M_{4B} can be adjusted to the desired value of the gain without affecting the value of either ω_0 or Q_0.

9.2   Basics of OTA-C filters

The G_m-C (transconductance-capacitor), also sometimes called OTA-C [15], is the most popular technique employed for the integration of continuous-time filters in CMOS technologies [16]. The basic elements of G_m-C circuits are voltage-to-current converters, or transconductors, and capacitors. Ideally, the transconductor is an infinite-bandwidth circuit that converts an input voltage linearly into an output current with input and output impedances ideally infinite, as shown in Figure 9.11(a). A transconductor can be used as a (grounded) resistor, as shown in Figure 9.11(c). The block in Figure 9.11(d) gives an output voltage that is, ideally, proportional to the integral of the input voltage. The transconductor converts the input voltage into an output current, which, in turn, is integrated in capacitor C.

  G_m-C filters make use of grounded capacitors, which can be implemented more easily than can floating capacitors. Gate capacitors can be used in this technique since a dc voltage can be applied to the ac-grounded capacitor terminal for appropriate biasing. On the other hand, the integrating capacitor (see Figure 9.11(d)) is directly affected by the sum of all parasitic capacitances in parallel with C.

  G_m-C filters, similarly to MOSFET-C filters, suffer from high variability owing to temperature variations, process deviations, and aging; thus, some kind of tuning is required in order to keep the frequency response within its specifications. Fortunately, the frequency response of OTA-C filters can be controlled by the transconductance of the V-to-I converters and/or the capacitances. The usual form of tuning the transfer function of a filter is by adjusting the bias current of the transconductors. In the next sections, we will introduce some simple transconductor schemes and filter examples, and will study the influence of the non-idealities of both the transconductor and the capacitor on the performance of the integrator.

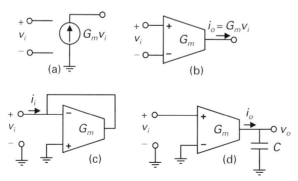

Fig. 9.11   (a) The small-signal equivalent circuit of the ideal transconductor. (b) Its symbol. (c) The use of a transconductor as a resistor ($i_i/v_i = G_m$). (d) The G_m-C integrator.

9.2.1   Transconductors

The main component of G_m-C filters is the adjustable transconductor. An abundance of topologies in the technical literature is available for CMOS transconductors. The interested reader is referred to [16] for an extensive list of references on G_m-C filters and their applications, together with transconductor topologies and linearization techniques. We will describe just some of the simplest architectures of CMOS voltage-to-current converters. Basically, a transconductor, a device with high input and output impedances, is a combination of transistors aimed at obtaining an adjustable linear voltage-to-current conversion.

Figure 9.12(a) is the scheme of the simplest version of a transconductor, composed of a transistor and a current source that provides the appropriate bias for the transistor. Assuming that the transistor operates in saturation, a condition required for high output impedance, the relationship between the transistor current and the input voltage is, assuming the slope factor n to be constant, given by

$$\frac{V_{IN} - V_{T0}}{n\phi_t} = \sqrt{1 + \frac{I_B - i_O}{I_S}} - 2 + \ln\left(\sqrt{1 + \frac{I_B - i_O}{I_S}} - 1\right), \qquad (9.2.1)$$

where the drain current $I_D = I_B - i_O$. In the scheme of Figure 9.12(a), an increase in the input voltage causes a decrease in the output current; thus, this is an inverting transconductor. Note that the relationship between current and voltage is non-linear. The input voltage range for low distortion is very narrow and independent of the inversion level I_B/I_S for weak inversion, whereas it increases with the square root of the inversion level for strong inversion (see Problem 9.4).

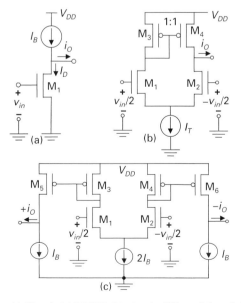

Fig. 9.12   (a) Simple MOSFET, (b) simple differential, and (c) fully-differential transconductors.

The circuit in Figure 9.12(b) shows a transconductor with a differential input. In this case, the linearity range is improved by a factor of two compared with the transconductor in Figure 9.12(a) for the same inversion level of the input transistors. In Figure 9.12(c), we have a fully-differential transconductor. In many applications, particularly those in low-voltage technologies, fully-differential topologies are widely employed. Although the topologies in Figure 9.12 can be highly non-linear, all of them can be used in applications for which the signal amplitude is low [17]–[20]. Some schemes intended to increase the linearity of the transconductor will be shown next.

A plethora of techniques for linearization of transconductors is based on the cancellation of non-linear terms. Most of them exploit the square law of the voltage-to-current characteristic of MOS transistors operating in strong inversion and combine transistors to eliminate the quadratic dependence of the current on the input voltage and to keep only the linear term [21]–[26].

In other transconductors, the input signal is either split into several sections [27] or attenuated at the input devices as a result of the use of a source degeneration structure [28], [29], resulting in an increased input range. A good choice to improve linearity is the use of balanced topologies, in which the even-order harmonics would be eliminated if matching were perfect. The authors of the majority of technical papers on transconductors employ fully-differential topologies as a way to reduce distortion.

Figure 9.13 illustrates a V–I converter in which the input voltage simply divides equally between two source-coupled input pairs [27]. In this case, the transconductance is divided by a factor of 2 compared with the conventional differential pair. On the other hand, the magnitude of the second harmonic is divided by a factor of 4. This circuit, besides providing improved linearity, is also useful for the detection of common-mode voltages (see Problem 9.5).

Figure 9.14 shows an interesting implementation of a fully balanced transconductor with source degeneration realized through voltage-controlled transistors M_{2A} and M_{2B} [29]. The authors of [29] analyze the circuit in Figure 9.14 using the square-law MOSFET model and verify how the linearity of the voltage-to-current conversion is affected by the ratio S_1/S_2 and conclude that $S_1/S_2 \cong 7$ gives the best linearity performance.

The transconductance of the circuit in Figure 9.14 [29] is

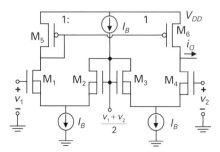

Fig. 9.13   A transconductor that divides the input voltage between two source-coupled pairs [27].

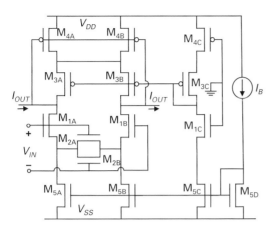

Fig. 9.14 A fully balanced transconductor with source degeneration [29].

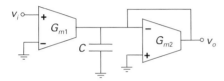

Fig. 9.15 A first-order low-pass G_m-C filter.

$$G_m = \frac{G_{m(S_2 \to \infty)}}{1 + S_1/(4S_2)}. \tag{9.2.2}$$

As compared with the simple differential pair, the transconductance of the circuit in Figure 9.14 is lower by a factor of $1 + S_1/(4S_2)$. The common-mode output voltage is stabilized by transistors M_{4A} and M_{4B}, which operate in the triode region, as previously explained in Chapter 8 (see Figure 8.37).

Several factors can degrade the performance of transconductors and, consequently, of the filters in which they are embedded. Important factors that affect the performance of transconductors are offset voltage, non-linearities, limited bandwidth, parasitic components, finite input and output impedances, and noise.

The offset voltage of a transconductor, as was previously examined for operational amplifiers, is one of the major problems in G_m-C filters. Some potential effects of the offset voltage of the transconductor are a reduction in voltage swing, a change in the effective transconductance, and an increase in harmonic distortion. In fact, each node of a G_m-C filter is affected by the offset of each transconductor multiplied by the dc transfer function from the input of the offset-generating transconductor to the node under consideration. As a result, if the designer is not aware of this problem, the dc input voltage of a transconductor can even be, in some cases, higher than the maximum allowable differential input voltage, giving rise to high distortion and a transconductance lower than the design value. A simple example to illustrate this issue is given in association with the first-order low-pass filter in Figure 9.15, for which the (ideal) transfer function is

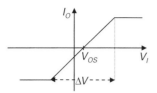

Fig. E9.4   The transfer characteristic of the transconductors in Figure 9.15.

$$\frac{V_O}{V_I}(s) = \frac{G_{m1}/G_{m2}}{1 + sC/G_{m2}}. \tag{9.2.3}$$

The dc output voltage V_{O1} due to V_{OS1}, the offset voltage of the first transconductor, is $V_{O1} = (G_{m1}/G_{m2})V_{OS1}$, i.e. the dc gain times the offset voltage. As an example, assume that $V_{OS1} = 5$ mV and that the dc gain $G_{m1}/G_{m2} = 20$. In this case, the differential input voltage for the second transconductor would be $V_{O1} = 20 \times 5 = 100$ mV, a value that, in some cases, can drive the transconductor into a highly non-linear region, thus increasing harmonic distortion and lowering the transconductance.

Example 9.4

Assume that the voltage-to-current transfer function of the transconductors in Figure 9.15 can be approximated by that given in Figure E9.4. The nominal values of the components are $G_{m1} = 1$ μA/V, $G_{m2} = 0.1$ μA/V, $C = 0.2$ pF, and $\Delta V = 100$ mV. Calculate the -3 dB frequency of the filter and the maximum offset voltage of transconductor 1 for a peak-to-peak input voltage of $(G_{m2}/G_{m1})\Delta V/2 = 5$ mV.

Answer

According to (9.2.3), the -3 dB frequency is

$$\omega_{-3dB} = G_{m2}/C = 0.1 \times 10^6/0.2 = 0.5 \text{ Mrad/s}.$$

The combined effect of the input offset voltages V_{OS1} and V_{OS2} on the output voltage is $(G_{m1}/G_{m2})V_{OS1} + V_{OS2} \cong 10V_{OS1}$. This approximation is valid since the dc output voltage is mostly affected by transconductor 1 for comparable values of the offset voltages. Using this approximation, the input of transconductor 2 is within its linear range when

$$-\Delta V/2 \leq V_O \leq \Delta V/2.$$

Let us assume that the frequency of the input signal is below the -3 dB frequency. In this case, we have

$$V_O \cong (G_{m1}/G_{m2})(V_I - V_{OS1}) = 10(V_I - V_{OS1}).$$

Using this expression together with the previous one and when $V_I = V_P \cos(\omega_0 t)$, with $V_P = 5/2$ mV, we find that $|V_{OS1}| \leq 2.5$ mV.

The distortion in transconductors is another degradation factor of the filter performance. Basically, the second- and third-order harmonic components are of major concern. Obviously, the distortion is essentially dependent on the transistor non-linearities and on the transconductor topology. As a general rule, the linearity range of a transconductor is not affected by the inversion level in weak inversion and increases roughly with the square root of the inversion level in strong inversion. Fully balanced structures are preferable to single-ended topologies, since they are capable of reducing significantly the second-order harmonic. The price to be paid is the need for a CMFB circuit to stabilize the common-mode dc output voltage.

Owing to the existence of many variants of transconductor topologies, we will not calculate here the harmonic distortion. The interested reader can use the UICM for each particular topology to give a rough estimate of the harmonic distortion, similarly to the derivation of harmonic distortion calculated in Appendix A9.1 for the MOSFET acting as a resistor. For the sake of simplicity, one can assume that neither the mobility nor the slope factor is dependent on the gate voltage, which can sometimes lead to relatively poor approximations for calculating the harmonic distortion.

Another fundamental drawback associated with G_m-C filters is the finite transconductor bandwidth, which can be limited both by the distributed nature of the MOSFET, as previously explained in the section on MOSFET-C filters, and by internal nodes, which can attenuate the signal transferred at higher frequencies. In this section we will analyze the effect of the frequency-dependent transconductance on the performance of the integrator, a basic building block of G_m-C filters. In an IC design, the parameters responsible for the frequency-dependent response of the transconductor must be properly accounted for so that the designer can, at least approximately, correct or compensate for this error. To explain the effects of the parasitic components and limited bandwidth of the transconductor on filters, we now proceed to analyzing the G_m-C integrator. In fact, it is very important to study the integrator since, in many filters, the main building blocks are integrators. In fact, the minimum dc gain and phase specifications required for the integrators can be derived from the filter specifications [30].

9.2.2 G_m-C integrators

Figure 9.16 shows schematically a G_m-C integrator that includes a frequency-dependent transconductance and other elements that degrade its performance. Here, G_o denotes the output conductance and C_P is a parasitic capacitance, which includes the capacitance associated with the transconductor output, the input capacitance of the following stage, and the parasitic capacitance associated with the top plate of the integrating capacitor C. The transfer function of the ideal integrator is

$$\frac{V_O}{V_I}(s) = \frac{G_{m0}}{sC},$$
(9.2.4)

where G_{m0} is the low-frequency transconductance, which, in the ideal integrator, is independent of frequency. The unity-gain frequency of the ideal integrator is exclusively

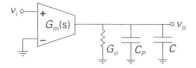

Fig. 9.16   A G_m-C integrator and parasitic elements.

determined by the transconductance and the load capacitance C. In practice, however, the transfer function becomes

$$\frac{V_O}{V_I}(s) = \frac{G_{m0}F(s)}{G_o + s(C + C_P)},\tag{9.2.5}$$

where $F(s)$ represents the dependence of the transconductance on frequency. Since the singularities of $F(s)$ must be located at frequencies much higher than the integrator unity-gain frequency, (9.2.5) can be simplified to

$$\frac{V_O}{V_{IN}}(s) \cong \frac{G_{m0}}{(C + C_P)[s + G_o/(C + C_P)](1 + \tau_{eq}s)}\tag{9.2.6}$$

for frequencies $\omega \leq G_{m0}/(C + C_P)$. In (9.2.6), τ_{eq} represents the effects of the singularities of the transconductance on the frequency response of the integrator. This is an acceptable approximation at least for frequencies not exceeding the unity-gain frequency of the integrator. If the singularities are not located at frequencies much higher than the integrator unity-gain frequency, the performance of the integrator will be highly affected by the singularities of the transconductor. At the unity-gain frequency, the phase deviation given by (9.2.6) with respect to the ideal deviation of $-\pi/2$ rad at the unity-gain frequency is

$$\Delta\phi = \tan^{-1}\left(\frac{G_0}{\omega_u(C + C_P)}\right) - \tan^{-1}\left(\omega_u\tau_{eq}\right) \cong \frac{G_0}{G_m} - \frac{G_m}{C + C_P}\tau_{eq}.\tag{9.2.7}$$

Note the similarities between expressions (9.2.7) and (9.1.16). Considering that the phase deviation is the reciprocal of the integrator Q-factor, both equations contain a first term equal to the inverse of a dc gain plus a second term equal to the phase lag created by the transistor's distributed RC effects and the finite unity-gain frequency of the op amp in the case of a MOSFET-C or by the transconductor delay at the unity-gain frequency and the parasitic capacitance in the case of a G_m-C.

Example 9.5

Assume that $W/L = 120\ \mu m/4\ \mu m$, $n = 1.2$, $\phi_t = 25$ mV, $\mu = 400\ cm^2/V$ per s, $I_{SH} = 100$ nA, and the transistor Early voltage per unit length $V_E = 5$ V/μm for the circuit in Figure E9.5. (a) Calculate the unity-gain frequency of the ideal integrator. (b) Calculate the phase deviation relative to the ideal inverting integrator due to both the distributed effect of the MOSFET and the finite gain of the transconductor. (c) What would be the unity-gain frequency in the case of a parasitic capacitance $C_P = 0.25C_L$, and what would be the value

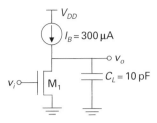

Fig. E9.5   A G_m-C integrator.

of the bias current required to restore the nominal value of the unity-gain frequency? Assume that the current source is ideal and the transistor is operating in saturation.

Answer

(a) The unity-gain frequency of the ideal integrator is $\omega_u = g_m/C_L$. To calculate the transconductance g_m we have to calculate the inversion level i_f, which is given by

$$i_f = I_B/I_S = I_B/(SI_{SH}) = 300/(30 \times 0.1) = 100.$$

Using the transconductance-to-current ratio we find that

$$g_m = \frac{2I_B}{n\phi_t(\sqrt{1 + i_f} + 1)} = \frac{600 \times 10^6}{30 \times 10^{-3}(\sqrt{1 + 100} + 1)} = 1.8 \text{ mA/V}.$$

Consequently, $\omega_u = 1.8 \text{ mA/V}/10 \text{ pF} = 180 \text{ Mrad/s}$.

(b) We use the expression for the drain-to-gate admittance

$$y_{dg} \cong g_m(1 - j\omega\tau_1),$$

with

$$\tau_1 = \frac{4}{15}\frac{L^2}{\mu\phi_t\sqrt{1 + i_f}}\frac{1 + 3\alpha + \alpha^2}{(1 + \alpha)^3} \cong \frac{4}{15}\frac{(4 \times 10^{-6})^2}{400 \times 10^{-4} \times 25 \times 10^{-3}\sqrt{101}} \cong 425 \text{ ps}$$

to find the contribution of the distributed MOSFET model to the phase deviation in the transconductance. Note that the rational function in α in the formula above is $\cong 1$. The phase shift of this integrator at the unity-gain frequency compared with that of the ideal integrator is

$$\Delta\phi = -\omega_u\tau_1 + [I_B/(V_EL)]/g_m$$
$$= -180 \times 10^6 \times 425 \times 10^{-12} + [0.3 \times 10^{-3}/(5 \times 4)]/(1.8 \times 10^{-3})$$
$$= -0.068 \text{ rad}.$$

The term in square brackets in the formula above is the transistor output conductance. The result we obtained for the phase deviation corresponds to an integrator Q-factor of approximately -14.7.

(c) When a parasitic capacitance of $0.25C_L$ is included in parallel with C_L, the unity-gain frequency will reduce to 80% (1/1.25) of the value calculated considering that the parasitic capacitance is zero. To keep the unity-gain frequency unchanged, the transconductance would have to increase by 25%; thus $g_m = 1.25 g_m|_{if=100}$. In this case, the inversion level must be increased to the value given below, in order to satisfy the unity-gain bandwidth requirement:

$$\sqrt{1 + i_f} - 1 = \left(g_m / g_m|_{i_f=100} \right)\left(\sqrt{1 + 100} - 1 \right) \rightarrow i_f \cong 151.$$

Consequently, the bias current must increase to $I_B = (151/100) \times 300 = 450\,\mu\text{A}$, which amounts to a 50% increase in power consumption.

Since the frequency response of high-frequency filters is very sensitive to phase errors in integrators, it is sometimes advisable to use some kind of phase compensation in order to keep the phase deviation within limits that will not significantly affect the filter response. However, the phase errors caused by parasitic components or parasitic effects in active devices are, in general, too difficult to predict and, therefore, to compensate for or eliminate by design [31]. Thus, high-frequency filters usually employ some phase compensation, either at the transconductor [32], [33] or at the integrating capacitor [34]. The compensation method introduced in [34] is simply to connect an MOS transistor, which can be tuned to compensate for the relatively unpredictable phase lag introduced by the transconductor (see Problem 9.7), in series with the integrating capacitor. Another form of compensation of the phase lag is, as indicated in [4], the use of the series resistance of the integrating capacitors to create a phase lead.

9.2.3   Signal-to-noise ratio, dynamic range, and power

To develop some concepts in this section we recall some results of the fundamental noise theory discussed in Chapter 4. The application of the *energy equipartition principle* (see Problem 4.4) to the RC circuit in Figure 9.17 leads to the following mean-square value for the output voltage noise:

$$\overline{v_{on}^2} = kT/C. \tag{9.2.8}$$

In (9.2.8) we have included the subscript n (for noise) in the symbol for the output voltage to emphasize that (9.2.8) is the noise contribution to the output voltage, that is, it

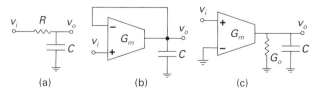

(a)   (b)   (c)

Fig. 9.17   (a) An RC circuit, (b) a G_m-C lossy integrator with dc gain of unity, and (c) a G_m-C lossy integrator with dc gain of G_m/G_o.

represents the mean-square value of the output-voltage noise for $v_i = 0$. Now let us consider that the instances of noise of the transconductors in Figures 9.17(b) and (c) have a white PSD given by

$$\overline{i_{on}^2}/\Delta f = 4kT\gamma G_m, \tag{9.2.9}$$

where γ is the excess noise factor, which usually has a minimal value of unity[2] [35]. In this textbook, as in [35], [36], we consider only the fundamental thermal noise, i.e. we assume that the effect of the $1/f$ noise is negligible. Under this assumption, the output noise power in the circuit of Figure 9.17(b) is given by

$$\overline{v_{on}^2} = \int_0^\infty \frac{\overline{i_{on}^2}}{\Delta f}|H(j\omega)|^2\,df = \int_0^\infty 4kT\gamma G_m \frac{1}{G_m^2 + (\omega C)^2}\frac{d\omega}{2\pi} = \gamma\frac{kT}{C}, \tag{9.2.10}$$

where $H(j\omega)$ is the transfer function that represents the conversion of the output noise current i_{on} of the transconductor into the output noise voltage. To derive the output noise voltage of the low-pass filter in Figure 9.17(c), we assume that $G_m \gg G_o$. In this case, the dominant noise source is the transconductor (regardless of whether G_o is a resistor or another transconductor). Using the same procedure as that used to derive (9.2.10) and assuming that G_o contributes negligibly to the output noise, we find that

$$\overline{v_{on}^2} = \int_0^\infty 4kT\gamma G_m \frac{1}{G_o^2 + (\omega C)^2}\frac{d\omega}{2\pi} = \gamma\frac{G_m}{G_o}\frac{kT}{C}. \tag{9.2.11}$$

The signal-to-noise ratio (SNR) is the ratio of the power of the output signal to the power of the noise signal, i.e.

$$\text{SNR} = \overline{v_o^2}/\overline{v_{on}^2} = \left(V_{opp}^2/8\right)/\overline{v_{on}^2}, \tag{9.2.12}$$

where the term in parentheses is the power of a sinusoidal signal with a peak-to-peak voltage of V_{opp}. In practice, the output voltage is generally limited to a value such that the distortion is acceptable for the intended application.

We define here the dynamic range DR of a circuit as the maximum achievable SNR, that is

$$\text{DR} = \text{SNR}_{max} = \overline{v_{o,max}^2}/\overline{v_{on}^2} = \left(V_{opp,max}^2/8\right)/\overline{v_{on}^2}. \tag{9.2.13}$$

The dynamic range is the ratio of the maximum output signal power for distortion below an acceptable level to the noise power. In our definition we consider that the noise level is independent of the signal level (quasi-equilibrium conditions). In syllabic companding, a technique in which the signal is expanded and compressed on the basis of the envelope of the signal, the definitions of dynamic range and maximal SNR are not

---

[2] For the case in which the transconductor is a simple long-channel MOS transistor in saturation, γ is $n/2$ in weak inversion and $2n/3$ in strong inversion. More elaborate transconductors have values that are typically slightly higher than those in a simple transistor transconductor.

coincident [37]. Note that the maximal output signal can be limited by the linearity range of any of the input, internal, or output nodes. In the example that follows the dynamic range is limited by the output node.

Example 9.6

Assume the following values for the components in Figure 9.15: $G_{m1} = 1\ \mu A/V$, $G_{m2} = 0.1\ \mu A/V$, $C = 0.2\ pF$, and $\gamma = 1$. Calculate the dynamic range when the maximal differential input voltage for acceptable distortion is (a) 50 mV for both transconductors and (b) 50 mV for transconductor 1 and 200 mV for transconductor 2. Assume that the offset voltage of both transconductors is zero.

Answer

(a) The mean-square value of the noise voltage for the low-pass filter (see (9.2.11)) is

$$\overline{v_{on}^2} = \gamma \frac{G_{m1} + G_{m2}}{G_{m2}} \frac{kT}{C} = 11 \frac{kT}{C}.$$

The linear range is limited to a 50-mV peak signal at transconductor 2. Therefore, the maximum signal swing is $V_{opp,max} = 100\ mV$. For a sine wave, the DR is

$$DR = \left(V_{opp,max}^2/8 \right) \Big/ \overline{v_{on}^2} = \left(V_{opp,max}^2/8 \right) \Big/ (11kT/C) \cong 5.5 \times 10^3$$

or, equivalently, $DR_{dB} = 10 \log DR = 37.4\ dB$. Note that the maximum peak-to-peak input signal is 10 mV in this case.

(b) The maximum output voltage is $V_{opp,max} = 400\ mV$. In comparison with the previous case, the dynamic range is greater by a factor of 16; this is equivalent to an increase of 12 dB. Thus, $DR_{dB} = 37.4 + 12 = 49.4\ dB$. This improvement is obtained with an increase in the linear range for transconductor 2, which, in turn, is generally obtained through an increase in power consumption.

Now, using simple models, as in [37], [38], we will determine how the power P delivered by the source of energy is related to the required dynamic range and bandwidth of an analog circuit. Let us consider the simple example of the unity-gain low-pass filter shown in Figure 9.18. As in [37], we assume that the transconductor operates in class B, that is, during capacitor charging the current through the negative supply is zero, whereas during discharging the current through the positive supply is zero. In this case, writing $V_O = (V_{opp}/2)\sin(\omega t)$, the instantaneous power p at the positive supply is

$$p = \frac{V_{DD}}{2} I_{D+} = \frac{V_{DD}}{2} I_C = \frac{V_{DD}}{2} C \frac{dV_O}{dt} = \frac{V_{DD}}{2} \omega C \frac{V_{opp}}{2} \cos(\omega t). \tag{9.2.14}$$

The energy delivered by the positive power supply is thus

$$\int_{-T/4}^{T/4} \frac{V_{DD}}{2} \omega C \frac{V_{opp}}{2} \cos(\omega t) dt = C \frac{V_{DD} V_{opp}}{4} \sin \theta \Big|_{-\pi/2}^{\pi/2} = C \frac{V_{DD} V_{opp}}{2}. \tag{9.2.15}$$

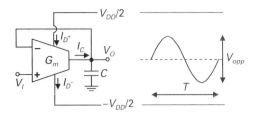

Fig. 9.18   A G_m-C unity-gain low-pass filter.

The average power P delivered by the positive source is

$$P = \frac{1}{T/2} \frac{CV_{DD}V_{opp}}{2} = fCV_{DD}V_{opp}, \tag{9.2.16}$$

which is the same as the average power delivered by the negative power supply. Note that, for $V_{opp} = V_{DD}$, the result given by (9.2.16) is identical to that for power dissipation in the CMOS static inverter. Using (9.2.10), which is valid for the low-pass filter under analysis with $\gamma = 1$, and (9.2.12), we find the SNR as

$$\text{SNR} = \left(V_{opp}^2/8\right)\Big/\overline{v_{on}^2} = \left(V_{opp}^2/8\right)\Big/(kT/C). \tag{9.2.17}$$

Now, by combining the results of (9.2.16) and (9.2.17), we find that

$$P = 8\frac{V_{DD}}{V_{opp}}kTf \cdot \text{SNR}. \tag{9.2.18}$$

The power consumption given by (9.2.18) is important as a criterion by which to compare various circuit solutions in terms of power efficiency [37]. In the most favorable (and hypothetical) scenario, the output-voltage (and common-mode input-voltage in the particular case of Figure 9.18) swing equals the power-supply voltage with acceptable distortion, i.e. $V_{opp} = V_{DD}$. In this case, the power delivered by the source is, at least, the minimum value [37], [38] given by

$$P_{\min} = P|_{V_{opp}=V_{DD}} = 8kTf \cdot \text{SNR} \tag{9.2.19}$$

for a given temperature, frequency, and SNR. As noted in [37], [38], the frequency f is the bandwidth of this low-pass filter. In conclusion, the power delivered by the energy source is proportional to the dynamic range required for the application.

It is worth noting that the application of the same reasoning to calculate the minimum power for the circuit in Figure 9.15 leads to the result

$$P = 8kT\Delta f\frac{V_{DD}}{V_{opp}} \cdot \text{SNR} \cdot (1 + A_{V0}), \tag{9.2.20}$$

where $A_{V0} = G_{m1}/G_{m2}$ is the low-frequency gain and $\Delta f = G_{m2}/(2\pi C)$ is the bandwidth

For readers who wish to deepen their knowledge on trade offs among power consumption, SNR, and signal bandwidth, we recommend the reading of [37], an excellent reference in which the power efficiency of some commonly employed circuits is analyzed.

9.2.4   Filter examples

An interesting example of a low-pass filter for signals below 1 Hz is presented in [20] and reproduced in Figure 9.19. The nominal parameters are $G_{m4} = 90$ pS and $C = 50$ pF.

The plot in Figure 9.19 shows the frequency response of a low-pass filter based on an integrator. The -3 dB cut-off frequency was measured as 0.302 Hz. An independently powered unity-gain buffer was also incorporated into the circuit to drive the output pad. The transconductor was implemented as a symmetric differential core followed by a current divider composed of a parallel–series association of transistors, as shown in Figure 9.20

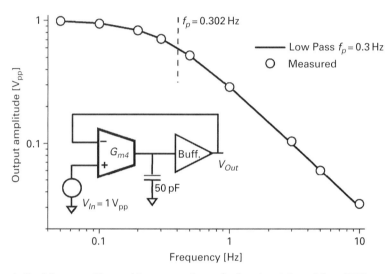

Fig. 9.19   A G_m-C low-pass filter and its measured transfer function (adapted from [20]).

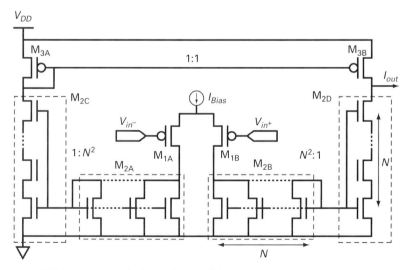

Fig. 9.20   A p-MOS-input symmetric OTA with parallel–series current division to reduce transconductance without loss in the linear range.

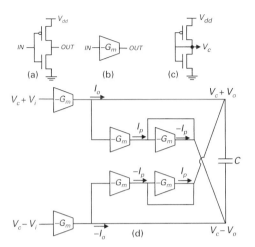

Fig. 9.21   (a) A CMOS inverter and (b) its symbol. (c) The circuit for generation of the common-mode voltage. (d) The complete transconductance element used in an integrator.

[19], [20]. The linear range of the differential input pair is 500 mV and its transconductance is around 69 nS. A division factor of 784 : 1 for transconductance attenuation was achieved employing two current mirrors with 28 identical parallel-connected transistors at the input and 28 identical series-connected transistors at the output. The resulting transconductance is $G_{m4} = 69/784 \, \text{nS} = 90 \, \text{pS}$.

A transconductance element for the realization of high-frequency filters was presented in [4], [39]. The transconductor, reproduced in Figure 9.21, is a fully-differential structure based on CMOS inverters. The voltage-to-current conversion is performed by two inverters driven by a differential input voltage balanced around a common-mode level equal to V_c [39]. It can be shown (see Problem 9.8) that in strong inversion and saturation of both p- and n-channel devices the output current is, to a first-order approximation, proportional to the input voltage when the current scaling factors β_n and β_p of the n- and p-transistors are matched ($\beta_n = \beta_p$). In this case, the transconductance of the CMOS inverter is

$$G_m = \beta(V_{dd} - V_{T0n} + V_{T0p}), \quad \beta = \mu C'_{ox}(W/L)/n. \tag{9.2.21}$$

As can be seen in (9.2.21), the transconductance can be adjusted by means of the local supply voltage V_{dd}.

The common-mode voltage V_c is generated by an inverter with input and output short-circuited. Note that, when the input voltage of the CMOS inverter is V_c, the output current of the transconductor is zero.

The complete transconductor is composed of six inverters, two of them responsible for the voltage-to-current conversion at the input and the remaining four for controlling the common-mode voltage. The dc gain of the G_m-C integrator is limited only by mismatch; a relative transconductance mismatch of less than 0.5% results in a dc gain larger than 200 [40].

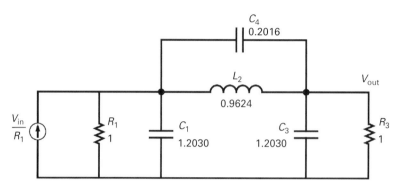

Fig. 9.22   A passive prototype filter for the generation of the G_m-C filter of Figure 9.23 [40].

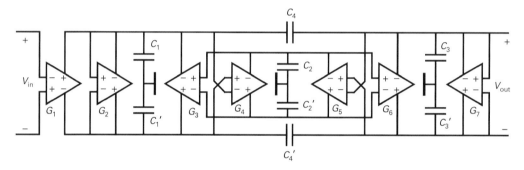

Fig. 9.23   Active implementation of the prototype filter in Figure 9.22 [40].

The transconductance of Figure 9.21(d) was employed in a third-order elliptic filter [40], which was derived from the passive ladder filter in Figure 9.22. The scheme of the G_m-C filter is shown in Figure 9.23. The active filter uses a gyrator $(G_3–G_6)$ loaded with a capacitor pair (C_2, C_2') to simulate the floating inductor L_2 of Figure 9.22. The usefulness of the gyrator and its realization on the basis of transconductors will be explained next. More details about the filter implementation can be found elsewhere [40].

In many cases, the design of analog integrated filters is based on classical LC filters with resistive source and load terminations [2], [41], [42]. An example of such an LC circuit is shown in Figure 9.22, where the filter components, apart from source and load terminations, are lossless. Designs based on LC filters, which lead to circuits with very low sensitivity to component tolerances, are the most appropriate for successful implementation of high-selectivity filters [41]. When inductors are either too bulky or the desired values are unavailable (or require very large areas), inductors can be eliminated with the help of gyrators. A gyrator is a two-port device (see Figure 9.24) described [2] by the equations

$$I_1 = -G_m V_2 \quad \text{and} \quad I_2 = G_m V_1. \tag{9.2.22}$$

The implementation of a gyrator using transconductors, together with the simulation of grounded and floating inductors using gyrator-C circuits, is shown in Figure 9.23. It is

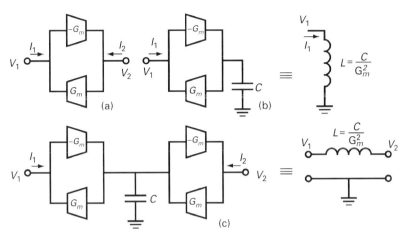

Fig. 9.24   (a) Implementation of a gyrator with transconductors. (b) Realization of a grounded inductor. (c) Realization of a floating inductor.

worth noting that the inductor value assumes an idealized model for the transconductors. In a real implementation, one must be aware of the input capacitance, output conductance, and phase shift of the transconductors, as well as mismatch between them [41]. In addition, transconductors also introduce noise into the filter. It has been demonstrated [41] that the noise power of an LC-tank where the inductor is simulated with a gyrator-C circuit is Q times (Q is the quality factor) higher than that of its passive counterpart.

9.3   Digitally-programmable continuous-time filters

Programmable filters are employed in several applications, such as storage systems [43]–[45], multistandard radio receivers [46]–[48], and cochlear implants [49]. Continuous-time filters are programmed by changing capacitances, or transconductances (or resistances in MOSFET-C filters), or both. The components can be controlled either continuously or in a quantized fashion. We will show here a simple component, the MOSFET-only current divider (MOCD), with which the parameters of a MOSFET-C filter [50], [51] can be digitally programmed.

The scheme of the MOCD [52], which is similar to the traditional R-2R network, is depicted in Figure 9.25. For an array of identical transistors, the current is successively halved through the branches of the MOCD. The output current is thus a fraction of the input current selected by the on state of the left-hand-side transistors in the parallel connections. This programmable current divider has two interesting properties: (i) the MOSFETs operate simultaneously as elements of the divider network and as switches and (ii) the resistance of the current attenuator is independent of both the number of bits and the attenuation factor. The high linearity of this division technique has been proved adequate for analog signal processing [52]. Note that for proper operation of the MOCD lines αI (the output line) and $(1 - \alpha)I$ (the dump line) must be at the same potential.

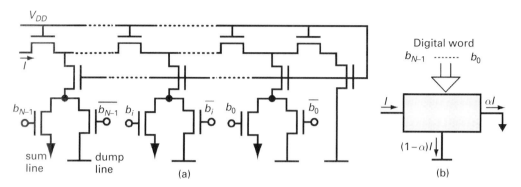

Fig. 9.25   A MOSFET-only (binary) current divider and its symbol.

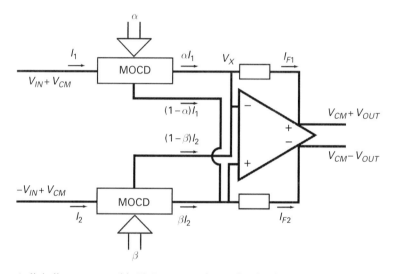

Fig. 9.26   A digitally programmable V–I converter for application in MOSFET-C filters.

The scheme of the digitally controlled V–I converter for application in MOSFET-C filters is represented in Figure 9.26. Note that the dump and output lines of the MOCDs are (virtually) at the same potential. Assuming that the MOCDs are matched and programmed with equal binary words, i.e. with equal attenuation factors ($\beta = \alpha$), the fundamental component of the difference between the output currents is given by

$$I_{F1} - I_{F2} = 2(2\alpha - 1)g_{ms}V_{IN}, \tag{9.3.1}$$

where g_{ms} is the conductance of the MOCD (equal to half the transconductance of a single transistor of the array) calculated for $V_{DS} = 0$. The attenuation factor can vary in the range from approximately 0 to $1 - 2^{-N}$ (N is the number of bits). Thus, the digitally programmable factor ($2\alpha - 1$) can vary between approximately 1 and -1. If the feedback components are capacitors, then the structure in Figure 9.26 can be either an inverting ($\alpha > 1/2$) or a non-inverting ($\alpha < 1/2$) integrator with a digitally controlled unity-gain frequency equal to $2(2\alpha - 1)g_{ms}/C$.

Examples of programmable MOSFET-C and active-*RC* filters using MOCDs are presented in [51] and [46], respectively. Even though the programmability of continuous-time filters is conceptually simple to implement, an important problem that arises from programmability is that of maintaining adequate dynamic range and frequency-response accuracy across the filter tuning range [9], [44].

9.4 On-chip tuning schemes

The transfer function of a filter is dependent on pole and zero frequencies, and gain constants. These parameters, in turn, are dependent on the accuracy of element values, which, in an integrated implementation, suffer deviations from their nominal values due to process and thermal variations, and aging. Also, the frequency response is affected by parasitic components the values of which are relatively unpredictable. Therefore, some kind of method to adjust the filter parameters must be used to obtain an accurate frequency response.

Two methods to control the accuracy of a filter are employed, namely direct and indirect tuning schemes [2], [7], [9]. In the direct tuning scheme, the filter that processes the signal is directly tuned. When the operation of the filter can be interrupted, the scheme in Figure 9.27(a) can be used. In the interrupting interval, the signal is disconnected and a

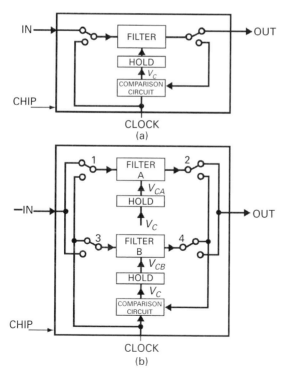

Fig. 9.27 On-chip direct tuning schemes: (a) interrupted filtering and (b) uninterrupted filtering (adapted from [7], [53]).

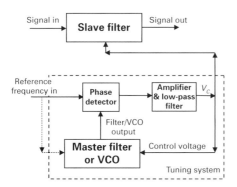

Fig. 9.28   An on-chip indirect tuning scheme with either a voltage-controlled filter or a voltage-controlled oscillator as the master (adapted from [54]). The dotted line indicates a connection for the case of the master filter only, but not for the case of the master VCO.

reference signal (clock) is applied to the filter input. A phase comparator compares the phase of the filter output with that of the clock signal and adjusts the control voltage V_C until the phase difference between the filter output and the clock becomes equal to a pre-established value. In Figure 9.27(b), the two filters are connected alternately in the signal path to avoid interruption in filtering; while the first one is performing the filtering function, the second one is being adjusted, and vice versa.

Another direct tuning method based on adaptive tuning techniques is described in [9]. In it, a tuning processor infers the response of the filter from the statistics of the filter input and output, and a tuning algorithm adjusts the filter coefficients in order to obtain the desired response. The price to be paid is that the tuning processor is generally very complex.

The basic idea of an indirect tuning scheme is illustrated in Figure 9.28 [54], [55]. In this approach, the main (slave) filter is permanently connected. The tuning system is composed of either a master filter or a master voltage-controlled oscillator (VCO), made of structures matched to those in the main filter, and the phase detector, amplifier, and low-pass filter. The phase detector compares the output of the VCO (or master filter) against a clock signal and generates an output signal that is low-pass filtered to generate V_C, the control voltage of the VCO (or master filter) and slave filter. Both the control voltage V_C and the VCO oscillation frequency (or the frequency response of the master filter) become constant when the oscillator (or master filter) tracks the clock signal. In this scheme, the tuning system adjusts the VCO frequency (or master filter response) directly and the main filter indirectly.

The indirect tuning technique described by Figure 9.28 could, at least in principle, work very well for the adjustment of the time constants of the filter, which are equal to C/G_m quotients. Theoretically, the Q-factor (or selectivity) of the filter is dependent on the ratio of the values of similar components only and, thus, theoretically, is sensitive to mismatch only. However, in a practical filter, the Q-factor is sensitive to parasitic components and highly sensitive to mismatch for high-Q filters. Therefore, the scheme in Figure 9.28 is appropriate for filters with relatively low quality factors, which are mainly determined by nominal ratios of similar components. For high-Q filters, a more elaborate scheme, called a vector-locked loop, includes not only the loop for frequency stabilization but also a second loop [1], [2], [9], [33] for stabilization of the quality factor of the filter.

Appendix

A9.1 Distortion of the MOSFET operating as a resistor

In this appendix, we compute the distortion in the I–V characteristic of the MOSFET shown in Figure 9.1, where voltages are referred to the substrate. Since $V_S = V_Q$ and V_G are constant, the forward current I_F is also constant, whereas the reverse current I_R is a function of the input voltage. The normalized drain current i_d can be expanded in power series in v_{in} as

$$i_d = \frac{I_D}{I_S} = i_f - i_r = \sum_{j=1}^{\infty} k_j v_{in}^j, \quad k_j = -\frac{1}{j!}\frac{d^j i_r}{dv_{in}^j}\bigg|_{i_r=i_f}. \tag{A9.1.1}$$

Recalling that

$$g_{md} = \frac{dI_D}{dV_D} = \frac{dI_D}{dV_{in}} = \frac{2I_S}{\phi_t}\left(\sqrt{1+i_r}-1\right), \tag{A9.1.2}$$

the coefficients k_j in (A9.1.1) can be readily calculated as

$$k_1 = \frac{2}{\phi_t}\left(\sqrt{1+i_f}-1\right), \tag{A9.1.3}$$

$$k_2 = -\frac{1}{\phi_t^2}\frac{\sqrt{1+i_f}-1}{\sqrt{1+i_f}}, \tag{A9.1.4}$$

$$k_3 = -\frac{1}{3\phi_t^3}\frac{\sqrt{1+i_f}-1}{\left(\sqrt{1+i_f}\right)^3}. \tag{A9.1.5}$$

Assuming that $V_{in} = V_M \sin(\omega t)$, we can immediately calculate the fundamental component of the current from

$$I_M = g_{md} V_M = \frac{2I_S}{\phi_t}\left(\sqrt{1+i_f}-1\right)V_M. \tag{A9.1.6}$$

The second- and third-order harmonic distortions of the current are given by

$$HD2 = \left|\frac{k_2}{2k_1}\right|V_M = \frac{V_M}{4\phi_t}\frac{1}{\sqrt{1+i_f}}, \tag{A9.1.7}$$

$$HD3 = \left|\frac{k_3}{4k_1}\right|V_M^2 = \frac{1}{24}\left(\frac{V_M}{\phi_t}\right)^2 \frac{1}{\left(\sqrt{1+i_f}\right)^3}. \qquad (A9.1.8)$$

The harmonic distortion is thus computed from (A9.1.7) and (A9.1.8) in terms of the normalized saturation current (i_f). The results of the second- and third-order harmonic distortions are shown in Figure A9.1.1 and A9.1.2 for various gate (V_G) and common-mode (V_Q) voltages.

The dominant deviation from linearity comes from the second-order term. The theoretical second-order distortion fits the experimental results very well. The error that results on calculating HD3 from (A9.1.7) is very large. HD3 increases considerably when the drain voltage approaches the saturation voltage. In fact, the analytical determination of third-order terms would need an I–V MOSFET model that is accurate up to the

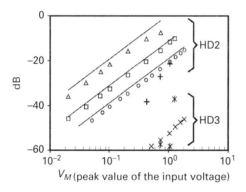

Fig. A9.1.1   Measured and theoretical harmonic distortion (in dB) of a transistor, $W/L = 18\,\mu m/5\,\mu m$, from a 2-μm technology: $\triangle$, $+$, $V_G = 1.2$ V and $V_Q = 0.1$ V ($i_f = 80$); $\square$, $*$, $V_G = 2.5$ V and $V_Q = 0.4$ V ($i_f = 1200$); and o, $\times$, $V_G = 5.0$ V and $V_Q = 1.1$ V ($i_f = 7200$).

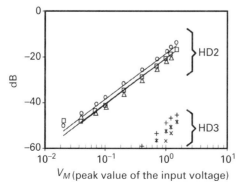

Fig. A9.1.2   Measured and theoretical harmonic distortion (in dB) of a transistor, $V_G = 5$ V, $W/L = 18\,\mu m/5\,\mu m$, from a 2-μm technology: o, $\times$, $V_Q = 1.9$ V ($i_f = 6600$); $\square$, $*$, $V_Q = 1.1$ V ($i_f = 10\,500$); and $\triangle$, $+$, $V_Q = 0.5$ V ($i_f = 11\,500$).

third-order derivative, and thus should include variations in mobility and slope factor in addition to short-channel effects.

Problems

9.1 Demonstrate that, in strong inversion, the differential output current $I_{OD}=I_{12}-I_{34}$ of the circuit in Figure 9.3 is given by

$$I_{12}-I_{34}=2\mu C'_{ox}\frac{W}{L}(V_A-V_B)V_{in}.$$

Assume that both n and μ are independent of the gate voltage and that all transistors are matched.

9.2 Consider the MOSFET modeled as a transmission line, as shown in Figure P9.2.

(a) Demonstrate that

$$I(y,s)=\frac{V_{IN}(s)}{rL}\frac{\sqrt{\tau_L s}L}{\sinh(\sqrt{\tau_L s}L)}\cosh\left[\sqrt{\tau_L s}(L-y)\right],$$

where $\tau_L=rc$ is the time constant per unit length squared.

(b) What is the transadmittance $y_{ds}=I(L,s)/V_{IN}(s)$?

(c) Demonstrate that for frequency values at which $\omega\tau_L L^2\ll 1$ we have

$$y_{ds}=\frac{1}{rL}\frac{1}{1+s\tau_L L^2/6}\cong\frac{1-s\tau_L L^2/6}{rL}. \qquad (P9.2.1)$$

(d) Verify that the values of r and c for $V_{DS}=0$ are, respectively,

$$r=\frac{1}{g_{ms}L}=\frac{-1}{\mu WQ'_{IS}}=\frac{1}{\mu WnC'_{ox}\phi_t q'_{IS}}$$

and

$$c=\frac{C_{sg}+C_{dg}+C_{sb}+C_{db}}{L}\bigg|_{V_{DS}=0}$$

$$=\frac{C_{gs}+C_{gd}+C_{bs}+C_{bd}}{L}\bigg|_{V_{DS}=0}=nC'_{ox}W\frac{q'_{IS}}{q'_{IS}+1}.$$

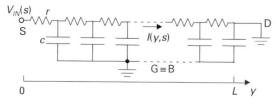

Fig. P9.2   A MOSFET modeled as a transmission line, with r and c equal to the resistance and capacitance per unit length, respectively.

Therefore,

$$\tau_L = rc = \frac{1}{\mu\phi_t(q'_{IS} + 1)} = \frac{1}{\mu\phi_t\sqrt{i_f + 1}}$$

and (P9.2.1) gives the same result as (9.1.8).

9.3 For the specifications of Example 9.3, find the integrating capacitor which leads to the "best" value of Q_I. What is this value?

9.4 Consider the transconductor in Figure 9.12(a). (a) Calculate the input voltage V_I for which the output current is zero. (b) Calculate the small-signal transconductance around $i_O = 0$. (c) Calculate the approximate deviation of the input voltage around the value calculated in (a) for which the transconductance deviates by $\pm 5\%$ around the value calculated in (b). (d) Assuming that $n = 1.2$, $V_{T0} = 0.6$ V, and $\phi_t = 25$ mV, determine the values in (a), (b), and (c) for $I_B/I_S = 0.1$, 1, 10, 100, and 1000.

9.5 For the circuit in Figure 9.13 demonstrate that (a) the transconductance is equal to half that of a single differential pair biased with the same I_B, (b) the magnitude of the second-harmonic component of the current is a quarter that of a single differential pair with the same I_B, and (c) the voltage at the intermediate node is equal to the common-mode voltage of v_1 and v_2. (d) Assuming that the substrate of all n-channel devices is connected to ground, describe how the voltage at the intermediate node is modified. In (c) and (d) neglect the influence of the drain voltage on the current.

9.6 Demonstrate that Equation (9.2.2) is valid for the source-degenerated transconductor in Figure 9.14.

9.7 Show that a resistor in series with the capacitor in an OTA-C integrator can compensate for the phase lag introduced by the transconductor. What is the value required for the resistor, assuming that the uncompensated phase deviation at the integrator unity-gain frequency is equal to $-\phi_u$?

9.8 (a) Determine the dependence of the output current on the input voltage of a CMOS inverter when both the p-channel device and the n-channel device operate in strong inversion and saturation. (b) Derive the condition for cancellation of the second-order harmonic distortion. (c) What are the common-mode voltage and the transconductance for the condition derived in (b)?

9.9 Analyze the effects of each of the following non-idealities of the transconductors on the inductance simulated by the gyrator-C circuit in Figure 9.24: (a) the output conductance, (b) the input capacitance, (c) the mismatch between transconductances, and (d) the transconductance phase lag. Compare your results with those of reference [41].

References

[1] K. R. Laker and W. M. C. Sansen, *Design of Analog Integrated Circuits and Systems*, New York: McGraw-Hill, 1994.

[2] R. Schaumann and M. E. Van Valkenburk, *Design of Analog Filters*, New York: Oxford University Press, 2000.

[3] Y. Sun (ed.), *Design of High Frequency Integrated Analogue Filters*, London: The Institution of Electrical Engineers, 2002.

[4] B. Nauta, *Analog CMOS Filters for Very High Frequencies*, Boston, MA: Kluwer, 1993.

[5] P. V. Ananda Mohan, *Current-Mode VLSI Analog Filters – Design and Applications*, Boston, MA: Birkhäuser, 2003.

[6] M. Banu and Y. P. Tsividis, "Fully integrated active *RC* filters in MOS technology," *IEEE Journal of Solid-State Circuits*, vol. **18**, no 6, pp. 644–651, Dec. 1983.

[7] Y. P. Tsividis, M. Banu, and J. Khoury, "Continuous-time MOSFET-C filters in VLSI," *IEEE Transactions on Circuits and Systems*, vol. **33**, no. 2, pp. 125–140, Feb. 1986.

[8] M. Banu and Y. Tsividis, "Detailed analysis of nonidealities in MOS fully integrated active *RC* filters based on balanced networks," *IEE Proceedings*, vol. **131**, part G, no. 5, pp. 190–196, Oct. 1984.

[9] S. Pavan and Y. Tsividis, *High Frequency Continuous Time Filters in Digital CMOS Processes*, Boston, MA: Kluwer, 2000.

[10] M. Banu and Y. Tsividis, "The MOSFET-C technique: designing power efficient, high frequency filters," Chapter 2 of *Design of High Frequency Integrated Analogue Filters*, ed. Yichuang Sun, London: The Institution of Electrical Engineers, 2002, pp. 35–55.

[11] M. C. Schneider, C. Galup-Montoro, S. M. Acosta, and A. I. A. Cunha, "Distortion analysis of MOSFETs for application in MOSFET-C circuits," *Proceedings of ISCAS* 1998, Monterey, CA, June 1998.

[12] Z. Czarnul, "Modification of Banu–Tsividis continuous-time integrator structure," *IEEE Transactions on Circuits and Systems*, vol. **33**, no. 7, pp. 714–716, July 1985.

[13] A. T. Behr, M. C. Schneider, S. Noceti Filho, and C. G. Montoro, "Harmonic distortion caused by capacitors implemented with MOSFET gates," *IEEE Journal of Solid-State Circuits*, vol. **27**, no. 10, pp. 1470–1475, Oct. 1992.

[14] Y. P. Tsividis, "Integrated continuous-time filter design – an overview," *IEEE Journal of Solid-State Circuits*, vol. **29**, no. 3, pp. 166–176, Mar. 1994.

[15] R. L. Geiger and E. Sánchez-Sinencio, "Active filter design using operational transconductance amplifiers: a tutorial," *IEEE Circuits and Devices Magazine*, vol. **1**, no. 2, pp. 20–32, Mar. 1985.

[16] E. Sánchez-Sinencio and J. Silva-Martínez, "CMOS transconductance amplifiers, architectures and active filters: a tutorial," *IEE Proceedings on Circuits Devices and Systems*, vol. **147**, no. 1, pp. 3–12, Feb. 2000.

[17] H. Khorramabadi and P. R. Gray, "High-frequency CMOS continuous-time filters," *IEEE Journal of Solid-State Circuits*, vol. **19**, no. 6, pp. 939–948, Dec. 1984.

[18] M. Steyaert, P. Kinget, and W. Sansen, "Full integration of extremely large time constants in CMOS," *Electronics Letters*, vol. **27**, no. 10, pp. 790–791, May 9, 1991.

[19] A. Arnaud and C. Galup-Montoro, "Pico-A/V CMOS transconductors using series-parallel current division," *Electronics Letters*, vol. **39**, no. 18, pp. 1295–1296, Sep. 4, 2003.

[20] A. Arnaud, R. Fiorelli and C. Galup-Montoro, "Nanowatt, sub-nS OTAs, with sub-10-mV input offset using series-parallel current mirrors," *IEEE Journal of Solid-State Circuits*, vol. **41**, no. 9, pp. 2009–2018, Sep. 2006.

[21] A. Nedungadi and T. R. Viswanathan, "Design of linear CMOS transconductance elements," *IEEE Transactions on Circuits and Systems*, vol. **31**, no. 10, pp. 891–894, Oct. 1984.

[22] C.-S. Park and R. Schaumann, "A high-frequency CMOS linear transconductance element," *IEEE Transactions on Circuits and Systems*, vol. **33**, no. 1, pp. 1132–1138, Nov. 1986.

[23] K. Bult and H. Wallinga, "A class of analog CMOS circuits based on the square-law characteristic of an MOS transistor in saturation," *IEEE Journal of Solid-State Circuits*, vol. **22**, no. 3, pp. 357–365, June 1987.

[24] E. Seevinck and R. F Wassennar, "A versatile CMOS linear transconductor/square-law function circuit," *IEEE Journal of Solid-State Circuits*, vol. **22**, no. 3, pp. 366–377, June 1987.

[25] S. Noceti Filho, M. C. Schneider, and R. N. G. Robert, "New CMOS OTA for fully integrated continuous-time circuit applications," *Electronics Letters*, vol. **25**, no. 24, pp. 1674–1675, Nov. 23, 1989.

[26] G. Wilson and P. K. Chan, "Low-distortion CMOS transconductor," *Electronics Letters*, vol. **26**, no. 11, pp. 720–722, May 24, 1990.

[27] R. R. Torrance, T. R. Viswanathan, and J. V. Hanson, "CMOS voltage to current transducers," *IEEE Transactions on Circuits and Systems*, vol. **32**, no. 11, pp. 1097–1104, Nov. 1985.

[28] Y. Tsividis, Z. Czarnul, and S. C. Fang, "MOS transconductors and integrators with high linearity," *Electronics Letters*, vol. **22**, no. 5, pp. 245–246, Feb. 27, 1986.

[29] F. Krummenacher and N. Joehl, "A 4-MHz CMOS continuous-time filter with on-chip automatic tuning," *IEEE Journal of Solid-State Circuits*, vol. **23**, no. 3, pp. 750–758, June 1988.

[30] W. J. A. de Heij, E. Seevinck, and K. Hoen, "Practical formulation of the relation between filter specifications and the requirements for integrator circuits," *IEEE Transactions on Circuits and Systems*, vol. **36**, no. 8, pp. 1124–1128, Aug. 1989.

[31] R. Schaumann, M. S. Ghausi, and K. Laker, *Design of Analog Filters: Passive, Active RC, and Switched Capacitor*, Englewood Cliffs, NJ: Prentice Hall, 1990.

[32] E. J. van der Zwan, E. A. M. Klumperink, and E. Seevinck, "A CMOS OTA for HF filters with programmable transfer function," *IEEE Journal of Solid-State Circuits*, vol. **26**, no. 11, pp. 1720–1723, Nov. 1991.

[33] V. Gopinathan, Y. P. Tsividis, K.-S. Tan, and R. K. Hester, "Design considerations for high-frequency continuous-time filters and implementation of an antialiasing filter for digital video," *IEEE Journal of Solid-State Circuits*, vol. **25**, no. 6, pp. 1368–1378, Dec. 1990.

[34] M. I. Ali, M. Howe, E. Sánchez-Sinencio, and J. Ramírez Angulo, "BiCMOS low-distortion tunable OTA for continuous-time filters," *IEEE Transactions on Circuits and Systems I*, vol. **40**, no. 1, pp. 43–49, Jan. 1993.

[35] G. Groenewold, "Optimal dynamic range integrators," *IEEE Transactions on Circuits and Systems I*, vol. **39**, no. 8, pp. 614–627, Aug. 1992.

[36] Y. Palaskas and Y. Tsividis, "Dynamic range optimization of weakly nonlinear, fully balanced G_m–C filters with power dissipation constraints," *IEEE Transactions on Circuits and Systems II*, vol. **50**, no. 10, pp. 714–727, Oct. 2003.

[37] E. A. Vittoz and Y. P. Tsividis, "Frequency-dynamic range-power," Chapter 10 in *Trade-Offs in Analog Circuit Design: The Designers Companion*, ed. C. Toumazou, G. Moschytz, and B. Gilbert, Boston, MA: Kluwer, 2002, pp. 283–313.

[38] B. J. Hosticka, "Performance comparison of analog and digital circuits," *Proceedings of the IEEE*, vol. **73**, no. 1, pp. 25–29, Jan. 1985.

[39] B. Nauta and E. Seevinck, "Linear CMOS transconductance element for VHF filters," *Electronics Letters*, vol. **25**, no. 7, pp. 448–450, Mar. 30, 1989.

[40] B. Nauta, "A CMOS transconductance-C filter technique for very high frequencies," *IEEE Journal of Solid-State Circuits*, vol. **27**, no. 2, pp. 142–153, Feb. 1992.

[41] Y.-T. Wang and A. A. Abidi, "CMOS active filter design at very high frequencies," *IEEE Journal of Solid-State Circuits*, vol. **25**, no. 6, pp. 1562–1574, Dec. 1990.

[42] H. Voorman and H. Veenstra, "Tunable high-frequency G_m–C filters," *IEEE Journal of Solid-State Circuits*, vol. **35**, no. 8, pp. 1097–1108, Aug. 2000.

[43] I. Mehr and D. R. Weland, "A CMOS continuous-time G_m–C filter for PRML read channel applications at 150 Mb/s and beyond," *IEEE Journal of Solid-State Circuits*, vol. **32**, no. 4, pp. 499–513, Apr. 1997.

[44] S. Pavan, Y. P. Tsividis, and K. Nagaraj, "Widely programmable high-frequency continuous-time filters in digital CMOS technology," *IEEE Journal of Solid-State Circuits*, vol. **35**, no. 4, pp. 503–511, Apr. 2000.

[45] G. Bollati, S. Marchese, M. Demicheli, and R. Castello, "An eighth-order CMOS low-pass filter with 30–120 MHz tuning range and programmable boost," *IEEE Journal of Solid-State Circuits*, vol. **36**, no. 7, pp. 1056–1066, July 2001.

[46] H. A. Alzaher, H. D. Elwan, and M. Ismail, "A CMOS highly linear channel-select filter for 3G multistandard integrated wireless receivers," *IEEE Journal of Solid-State Circuits*, vol. **37**, no. 1, pp. 27–37, Jan. 2002.

[47] V. Giannini, J. Craninckx, S. D'Amico, and A. Baschirotto, "Flexible baseband analog circuits for software-defined radio front-ends," *IEEE Journal of Solid-State Circuits*, vol. **42**, no. 7, pp. 1501–1512, July 2007.

[48] D. Chamla, A. Kaiser, A. Cathelin, and D. Belot, "A switchable-order G_m–C baseband filter with wide digital tuning for configurable radio receivers," *IEEE Journal of Solid-State Circuits*, vol. **42**, no. 7, pp. 1513–1521, July 2007.

[49] C. D. Salthouse and R. Sarpeshkar, "A practical micropower programmable bandpass filter for use in bionic ears," *IEEE Journal of Solid-State Circuits*, vol. **38**, no. 1, pp. 63–70, Jan. 2003.

[50] M. C. Schneider, C. Galup-Montoro, and S. Noceti Filho, "A digitally programmable V–I converter for application in MOSFET-C filters," *Electronics Letters*, vol. **31**, no. 18, pp. 1526–1527, Aug. 31, 1995.

[51] M. L. W. Cunha, S. Noceti Filho, and M. C. Schneider, "Automatic tuning of MOSFET-C filters using digitally programmable current attenuators," *Proceedings of IEEE ISCAS 1997*, pp. 329–332, 1997.

[52] K. Bult and G. J. G. M. Geelen, "An inherently linear and compact MOST-only current division," *IEEE Journal of Solid-State Circuits*, vol. **27**, no. 12, pp. 1730–1735, Dec. 1992.

[53] Y. Tsividis, "Self-tuned filters," *Electronics Letters*, vol. **17**, no. 12, pp. 406–407, June 11, 1981.

[54] Y. P. Tsividis and V. Gopinathan, "Continuous-time filters," in Chapter 6 in *Design of Analog-Digital VLSI Circuits for Telecommunications and Signal Processing*, ed. J. E. Franca and Y. Tsividis, Englewood Cliffs, NJ: Prentice-Hall, 1994, pp. 177–211.

[55] M. Banu and Y. P. Tsividis, "An elliptic continuous-time CMOS filter with on-chip automatic tuning," *IEEE Journal of Solid-State Circuits*, vol. **20**, no. 6, pp. 1114–1121, Dec. 1985.

10 Fundamentals of sampled-data circuits

The first part of this chapter is dedicated to the MOS sample-and-hold circuit (S/H). The main physical limitations of the MOS S/H, namely switch on-resistance, thermal noise, and switch charge injection, are reviewed. Distortion produced by the non-linear switch resistance is modeled and some circuit-linearization techniques for MOS switches are introduced as well as low-voltage circuit techniques. Finally, clock-jitter effects on the MOS S/H are modeled and these effects in analog-to-digital converters are commented on. The second part of the chapter presents the basics of switched-capacitor (SC) circuits. Parasitic insensitive integrators and second-order filters are reviewed. Amplifier specifications for SC circuits are detailed, as well as some circuit techniques to reduce the effects of op-amp imperfections. The important topic of SC circuits fully compatible with digital MOS technology ends the section on SC circuits. The last part of the chapter briefly introduces alternative switched circuits, specifically switched-MOSFET and switched-current filters.

10.1 MOS sample-and-hold circuits

Sample-and-hold (or track-and-hold) circuits are ubiquitous in signal-processing circuits. They are used at the front end of analog-to-digital (A/D) converters and at the back end of digital-to-analog (D/A) converters. The signal-to-noise ratio of an A/D converter is usually limited by the performance of the sample-and-hold block. Furthermore, precision analog integrated circuits are mostly implemented by sampled-data circuits in which the basic building block is the S/H.

10.1.1 Sample-and-hold basics

An ideal S/H is depicted in Figure 10.1. During the sampling mode, the switch S is on and the output voltage tracks the input voltage. At the moment the switch turns off, the input voltage is stored as a charge on the hold capacitor. Although the circuit discussed here implements a track-and-hold function, from now on we will adopt the term sample-and-hold, which is the most commonly used in the literature.

To avoid disturbing the signal source with the charging current of the capacitor, a front-end buffer is often added before the sample-and-hold module. Additionally, a buffer is also frequently placed at the output to reduce the loss of the stored charge due to leakage

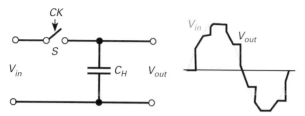

Fig. 10.1   A basic MOS sample-and-hold circuit (adapted from [1]).

currents. Since the sampling switch and the buffers present several non-idealities, more elaborate sample-and-hold architectures are often used for A/D front ends (see for example [1]–[3]).

In the next sections, the main physical limitations of the S/H are summarized, beginning with the most fundamental, i.e. thermal noise.

Example 10.1. Frequency limitations of a non-sampling A/D converter [3].

In the absence of a sample-and-hold block in the front end of an A/D converter, the input signal should not vary by more than 1 LSB (least significant bit) during the conversion time in order to maintain the error within the intended accuracy limit. Assume that the input to the converter is a sine wave with peak-to-peak amplitude equal to the full-scale voltage V_{FS}, i.e.

$$v(t) = \frac{V_{FS}}{2} \sin(2\pi ft). \tag{10.1.1}$$

The maximum rate of change of the input signal is

$$\left.\frac{dv}{dt}\right|_{\max} = 2\pi f \frac{V_{FS}}{2}. \tag{10.1.2}$$

During the conversion time, the maximum acceptable voltage variation is

$$dv = 1\,\mathrm{LSB} = \frac{V_{FS}}{2^B}. \tag{10.1.3}$$

Thus, the maximum frequency for a sine wave which can be converted with an error less than 1 LSB is, from (10.1.2) and (10.1.3), given by

$$f_{\max} = \frac{1}{\pi V_{FS}} \frac{V_{FS}}{2^B} \frac{f_s}{dt} = \frac{f_s}{\pi 2^B}, \tag{10.1.4}$$

where we have assumed, for simplicity, that the conversion time dt is equal to the inverse of the sampling frequency f_s, rather than less, as is the case in reality. Considering a 12-bit converter with sampling rate of 100 kS/s, the maximum frequency given by (10.1.4) is only 7.8 Hz. Clearly, this maximum frequency is orders of magnitude lower than the Nyquist rate of 50 kHz. To avoid this severe frequency limitation, a sample-and-hold function is added at the front end of the converter. The timing of the encoder is such that it performs the conversion during the hold time in which the signal is constant.

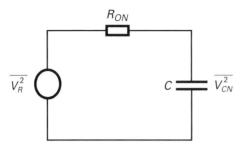

Fig. 10.2   The equivalent circuit of the S/H with the switch on and $v_i = 0$.

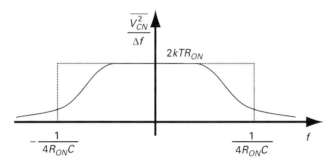

Fig. 10.3   The power-spectral density of the noise voltage across the capacitor of the circuit in Figure 10.2.

10.1.2   Thermal noise

To analyze thermal noise in an S/H, let us first consider that the switch is on and that the input signal v_{in} is zero, as shown in Figure 10.2, where R_{ON} represents the switch on-resistance and $\overline{v_R^2} = 4kTR_{ON}\Delta f$ is the voltage generator representing its thermal noise. The power-spectral density of the noise voltage on the capacitor is given by

$$\frac{\overline{v_{CN}^2}}{\Delta f} = \frac{\overline{v_R^2}}{\Delta f} \left| \frac{\frac{1}{j\omega C}}{R_{ON} + \frac{1}{j\omega C}} \right|^2 = \frac{4kTR_{ON}}{1 + (2\pi f\tau_{ON})^2}, \tag{10.1.5}$$

where $\tau_{ON} = R_{ON}C$ is the time constant of the S/H during the on time. The mean-square noise voltage on the capacitor, which was calculated in Problem 4.4, is given by

$$\overline{v_{CN}^2} = 4kTR_{ON} \int_0^\infty \frac{df}{1 + (2\pi f\tau_{ON})^2} = \frac{kT}{C}. \tag{10.1.6}$$

The (bilateral) power spectral density $\overline{v_{CN}^2}/\Delta f$ is shown in Figure 10.3. The total area under the curve, which is kT/C, is equal to that of an equivalent ideal unity-gain low-pass filter for which the input is the resistor thermal noise and the bandwidth is $f_{NB} = 1/(4R_{ON}C)$, as represented by the dotted rectangle; f_{NB} represents the noise bandwidth.

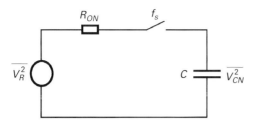

Fig. 10.4   The equivalent circuit of the MOS S/H with the switch off and $v_i = 0$.

The exact frequency-domain analysis of the track-and-hold circuit of Figure 10.4 is somewhat involved, but, to understand the effect of sampling on thermal noise let us consider, for simplicity, the idealized case of sampling by a periodic pulse train $s(t)$ given by

$$s(t) = \sum_{-\infty}^{\infty} \delta(t - nT_s), \qquad (10.1.7)$$

where $\delta(t)$ represents the unit impulse or Dirac delta function. The result of sampling a continuous-time function $x(t)$ is given by the product

$$x_s = x(t)s(t). \qquad (10.1.8)$$

The spectrum of the sampled signal is given by the classic Nyquist formula [4]

$$X_s(j2\pi f) = \frac{1}{T_s} \sum_{m=-\infty}^{+\infty} X(j2\pi f - jm2\pi f_s), \qquad (10.1.9)$$

where $X(j2\pi f)$, shown in Figure 10.5(a), is the spectrum of the continuous-time signal $x(t)$ and f_s is the sampling frequency. The spectrum of the sampled signal is the sum of the shifted (by $\pm m f_s$) spectra of the continuous-time signal, as shown in Figure 10.5(b).

For a continuous-time signal spectrum limited to $f_s/2$, the replicas of its frequency spectrum shifted by $\pm m f_s$ do not overlap. On the other hand, aliasing or folding occurs if the signal is not band-limited to $f_s/2$, as shown in Figure 10.5(b). The aliasing of the signal spectra is avoided by using anti-aliasing filters, but the aliasing of thermal noise is inevitable in an S/H.

In effect, since $\tau_{ON} = R_{ON}C$ is the time constant of the sample-and-hold module during the on time, the switch must be closed for a period of several time constants to allow charging of the hold capacitor within an error band to the input voltage. Let us assume that $T_s/2 > 6\tau_{ON}$. Hence, the noise bandwidth satisfies $f_{NB} = 1/(4\tau_{ON}) > 6(f_s/2)$; consequently, noise is aliased into all portions of the spectrum and, in particular, into the signal band.

Figure 10.6 shows a simplified representation of thermal-noise aliasing in which the thermal-noise power-spectral density is assumed constant over the noise bandwidth. As is clear from Figure 10.6, the resistor noise power density is multiplied by a factor of $2f_{NB}/f_s$. Thus,

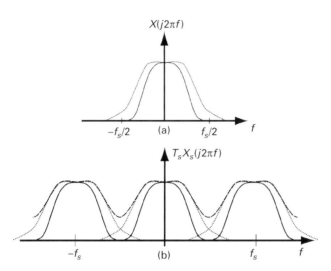

Fig. 10.5   Signal spectra: (a) for two continuous-time signals, band-limited to $f_s/2$ and not band-limited to $f_s/2$; and (b) for the two signals obtained by sampling the two continuous signals at frequency f_s, showing $(--)$ the aliased spectrum of a sampled signal that is not band-limited to $f_s/2$.

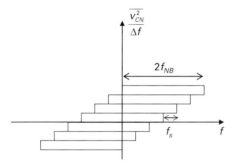

Fig. 10.6   A simplified representation of the aliasing of thermal noise due to sampling for the case $2f_{NB}/f_s = 6$.

$$2kTR_{ON} \cdot \frac{2f_{NB}}{f_s} = \frac{4kTR_{ON}}{f_s} \frac{1}{4R_{ON}C} = \frac{kT}{Cf_s}. \qquad (10.1.10)$$

Thermal-noise aliasing appears as an unwanted behavior, but it is physically unavoidable. The fully aliased thermal noise in the (useful) Nyquist bandwidth from $-f_s/2$ to $f_s/2$ is

$$f_s \frac{kT}{Cf_s} = \frac{kT}{C}, \qquad (10.1.11)$$

which is the total thermal noise on the capacitor of a continuous-time RC circuit. Consequently, according to the rules of physics, the related continuous-time and sampled-data circuits have similar thermal noise in their useful bandwidths. A direct physical interpretation of the aliasing of thermal noise in the S/H follows from the equivalence between a resistor and a switched capacitor (see Problem 10.1).

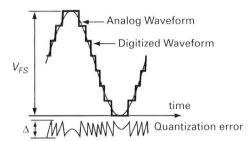

Fig. 10.7 The quantization error of a digitized analog waveform. V_{FS} is the full-scale voltage range and Δ is the size of the LSB (adapted from [6]).

Example 10.2. Comparing thermal and quantization noise [5].

Denoting Δ as the size of 1 LSB, and assuming a constant probability density of $1/\Delta$ for the quantization noise e (Figure 10.7) in the interval $-\Delta/2$ to $\Delta/2$, the rms value of the quantization noise is given by

$$e_{rms}^2 = \int_{-\Delta/2}^{\Delta/2} e^2 \left(\frac{1}{\Delta}\right) de = \frac{\Delta^2}{12}. \tag{10.1.12}$$

Thus, the condition of less thermal than quantization noise follows from (10.1.11) and (10.1.12) as

$$\frac{kT}{C} \leq \frac{\Delta^2}{12} \tag{10.1.13}$$

or, equivalently,

$$C \geq 12kT\left(\frac{2^B}{V_{FS}}\right)^2. \tag{10.1.14}$$

For $V_{FS} = 1$ V and $T = 300$ K we obtain the following minimum capacitances required to keep the thermal noise below the quantization noise:

Number of bits (B)	Capacitance (C)
8	3.3 fF
12	0.83 pF
14	13.3 pF
16	213 pF
20	55 nF

10.1.3 Switch on-resistance

In this section, we assume that the switch on-resistance is constant, independently of the value of the input voltage, and in the next section we will remove this restriction, considering the real case in which the MOSFET on-resistance is non-linear.

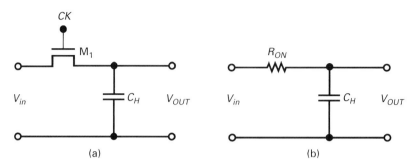

Fig. 10.8   (a) A basic MOS S/H. (b) The equivalent circuit in the sampling mode (adapted from [1]).

Let us assume that the simple S/H of Figure 10.8 has as input voltage $v_i = A\cos(\omega t)$ and that the time constant $\tau_0 = R_{ON}C$ of the S/H in the sampling mode satisfies two conditions, namely $\tau_0 \ll T_s$ and $\omega\tau_0 \ll 1$, where T_s is the clock period. In effect, with a switch on-time of $T_s/2$, an accurate settling (e.g. 0.1%) requires $\tau_0 \leq (T_s/2)/7$ and, thus, the first condition is satisfied. In addition, since the input waveform must satisfy the Nyquist condition ($\omega/(2\pi) \leq f_s/2$ or $\omega T_s \leq \pi$), the second condition follows. During the tracking period, the linear S/H satisfies

$$\frac{dv_o}{dt} + \frac{v_o}{\tau_0} = \frac{A}{\tau_0}\cos(\omega t). \tag{10.1.15}$$

On choosing the time origin $t=0$ at the beginning of the tracking time, the initial condition for the output voltage can be approximated by $v_o(0) = v_i(-T_s/2)$, if we neglect the small settling error.

The solution to (10.1.15) is the sum of the forced and natural responses. For $\omega\tau_0 \ll 1$ (see Problem 10.2), the forced response can be approximated by

$$v_o(t) \cong A\cos[\omega(t - \tau_0)]. \tag{10.1.16}$$

The solution of (10.1.15) that satisfies the initial condition is

$$v_0(t) \cong A\cos\omega(t - \tau_0) + A\left[\cos\left(\frac{\omega T_s}{2}\right) - \cos(\omega\tau_0)\right]e^{-t/\tau_0}. \tag{10.1.17}$$

Since $\omega\tau_0 \ll 1$ and $\omega T_s/2 \ll 1$, we can neglect the natural response and, even if these conditions are not fulfilled (sampling near the Nyquist limit), we can neglect it, except at the very beginning of the tracking time. Thus, in the tracking mode we can approximate the output signal with a delayed version of the input signal given by

$$v_o(t) \cong v_i(t - \tau_0). \tag{10.1.18}$$

10.1.4   Sampling distortion due to switch on-resistance

For the real MOS S/H, the switch on-resistance varies with the input voltage. Considering an n-MOSFET, as in Figure 10.8, the on-conductance of the transistor decreases for increasing values of the input voltage, as shown in Figure 10.9.

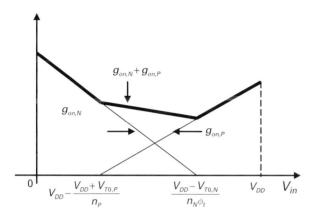

Fig. 10.9   Variation of the on-conductance of the n-MOS, p-MOS, and CMOS switches with the input voltage.

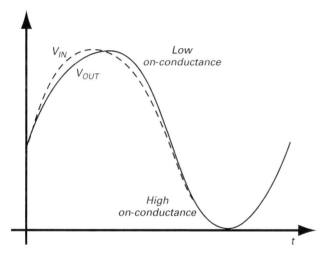

Fig. 10.10   Illustration of the distortion produced by the input-dependent delay of the MOS S/H in the tracking mode (adapted from [1]).

Figure 10.10 shows the distortion of a high-frequency sine wave due to the input-dependent on-resistance of the MOS switch in the S/H of Figure 10.8. The input signal (not satisfying the Nyquist condition) experiences an input-dependent phase delay, which increases for increasing values of the input voltage, because of the increase in the n-MOSFET on-resistance. This input-dependent phase shift produces harmonic distortion. A simple analytical model [7] for this distortion is summarized in Appendix A10.1.

Since the on-resistance of the p-MOS transistor decreases for increasing values of the input voltage, as shown in Figure 10.9, the complementary CMOS switch shown in Figure 10.11 is sometimes used to reduce the distortion due to the on-resistance non-linearity. The appropriate operation of the CMOS switch is hampered by the low supply

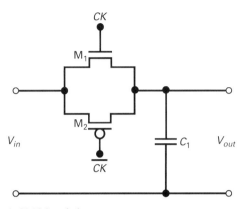

Fig. 10.11   A CMOS switch.

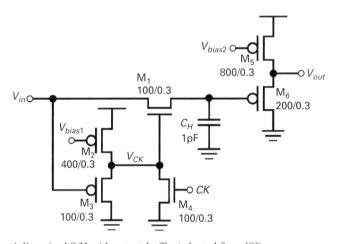

Fig. 10.12   A linearized S/H with output buffer (adapted from [8]).

voltages employed in more advanced technologies, as will be explained in Section 10.1.7 on low-voltage S/Hs.

10.1.5   Linearization of the MOS sampling switch

It is possible to reduce the sampling errors due to the variable MOS on-conductance by sampling the signal at a constant gate-to-source voltage across the switch.

The S/H of Figure 10.12 [8] keeps the gate-to-source voltage of the sampling transistor (M_1) constant during the tracking period. In effect, transistors M_2 and M_3 implement a source follower that applies the input voltage shifted by a constant voltage to the gate of the tracking transistor M_1. A sample of the input signal is stored on capacitor C_H when the clock signal is high, turning on M_4 and, consequently, opening M_1. Owing to the constant-current source M_2, M_4 closes at a constant current independently of the input voltage. Thus, M_1, which is opened through the action of M_4, samples the input voltage at

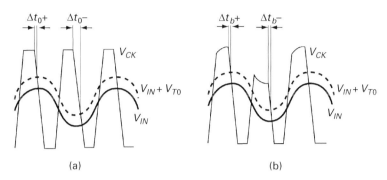

Fig. 10.13   Sampling-instant variation for (a) an ordinary S/H and (b) a linearized S/H (adapted from [8]).

a time independent of the value of the input voltage. Finally, transistors M_5 and M_6 implement an output buffer. Figure 10.13 compares the sampling-instant variation of the ordinary sampler and the linearized sampler, assuming that there is no body effect ($n = 1$).

Neglecting the body effect, the MOS switch samples the input voltage when $V_{GS1} = V_{T0}$. Since $V_{GS1} = V_{CK} - V_{in}$, sampling occurs at the instants when $V_{CK} = V_{in} + V_{T0}$, which are represented in Figure 10.13 by the intersection points of the dashed line with the clock signal.

Clearly, with the constant-amplitude clock applied to the gate of the MOS switch, there can be a large variation in the interval between two sampling instants as shown in Figure 10.13(a). On the other hand, if the amplitude of the clock signal follows the input voltage, as shown in Figure 10.13(b), there will be no variation in the interval between consecutive sampling instants. Summarizing, sampling at constant gate-to-source voltages reduces the input-dependent sampling instant and the switch on-resistance variations.

A constant gate-to-source voltage does not ensure that the switch resistance is constant, as is clear from (9.1.2), rewritten below as

$$R_{ON} = \frac{1}{\mu C'_{ox} \frac{W}{L}(V_G - V_{T0} - nV_S)} = \frac{1}{\beta[V_{GS} - V_{T0} - (n-1)V_S]}. \qquad (10.1.19)$$

Owing to the body effect ($n > 1$), a variable source voltage V_S ($= V_{in}$) causes the on-resistance to change even when the gate-to-source voltage is constant.

To completely eliminate the distortion introduced by the MOS switch, one can use a sample-and-hold topology that allows the switch to operate at constant voltage, as will be shown in Section 10.4.

10.1.6   Charge injection by the switch

The turning off of the transistor switch is inevitably accompanied by a release of charge that produces some disturbance in the charge stored on the sampling capacitor and, thus, an error in the sampled voltage. The main cause of charge injection is the release of carriers from the channel when the transistor turns off, but there is also some charge

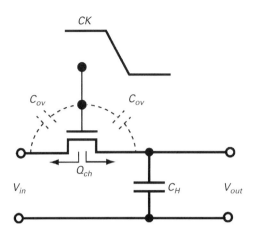

Fig. 10.14   MOS switch charge injection (adapted from [1]).

injection into the hold capacitor due to the coupling of the clock signal to the hold capacitor through the gate–drain overlap capacitance (Figure 10.14).

Charge injection, also commonly called clock feedthrough, is one of the major limitations to the accuracy of the MOS S/H. Analytical models have been developed to model charge injection [9], but the resulting formulas are very complicated even for the simplest structures. For this reason, we will present only a simple formula [10], [11], which gives the order of magnitude of charge injection, and we will summarize design strategies [12] based on numerical simulations, which are aimed at minimizing the charge-injection errors.

We assume that, for appropriate settling, the sampling clock half-period must be larger than five time constants, therefore

$$\frac{T_s}{2} > 5 R_{ON} C_H, \tag{10.1.20}$$

where T_s is the sampling period, R_{ON} is the on-resistance of the n-MOSFET switch, and C_H is the holding capacitance. We can rewrite (10.1.20) as

$$f_s < \frac{1}{10 R_{ON} C_H}, \tag{10.1.21}$$

where f_s is the (sampling) clock frequency.

For a MOSFET in the linear region, the on-resistance is the inverse of the source (or drain) transconductance, and it is given by

$$R_{ON} = \frac{1}{(W/L)\mu |Q_I'|}, \tag{10.1.22}$$

where Q_I' is the inversion charge per unit area. In this case, it is more convenient to write the transconductance in terms of the total inversion channel charge $Q_I = WLQ_I'$ as

$$R_{ON} = \frac{L^2}{\mu |Q_I|}. \tag{10.1.23}$$

The fraction of the total channel charge Q_I released to the hold capacitor is essentially dependent on the ratio between the signal source and hold capacitances, and on the switching speed [12]. If the clock waveform is very fast, the channel charge will flow equally through the source and drain terminals [12]. We will assume that this is the case. If it is not, the derived formula is nevertheless useful as an estimate of the order of magnitude of the error due to charge injection. Thus, assuming that half of the total inversion channel charge is injected into the hold capacitor C_H, the sampling voltage error is given by

$$\Delta V = \frac{|Q_I|}{2C_H}.$$ (10.1.24)

On combining (10.1.21), (10.1.23), and (10.1.24) we obtain

$$f_s < \frac{1}{10R_{ON}C_H} = \frac{\mu|Q_I|}{10L^2C_H} = \frac{\mu\Delta V}{5L^2}.$$ (10.1.25)

Clearly, since the frequency limit is inversely proportional to L^2, switches must be implemented with the minimum available channel length of the technology.

Example 10.3

Assuming a maximum voltage error of 1 mV due to charge injection, calculate the maximum clock frequencies for effective channel lengths of 1 µm, 0.316 µm, and 100 nm.

Answer

Using (10.1.25) and assuming an n-MOS switch with $\mu = 500$ cm^2/V per s, we obtain the following values for f_s: 10 MHz, 100 MHz, and 1 GHz, for channel lengths 1 µm, 0.316 µm, and 100 nm, respectively.

Since the frequency limit is inversely proportional to L^2, there is a substantial increase in the maximum speed for advanced technologies.

10.1.6.1 Reducing injection errors

Numerical modeling of the MOS S/H [12] shows that the splitting of the channel charge between the signal source and the sampling capacitor (see Figure 10.15) is dependent on the ratio between the signal source capacitance C_P and the hold capacitance C_H, and on a dimensionless characteristic switching parameter B defined as

$$B = \sqrt{\frac{C_g}{2C_H}}\omega_T t_{FALL},$$ (10.1.26)

where $C_g = WLC'_{ox}$ is the total gate capacitance, t_{FALL} is the fall time of the clock signal from V_{GON} to the threshold voltage of the MOS switch, and $\omega_T = [2\mu/(nL^2)]$ $(V_{GON} - V_{T0} - nV_{in})$ is the intrinsic transition frequency of the saturated on-transistor in strong inversion.

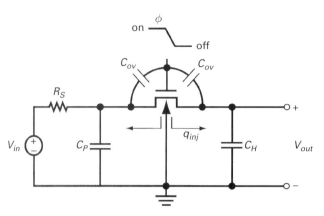

Fig. 10.15   Charge injection in the basic MOS S/H (adapted from [13]).

Numerical simulations [12] show that for fast switching, namely $B \ll 1$, half of the channel charge is injected into the sampling capacitor and the other half into the signal source, independently of the capacitance ratio C_P/C_H. On the other hand, for very long fall times ($B \gg 1$), the partitioning of the channel charge is dependent only on the ratio C_P/C_H. Finally, for intermediate cases, the charge division is dependent both on C_P/C_H and on B.

It is worth noting that for a CMOS switch there is a partial compensation of the released charges, but this compensation is not efficient since it is dependent on the input voltage V_{in} and on which of the n-MOS and p-MOS transistors switches off first. For this reason, compensation of charge injection is achieved using the same type of transistor.

A practical way to reduce charge injection into the hold capacitor is by using fast switching ($B \ll 1$) and achieving charge compensation using a single half-sized dummy switch as shown in Figure 10.16(a). Clearly, mismatch between the main and dummy switches and between clock signals imposes a limit on the charge compensation. Further improvement in the S/H precision can be obtained by using circuit techniques to reject the unavoidable injection errors, as shown in the next section.

10.1.6.2   Rejecting injection errors

Since charge injection is dependent on the input signal, it produces signal distortion. The use of the linearized switch introduced in Section 10.1.5 reduces the dependence of charge injection on the input voltage and, thus, the distortion.

As will be shown in Section 10.2.6, for circuits containing several switches, such as comparators and switched-capacitor circuits, turning off certain switches first ensures that the charge-injection errors of the overall circuit are due only to those switches. Then, for these circuits, it is possible to make the charge injection independent of the signal by closing first the switches that operate at a constant voltage. In this case, the effect of the charge injected by the switches appears simply as an offset voltage.

Another strategy to reduce distortion that can be applied in conjunction with other distortion-reduction techniques is the use of fully-differential circuits [13], [14], as shown

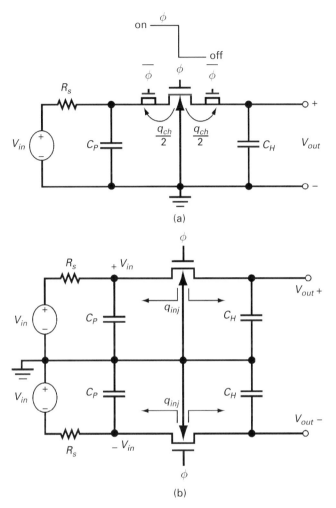

Fig. 10.16 Charge-injection-cancellation techniques: (a) short fall time of the clock and half-sized dummy switches; and (b) fully-differential structure (adapted from [13]).

in Figure 10.16(b). Clearly, a fully-differential balanced topology does not eliminate the effects of charge injection, since only for the case of equal charges injected by the switches will the differential output voltage be insensitive to charge injection. In the matched case, the charge injected by the switches would manifest itself as a common-mode output voltage.

10.1.7 Low-voltage sample-and-hold circuits

The continuous trend toward low-voltage circuits is a consequence of the advance of CMOS technologies. The supply voltage must be reduced when device dimensions shrink, in order to limit the maximum values of the electric field and thereby guarantee the long-term reliability of the chip. Another reason to reduce the supply voltage is the

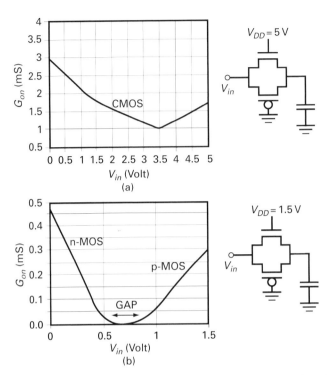

Fig. 10.17   The on-conductance of a CMOS switch for two different supply voltages: (a) $V_{DD}=5$ V and
(b) $V_{DD}=1.5$ V (adapted from [15]).

resulting reduction in the dynamic power consumption of digital circuits, which is bene-
ficial for portable electronics. A critical problem for low-voltage analog circuits is the
difficulty involved in implementing MOS switches. The on-conductance of the CMOS
switch varies with the signal voltage, as shown in Figure 10.17, where the switch
conductance is plotted against the input signal for two values of the supply voltage. In
the topmost case, for the higher supply voltage, the parallel connection of the n-MOS and
p-MOS transistors gives an acceptable minimum conductance for intermediate values of
the input signal. On the other hand, for the lower supply voltage, there is a range of V_{in} for
which the switch does not conduct. In fact, in the conduction gap the two transistors are still
conducting, but in weak inversion; thus, the switch has an extremely low conductance.

The conductance of the n-MOS transistor of the CMOS switch operating in strong
inversion with $V_G = V_{DD}$ and $V_S = V_{in}$ is given by

$$G_n = \beta_n(V_{DD} - V_{T0n} - n_n V_{in}). \tag{10.1.27}$$

Equation (10.1.27) is equivalent to (10.1.19), but rewritten including the subscript n to
emphasize that the parameters are specific to an n-channel transistor.

As explained in Example 2.8, for p-channel devices the bulk is usually at the most
positive potential, V_{DD} in this case. The gate and source potentials referenced to the
p-channel bulk are

$$V_G = 0 - V_{DD}; \qquad V_S = V_{in} - V_{DD}. \tag{10.1.28}$$

Thus, the conductance of the p-channel transistor is given by

$$G_p = \beta_p \left(V_{DD} - |V_{T0p}| - n_p (V_{DD} - V_{in}) \right). \tag{10.1.29}$$

From (10.1.27) and (10.1.29) it follows that, to guarantee the operation of the two transistors in strong inversion, the input voltage must be within the interval

$$\frac{V_{DD} - V_{T0n}}{n_n} > V_{in} > V_{DD} - \frac{V_{DD} - |V_{T0p}|}{n_p}. \tag{10.1.30}$$

Clearly, condition (10.1.30) can be satisfied only for V_{DD} above a minimum value [16] given by

$$V_{DDcrit} = \frac{n_p V_{T0n} + n_n |V_{T0p}|}{n_n + n_p - n_n n_p}. \tag{10.1.31}$$

Neglecting the body effect, (10.1.30) and (10.1.31) reduce to

$$V_{DD} - V_{T0n} > v_i > |V_{T0p}|, \tag{10.1.32}$$

$$V_{DDcrit} = V_{T0n} + |V_{T0p}|. \tag{10.1.33}$$

It is worth noting that, due to the body effect, the minimum value of V_{DD} given by (10.1.31) can be somewhat higher than $V_{T0n} + |V_{T0p}|$.

A simple way to deal with the conduction gap is to use p- and n-channel switches, each with an adequate range of V_{in}, as shown in Figure 10.18(a) [17]. To operate properly, n- and p-channel switches must be connected to nodes with dc voltages close to ground and V_{DD}, respectively. A low-voltage S/H implemented with single n- and p-MOS switches is shown in Figure 10.18(b) [17]. During ϕ_1, the hold capacitor samples the input signal, which is close to zero volts referenced to V_{DD}. During ϕ_2, C_H is connected so as to give feedback to the op amp, providing an output sample $V_{out} = V_{DD} - V_{in}$. Clearly, as shown in Figure 10.18(a), the penalty is the reduced available voltage swing.

Because low supply voltage is the mainstream of electronics, several techniques have been developed to achieve high-performance S/Hs with reduced supply voltages. Certain enhancements to the CMOS technology, such as low-threshold voltage devices, are sometimes used. For standard CMOS technologies, the main alternatives are clock-voltage multiplication and clock bootstrapping.

Voltage-multiplication techniques introduce reliability problems because some devices operate with voltages above the nominal voltage of the technology. A bootstrap technique developed in [18] satisfies reliability constraints by avoiding having terminal-to-terminal transistor voltages exceed the supply voltage.

As shown in Figure 10.19(a), in the on state of M_1, a voltage equal to V_{DD} stored on capacitor C_1 is applied across the gate-to-source terminals of the MOSFET switch. In the off state, the gate of the switch is grounded and capacitor C_1 is recharged to V_{DD}. The

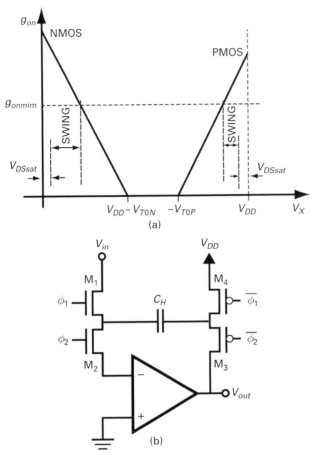

(a)

(b)

Fig. 10.18   (a) Available output swing obtained by dc-shifting the input signal applied to the n- and p-MOS switches (V_{DSsat} is the voltage margin to either V_{DD} or ground required for the proper operation of the blocks, e.g. amplifiers, connected to the switches). (b) The low-voltage S/H that provides dc bias for proper operation of both switches (adapted from [17]).

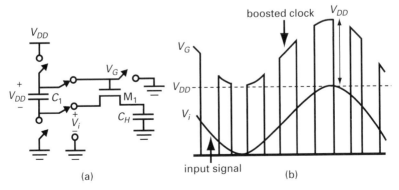

(a)

(b)

Fig. 10.19   A bootstrapped MOS switch: (a) simplified schematic diagram and (b) input (source) and clock (gate) signals (adapted from [18]).

resulting gate signal is shown in Figure 10.19(b). The actual circuit [18] is more complicated; in particular, a specific clock multiplier circuit is required in order to drive the switches which allow the recharging of capacitor C_1 to V_{DD}. Since the gate-to-source voltage is constant, the switch linearity is improved, but, similarly to the circuit in Section 10.1.5, the variations in the on-resistance due to the body effect cannot be eliminated. For an analog circuit containing many switches, such as a switched-capacitor filter, the area and power overhead due to the bootstrap circuits can be substantial [18] because each switch in the signal path requires one bootstrap circuit.

10.1.8   Jitter analysis

Ideally, the sampling instants at which the switch of the S/H turns off are spaced uniformly in time by the interval $T = 1/f_s$. However, as we have shown in Section 10.1.5, variation in the transistor source voltage causes an input-dependent sampling instant. There are other causes of uncertainty in the sampling period. A useful model considers that the sampling period T is a random variable the standard deviation of which is called (aperture) jitter or aperture uncertainty τ_a, measured in root-mean-square (rms) seconds. Typical clocks have an aperture jitter of the order of 100 ps rms, and high-quality clocks have a jitter of the order of 1 ps rms. The dominant cause of clock jitter is usually the power-supply noise produced by the switching of the (many) different circuits on the chip. Assuming that the delay of an inverter is 100 ps and that the delay varies 20%, for a 1 V variation in the power-supply voltage, a 200 mV noise would produce 4 ps of jitter [5]. An expression for the voltage error caused by clock jitter can be easily derived in terms of τ_a. For dc signals, the clock jitter clearly does not produce voltage errors. For sinusoidal waves, the worst case for the voltage error occurs for a signal with full-scale voltage swing and with the highest frequency in the Nyquist band ($f_s/2$), namely

$$v(t) = \frac{V_{FS}}{2} \sin\left(2\pi\frac{f_s}{2}t\right).$$
(10.1.34)

The maximum voltage error will occur for a signal sampled at its maximum slope, which is

$$\frac{V_{FS}}{2}\pi f_s.$$
(10.1.35)

Thus, the rms voltage error is given by the product of the maximum slope of the sampled signal and τ_a, that is,

$$v_{rms} = \pi f_s V_{FS}\tau_a/2.$$
(10.1.36)

Finally, the signal-to-noise ratio (SNR) of the S/H due to clock jitter is given by

$$\text{SNR} = 10\log_{10}\left(\frac{\frac{1}{2}\left(\frac{V_{FS}}{2}\right)^2}{\left(\pi f_s V_{FS}\frac{\tau_a}{2}\right)^2}\right) = -20\log_{10}\left(\sqrt{2}\pi f_s\tau_a\right).$$
(10.1.37)

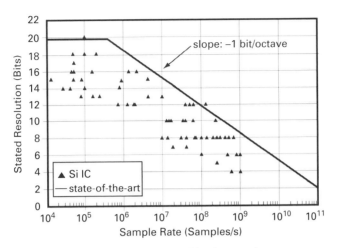

Fig. 10.20 Resolution, in number of bits, as stated by the manufacturer, versus sampling rate, for A/D converters implemented in silicon (adapted from [6]).

10.1.9   Tradeoff between resolution and sampling rate in analog-to-digital converters

The main parameters that characterize A/D converters are resolution and speed. For low speed an A/D converter's resolution is limited by thermal noise, whereas for sampling rates between megasamples and gigasamples per second, the resolution is limited by the aperture jitter of the front-end S/H [6].

Equating the noise voltage associated with the aperture jitter given by (10.1.36) with the quantization noise in (10.1.12) allows us to write

$$\frac{\Delta}{\sqrt{12}} = \pi f_s V_{FS} \frac{\tau_a}{2}. \tag{10.1.38}$$

Denoting as B_{jitter} the maximum resolution in number of bits due to clock jitter, we can write

$$\Delta = \frac{V_{FS}}{2^{B_{jitter}}} \tag{10.1.39}$$

and, from (10.1.38) and (10.1.39), it follows that

$$B_{jitter} = \log_2 \left(\frac{1}{\sqrt{3}\pi f_s \tau_a} \right). \tag{10.1.40}$$

Thus, the bit resolution falls off by one bit for every doubling of the sampling rate.

Figure 10.20 shows the bit resolution versus the sampling rate for a large number of A/D converters. It clearly illustrates the trend of a 1-bit loss in resolution for every doubling of the sampling frequency. The -1-bit per octave slope is related to the sample-to-sample variation of the instant of time at which sampling occurs.

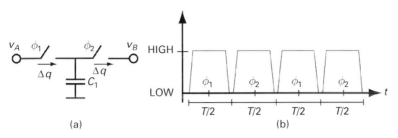

Fig. 10.21 The basic switched capacitor and non-overlapping clock signals.

10.2 Basics of switched-capacitor filters

The switched-capacitor (SC) circuit technique [19], [20] allowed the development of analog MOS integrated circuits in the late 1970s and it is still one of the most widely used analog-circuit techniques.

10.2.1 Basic principles of operation of switched-capacitor circuits

The SC technique is based on the exchange of charge between capacitors. The SC circuits are discrete-time charge processors composed of capacitors, switches, and operational amplifiers.

To start with a simple example of an application of the basic principle, let us show how to substitute a switched capacitor for a resistor. The elementary SC circuit illustrated in Figure 10.21 consists of a grounded capacitor and two switches (realized by MOS transistors) controlled by two non-overlapping clock signals ϕ_1 and ϕ_2. Assuming that the switches are n-channel transistors, they turn on when the clock signals applied to the gates are high. Since the two clock signals are never high at the same time, the two switches are never simultaneously on; thus, there is no conducting path between nodes A and B at any time.

Assuming, for simplicity, that v_A and v_B are dc voltages, and neglecting the on-resistances of the MOS switches, capacitor C_1 charges to voltage v_A during ϕ_1, and to voltage v_B during ϕ_2. Thus, the charges held on capacitor C_1 are

$$q_1 = C_1 v_A \tag{10.2.1}$$

and

$$q_2 = C_1 v_B \tag{10.2.2}$$

during phases ϕ_1 and ϕ_2, respectively. Consequently, in one clock period, a charge

$$\Delta q = q_1 - q_2 = C_1(v_A - v_B) \tag{10.2.3}$$

is transferred from node A to node B. Since this charge transfer occurs in each clock period, an average current

$$i_{av} = \Delta q / \Delta T = (C_1/T)(v_A - v_B) \tag{10.2.4}$$

flows between nodes A and B. Thus, on average, the switched capacitor behaves as a resistor with its resistance value given by

$$R_1 = T/C_1 = 1/(C_1 f_s), \tag{10.2.5}$$

where f_s is the clock frequency. Summarizing, the equivalent resistor of a switched capacitor is proportional to the clock period and inversely proportional to the capacitance. Thus, combining a low-frequency clock with low capacitance values is a straightforward way to implement high resistance values.

Example 10.4

Calculate the capacitance values of switched capacitors equivalent to 1- and 10-MΩ resistors for a clock frequency of 1 MHz and estimate the area needed to build them.

Answer

Applying (10.2.5), it follows that 1-pF and 0.1-pF capacitors, switched at 1 MHz, simulate resistors of 1 and 10 MΩ, respectively. Assuming that the capacitance per unit area is 0.5 fF/μm^2, the areas occupied by the 1- and 0.1-pF capacitors are 2000 and 200 μm^2, respectively.

The above example clearly shows that the SC technique is very efficient for the implementation of high resistance values, which are useful for the realization of low-frequency analog filters.

The equivalence of the switched capacitor with a resistor is a very useful concept, allowing the rapid synthesis of SC filters from continuous-time prototypes. If the clock frequency is much higher than the maximum signal frequency, the continuous-time model of the switched capacitor can provide an adequate representation of it; otherwise, an accurate time (frequency)-domain model of the switched circuit must be used. Let us consider as an example the first-order low-pass SC filter shown in Figure 10.22. In this figure, the sampling periods corresponding to the clock phase ϕ_2 are denoted as integers and phase 2 is assumed to be a half clock period shifted in relation to phase 1. A straightforward analysis of the circuit follows, assuming that the capacitors are fully charged at the end of the sampling periods.

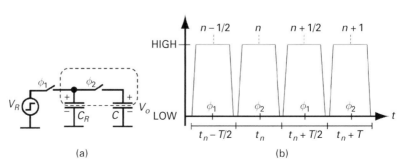

Fig. 10.22   A first-order low-pass SC filter.

During ϕ_1, C_R is charged to voltage v_R and C holds the charge stored during the previous clock half cycle. During ϕ_2 (the switch controlled by ϕ_1 is open), the charge inside the closed surface within dashed lines is conserved; thus

$$C_R v_R(t_n - T/2) + C v_o(t_n - T/2) = (C_R + C)v_o(t_n). \tag{10.2.6}$$

Since v_o changes during ϕ_2 only, we can write

$$v_o(t_n - T/2) = v_o(t_n - T). \tag{10.2.7}$$

Assuming that V_R is also a signal sampled in ϕ_2, it follows that

$$v_R(t_n - T/2) = v_R(t_n - T). \tag{10.2.8}$$

Combining (10.2.6), (10.2.7), and (10.2.8) yields

$$C_R v_R(t_n - T) + C v_o(t_n - T) = (C_R + C)v_o(t_n). \tag{10.2.9}$$

The standard procedure to deal with recursive relations such as that in (10.2.9) is to use the z-transform, the main properties of which are summarized in Problems 10.6 and 10.7. Taking the z-transform of (10.2.9), it follows that

$$C_R V_R^{\phi_2} z^{-1} + C V_o^{\phi_2} z^{-1} = (C_R + C)V_o^{\phi_2} \tag{10.2.10}$$

or

$$\frac{V_o^{\phi_2}}{V_R^{\phi_2}} = \frac{C_R}{C_R + C} \frac{z^{-1}}{1 - z^{-1}C/(C_R + C)}. \tag{10.2.11}$$

Example 10.5

Calculate the response of the first-order low-pass SC filter of Figure 10.22 to a voltage step of amplitude V_R.

Answer

The z-transform of the step function (see Problem 10.6) is given by

$$U(z) = \frac{V_R}{1 - z^{-1}}, \tag{10.2.12}$$

where V_R is the amplitude of the input voltage step. The z-transform of the output voltage is obtained by the combination of (10.2.11) and (10.2.12), which yields

$$
\begin{aligned}
V_o(z) &= \frac{C_R}{C_R + C} \frac{z^{-1}}{1 - z^{-1}C/(C_R + C)} \frac{V_R}{1 - z^{-1}} \\
&= V_R \left(\frac{1}{1 - z^{-1}} - \frac{1}{1 - z^{-1}C/(C_R + C)} \right).
\end{aligned} \tag{10.2.13}
$$

Hence, from (10.2.13) and the results of Problem 10.6, we have

$$v_o(nT) = V_R \left[1 - \left(\frac{C}{C + C_R} \right)^n \right] u(nT) = V_R \left(1 - e^{-\alpha nT} \right) u(nT), \tag{10.2.14}$$

where

$$\alpha T = \ln[(C + C_R)/C] \qquad (10.2.15)$$

As shown in Example 10.5, the time constant of the low-pass SC filter is

$$\tau = 1/\alpha = T/\ln(1 + C_R/C). \qquad (10.2.16)$$

Thus, the time constant is proportional to the clock period and is also dependent on the capacitance ratio. The clock period is usually established by a quartz-crystal-controlled oscillator, which is very accurate and stable. The capacitance ratio can also be made accurate and stable using some simple matching rules, as explained in Section 3.4.3.1. Summarizing, the time constants of SC filters can be made very stable and accurate (of the order of 0.1% [4]), and this is the main reason for the popularity of SC circuits since their inception in the late 1970s.

Let us analyze two important limit cases of the low-pass time constant. For $C_R/C \ll 1$, it follows that $\ln(1 + C_R/C) \approx C_R/C$ and, thus,

$$\tau = T(C/C_R). \qquad (10.2.17)$$

Clearly, we could have derived (10.2.17) directly using the equivalence between a switched capacitor and a resistor given by (10.2.5). The advantage of the general relation (10.2.15) is that it holds even for $C_R/C \gg 1$. In this case, $\tau = T/\ln(1 + C_R/C) \approx T/\ln(C_R/C)$ and the time constant is lower than the clock period. In effect, when $C_R \gg C$, the capacitor C receives most of its (final) charge in just one clock cycle.

The frequency response of the first-order low-pass SC filter, obtained by substituting $e^{j\omega T}$ for z in the z-transform given by (10.2.11), is

$$H(e^{j\omega T}) = \left. \frac{z^{-1}}{1 + C/C_R(1 - z^{-1})} \right|_{z=e^{j\omega T}} = \frac{1}{e^{j\omega T}(1 + C/C_R) - C/C_R}. \qquad (10.2.18)$$

For $\omega T \ll 1$, we can approximate the imaginary exponential to first-order as $e^{j\omega T} \approx 1 + j\omega T$; thus (10.2.18) can be rewritten as

$$H(e^{j\omega T}) = \frac{1}{1 + j\omega T(1 + C/C_R)}. \qquad (10.2.19)$$

Note that, for frequencies much lower than the Nyquist frequency $f_s/2$, the discrete-time low-pass filter of Figure 10.22 operates as a continuous-time filter composed of a capacitance equal to $C_R + C$ and a resistance equal to T/C_R. Additionally, when $\omega T (1 + C/C_R) \ll 1$, the output of the SC filter is a replica of the input with negligible attenuation and a delay equal to $T(1 + C/C_R)$.

10.2.2   Switched-capacitor integrators

The basic building block of SC filters is the integrator, which can be obtained directly from the continuous-time integrator shown in Figure 10.23(a) by substituting a switched capacitor for the input resistor R_1, as shown in Figure 10.23(b).

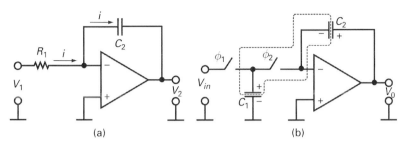

Fig. 10.23 (a) Continuous-time and (b) parasitic-sensitive switched-capacitor integrators.

The time-domain analysis of the SC integrator in Figure 10.23 is straightforward if one assumes that the capacitors are fully charged at the end of the clock phases and the operational amplifier is ideal. During phase ϕ_1, the input source charges capacitor C_1 to v_{in} while the integrating capacitor C_2 conserves the charge that was stored during the preceding cycle of phase ϕ_2. Thus, using the notation indicated in Figure 10.22 for the clock phases, we can write

$$q_1(n - 1/2) = C_1 v_{in}(n - 1/2),\qquad(10.2.20)$$

$$v_o(n - 1/2) = v_o(n - 1).\qquad(10.2.21)$$

During phase ϕ_2, the charge inside the region enclosed by the dashed border is conserved; thus

$$q_1(n) - q_2(n) = q_1(n - 1/2) - q_2(n - 1/2)\qquad(10.2.22)$$

or, equivalently,

$$C_2[v_o(n) - v_o(n - 1/2)] = q_2(n) - q_2(n - 1/2) = -q_1(n - 1/2),\qquad(10.2.23)$$

since q_1 is zero during ϕ_2 ($q_1(n)=0$).

On substituting (10.2.20) and (10.2.21) into (10.2.22) we obtain

$$v_o(n) = v_o(n - 1) - (C_1/C_2)v_{in}(n - 1/2).\qquad(10.2.24)$$

To write the expression equivalent to (10.2.24) in the z-domain, we recall that delays of one clock period and half a clock period appear as factors of z^{-1} and $z^{-1/2}$, respectively, which multiply the z-transform of the input signal. Thus, the z-domain equivalent to (10.2.24) is

$$V_o^{\phi_2}(z) = z^{-1} V_o^{\phi_2}(z) - (C_1/C_2)z^{-1/2}V_{in}^{\phi_1}(z),\qquad(10.2.25)$$

from which the transfer function is obtained as

$$H_{21}(z) = \frac{V_o^{\phi_2}}{V_{in}^{\phi_1}} = -\frac{C_1}{C_2}\frac{z^{-1/2}}{1 - z^{-1}}.\qquad(10.2.26)$$

The notation H_{21} indicates that the z-domain transfer function is the ratio of the z-transform of the output signal during ϕ_2 to the z-transform of the input signal available to the integrator during ϕ_1. Note that, for this specific integrator, $H_{12} = H_{22} = 0$, since the value of the input signal during ϕ_2 does not affect the integrator output.

Finally, the integrator frequency response is

$$H(z = e^{j\omega T}) = -\frac{C_1}{C_2} \frac{e^{-j\omega T/2}}{1 - e^{-j\omega T}} = -\frac{C_1}{C_2} \frac{1}{j\omega T} \frac{\omega T/2}{\sin(\omega T/2)}. \tag{10.2.27}$$

Equation (10.2.27) can be viewed as the transfer function of a continuous-time integrator divided by $(\sin x)/x$. The $(\sin x)/x$ "error" is, in fact, not surprising, since it is common to all sampled-data signals [1]–[4]. Clearly, for $\omega T \ll 1$, expression (10.2.27) reduces to that of the ideal continuous-time integrator, which is

$$H(z = e^{j\omega T})\big|_{\omega T \ll 1} \cong -\frac{C_1}{C_2} \frac{1}{j\omega T}. \tag{10.2.28}$$

It is important to note that, when the output voltage is sampled in phase ϕ_1, another half-clock-period delay occurs in the signal path. The transfer function, for this case, is

$$H_{11}(z) = \frac{V_o^{\phi_1}}{V_{in}^{\phi_1}} = z^{-1/2} H(z) = -\frac{C_1}{C_2} \frac{z^{-1}}{1 - z^{-1}}. \tag{10.2.29}$$

In the above analysis we neglect the effects of the parasitic capacitances of both C_1 and C_2. In fact, there are relatively large parasitic capacitances from the bottom plates of the capacitors to the underlying conductive layer (substrate), which is usually grounded. Additionally, there are parasitic (non-linear) capacitances associated with the MOSFET switches. For the parasitic-sensitive integrator of Figure 10.23(b), the only relevant parasitic capacitances are those from the top plate of C_1 to ground, since they are in parallel with C_1, increasing its (ideal) value. The parasitic capacitances from the op-amp inverting node to ground as well as those from the op-amp output to ground are irrelevant for an ideal amplifier. In effect, the parasitic capacitances tied to the virtual ground remain discharged and, thus, they do not affect the transfer function of the integrator. Since the parasitic capacitances connected to the op-amp output are charged by an (ideal) voltage source, they do not contribute either to the charge processing or to the transfer function. To overcome the effect of parasitic capacitances, which are not well controlled, parasitic-insensitive integrators [4], [11] were introduced early in the development of SC circuits. The idea behind parasitic-insensitive SC circuits is to employ topologies in which the parasitic capacitances are switched between voltage sources (including ground) or between ground and virtual ground, avoiding connections such as that of the top plate of capacitor C_1 in Figure 10.23(b).

The circuits with floating-input capacitors shown in Figure 10.24 are two examples of parasitic-insensitive integrators. During phase ϕ_2, capacitor C_1 of the circuit of Figure 10.24(a) is charged to the input voltage. During phase ϕ_1, this capacitor is discharged, its charge being transferred to the integrating capacitor C_2. Since discharge occurs through the terminal that was previously grounded, this reverse connection injects into the output node a charge of the same polarity as that at the input node, emulating a negative resistor. Thus, the circuit in Figure 10.24(a) implements a non-inverting

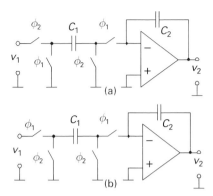

Fig. 10.24   (a) Non-inverting and (b) inverting parasitic-insensitive integrators.

integrator. The two available transfer functions, derived as in the case of the parasitic-sensitive integrator, are

$$H_{12}(z) = \frac{V_2^{\phi_1}(z)}{V_1^{\phi_2}(z)} = +\frac{C_1}{C_2}\frac{z^{-1/2}}{1-z^{-1}}, \tag{10.2.30}$$

$$H_{22}(z) = \frac{V_2^{\phi_2}(z)}{V_1^{\phi_2}(z)} = +\frac{C_1}{C_2}\frac{z^{-1}}{1-z^{-1}}. \tag{10.2.31}$$

Finally, for the inverting parasitic-insensitive integrator of Figure 10.24(b), the transfer functions are

$$H_{11}(z) = \frac{V_2^{\phi_1}(z)}{V_1^{\phi_1}(z)} = -\frac{C_1}{C_2}\frac{1}{1-z^{-1}}, \tag{10.2.32}$$

$$H_{21}(z) = \frac{V_2^{\phi_2}(z)}{V_1^{\phi_1}(z)} = -\frac{C_1}{C_2}\frac{z^{-1/2}}{1-z^{-1}}. \tag{10.2.33}$$

It is interesting to note that, for Figure 10.24(b), a delay-free integrator is available when the input and output are both sampled in phase ϕ_1.

10.2.3   Offset compensation

Switched-capacitor circuits require the inverting op-amp inputs to behave as virtual grounds. CMOS operational amplifiers usually have input-referred offset voltages of the order of several mV. In addition to the dc offset voltage, the finite voltage gain of the amplifier and both thermal and $1/f$ noise contribute to the non-zero voltage at the inverting input of the op amp. Offset and noise reduce the dynamic range of the circuits, which is particularly harmful for low-voltage circuits. Fortunately, there are circuit techniques that can ideally cancel out the effect of dc offsets and greatly reduce the effect of low-frequency disturbances such as $1/f$ noise. The effect of the op amp's finite

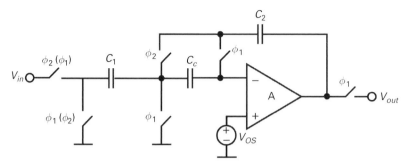

Fig. 10.25   An SC integrator with offset-storing capacitor (adapted from [13]).

gain can also be considered a low-frequency effect if the signal is highly oversampled, that is, if the signal variations in a clock period are minimized.

The basic idea of offset compensation is to store the offset voltage during a calibration phase and to subtract it from the signal in the next phase [13]. In this way, the effect of the offset at the circuit output is ideally cancelled out.

The integrator shown in Figure 10.25 has an additional capacitor C_C that stores the offset voltage V_{os} during phase ϕ_1. Next, during phase ϕ_2, capacitor C_C, which is charged to V_{os}, is connected in series with the input capacitor C_1, thus, canceling out the effect of V_{os} at the op-amp output. In fact, capacitor C_C stores the voltage of the op amp's inverting input during phase ϕ_1; thus, during ϕ_2, slowly varying voltages will be (partially) cancelled out, such as $1/f$ noise and the attenuated copy of the input voltage at the op amp's inverting input due to the finite gain, for highly oversampled signals.

There are numerous circuit variations that carry out offset cancellation in SC integrators, amplifiers, and S/Hs. Some techniques receive specific names, such as auto-zero and correlated double sampling. The description of some more accurate circuits for reduction of offset and low-frequency effects is beyond the scope of this text; the interested reader is referred to the tutorial in [13] for a detailed presentation.

10.2.4   Biquad filters

Using the integrators introduced in the previous sections, it is possible to implement a great variety of high-order filters. Of particular importance are the generic second-order filters, commonly designated biquads. Maybe the simplest way to synthesize an SC filter is to emulate a continuous-time structure [4], [11] and take advantage of the (approximate) equivalence between a resistor and a switched capacitor.

The starting point is the continuous-time biquadratic transfer function given below

$$H_a(s) = \frac{V_o(s)}{V_i(s)} = -\frac{k_2 s^2 + k_1 s + k_0}{s^2 + (\omega_0/Q)s + \omega_0^2}, \qquad (10.2.34)$$

where s is the complex frequency, ω_0 the pole frequency, and Q the pole quality factor. The equation above can be rewritten in order to obtain the output voltage as a linear combination of integrals of the input and output voltages. Let us now rewrite (10.2.34) as

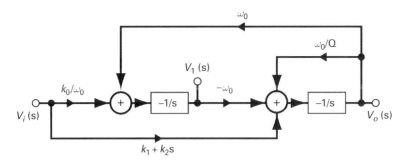

Fig. 10.26 **Fig. 10.26** The signal-flow graph of a continuous-time biquad (adapted from [11]).

$$V_o(s)\left(s^2 + \frac{\omega_0}{Q}s + \omega_0^2\right) = -\left(k_2 s^2 + k_1 s + k_0\right)V_i(s) \tag{10.2.35}$$

or, equivalently, as

$$s^2 V_o(s) = -\left(k_2 s^2 + k_1 s + k_0\right)V_i(s) - \left(\frac{\omega_0}{Q}s + \omega_0^2\right)V_o(s). \tag{10.2.36}$$

Finally, on dividing both sides by s^2 and rearranging, the output can be obtained from two integrators as

$$V_o(s) = -\frac{1}{s}\left[(k_1 + k_2 s)V_i(s) + \frac{\omega_0}{Q}V_o(s) - \omega_0 V_1(s)\right] \tag{10.2.37}$$

and

$$V_1(s) = -\frac{1}{s}\left(\frac{k_o}{\omega_o}V_i(s) + \omega_o V_o(s)\right). \tag{10.2.38}$$

Figure 10.26 shows the signal-flow graph of Equations (10.2.37) and (10.2.38). It contains two inverting integrators $(-1/s)$, two summers at the input of the integrators, and various coupling branches. An active-RC configuration of the signal-flow graph using both positive and negative resistors is shown in Figure 10.27. In effect, for a switched-capacitor emulation of the continuous-time filter, the parasitic-insensitive SC topologies allow the emulation of positive and negative resistors, as shown in Section 10.2.2.

An SC emulation of the active biquad is shown in Figure 10.28, where the resistors have been substituted with parasitic-insensitive switched capacitors. The topology in Figure 10.28, which has a minimum number of switches, was obtained first by substituting for each resistor a floating switched capacitor and its four associated switches, and then eliminating the redundant switches. The capacitor labels are their values with respect to a unity capacitor. Since, in general, $\omega_0 T \ll 1$ and the integrating capacitors were chosen to be of unit value, the ratio of the largest to the smallest capacitances, the so-called capacitance spread, is not less than $1/(\omega_0 T)$.

When the quality factor of the filter is high $(Q \gg 1)$, the capacitance spread becomes $Q/(\omega_0 T)$. Clearly, a very large capacitance spread, say 1000, is not practical, and for this

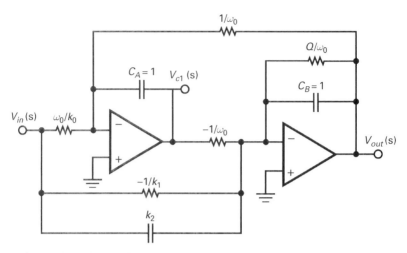

Fig. 10.27   Active-RC biquad realization of the signal-flow graph of Figure 10.26 (adapted from [11]).

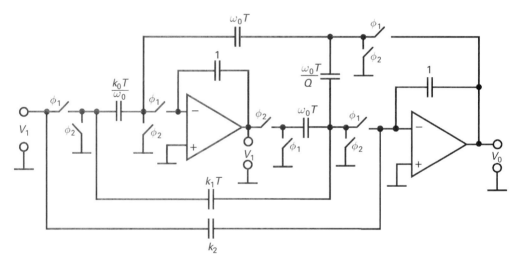

Fig. 10.28   A switched-capacitor configuration equivalent to the continuous-time biquad of Figure 10.27.

reason the above topology is used only for low-Q filters. To implement high-Q filters, other topologies are preferred. In effect, there are many possible configurations of the biquadratic function using two integrators [21].

To implement high-order filters, several biquads can be cascaded. Other available strategies for the construction of high-order filters can be found in [4], [22], [23].

Finally, it should be noted that a continuous-time prototype was used only for convenience, in order to obtain the SC biquad topology. The frequency response for the circuit of Figure 10.28 can accurately be determined using the discrete-time analysis introduced in Section 10.2.1 and detailed in [4], [10].

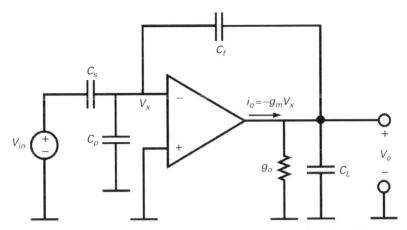

Fig. 10.29 The equivalent circuit for calculating the static error and the settling time of an op amp in a capacitive-feedback configuration [5].

10.2.5 Amplifier specifications

The operation of the SC integrator was described in Section 10.2.2 assuming the op amp to be ideal. In the actual design of SC circuits, the op-amp non-idealities must be considered carefully, since they are responsible for the main limitations of SC circuits.

In this section, we will consider the effect of the most important op-amp non-idealities other than offset, namely finite gain and unity-gain frequency. For the effects of other op-amp non-idealities, such as finite slew rate and electrical noise, the interested reader can consult the SC-filter literature [4], [22].

The speed/accuracy metrics for SC filters are related to the settling behavior of the operational amplifier. To obtain simple design rules we will analyze the step response of an op amp with capacitive feedback, such as the one shown in Figure 10.29. The op amp is represented as a voltage-controlled current source for which the transconductance is g_m and the output conductance is g_o. C_p represents all the capacitances between node v_x and the ac ground. We will first show how the amplifier gain influences the settling accuracy and then calculate the effect of the op-amp bandwidth on the settling time. Thus, we will analyze static and dynamic errors separately; the total error can then be calculated as the superposition of all the individual errors.

Charge balance at the inverting node of the op amp in Figure 10.29 yields

$$(C_s + C_p + C_f)v_x = C_s v_i + C_f v_o. \tag{10.2.39}$$

For low-frequency operation we have

$$V_x = -\frac{V_o}{A_{V0}}, \tag{10.2.40}$$

where $A_{V0} = g_m/g_o$ is the low-frequency voltage gain. Combining (10.2.39) and (10.2.40) allows us to write the static voltage gain of the stage as

$$\frac{v_o}{v_i} = -\frac{C_s}{C_f}\frac{1}{1 + 1/(\alpha A_{V0})}, \qquad (10.2.41)$$

where the feedback factor α is given by

$$\alpha = \frac{C_f}{C_s + C_p + C_f}. \qquad (10.2.42)$$

Example 10.6

Calculate the minimum amplifier gain for a static error of 0.1%, assuming $C_f = 1\,\text{pF}$, $C_s = 4\,\text{pF}$, and $C_p = 1\,\text{pF}$.

Answer

For a small error (0.1%) we can approximate (10.2.41) as

$$\frac{v_o}{v_i} \cong -\frac{C_s}{C_f}\left(1 - \frac{1}{\alpha A_{V0}}\right). \qquad (10.2.43)$$

Thus, $\alpha A_{V0} > 1000$. Since $\alpha = 1/(4+1+1) = 1/6$, $A_{V0} > 6000$.

To model the settling time we will neglect the effect of the op-amp output resistance ($g_o = 0$), which provides an acceptable approximation for the CMOS op amps used in SC filters. The Kirchoff current law applied to the inverting input and output nodes of the op amp yields

$$s(C_s + C_p + C_f)v_x = sC_s v_i + sC_f v_o, \qquad (10.2.44)$$

$$s(C_f + C_L)v_o + g_m v_x = sC_f v_x. \qquad (10.2.45)$$

On combining (10.2.44) and (10.2.45) we obtain the voltage transfer function as [5]

$$\frac{v_o}{v_i} = -\frac{C_s}{C_f}\frac{1 - \dfrac{sC_f}{g_m}}{1 + \dfrac{sC_{Lef}}{\alpha g_m}}, \qquad (10.2.46)$$

where the feedback factor α is given by (10.2.42) and the effective load is

$$C_{Lef} = C_L + (1 - \alpha)C_f. \qquad (10.2.47)$$

Neglecting the zero of the transfer function in (10.2.46), since it is usually at a frequency much higher than the pole frequency, we approximate the transfer function of the feedback op amp as

$$\frac{v_o}{v_i} \cong -\frac{C_s}{C_f}\frac{1}{1 + s\tau}, \qquad (10.2.48)$$

with τ given by

$$\tau = \frac{C_{Lef}}{\alpha g_m}. \qquad (10.2.49)$$

Since the step response is given by a simple exponential with time constant τ, the error calculated at the settling time t_s is given by

$$\varepsilon = \frac{v_o(t \to \infty) - v_o(t = t_s)}{v_o(t \to \infty)} = e^{-t_s/\tau}. \tag{10.2.50}$$

Thus,

$$t_s = \tau \ln\left(\frac{1}{\varepsilon}\right). \tag{10.2.51}$$

Since the unity-gain frequency GB of the op amp is

$$GB = \frac{g_m}{2\pi C_{Lef}}, \tag{10.2.52}$$

the settling condition $t_s < T/2$ can be written, by combining (10.2.49), (10.2.51), and (10.2.52), as [5]

$$GB > \frac{f_s}{\pi\alpha} \ln\left(\frac{1}{\varepsilon}\right). \tag{2.53}$$

where $f_s = 1/T$ is the clock frequency. For $\alpha = 1$ and $\varepsilon = 0.1\%$, $GB > 2.2 f_s$.

Example 10.7 [5]

Calculate the minimum transconductance necessary to obtain a settling time of 10 ns with an error of $\varepsilon = 0.1\%$, assuming $C_f = 1\,\mathrm{pF}$, $C_s = 4\,\mathrm{pF}$, and $C_p = 1\,\mathrm{pF}$.

Answer

$$\alpha = \frac{C_f}{C_s + C_p + C_f} = \frac{1}{4 + 1 + 1} \simeq 0.17,$$
$$C_{Lef} = C_L + (1 - \alpha)C_f = 5 + 0.83 = 5.83\,\mathrm{pF}.$$

The ratio between the zero (g_m/C_f) and pole ($\alpha g_m/C_{Lef}$) frequencies is ($C_{Lef}/(\alpha C_f)$) = $5.83/0.17 \gg 1$; thus, neglecting the effect of the zero on the settling time is reasonable. With this simplification we have

$$\tau = t_s \Big/ \ln\left(\frac{1}{\varepsilon}\right) = 10/6.9 = 1.45\,\mathrm{ns}.$$

Finally, $g_m = C_{Lef}/(\alpha\tau) = 5.83 \times 10^{-12}/(0.17 \times 1.45 \times 10^{-9}) = 24\,\mathrm{mA/V}$.

10.2.6 Low-distortion switched-capacitor filters

The widespread use of SC filters is due to their excellent analog performance. They are very stable and accurate, and can be designed for very low distortion and high dynamic

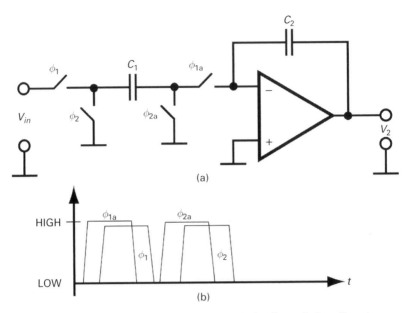

Fig. 10.30　A switched-capacitor integrator with four-phase clock scheme for low distortion.

range. Currently, there is no alternative analog integrated-circuit technique with all these qualities. In the early days of the SC technique, the non-linear parasitic capacitors were the major source of distortion. With the introduction of parasitic-insensitive topologies, the signal-dependent charge injection of MOSFET switches became the major source of distortion. Fortunately, an improved four-phase clocking scheme [24] allows us to obtain very-low-distortion filters; circuits with total harmonic distortion (THD) of −80 dB or below for signal swing close to the supply rails have been reported [13], [25].

The basic idea of the four-phase clock scheme is to advance slightly the times at which the switches connected to constant-voltage nodes (ground or virtual ground) open. As can be seen in Figure 10.30, the switches controlled by phases ϕ_{1a} and ϕ_{2a} turn off before the other two switches. Since the former switches are connected to constant-voltage nodes, the charge injected into the capacitors by turning them off is constant, thus producing dc offset but no distortion. When the switches controlled by phases ϕ_1 and ϕ_2 turn off, they cannot inject charge into the integrating capacitor since the current path is already blocked.

Similarly to the case of continuous-time filters, the use of fully-differential topologies is a must for low-distortion circuits since even harmonics are canceled out. Fully-differential structures are also adopted because they reject common-mode noise such as substrate noise, thus allowing high-dynamic-range circuits to be obtained. To design very-low-distortion filters all the main causes of distortion must be circumvented, including capacitor and op-amp non-linearities. For low-distortion and high-dynamic-range SC circuits, the reader is referred to [13], [25].

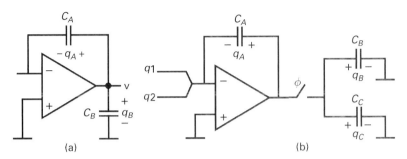

Fig. 10.31 (a) An elementary charge mirror and (b) basic SC signal-processing blocks (adapted from [26]).

10.3 Switched-capacitor circuits as charge processors

In this section we show that, in some cases, it is advantageous to regard SC circuits as charge processors [26]. Firstly, we establish under which conditions SC circuits containing non-linear capacitors can act as linear charge processors. Secondly, the conditions to obtain linear and accurate input-to-output voltage relationships in networks containing non-linear capacitors are derived. Finally, the alternatives for capacitor implementations are reviewed to make it clear that SC filters can be implemented in any digital MOS process.

To consider SC circuits as charge, rather than voltage, processors, we return to the basic SC cell and describe it as a charge mirror. Let us assume that the capacitors in Figure 10.31 are non-linear but have the same physical structure and that the op amp is ideal. The relationships between the stored charges and the op-amp output voltage can be written as

$$q_A = C_A v = C_{A0} f(v) v, \tag{10.3.1}$$

$$q_B = C_B v = C_{B0} f(v) v, \tag{10.3.2}$$

where $f(v)$ is a dimensionless function describing the non-linearity of the capacitors. This non-linearity is dependent on the structure of the capacitors and can be considered to be equal for capacitors that have the same structure and are physically close to each other on a chip. Parameters C_{A0} and C_{B0} are the values of the capacitances for a signal voltage v equal to zero, and are proportional to the areas of the capacitors.

From (10.3.1) and (10.3.2) it follows that

$$\frac{q_B}{q_A} = \frac{C_{B0}}{C_{A0}} = \frac{\text{area}(C_B)}{\text{area}(C_A)}. \tag{10.3.3}$$

Thus, the basic block in Figure 10.31(a) performs as a (linear) charge mirror with gain defined by a geometric ratio and, consequently, independent of the technology.

The basic building blocks commonly employed in SC networks are shown in Figure 10.31(b). Assuming that the capacitors have matched non-linearities and the op amp is ideal, each output charge (q_B, q_C) is a delayed and scaled image of the sum of the input charges (q_1, q_2).

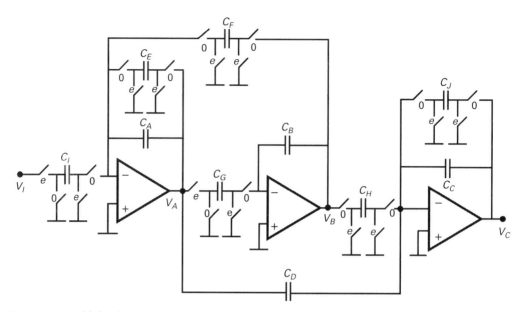

Fig. 10.32   A third-order low-pass SC filter (adapted from [26]).

From these considerations, it is clear that SC circuits employing non-linear capacitors can maintain linearity from input to output. If the external quantity is a voltage signal, linear input V–Q and output Q–V converters are then required, as detailed in the next section.

10.3.1   Realization of linear voltage processors

Let us consider as an example the third-order low-pass SC filter shown in Figure 10.32. As do most of the practical SC filters, this network satisfies the following topological constraint: after the settling time corresponding to each clock phase, the voltage across any capacitor is dependent only on a single node voltage. In other words, each capacitor has one of its terminals connected (or switched) to ground (or to virtual ground) and the other connected (or switched) to a voltage source (including ground). For the analysis that follows, we consider that the op amps and the switches are ideal.

Assuming that the capacitors are linear, the charge-conservation equations at the virtual grounds of the op amps are

$$-C_{I0}v_I(nT - T/2) + C_{E0}v_A(nT) + C_{F0}v_B(nT)$$
$$= -C_{A0}v_A(nT) + C_{A0}v_A(nT - T), \qquad (10.3.4)$$

$$-C_{G0}v_A(nT - T) = -C_{B0}v_B(nT) + C_{B0}v_B(nT - T), \qquad (10.3.5)$$

$$C_{H0}v_B(nT) + C_{J0}v_C(nT) + C_{D0}v_A(nT) - C_{D0}v_A(nT - T)$$
$$= -C_{C0}v_C(nT) + C_{C0}v_C(nT - T). \qquad (10.3.6)$$

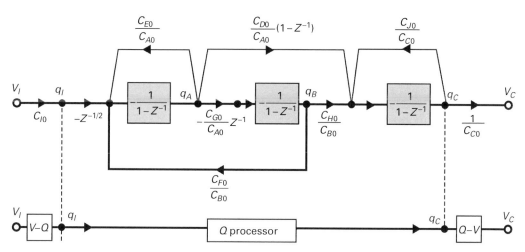

Fig. 10.33 The signal-flow graph of the SC filter in Figure 10.32.

Rewriting the above equations in terms of the charges stored on the capacitors yields

$$-q_I(nT - T/2) + \frac{C_{E0}}{C_{A0}}q_A(nT) + \frac{C_{F0}}{C_{B0}}q_B(nT)$$
$$= -q_A(nT) + q_A(nT - T), \qquad (10.3.7)$$

$$-\frac{C_{G0}}{C_{A0}}q_A(nT - T) = -q_B(nT) + q_B(nT - T), \qquad (10.3.8)$$

$$\frac{C_{H0}}{C_{B0}}q_B(nT) + \frac{C_{J0}}{C_{C0}}q_C(nT) + \frac{C_{D0}}{C_{A0}}q_A(nT) - \frac{C_{D0}}{C_{A0}}q_A(nT - T)$$
$$= -q_C(nT) + q_C(nT - T). \qquad (10.3.9)$$

The coefficients of the equations above correspond to charge-mirror gains [26]. For instance, the charge in capacitor C_A is mirrored to capacitors C_E, C_G, and C_D with gains C_{E0}/C_{A0}, C_{G0}/C_{A0}, and C_{D0}/C_{A0}, respectively.

Equations (10.3.7) through (10.3.9) are represented by the signal flow of the Q-processor in Figure 10.33, where the branch transmittances are proportional to the charge-mirror gains. This figure also shows the input V–Q and output Q–V conversions of the SC filter.

Let us consider now that the filter is built using non-linear capacitors with identical non-linearity and the same area ratios as the linear capacitors. The charge-mirror gains will be the same as those for the linear network, and, since the coefficients of Equations (10.3.7)–(10.3.9) correspond to the charge-mirror gains, the very same equations will represent the non-linear network with matched non-linearities.

Summarizing, provided that some simple conditions are satisfied, an SC network containing non-linear capacitors can realize linear charge processing. Furthermore, the expression of its charge transfer function will be equal to that of the same network with

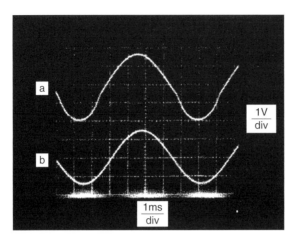

Fig. 10.34   Measured output waveforms at (a) an intermediate node and (b) the output node of an SC filter (adapted from [26]).

linear capacitors. Another important conclusion is that this charge transfer function is proportional to the voltage transfer function of the linear network. Provided that the capacitors connected to the input and to the network output nodes are linear, the input $V–Q$ and output $Q–V$ conversions will be linear, as will the input-to-output voltage transfer function. As an example, Figure 10.34 shows the waveforms at an internal node and at the output node (for a sinusoidal input) of an SC filter built with linear capacitors connected to the input and output nodes. All the other capacitors of the filter are non-linear with matched non-linearities. As is clear in Figure 10.34, in spite of the strong distortion of the voltage signal at the intermediate node, the output signal is a fine sinusoid.

10.3.2   Implementation issues

For the implementation of the non-linear (internal) capacitors, the natural choice is to use a MOSFET gate structure. An important feature of this structure is the use of the thin gate oxide as the dielectric. Compared with the commonly employed double-poly capacitors, gate capacitors in VLSI processes present a larger capacitance per unit area and have better matching properties. Moreover, gate capacitors are fully compatible with any digital MOS process, whereas linear capacitors are costly to implement in VLSI processes. To implement the few linear capacitors required, two main alternatives are available. One of them is the use of MIM capacitors or another type of available linear capacitor. Even though MIM capacitors have low specific capacitances, they represent an attractive solution to implementing few capacitors.

In many practical situations, the total area occupied by the capacitors in a filter implemented using MOSFET gates as internal capacitors and MIM capacitors in the input/output branches will be smaller than that which would be required if double-poly capacitors were to be used for the entire filter. The other alternative is to construct the input/output capacitors entirely from MOS transistors and to apply a dc bias [27], [28] or combine two or more non-linear capacitors to partially cancel out their non-linearities [29]. There are several ways to create MOSFET-based capacitors, and these devices can be combined using series and/or parallel combinations for non-linearity compensation. Using the series-compensation technique, SC filters with very high linearity, comparable to that achieved using highly linear poly-capacitors, can be realized [29].

Finally, if the topology chosen for the SC filter has too many capacitors connected to the input and output nodes, the linear V–Q and Q–V conversions can be realized by using amplifier stages at both ends. This brings down the required number of linear capacitors to two or three.

10.4   Alternative switched-circuit techniques

Sampled-data analog signal processing requires four basic operations: inversion, addition, multiplication, and delay. In SC circuits these operations are realized by switched charge mirrors, as explained in Section 10.3. Another alternative is switched current circuits, in which the four basic operations are carried out by switched current mirrors [30]–[32]. These mirrors perform a non-linear I–V conversion followed by its inverse V–I conversion, as shown in Figure 10.35(a), leading to a linear relationship between input and output currents.

In the circuit shown in Figure 10.35(b), assuming the op amp to be ideal, transistors M_1 and M_2 are both biased with the same set of voltages. Therefore, neglecting transistor mismatch, the output current i_o is an inverted replica of the input current i_{in}, i.e. $i_o = -Ai_{in}$, where $A = (W/L)_2/(W/L)_1$. The S/H shown in Figure 10.35(b) operates as follows [15].

- Track mode. The input current is fed to the cell when the switch is closed. The current is memorized as a voltage across the holding capacitor. It should be emphasized that linear capacitors are not needed to store the data.
- Hold mode. After the switch has opened, the voltage is held on the capacitor. The output current is equal to that in the previous clock phase. An important property of the S/H shown in Figure 10.35(b) is that the switch operates at a constant voltage. Therefore, the conduction gap of the switches at low supply voltages is avoided. Since the dc operating point of the switches is constant, both the charge injected into the holding capacitor and the settling time of the S/H are signal-independent.

The bias voltage (V_B), which allows the highest current swing, is generated by the circuit shown in Figure 10.35(c), where M_A and M_B are identical transistors. Figure 10.35(d) shows a plot of the conductance of the n-MOS switch for an input voltage varying between 0 and V_{DD}. Note that the op amp in the feedback configuration imposes operation of the switch at a constant voltage (equal to V_B), thus resulting in a conductance of around 0.28 mS for the example shown.

The current S/H of Figure 10.35(b) can be easily modified to implement digitally programmable analog filters, substituting the MOSFET-only current divider (MOCD) described in Chapter 9, and shown in Figure 10.36, for transistor M_2.

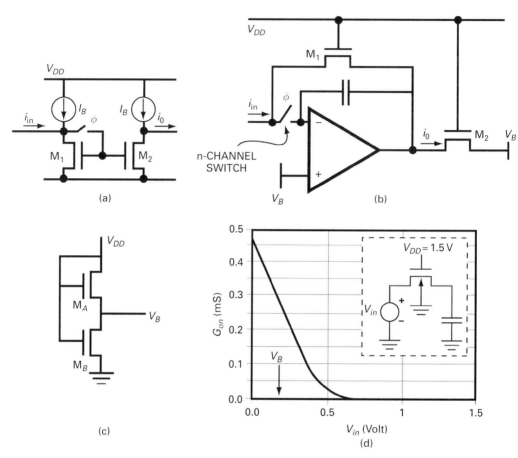

(d)

Fig. 10.35   Switched-current mirrors: (a) using saturated transistors operating as current sources; and (b) using an op amp and MOSFETs in the triode region. In (c) the bias circuit (M_A and M_B are identical transistors) for the current mirror in (b) is shown, and in (d) is shown the on-conductance of the n-MOS switch (adapted from [15]).

As an example, the circular finite-impulse-response (FIR) structure proposed in [33], which is programmable with MOCDs, was implemented using the current S/H shown in Figure 10.35(b). The circular FIR filter, which is shown in Figure 10.37, operates as follows.

- In the ck1 phase, the S/H 1 is in the sample mode (active), and the other S/Hs are in the hold mode. The filter coefficients and the stored signals are distributed as shown in Figure 10.37(a) and (b), respectively.
- In the ck2 phase, S/H 2 is active, and the stored values of all S/Hs, except the second, remain unchanged. However, all tap weights must be shifted, as shown in Figure 10.37, to obtain

$$H(z) = h_1 + h_2 z^{-1} + h_3 z^{-2} + h_4 z^{-3}. \tag{10.4.1}$$

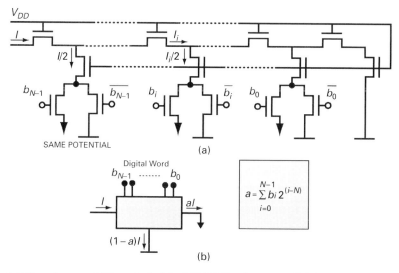

(a)

(b)

$$a = \sum_{i=0}^{N-1} b_i 2^{(i-N)}$$

Fig. 10.36   (a) The scheme and (b) symbol of the MOSFET-only current divider (adapted from [15]).

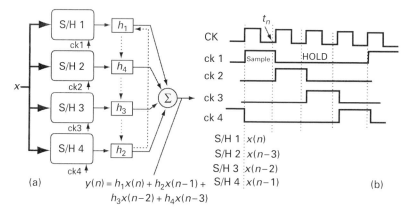

(a)

$y(n) = h_1 x(n) + h_2 x(n-1) +$
$h_3 x(n-2) + h_4 x(n-3)$

(b)

Fig. 10.37   (a) A circular FIR filter (coefficients during ck1) and (b) clock waveforms (adapted from [15]).

The circulation process continues in each clock cycle, thus cycling the digital coefficients through all the multipliers. This process simulates a tapped delay line without the sampled signal passing through series delay elements in each clock cycle. Consequently, this structure has the advantage of avoiding the propagation of both the offset voltage and noise from each cell to the next (multiple resampling errors).

The S/H of Figure 10.35(b) and the programmable MOCD in Figure 10.36 were used for the implementation of a four-tap circular FIR filter, the schematic diagram of which is shown in Figure 10.38 [15]. The input section performs a linear voltage-to-current conversion, while the output S/H acts, simultaneously, as a summer block and a linear current-to-voltage converter.

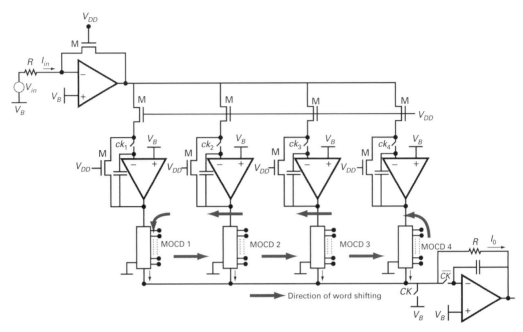

Fig. 10.38   The switched-MOSFET scheme of the circular FIR filter (adapted from [15]).

Appendix

A10.1   Modeling the sampling distortion due to the non-linearity of the switch on-resistance

For the n-MOS transistor operating in strong inversion, as is currently the case in almost all S/Hs, the on-resistance in the triode region (with $V_{DS}=0$) is given by

$$R_{ON} = \frac{1}{\mu C'_{ox} \frac{W}{L}(V_G - V_{T0} - nV_S)} = \frac{1}{\beta(V_G - V_{T0} - nV_S)}. \qquad (A10.1.1)$$

In the tracking mode, the gate is connected to the power supply V_{DD} and the source to the input signal. Assuming that the quiescent voltage at the source is V_{SQ}, it follows that

$$V_G - V_{T0} - nV_S = V_{DD} - V_{T0} - nV_{SQ} - nv_i = V_{ovQ} - nv_i. \qquad (A10.1.2)$$

The time constant of the S/H in the tracking mode is now signal-dependent and given by

$$\tau = R_{ON}C = \frac{C}{\beta(V_{ovQ} - nv_i)}. \qquad (A10.1.3)$$

We now assume that the output signal is given by the delayed input signal as shown in Section 10.1.3, but considering the signal-dependent delay, as in (A10.1.3).

With the assumption of a small delay, a closed-form expression for the harmonic distortion is easily obtained [7]. The output signal of the S/H in the tracking mode can be approximated by its first-order Taylor power expansion

$$v_o(t) = v_i(t - \tau(v_i)) \cong v_i(t) - \tau(v_i)\frac{dv_i}{dt} \qquad \text{(A10.1.4)}$$

with

$$\tau(v_i) \cong \tau_o + \tau_1 v_i + \tau_2 v_i^2. \qquad \text{(A10.1.5)}$$

Expanding now the switch on-resistance to second-order,

$$R_{ON} \cong R_0\left[1 + \frac{nv_i}{V_{ovQ}} + \left(\frac{nv_i}{V_{ovQ}}\right)^2\right], \qquad \text{(A10.1.6)}$$

where

$$R_0 = \left(\beta V_{ovQ}\right)^{-1}, \qquad \text{(A10.1.7)}$$

and the coefficients in (A10.1.5) are given by

$$\tau_o = R_o C, \qquad \text{(A10.1.8)}$$

$$\tau_1 = \frac{nC}{\beta V_{ovQ}^2}, \qquad \text{(A10.1.9)}$$

$$\tau_2 = \frac{n^2 C}{\beta V_{ovQ}^3}. \qquad \text{(A10.1.10)}$$

Assuming a sinusoidal input signal $v_i = A\sin(\omega t)$, we can calculate the first two components of the harmonic distortion by combining (A10.1.4) through (A10.1.10), yielding

$$HD2 = \frac{\tau_1 \omega}{2} A = \frac{nC\omega}{2\beta V_{ovQ}^2} A, \qquad \text{(A10.1.11)}$$

$$HD3 = \frac{\tau_2 \omega}{4} A^2 = \frac{n^2 C\omega}{4\beta V_{ovQ}^3} A^2. \qquad \text{(A10.1.12)}$$

As shown in Figure A10.1.1, the components of different orders of the harmonic distortion have a linear dependence on the input frequency, as predicted by (A10.1.11) and (A10.1.12).

A similar model [7] can be applied to a transmission gate in which the on-resistance is given by the parallel connection of the on-resistances of the n-MOS and p-MOS transistors.

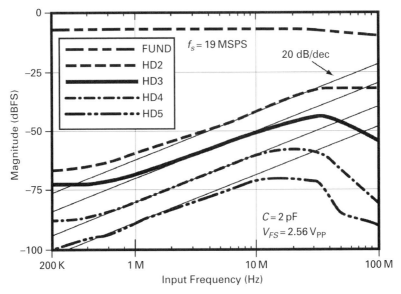

Fig. A10.1.1   Measured and modeled harmonic distortion versus input signal frequency for a sample-and-hold test circuit fabricated in 0.25-μm technology operating with 3.3 V. The sampling frequency is 19 MS/s (adapted from [34]).

Problems

10.1 Apply the Nyquist noise formula to the equivalent resistance $1/(Cf_s)$ of a switched capacitor and compare the obtained (bilateral) power-spectral density with expression (10.1.10) for the aliased thermal noise of the MOS S/H. Comment on the result.

10.2 Show that, for a linear S/H, we can approximate the solution of Equation (10.1.15) for the tracking period, repeated below,

$$\frac{dv_o}{dt} + \frac{v_o}{\tau_0} = \frac{A}{\tau_0}\cos(\omega t), \tag{P10.1.1}$$

as

$$v_o(t) \cong A\cos[\omega(t - \tau_0)] \tag{P10.1.2}$$

for $\tau_0 \ll T_s$ and $\omega\tau_0 \ll 1$.

Hint: the general solution of (P10.1.1) can be written as

$$v_o = A\frac{e^{-t/\tau_0}}{\tau_0}\int e^{t/\tau_0}\cos(\omega t)dt + Ke^{-t/\tau_0},$$

where K is a constant to be determined.

10.3 For an MOS S/H, plot HD2 and HD3 due to the non-linear on-resistance versus the signal frequency in the interval $1\,\text{MHz} < f < 100\,\text{MHz}$, assuming $v_i(t) = A\sin(2\pi ft) + 1\text{V}$, $A = 750\,\text{mV}$, $C_H = 1\,\text{pF}$, $V_{DD} = 3.3\,\text{V}$, $W = 50\,\mu\text{m}$, $L = 0.2\,\mu\text{m}$, $V_{T0} = 0.6\,\text{V}$, $C'_{ox} = 4.5\,\text{fF}/\mu\text{m}^2$, $\mu = 400\,\text{cm}^2/\text{V}$ per s, and $n = 1.2$. Comment on the results.

10.4 The transit time τ of a MOSFET is defined as

$$\tau = \frac{|Q_I|}{I_D}, \tag{P10.4.1}$$

where Q_I is the total carrier charge stored in the channel and I_D is the drain current.

(a) Show that, in the linear region, for $V_{DS} \cong 0$, we have

$$\tau \cong \frac{L^2}{\mu V_{DS}}, \tag{P10.4.2}$$

where μ is the carrier mobility and L the channel length.

(b) Using (P10.4.2), show that expression (10.1.25), which gives the maximum sampling frequency for a given charge-injection-related error, can be rewritten as

$$f_s < \frac{1}{5} \frac{1}{\text{transit time}}. \tag{P10.4.3}$$

Hint: the voltage V_{DS} of the transistor when it turns off is the voltage sampling error ΔV.

10.5 (a) Show that expression (15c) of reference [12] can be rewritten as

$$B = \sqrt{\frac{C_g}{2C_h}} \, \omega_T t_{FALL}, \tag{P10.5.1}$$

where $C_g = WLC'_{ox}$ is the total gate capacitance, $\omega_T = [2\mu/(nL^2)](V_{GON} - V_{T0} - nV_{in})$ is the intrinsic transition frequency of the saturated on-transistor, and t_{FALL} is the fall time of the clock signal from V_{GON} to the threshold voltage of the MOS switch.

(b) For $C_g/C_h = 0.1$ and $t_{FALL} = 1\,\text{ns}$, calculate the transistor f_T required to obtain $B = 0.1$ (short clock fall time).

(c) Assuming $V_{in} = 0$, $V_{T0} = 0.6\,\text{V}$, $n = 1.2$, $\mu = 400\,\text{cm}^2/\text{V}$ per s and $L = 0.2\,\mu\text{m}$, determine the V_{GON} required to obtain the value of f_T determined in part (b).

10.6 For a semi-infinite sequence of samples $x(nT)$ $(n = 0, 1, \ldots)$, the z-transform is defined as

$$X(z) = \sum_{n=0}^{\infty} x(nT)z^{-n}. \tag{P10.6.1}$$

(a) Prove that the z-transform of Dirac's delta, $x(0) = 1$, otherwise $x(nT) = 0$, is

$$X(z) = 1. \tag{P10.6.2}$$

(b) Prove that the z-transform of the step function

$$u(nT) = 1, n \geq 0 \qquad \text{(P10.6.3)}$$

is given by

$$1 + z^{-1} + z^{-2} + z^{-3} + z^{-4} + \cdots = 1/(1 - z^{-1}) \qquad \text{(P10.6.4)}$$

for $|z| > 1$.

(c) Prove that the z-transform of the sampled exponential

$$x(nT) = e^{-anT} \qquad \text{(P10.6.5)}$$

is

$$X(z) = \frac{1}{1 - e^{-aT}z^{-1}} \qquad \text{(P10.6.6)}$$

for $|z| > e^{-aT}$.

(d) Prove that multiplication of the z-transform by z^{-1} corresponds to delaying the corresponding time sequence by one clock period.

10.7 The spectrum of the sampled signal

$$x_s(t) = \sum_{-\infty}^{+\infty} x(nT)\delta(t - nT)$$

is calculated as

$$F_s(j\omega) = \int_{-\infty}^{\infty} \sum_{-\infty}^{\infty} x(nT)\delta(t - nT)e^{-j\omega t}dt. \qquad \text{(P10.7.1)}$$

Show that

$$F_s(j\omega) = \sum_{-\infty}^{+\infty} x(nT)e^{-j\omega nT}. \qquad \text{(P10.7.2)}$$

and compare (P10.7.2) with the z-transform of $x(nT)$, to derive

$$F_s(j\omega) = X(z = e^{j\omega T}).$$

Hint: interchange the order of the summation and the integration in (P10.7.1).

10.8 Derive expressions (10.2.30) and (10.2.31) for the transfer functions of the parasitic-insensitive non-inverting integrator.

10.9 Derive the expressions for all the (normalized) capacitances of the biquad given in Figure 10.28.

10.10 (a) Prove that the scaling (by the same factor k) of **all** the capacitances connected to the **op-amp output** (Figure P10.10.1) maintains the same **charge** transfer function, but the voltage V_i at the op-amp output is scaled by k, i.e. $V_i \rightarrow V_i' = V_i/k$, where V_i (V_i') is the voltage in the original (scaled) network.

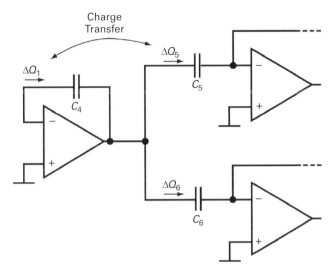

Fig. P10.10.1 The SC circuit viewed as a charge processor: **voltage scaling**.

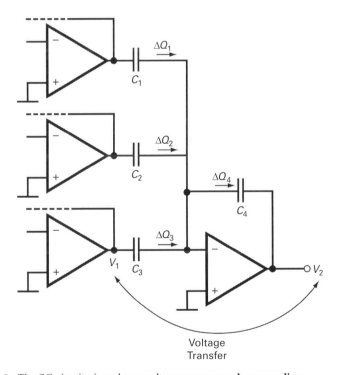

Fig. P10.10.2 The SC circuit viewed as a voltage processor: **charge scaling**.

(b) Prove that on scaling **all** of the capacitances connected to the **op-amp input** (Figure P10.10.2), the same **voltage** transfer function is maintained, but the charges (and capacitances) at the op-amp input are scaled by k: $\Delta Q_i \rightarrow \Delta Q_i' = \Delta Q_i/k$.

The interested reader may refer to Section 5.6 in [4] for the application of voltage and charge scaling to the optimization of dynamic range and capacitance spread in SC filters.

References

[1] B. Razavi, *Principles of Data Conversion System Design*, Piscataway, NJ: IEEE Press, 1995.

[2] R. van de Plassche, *Integrated Analog-to-Digital and Digital-to-Analog Converters*, Boston, MA: Kluwer, 1994.

[3] H. Zumbahlen, editor, *Linear Circuit Design Handbook*, Amsterdam: Elsevier/Newnes, 2008.

[4] R. Gregorian and G. Temes, *Analog MOS Integrated Circuits for Signal Processing*, New York: Wiley, 1986.

[5] A. M. Niknejad and B. E. Boser, *Analog Integrated Circuits, EECS 240 Course Notes*.

[6] R. H. Walden, "Analog-to-digital converter survey and analysis," *IEEE Journal of Selected Areas in Communications*, vol. **17**, no. 4, pp. 539–550, Apr. 1999.

[7] F. Centurelli, P. Monsurro, and A. Trifiletti, "A model for the distortion due to switch on-resistance in sample-and-hold circuits," *Proceedings of IEEE ISCAS* 2006, pp. 4787–4790.

[8] D. Jaconis and C. Svensson, "A 1 GHz linearized CMOS track-and-hold circuit," *Proceedings of IEEE ISCAS* 2002, vol. **5**, pp. 577–580.

[9] B. J. Sheu and C. M. Hu, "Switched induced error voltage on a switched capacitor," *IEEE Journal of Solid-State Circuits*, vol. **19**, no. 4, pp. 519–525, Aug. 1984.

[10] G. C. Temes, "Simple formula for estimation of minimum clock-feedthrough error voltage," *Electronics Letters*, vol. **22**, no. 20, pp. 1069–1070, Sep. 1986.

[11] D. A. Johns and K. Martin, *Analog Integrated Circuit Design*, New York: Wiley, 1997.

[12] G. Wegmann, E. A. Vittoz, and F. Rahali, "Charge injection in analog MOS switches," *IEEE Journal of Solid-State Circuits*, vol. **22**, no. 6, pp. 1091–1097, Dec. 1987.

[13] C. C. Enz and G. C. Temes, "Circuit techniques for reducing the effects of op-amp imperfections: autozeroing, correlated double sampling, and chopper stabilization," *Proceedings of the IEEE*, vol. **84**, no. 11, pp. 1584–1614, Nov. 1996.

[14] K.-L. Lee and R. G. Meyer, "Low-distortion switched-capacitor filter design techniques," *IEEE Journal of Solid-State Circuits*, vol. **20**, no. 6, pp. 1103–1113, Dec. 1985.

[15] F. A. Farag, C. Galup-Montoro, and M. C. Schneider, "Digitally programmable switched-current FIR filter for low-voltage applications," *IEEE Journal of Solid-State Circuits*, vol. **35**, no. 4, pp. 637–641, Apr. 2000.

[16] E. Vittoz, "Micropower techniques," in *Design of Analog-Digital VLSI Circuits for Telecommunications and Signal Processing*, 2nd edn., ed. J. E. Franca and Y. Tsividis, Englewood Cliffs, NJ: Prentice-Hall, 1994, pp. 53–96.

[17] A. Baschirotto, "A low-voltage sample-and-hold circuit in standard CMOS technology operating at 40 MS/s, *IEEE Transactions on Circuits and Systems II*, vol. **48**, no. 4, pp. 394–399, Apr. 2001.

[18] A. M. Abo and P. R. Gray, "A 1.5-V, 10-bit, 14.3-MS/s CMOS pipeline analog-to-digital converter," *IEEE Journal of Solid-State Circuits*, vol. **34**, no. 5, pp. 599–606, May 1999.

[19] B. J. Hosticka, R. W. Broderson, and P. R. Gray, "MOS sampled data recursive filters using switched capacitor integrators," *IEEE Journal of Solid-State Circuits*, vol. **12**, no. 6, pp. 600–608, Dec. 1977.

[20] J. T. Caves, M. A. Copeland, C. F. Rahim, and S. D. Rosenbaum, "Sampled analog filtering using switched capacitors as resistor equivalents," *IEEE Journal of Solid-State Circuits*, vol. **12**, no. 6, pp. 592–600, Dec. 1977.

[21] J. Bermudez and B. Bhattacharyya, "A systematic procedure for generation and design of parasitic insensitive SC biquads," *IEEE Transactions on Circuits and Systems*, vol. **32**, no. 8, pp. 767–783, Aug. 1985.

[22] K. R. Laker and W. M. C. Sansen, *Design of Analog Integrated Circuits and Systems*, New York: McGraw-Hill, 1994.

[23] R. Schaumann and M. E. Van Valkenburk, *Design of Analog Filters*, New York: Oxford University Press, 2000.

[24] D. G. Haigh and B. Singh, "A switching scheme for switched capacitor filters which reduces the effect of parasitic capacitances associated with switch control terminals," *Proceedings of IEEE ISCAS* 1983, pp. 586–589.

[25] G. Nicollini, A. Nagari, P. Confalonieri, and C. Crippa, "A −80 dB THD, 4 V_{pp} switched capacitor filter for 1.5 V battery-operated systems," *IEEE Journal of Solid-State Circuits*, vol. **31**, no. 8, pp. 1214–1219, Aug. 1996.

[26] J. C. M. Bermudez, M. C. Schneider, and C. G. Montoro, "Compatibility of switched capacitor filters with VLSI processes," *IEE Proceedings, Part G*, vol. **139**, no. 4, pp. 413–418, Aug. 1992.

[27] A. T. Behr, M. C. Schneider, S. Noceti Filho, and C. G. Montoro, "Harmonic distortion caused by capacitors implemented with MOSFET gates," *IEEE Journal of Solid-State Circuits*, vol. **27**, no. 10, pp. 1470–1475, Oct. 1992.

[28] M. C. Schneider, C. Galup-Montoro, and J. C. M. Bermudez, "Explicit formula for harmonic distortion in SC filters due to weakly nonlinear capacitors," *IEE Proceedings, Part G*, vol. **141**, no. 6, pp. 505–509, Dec. 1994.

[29] H. Yoshizawa, Y. Huang, P. F. Ferguson Jr., and G. C. Temes, "MOSFET-only switched-capacitor circuits in digital CMOS technology," *IEEE Journal of Solid-State Circuits*, vol. **34**, no. 6, pp. 734–747, June 1999.

[30] J. B. Hughes, N. C. Bird, and I. C. Macbeth, "Switched current – a new technique for analog sampled-data signal processing," *Proceedings of IEEE ISCAS* 1989, pp. 1584–1587.

[31] J. B. Hughes, I. C. Macbeth, and D. M. Pattullo, "Switched current filters," *IEE Proceedings, Part G*, vol. **137**, no. 2, pp. 156–162, Apr. 1990.

[32] T. S. Fiez, G. Liang, and D. J. Allstot, "Switched current circuit design issues," *IEEE Journal of Solid-State Circuits*, vol. **26**, no. 3, pp. 192–202, Mar. 1991.

[33] R. Barnett and R. Harjani, "A 200 MHz differential sampled data FIR filter for disk drive equalization," *Proceedings of IEEE ISCAS* 1996, vol. 1, pp. 429–432.

[34] T. W. Brown, T. S. Fiez, and M. Hakkarainen, "Prediction and characterization of frequency dependent MOS switch linearity and the design implications," *CICC Technical Digest*, pp. 237–240, Sep. 2006.

11 Overview of MOSFET models and parameter extraction for design

Compact models, which describe the electrical behavior of the passive and active devices on a chip, are the fundamental link between circuit designers and foundries. Compact models allow simulation of the circuit functionality before its fabrication, thus saving time and money. In CMOS technologies, the MOS transistor is the principal component; consequently, its model plays a decisive role in the analysis and design of integrated circuits.

Early compact MOSFET models rely on approximate solutions that are valid only in particular regions of operation, which are connected mathematically by smoothing functions. Because the threshold of strong inversion V_T is the key parameter in these regional models, they are also called V_T-based models. This regional approach leads to inaccuracy between regions and, consequently, this class of models is not accurate enough to represent the moderate-inversion region [1]. To overcome the limitations of V_T-based MOSFET models, a new class of models emerged, namely inversion-charge-based and surface-potential-based models [2].

This chapter provides an overview of the approaches taken by the developers of MOSFET models. After a brief review of V_T-based MOSFET models, we present a summary of some fundamental properties of advanced MOSFET models.

The accuracy of the transistor characteristics depends not only on an appropriate device model but also on the accuracy of its fundamental parameters. To complete the chapter we describe some procedures to extract fundamental parameters of MOSFET models, in particular those used in this textbook as design parameters. We will give special attention to the extraction of the specific current, slope factor, threshold voltage, carrier mobility, and Early voltage. Other parameters such as gate oxide thickness and junction and overlap capacitances are assumed to have the nominal values of the technology process.

11.1 MOSFET models for circuit simulation

Despite their limitations, V_T-based MOSFET models have been used successfully for much circuit-design work. Examples of threshold-voltage-based models are BSIM3 and BSIM4 (BSIM stands for Berkeley Short-Channel IGFET Model), from a series of models, and MOS Model 9. Limitations of V_T-based MOSFET models include asymmetries between source and drain, and discontinuities in derivatives.

To avoid the drawbacks of V_T-based MOSFET models, two alternative approaches, namely surface-potential-based (ϕ_s-based) [3]–[5] and inversion-charge-based (Q_I'-based) [6]–[8] models, have been employed more recently. In these two alternative models, the drain current and the terminal charges are indirect functions of the terminal voltages, and are obtained through either the surface potential or the inversion charge density. The ϕ_s-based and Q_I'-based models have a common background, but there exist some differences between them, motivating model developers to support one approach rather than the other. In this chapter we will, following the joint paper by model developers [2], briefly review some Q_I'-based and ϕ_s-based models. Because the complete transistor model, including the various physical effects relevant to advanced technologies, is rather complex, we will restrict our presentation to the core models.

11.1.1 Threshold-voltage-based models (BSIM3 and BSIM4)

Here we describe the equations used in BSIM3 for the calculation of the inversion charge density [9], which are also used in BSIM4 in a slightly modified format. The reader should note that we have modified the original notation to adapt the symbols used in the BSIM user's manual to those employed in this textbook.

Initially, we write the inversion charge density at the source for the case of weak inversion as that obtained using the charge-sheet model [10], which is given by

$$Q_{I,wi}' = -(n-1)C_{ox}'\phi_t \exp\left(\frac{V_{GS} - V_T}{n\phi_t}\right). \tag{11.1.1}$$

On the other hand, the inversion charge density in strong inversion is written as

$$Q_{I,si}' = -C_{ox}'(V_{GS} - V_T). \tag{11.1.2}$$

In BSIM3, the inversion charge density is, for any inversion level, given by

$$Q_I' = -C_{ox}'V_{gsteff}, \tag{11.1.3}$$

where V_{gsteff} is an interpolating function, which in BSIM3 is

$$V_{gsteff} = \frac{2n\phi_t \ln[1 + \exp[(V_{GS} - V_T)/(2n\phi_t)]]}{1 + \dfrac{2n}{n-1}\exp[-(V_{GS} - V_T)/(2n\phi_t)]}. \tag{11.1.4}$$

The interpolation function in (11.1.4) is plotted for values of n of 1.2 and 1.4 in Figure 11.1. It is simple to demonstrate (see Problem 11.1) that the application of (11.1.4) to (11.1.3) gives the asymptotic cases of weak and strong inversion in (11.1.1) and (11.1.2), respectively. In the BSIM equations, the slope factor n for a long-channel device with $V_{BS}=0$ is a constant given by

$$n = 1 + \gamma\left/\left(2\sqrt{2\phi_F}\right)\right., \tag{11.1.5}$$

where ϕ_F is the Fermi potential of the holes in the p-substrate.

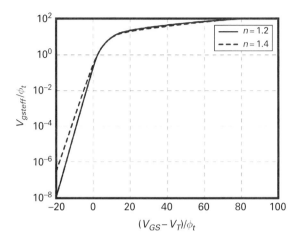

Fig. 11.1   The interpolating function of expression (11.1.4) for BSIM3 for $n = 1.2$ and $n = 1.4$.

Example 11.1

Show that, for $n \to 1$, the curves of expressions (11.1.1) and (11.1.2) intersect at $V_{GS} \cong V_T + n(n-1)\phi_t$.

Answer

For $n = 1$, the two curves intersect at $V_{GS} = V_T$. For n slightly higher than unity, we can use the approximation $\exp(x) \cong 1 + x$. On equating (11.1.1) and (11.1.2) we find

$$-(n-1)C'_{ox}\phi_t\left(1 + \frac{V_{GS} - V_T}{n\phi_t}\right) \cong -C'_{ox}(V_{GS} - V_T),$$

which gives $V_{GS} \cong V_T + n(n-1)\phi_t$.

11.1.2   Surface-potential-based models (HiSIM, MM11, and PSP)

The application of the gradual-channel and charge-sheet approximations yields the implicit relation for the surface potential given below:

$$(V_G - V_{FB} - \phi_s)^2 = \gamma^2 \phi_t [e^{-\phi_s/\phi_t} + \phi_s/\phi_t - 1$$
$$+ e^{-(2\phi_F + V_C)/\phi_t}\left(e^{\phi_s/\phi_t} - \phi_s/\phi_t - 1\right)]. \qquad (11.1.6)$$

The accurate solution of (11.1.6), which was once a big issue, is no longer a problem. Considerable progress has been made in the development of fast and accurate algorithms to find the solution for the surface potential in (11.1.6). After ϕ_s has been determined from (11.1.6), the charge-sheet approximation [10] (see Appendix A2.1) can be used to calculate the bulk charge density Q'_B as

$$Q'_B = -\text{sgn}(\phi_s)C'_{ox}\gamma\sqrt{\phi_s + \phi_t\left(e^{-\phi_s/\phi_t} - 1\right)}. \tag{11.1.7}$$

Expression (11.1.7) gives a continuous model from accumulation through depletion to inversion. Using the values of the surface potential and the depletion charge determined previously, the inversion charge density Q'_I can be calculated from the potential-balance equation as

$$Q'_I = -C'_{ox}\left(V_G - V_{FB} - \phi_s + \frac{Q'_B}{C'_{ox}}\right). \tag{11.1.8}$$

The main variables used to build (long-channel) surface-potential models are determined from Equations (11.1.6), (11.1.7), and (11.1.8).

A disadvantage of the original ϕ_s-based models was the use of intricate and lengthy expressions for the calculation of current, total charges, and noise [2]. To simplify these calculations, some ϕ_s-based compact models use linearization of the inversion charge density as a function of the surface potential in a similar way to Q'_I-based models [2]. In the following, we describe some of the approaches taken by the developers of the most representative ϕ_s-based models of the MOSFET at the time of writing.

Example 11.2

Find an approximation for the surface potential and the bulk charge in terms of V_G for the case of deep accumulation.

Answer

For the case of deep accumulation, $-\phi_s/\phi_t \gg 1$, thus, from (11.1.7), we can write

$$Q'_B \cong \gamma C'_{ox}\sqrt{\phi_t e^{-\phi_s/\phi_t}}.$$

From the potential-balance equation ($Q'_I = 0$), we have

$$V_G - V_{FB} = \phi_s - Q'_B/C'_{ox} \cong -Q'_B/C'_{ox}.$$

An estimate of the surface potential can be found from the approximations for the bulk charge density and potential-balance equations, leading to

$$\gamma\sqrt{\phi_t e^{-\phi_s/\phi_t}} \cong V_{FB} - V_G \rightarrow \phi_s \cong -2\phi_t \ln\left(\frac{V_{FB} - V_G}{\gamma\sqrt{\phi_t}}\right).$$

Using the potential-balance equation, we find that

$$Q'_B = C'_{ox}(\phi_s + V_{FB} - V_G) \cong C'_{ox}\left[-2\phi_t \ln\left(\frac{V_{FB} - V_G}{\gamma\sqrt{\phi_t}}\right) + V_{FB} - V_G\right].$$

11.1.2.1   The HiSIM model

HiSIM (the Hiroshima-University STARC IGFET Model [3]) calculates the potentials by solving the Poisson equation iteratively both at the source side and at the drain side. An accuracy of 10 pV has been achieved with fast algorithms. Such an extreme accuracy in the surface-potential calculations has turned out to be absolutely necessary for maintaining sufficiently accurate solutions for transcapacitance values, as well as for achieving stable circuit simulations [2].

HiSIM is based on the charge-sheet approximation, as are other surface-potential models, but does not use the linearization of the inversion charge density as a function of ϕ_s employed in the MM11 and PSP models. The drain current (for constant mobility) in HiSIM is given by

$$I_D = I_{drift} + I_{diff}, \tag{11.1.9}$$

$$I_{drift} = \mu_n \frac{W}{L} C'_{ox} \left[V'_G(\phi_{sL} - \phi_{s0}) - \frac{1}{2}(\phi^2_{sL} - \phi^2_{s0}) \right.$$
$$\left. - \frac{2}{3}\gamma[(\phi_{sL} - \phi_t)^{3/2} - (\phi_{s0} - \phi_t)^{3/2}] \right], \tag{11.1.10}$$

$$I_{diff} = \mu_n \frac{W}{L} C'_{ox} \phi_t \left\{ (\phi_{sL} - \phi_{s0}) + \gamma[(\phi_{sL} - \phi_t)^{1/2} - (\phi_{s0} - \phi_t)^{1/2}] \right\}. \tag{11.1.11}$$

Without the linear approximation of Q'_I as a function of ϕ_s, the model equations, particularly those for the intrinsic charges, are very complicated [11].

In HiSIM, short-channel effects are accounted for by employing the bias- and geometry-dependent lateral-gradient factor originally introduced in [12]. In this approach [3], the gate-to-source voltage V_{gs} is shifted by a value that is dependent on the gradient of the lateral electric field F_{yy}:

$$V'_G = V_{gs} + \Delta V'_G - V_{FB}, \tag{11.1.12}$$

where

$$\Delta V'_G = \frac{\varepsilon_S}{C'_{ox}} \sqrt{\frac{2\varepsilon_S}{qN_{sub}}[\phi_s(y) - V_{bs} - \phi_t]F_{yy}},$$

with

$$F_{yy} = \frac{dF_y}{dy}.$$

F_{yy}, the gradient of the lateral electric field, is assumed to be independent of position owing to a parabolic approximation of the electrostatic potential along the channel. F_{yy} is extracted from the measured threshold voltage V_{th} versus bias characteristics.

11.1.2.2   The MOS Model 11

MOS Model 11 (MM11) [4] is the successor to MOS Model 9 (MM9) [13]. Its development was aimed at fulfilling the requirements for a model that satisfies the accuracy demands of analog and RF circuit design, but with a computational complexity that allows its application in digital design.

To obtain efficient expressions for outputs such as current, charges, and noise, several approximations, mainly based on a linearization of the inversion charge as a function of ϕ_s, were developed. In MM11, this linearization is performed around the mean value of the source and drain surface potentials [4], which results in simpler expressions without loss of accuracy.

The expression for the drain current in MM11 is

$$I_{DS} = -\frac{W}{L}\mu_n(\bar{Q}'_I - n_c C'_{ox}\phi_t)(\phi_{sL} - \phi_{s0}). \tag{11.1.13}$$

The "average" charge density $\bar{Q}'_I$ and the effective capacitance $n_c C'_{ox}$ are both calculated using the mean value of the surface potential $(\phi_{sL} + \phi_{s0})/2$. Note that the first component of the drain current, that associated with the average charge density, corresponds to drift, whereas the second one, that associated with the charge density at pinch-off, corresponds to diffusion.

The MM11 approach ensures that the model symmetry with respect to source–drain interchange is maintained. Note that this approach is similar to the symmetric linearization used in SP [5], a surface-potential-based model, which made it easy to merge MM11 and SP into one model called PSP.

In MM11, a strong emphasis has been placed on distortion modeling. For an accurate description of distortion, the model should accurately describe the drain current and its higher-order derivatives (up to at least third order). MM11 was specially developed for this purpose. MM11 contains improved expressions for mobility reduction, velocity saturation, and various conductance effects [4]. The distortion modeling of MM11 has been tested extensively on several MOSFET technologies, and it gives an accurate description of modern CMOS technologies [4]. MM11 includes an accurate description of several important physical phenomena, such as poly-depletion, quantum-mechanical effects, gate tunneling current, and gate-induced drain leakage, besides taking into account the effects of pocket implants.

The expressions of MM11 for the source- and drain-associated charges, written below, use the inversion charge density as the key variable. As indicated in [14], a change of notation allows one to write the equations of MM11 for the source- and drain-associated charges exactly as they are written in Chapter 2 for a "charge-based model." This is to be expected since modern inversion-charge- and surface-potential-based models both make use of the same fundamental approximations:

$$Q_D = WL\left[\frac{6Q'^3_R + 12Q'_F Q'^2_R + 8Q'^2_F Q'_R + 4Q'^3_F}{15(Q'_F + Q'_R)^2} + \frac{n}{2}C'_{ox}\phi_t\right], \tag{11.1.14}$$

$$Q_S = WL\left[\frac{6Q'^3_F + 12Q'_R Q'^2_F + 8Q'^2_R Q'_F + 4Q'^3_R}{15(Q'_F + Q'_R)^2} + \frac{n}{2}C'_{ox}\phi_t\right]. \tag{11.1.15}$$

11.1.2.3   The PSP model

The PSP [15], [16] model is a surface-potential-based model obtained by merging and developing the best features of two ϕ_s-based models, namely SP, developed at Pennsylvania State University, and MM11, developed at Philips [15]. In 2005 the Compact Modeling Council elected PSP as the industrial standard compact model for MOSFETs. At the time of writing, PSP is being jointly developed by NXP Semiconductors (formerly part of Philips) and Arizona State University. The PSP model, intended for digital, analog, and RF design, contains all relevant physical effects (mobility reduction, velocity saturation, DIBL, gate current, etc.) needed to model present-day and upcoming deep-submicron bulk CMOS technologies. According to its developers, PSP gives an accurate description of currents, charges, and their first-order and higher-order derivatives, resulting in an accurate description of electrical distortion behavior [16]. PSP also describes accurately the noise behavior of MOSFETs and has an option for non-quasi-static (NQS) effects. Details on the structure of the model, basic equations for the intrinsic model, extrinsic components, and extraction procedures can be found in [15], [16].

In PSP, the surface potential is calculated using (11.1.6) or a slightly modified form of it. PSP uses a symmetric linearization method (SLM) in which the bulk and, consequently, the inversion charge densities are linear functions of the surface potential. The linearization is performed around the midpoint of the surface potential. The drain current is calculated using the inversion charge at the midpoint of the surface potential and the "average" gradient of the surface potential along the channel. As expected, the mobility is modified both by the effects of velocity saturation and by the transverse electric field.

The SLM also yields an explicit form of the dependence of the surface potential on the position along the channel, which, in turn, allows one to compute the source and drain terminal charges for the quasi-static model using the Ward–Dutton charge partitioning. The PSP model includes quantum-mechanical corrections and the effects of the poly-silicon depletion region. As regards the extrinsic model, the gate/source and gate/drain overlap regions and the gate and bulk currents are also described. PSP also includes the noise models which are indispensable for analog and RF designs.

11.1.3   Charge-based models (EKV, ACM, and BSIM5)

In inversion charge-based models, the equations for current, charges, and noise are expressed in terms of the inversion charge densities at the two ends of the channel. In their pioneering work [6], Maher and Mead showed that the drain current I_D can be expressed as a very simple function of the area densities of the inversion charges at source and drain (Q'_{IS} and Q'_{ID}). For a long-channel transistor, the expression for the drain current is

$$I_D = \frac{\mu W}{L}\left[\frac{Q'^2_{IS} - Q'^2_{ID}}{2nC'_{ox}} - \phi_t\left(Q'_{IS} - Q'_{ID}\right)\right],\qquad (11.1.16)$$

as shown previously in Chapter 2. Cunha *et al.* [8] derived expressions, also shown in Chapter 2, for the total charges and small-signal parameters as functions of the source and drain channel charge densities. Shur and collaborators proposed a single equation for the charge densities in terms of the terminal voltages [7], called the Unified Charge-Control Model (UCCM). An improved version of the UCCM presented in [7], which has been used in this textbook, is

$$V_P - V_C = \phi_t \left[\frac{Q'_{IP} - Q'_I}{nC'_{ox}\phi_t} + \ln\left(\frac{Q'_I}{Q'_{IP}}\right) \right]. \qquad (11.1.17)$$

Q'_I-based models rely on both the gradual-channel approximation and the linearization of the bulk and inversion charges with respect to the surface potential at a fixed gate bias.

11.1.3.1   The EKV model

The EKV model [17], [18] was initially aimed at the design of very-low-power analog integrated circuits, with the objective of obtaining a simple analytical model for all regions of operation. To this end, the EKV authors devised an interpolation function that should obey the asymptotic cases of weak and strong inversion. They also defined the forward and reverse inversion levels on the basis of decomposition of the drain current into forward and reverse components. Using this strategy, the EKV model was the first to introduce single-piece analytical expressions for the current, transconductances, intrinsic capacitances, non-quasi-static transadmittances, and noise that are valid in weak, moderate, and strong inversion and from the linear to the saturation region. All the small-signal parameters are written in terms of the forward and reverse inversion levels; this is appropriate for analog design since most analog circuits are current-biased.

In contrast to most MOSFET models, at the time of its introduction the EKV model exploited the inherent symmetry of the MOSFET by referring all the terminal voltages to the substrate. The EKV model evolved into a charge-based formulation when the interpolation function for the continuous g_m/I_D characteristic was substituted by a physics-based expression covering weak to strong inversion.

It should be noted here that the specific current I_{spec} in the EKV model is

$$I_{spec} = 2\mu_n C'_{ox} n\phi_t^2 \frac{W}{L}, \qquad (11.1.18)$$

which is four times larger than the specific current I_S in the ACM model.

The main equations of the ACM and EKV models are similar, but there are some differences concerning the pinch-off potential and the slope-factor definitions that have some impact on the precision of the inversion-charge modeling. Details of these differences, which are beyond the scope of this book, can be found elsewhere [14].

The EKV charge-based model has now evolved into a full-featured compact model that includes all the major effects that have to be accounted for in deep-submicron CMOS technologies (for details, see http://legwww.epfl.ch/ekv).

11.1.3.2   The ACM model

The initial motivation for the ACM modeling approach [8], [14], [19] came from an analog design in digital CMOS technology carried out in the late 1980s. The use of the MOS gate as a linear capacitor required the calculation of the weak non-linearities of the

MOS capacitor in accumulation and strong inversion. The classical strong-inversion approximation was clearly not appropriate, and improved capacitive models of the MOS gate valid for moderate inversion and accumulation were therefore developed [20].

The use of the new gate capacitor model to achieve a four-terminal MOS model accurate in both weak and moderate inversion was a natural step forward. Also, the necessity for a symmetric MOSFET model to describe the series association of transistors became clear at that time [21]. An appropriate MOSFET model was finally achieved in [8]. The symmetry of the transistor with respect to source and drain was obeyed. Rigorous definitions of pinch-off and threshold voltages, which are essential for consistent and accurate models, were given in [14]. The dc, ac, and non-quasi-static models were developed in [19]. All transistor parameters were given as very simple (rational) functions of the inversion charge densities at the channel boundaries. A computer-implemented version of the ACM model has been included in the SMASH circuit simulator since 1997 [22]. The ACM model has a hierarchical structure that facilitates the inclusion of various phenomena into the model. It was the first model to furnish simple explicit expressions for all the intrinsic capacitive coefficients. More recently, unified $1/f$-noise and mismatch models have been presented in [23] and [24], respectively.

The extraction of the main dc parameters of the ACM model used for circuit design will be presented in Section 11.2.

11.1.3.3   The BSIM5 model

The BSIM5 model was introduced in [25]. At the time of its introduction, the BSIM authors were striving to develop a new-generation MOSFET model fully based on physics that would be flexible enough for the inclusion of small-dimension effects. BSIM5 uses the standard charge-based expression for the current, expression (11.1.16), exactly as ACM does [25], [26]. The inversion charge density is also calculated through the UCCM given by (11.1.17), using a linearized expression for the pinch-off voltage in terms of the gate voltage. Poly-depletion and quantum effects are handled by using correction terms for the slope factor and the effective bulk potential. Velocity saturation, velocity overshoot, and source-velocity limits are modeled in a unified way.

11.2   Parameter extraction for first-order design

The correct determination of MOSFET parameters is fundamental for the meaningful analysis, design, and simulation of MOS circuits. The accuracy of the transistor characteristics depends not only on a good device model but also on the values of fundamental parameters. This section describes some procedures that can be used to extract fundamental dc parameters from current–voltage characteristics of MOSFETs. We will not discuss the extraction of capacitances since these parameters, be they intrinsic (such as the gate oxide capacitance) or extrinsic (such as overlap and junction capacitances), are common to different models and, therefore, can be easily accessed from the process parameters provided by the foundries. The flicker-noise parameter N_{ot} that we use in the model shown in Chapter 4 is related to the conventional parameter K_F [14] through the following relationship:

$$K_F = N_{ot} q^2 / C'_{ox}. \tag{11.2.1}$$

Finally, the MOSFET mismatch parameters that we use for circuit design, namely N_{oi} and A_{ISH}, can easily be found from the conventional mismatch parameters A_{VT} and A_β through the relations

$$A_{VT}^2 = 6(q^2/C_{ox}'^2)N_{oi}, \qquad A_\beta^2 = 2A_{ISH}^2 \tag{11.2.2}$$

previously presented in Chapter 4.

In view of the previous considerations, we will limit the discussion here to the extraction of dc parameters that have been used in this textbook for a first-order design, namely, the specific current, the threshold voltage, the carrier mobility, the slope factor, and the Early voltage. We will not compare the extraction procedures to be presented next with other methods available in the technical literature. Comparison of extraction methods can be found elsewhere (see [14] and the references therein). The purpose of this section is to show how the designer can extract the set of dc parameters required for first-order design of MOS circuits, either from experimental or from simulation results. We will present the main dc parameters extracted for 0.35- and 0.13-μm CMOS technologies, for both long- and short-channel n-MOS devices and for long-channel p-MOS transistors. We will mainly use an extraction procedure based on the variation of the drain current with the gate voltage. The extraction procedure should be applied cautiously to more advanced technologies due to the significant impact of physical effects, such as quantum confinement of carriers and velocity saturation, on the performance of deeply scaled devices.

11.2.1 Specific current and threshold voltage

The values of the threshold voltage and specific current are determined from the transconductance-to-current-ratio, or simply g_m/I_D, method [14], to be described next. This method has the advantage of being based on a physical property of the device, and thus being independent of a specific model. The procedure proposed for the extraction of V_T and I_S is performed over the weak- and moderate-inversion regions with small drain-to-source voltages so that short-channel effects are avoided. Series resistances of source and drain may be disregarded owing to the negligible voltage drops across them.

Neglecting the (small) dependence of I_S on the gate voltage, the transconductance-to-current ratio can be written as

$$\frac{g_m}{I_D} = \frac{1}{i_f - i_r}\left(\frac{\partial i_f}{\partial V_G} - \frac{\partial i_r}{\partial V_G}\right) = \frac{2}{n\phi_t\left(\sqrt{1+i_f} + \sqrt{1+i_r}\right)}. \tag{11.2.3}$$

For V_{DS} much lower than the thermal voltage ϕ_t, $i_f \cong i_r$ and (11.2.3) becomes

$$\frac{g_m}{I_D} \cong \frac{1}{n\phi_t\sqrt{1+i_f}}. \tag{11.2.4}$$

Now, if one assumes n to be almost constant over the measurement range, (11.2.4) can be written as

$$\frac{g_m}{I_D} \simeq \frac{g_m}{I_D}\bigg|_{max} \frac{1}{\sqrt{1 + i_f}}. \tag{11.2.5}$$

In expression (11.2.5), the maximum transconductance-to-current ratio is that deep in weak inversion ($i_f \to 0$). For $i_f = 3$, the transconductance-to-current ratio is half of its peak value. Equation (11.2.5) shows that the deviation of the transconductance-to-current ratio relative to the maximum value in weak inversion is dependent only on the inversion level.

Equation (11.2.5) is the basis of a very simple and quick method for determining both the threshold voltage and the specific current I_S using a single current–voltage characteristic, as will be shown next.

Using the UICM for both the source and the drain transistor terminals, and subtracting one of the resulting expressions from the other, we obtain

$$\frac{V_{DS}}{\phi_t} = \sqrt{1 + i_f} - \sqrt{1 + i_r} + \ln\left(\frac{\sqrt{1 + i_f} - 1}{\sqrt{1 + i_r} - 1}\right). \tag{11.2.6}$$

For low values of V_{DS}, $i_r \to i_f$ and expression (11.2.6) becomes

$$\frac{V_{DS}}{\phi_t} \simeq \frac{1}{2} \frac{i_f - i_r}{\sqrt{1 + i_f} - 1} = \frac{1}{2} \frac{I_D/I_S}{\sqrt{1 + i_f} - 1}. \tag{11.2.7}$$

The measurement of both V_T and I_S is taken from the g_m/I_D versus V_G characteristic according to expressions (11.2.5) and (11.2.7). The gate voltage at which the value of g_m/I_D drops to half of its peak value corresponds to $i_f = 3$. Recalling that for $i_f = 3$ and $V_S = 0$, the pinch-off voltage $V_P = 0$ or, equivalently, $V_G = V_{T0}$, the equilibrium threshold voltage V_{T0} is the value of V_G at which $g_m/I_D \cong 0.5(g_m/I_D)_{max}$. Now, using (11.2.7), we find that the value of the specific current is, for $i_f = 3$, calculated as

$$I_S \simeq \frac{\phi_t}{2V_{DS}} \frac{I_D}{\sqrt{1 + i_f} - 1}\bigg|_{i_f=3} = \frac{\phi_t}{2V_{DS}} I_D. \tag{11.2.8}$$

For $V_{DS}/\phi_t = 1/2$, we have $I_S \cong I_D$.[1]

To illustrate how to determine the threshold voltage and the specific current, we measure the drain current of a MOSFET as a function of the gate voltage for $V_{DS} = 13\,\text{mV}$, as shown in the inset of Figure 11.2, and calculate

$$\frac{g_m}{I_D} = \frac{dI_D}{I_D\,dV_G} = \frac{d\ln I_D}{dV_G}. \tag{11.2.9}$$

[1] For $V_{DS}/\phi_t = 1/2$ and $i_f = 3$, more accurate values for the transconductance-to-current ratio and the specific current I_S are 0.53 times the peak value of g_m/I_D and 1.13 times the measured current, respectively [27].

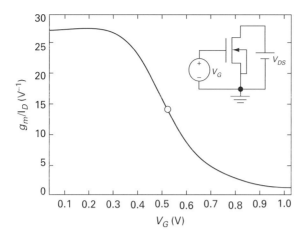

Fig. 11.2   The transconductance-to-current ratio of a MOSFET against gate voltage for $V_{DS} \cong \phi_t/2$ and $V_S = 0$. The inset shows the circuit employed to measure the dependence of the current on the gate voltage. The small circle indicates the point at which $g_m/I_D \cong 0.5(g_m/I_D)_{\max}$.

The g_m/I_D curve thus generated was used to find both V_{T0} and I_S. The circle in Figure 11.2 corresponds to the point at which g_m/I_D drops to around half (53%) of its peak value. The gate voltage thus measured is the equilibrium threshold voltage V_{T0} and the corresponding current is approximately the specific current I_S (in fact, the specific current is 1.13 times the drain current measured for $V_G = V_{T0}$).

Using the g_m/I_D ratio for a device composed of 32 n-channel transistors ($W = 8\,\mu\text{m}$, $L = 3.2\,\mu\text{m}$) in a 0.35-μm technology connected in parallel, for $V_{DS} = 13\,\text{mV}$ we find from experiment $V_{T0} = 551\,\text{mV}$ and $I_S = 6.84\,\mu\text{A}$, whereas from simulation using BSIM3v3 (level $= 53$) $V_{T0} = 552\,\text{mV}$ and $I_S = 7.95\,\mu\text{A}$. Of course, there is some difference between experimental and simulation results, which can be ascribed to deviations of the nominal parameters of the technology and/or some inadequacy of the model employed for the simulation.

11.2.2   The slope factor

After having determined the threshold voltage and the specific current (for $V_G = V_{T0}$), we show now how to determine the body-effect factor and the bulk Fermi potential from the slope factor n. A very simple methodology to determine the slope factor is described in [28] and presented next. Recalling that the UICM is given by

$$(V_P - V_S)/\phi_t = \sqrt{1 + i_f} - 2 + \ln\left(\sqrt{1 + i_f} - 1\right), \qquad (11.2.10)$$

then, for constant i_f, the difference between V_P and V_S is constant. In particular, for $i_f = 3$, the pinch-off voltage is equal to the source voltage. A simple and useful circuit for determining the dependence of V_P on V_G is shown in the inset of Figure 11.3 [28].

The MOSFET is biased with a current source equal to $3I_S$, with I_S being determined by the g_m/I_D characteristic. Since the transistor operates in saturation, the reverse current is

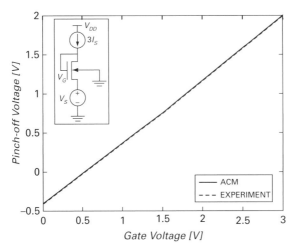

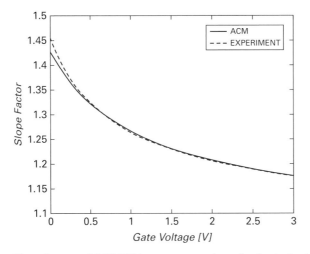

Fig. 11.3   Pinch-off voltage versus gate voltage of a device composed of 32 n-channel transistors ($W = 8\ \mu m$, $L = 3.2\ \mu m$) in parallel in 0.35-μm technology. The inset shows the circuit employed for the measurements. The solid line represents the ACM model (see details in the text) and the dashed line represents the experiment.

Fig. 11.4   Slope factor $n = 1/(dV_P/dV_G)$ versus gate voltage for the device in Figure 11.3.

much lower than the forward current; consequently, we have $i_f = I_D/I_S = 3$ and, as (11.2.10) shows, the source voltage equals the pinch-off voltage.

The graph in Figure 11.3 shows the source voltage, i.e., the pinch-off voltage, versus the gate voltage measured through the circuit shown in the insert. The solid line represents the ACM curve, which was traced after extraction of parameters γ and ϕ_F from measurements of the slope factor n, which is shown in Figure 11.4 as a function of the gate voltage.

The values of the experimental slope factor were then used to draw the graph $1/(n-1)^2$ versus V_P, as illustrated in Figure 11.5. From the formula $n = 1 + \gamma/(2\sqrt{2\phi_F + V_P})$ for the slope factor we find that

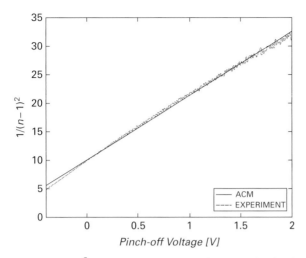

Fig. 11.5   Plot of $1/(n-1)^2$ versus pinch-off voltage for the device in Figure 11.3.

$$\frac{1}{(n-1)^2} = \frac{4V_P}{\gamma^2} + \frac{8\phi_F}{\gamma^2}. \tag{11.2.11}$$

Using the experimental data, the slope and the y-intercept of the interpolation line give $\gamma = 0.60 \text{ V}^{1/2}$ and $2\phi_F = 0.89 \text{ V}$, which are very close to the values provided by the foundry.

11.2.3   Mobility

To complete the extraction procedure for the fundamental dc parameters of the long-channel MOSFET, we now show how to extract the parameter associated with the dependence of the mobility on the transversal field. Once again, the transistor operates in the triode region. For low values of V_{DS}, the bulk and inversion charge densities along the channel are approximately constant. Thus, the dependence of the mobility on the transverse electric field is written as in (2.4.1), with equal charge densities at the source and drain ends of the channel, yielding

$$\mu = \frac{\mu_0}{1 - \alpha_\theta \left(\dfrac{Q'_B + \eta Q'_I}{\varepsilon_s} \right)}, \tag{11.2.12}$$

where $\eta \cong 1/2$ for electrons and $\eta \cong 1/3$ for holes. For the purpose of calculating the parameter α_θ, we determine the mobility variation for cases in which the depletion charge is much higher than the inversion charge density. In these cases, the depletion charge density is

$$Q'_B = Q'_{Ba} - Q'_I(n - 1)/n \cong Q'_{Ba} = -\gamma C'_{ox} \sqrt{2\phi_F + V_P}. \tag{11.2.13}$$

Substitution of (11.2.13) into (11.2.12) leads to

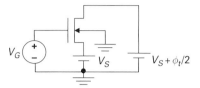

Fig. 11.6   Circuit for generating curves of transconductance-to-current ratio similar to that shown in Figure 11.2, for several values of the specific current. Similarly to the curve in Figure 11.2, the specific current is measured for $i_f = 3$, the inversion level at which the g_m/I_D ratio is 53% of its peak value.

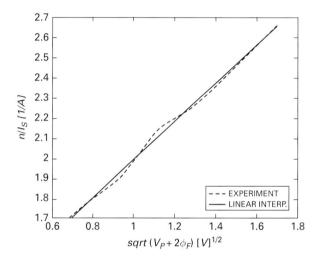

Fig. 11.7   A plot of n/I_S versus the square root of $(V_P + 2\phi_F)$ for extracting the mobility parameters.

$$\mu = \frac{\mu_0}{1 + \theta\sqrt{V_P + 2\phi_F}} \tag{11.2.14}$$

with $\theta = \alpha_\theta C'_{ox}\gamma/\varepsilon_s$ in $V^{-1/2}$.

The value of θ is extracted from measurements of the specific current as a function of the pinch-off voltage. Using the expression for the specific current and (11.2.14), we can write

$$\frac{n}{I_S} = \frac{1}{\mu C'_{ox}\phi_t^2[W/(2L)]} = \frac{1 + \theta\sqrt{V_P + 2\phi_F}}{\mu_0 C'_{ox}\phi_t^2[W/(2L)]}. \tag{11.2.15}$$

Expression (11.2.15) shows that the plot of n/I_S versus the square root of $(V_P + 2\phi_F)$ allows the determination of the values of both θ and the denominator of (11.2.15). The specific current for various values of gate voltage and, consequently, of V_P, was measured for source voltages of $-0.4, 0, 0.4, 0.8, 1.2$, and 2 V with the circuit shown in Figure 11.6.

The procedure for the determination of I_S is the same as that described for Figure 11.2. The experimental and fitted (ACM) values obtained for n/I_S are plotted in Figure 11.7 in terms of the square root of $(V_P + 2\phi_F)$. The values extracted for $\mu_0 C'_{ox}\phi_t^2[W/(2L)]$ and θ are 8.8 μA and 0.75 $V^{-1/2}$, respectively. Note that the value of the "zero-field mobility" μ_0

Table 11.1 Parameters extracted for a parallel association of 32 transistors ($W = 8\ \mu m$, $L = 3.2\ \mu m$) in a 0.35-μm CMOS technology

Parameter	V_{T0}	$2\phi_F$	γ	$\mu_0 C'_{ox}\phi_t^2[W/(2L)]$	θ
Value	0.552 V	0.89 V	0.60 $V^{1/2}$	8.8 μA	0.75 $V^{-1/2}$

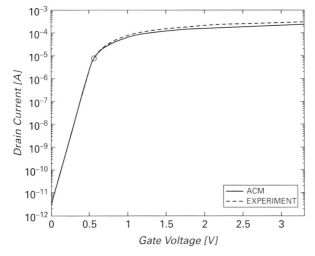

Fig. 11.8 I_D versus V_G plots of both experiment and the ACM model for a long-channel ($L = 3.2\ \mu m$) n-MOS transistor in a 0.35-μm CMOS technology, with $V_S = 0$ and $V_{DS} = 13$ mV. The maximum error for currents higher than 10^{-11} A is around 30% for $V_G = 3.3$ V.

can be extracted from $\mu_0 C'_{ox}\phi_t^2[W/(2L)]$ when the oxide capacitance and transistor dimensions are known.

Table 11.1 gives the extracted parameters of a long-channel n-MOS transistor in a 0.35-μm technology. The parameters were extracted using the methodology in this section.

11.3   Comparison between experiment and the ACM model in a 0.35-μm technology

For the sake of comparison, Figure 11.8 shows the experimental and ACM model plots of the drain current versus gate voltage for the previously described NMOS transistor. For the plot of the ACM curve we used the parameters in Table 11.1 and the following set of equations:

$$V_P = \left[\sqrt{V_G - V_{T0} + \left(\sqrt{2\phi_F} + \gamma/2\right)^2} - \gamma/2 \right]^2 - 2\phi_F, \tag{11.3.1}$$

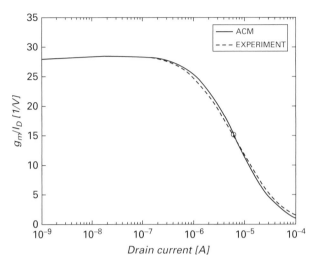

Fig. 11.9   Plots of experimental and modeled transconductance-to-current ratios versus drain current for the characteristics shown in Figure 11.8.

$$\frac{V_P - V_{S(D)}}{\phi_t} = \sqrt{1 + i_{f(r)}} - 2 + \ln\left(\sqrt{1 + i_f} - 1\right), \qquad (11.3.2)$$

$$I_D = I_S\left(i_f - i_r\right), \qquad (11.3.3)$$

$$I_S = \mu_0 C'_{ox} \phi_t^2 [W/(2L)] n(\mu/\mu_0). \qquad (11.3.4)$$

For the normalized mobility μ/μ_0 given by (11.2.12), the depletion charge is calculated as

$$Q'_B = Q'_{Ba} - Q'_I(n - 1)/n = -\gamma C'_{ox}\sqrt{2\phi_F + V_P} - Q'_I(n - 1)/n \qquad (11.3.5)$$

with Q'_I given by the UCCM.

Using (11.3.5) in (11.2.12), we find that

$$\frac{\mu_0}{\mu} = 1 + \theta\left[\sqrt{V_P + 2\phi_F} - \frac{Q'_I}{\gamma C'_{ox}}\left(\eta - \frac{n - 1}{n}\right)\right]. \qquad (11.3.6)$$

As can be noted in Figure 11.8, the ACM model fits very well the experimental results. For over seven decades of current, from 10^{-11} to 3×10^{-4} A, the percentage error for the current is lower than 30%.

In Figure 11.9 we compare the transconductance-to-current ratio of the experimental results and those obtained from the ACM model. Once again, the fitting of the experimental data to the ACM model is extremely good, except for currents approaching 10^{-4} A, for which the relative error in the transconductance-to-current ratio is moderately large.

Using the procedure described previously, we have also extracted the dc parameters of a short-channel n-MOS transistor in a 0.35-μm technology. The graphs of the measured and modeled drain current against gate voltage are shown in Figure 11.10. In this case,

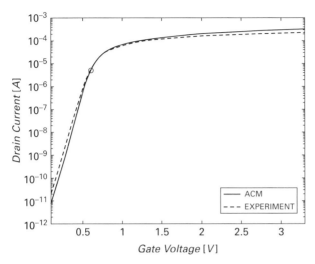

Fig. 11.10   A plot of the experimental and modeled currents versus gate voltage for a minimum-length n-MOS transistor in a 0.35-μm technology.

the model deviates from the experiment by a maximum factor of 2.5 for currents deep in weak inversion, which are around four or more decades below the threshold point indicated by the circle.

Example 11.3
Assume the following set of parameters of a p-channel transistor:

Parameter	V_{T0}	$2\phi_F$	γ	$\mu_0 C'_{ox}\phi_t^2[W/(2L)]$	θ
Value	-0.7 V	-0.81 V	0.40 V$^{1/2}$	$10\ \mu$A	0.5 V$^{-1/2}$

Estimate the variation in V_P, n, μ, and I_S for $0 < V_G < 3.3$ V, $V_B = 3.3$ V, and $V_D = V_S = 0$.

Answer
For the p-channel transistor we write the set of equations

$$-V_P \cong \left[\sqrt{-(V_{GB} - V_{T0}) + \left(\sqrt{-2\phi_F} + \gamma/2\right)^2} - \gamma/2\right]^2 + 2\phi_F,$$

$$n = 1 + \frac{\gamma}{2\sqrt{-(V_P + 2\phi_F)}},$$

$$\frac{\mu_0}{\mu} = 1 + \theta\left[\sqrt{-(V_P + 2\phi_F)} + \frac{Q'_I}{\gamma C'_{ox}}\left(\eta - \frac{n-1}{n}\right)\right]$$

from the counterpart equations that we have written for n-channel transistors. Using the formulas above, we find that the values of V_P, n, μ, and I_S lie within the ranges given in following the table:

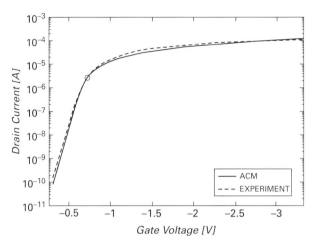

Fig. 11.11   Experimental and modeled currents versus gate voltage for a long-channel ($L = 3.2$ μm) p-MOS transistor in a 0.35-μm technology.

V_G (V)	V_{GB} (V)	V_P	n	μ/μ_0	I_S (μA)
0	−3.3	−2.26	1.11	0.38	4.2
3.3	0	0.55	1.76	0.80	14.1

For the calculation of the mobility, the inversion charge density is taken as zero for $V_{GB} = 0$ and as $Q'_I = -C'_{ox}(V_{GB} - V_{T0})$ for $V_{GB} = -3.3$ V. The value of η is assumed to be 1/3.

The I_D versus V_G experimental and model curves of a long-channel PMOS transistor for $V_{SD} = 13$ mV are shown in Figure 11.11. The fitting of the experimental curve to the model is very good, although not as good as that for the case of the long-channel n-MOS transistor. The maximum difference between the model and the experiment for a current span of seven decades is below 100%. The experimental and modeled transconductance-to-current ratios, which are shown in Figure 11.12, are in close agreement, with errors below 30% over almost the entire range, except for very high currents.

11.4   Comparison between simulation and the ACM model in a 0.13-μm technology

We will propose here a simple modification to the classical UCCM model to accommodate important changes that we have observed in the simulation results of MOSFETs in a 0.13-μm technology. We will make this modification without entering into details as to why the characteristics of deeply scaled devices can deviate significantly from the classical model. As explained in Chapter 2, the UCCM is based on the charge-sheet approximation (CSA). In effect, using CSA and Boltzmann statistics, the inversion charge

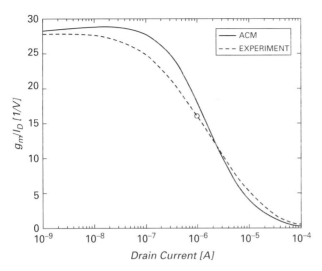

Fig. 11.12   Experimental and modeled transconductance-to-current ratios versus drain current for a long-channel p-MOS transistor ($L = 3.2\ \mu$m) in a 0.35-μm CMOS technology.

density located at the semiconductor surface is proportional to $\exp[(\phi_s - V_C)/\phi_t]$. The inversion capacitance reduces to

$$C_i' = -\frac{dQ_I'}{d\phi_s} = -\frac{Q_I'}{\phi_t}. \tag{11.4.1}$$

It is worth observing that (11.4.1) is accurate in the subthreshold region only. In effect, the resolution of the Poisson equation in the semiconductor shows that the logarithmic slope of the $Q_I'(\phi_s)$ curve varies from $1/\phi_t$ in weak inversion to $1/(2\phi_t)$ deep in strong inversion [14]. For thin oxides, the strong electric field at the interface makes quantum confinement of the inversion channel relevant, reducing the inversion capacitance [14]. To include quantum effects, as well as other effects that contribute to reducing the gate capacitance, we can simply modify the classical UCCM, approximating the inversion-charge capacitance as

$$C_i' = -\frac{Q_I'}{n_{ch}\phi_t}, \tag{11.4.2}$$

where $n_{ch} \geq 1$. The classical UCCM was derived in Chapter 2 assuming that $n_{ch} = 1$. Using once again the capacitive model in Figure 11.13, we obtain, after some algebra, the differential UCCM model

$$\frac{dQ_I'}{nC_{ox}'}\left(1 - \frac{nC_{ox}' n_{ch}\phi_t}{Q_I'}\right) = dV_C. \tag{11.4.3}$$

We can readily note that (11.4.3) uses the same formalism as the classical UCCM, but its interpretation is in order. Expression (11.4.3) can be viewed as the classical UCCM equation, but with the device operating at temperature $n_{ch}T$, rather than T. Thus, when the

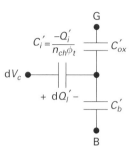

Fig. 11.13 A capacitive model of the three-terminal MOS structure with the inversion capacitance approximated as $C_i' = -Q_I'/(n_{ch}\phi_t)$.

MOSFET inversion capacitance in weak inversion follows (11.4.2), one can still use the classical UCCM formalism but with the following change

$$T \to n_{ch}T. \tag{11.4.4}$$

It is interesting to point out that the model parameters such as Q_{IP}' and I_{SH} that we have been using in this textbook should be modified accordingly, i.e.

$$Q_{IP}' = -nC_{ox}'\phi_t \to -nC_{ox}'n_{ch}\phi_t; \qquad I_{SH} = \mu C_{ox}'n\phi_t^2/2 \to \mu C_{ox}'n(n_{ch}\phi_t)^2/2. \tag{11.4.5}$$

After having discussed the non-classical UCCM, we will show how to extract the dc parameters of the ACM model and then draw a comparison between simulation results, which were obtained using the BSIM3v3 model, and the ACM model. Recall that the ACM model uses the UCCM for the relationship between inversion charges and terminal voltages.

Let us start with the determination of n_{ch}, which is obtained from the dependence of the drain current on the source voltage in weak inversion, according to the scheme shown in the inset of Figure 11.14. The value of n_{ch} is determined from the slope of the curve as follows:

$$\frac{1}{n_{ch}} = -\phi_t \frac{dI_D}{I_D \, dV_S} = -\phi_t \frac{d\ln I_D}{dV_S} = -2.3\phi_t \frac{d\log I_D}{dV_S}. \tag{11.4.6}$$

For a -70-mV variation in the source voltage per decade of current in the graph of Figure 11.14, we have $n_{ch} \cong 1.18$.

It should be mentioned here that the value of n_{ch} is slightly dependent on V_G, but we have considered it constant for the sake of simplicity. Using the procedure described previously, we find the dc parameters listed in Table 11.2. Two problems we have found with the simulations using the BSIM3v3 model are the non-physical non-monotonicity and the discontinuity in the derivative of the slope factor around the threshold voltage. Such problems most likely arise from the interpolation and/or smoothing functions employed in BSIM3 to model the MOS transistor. To extract the parameters γ and $2\phi_F$ from simulation results, we have used the values of n for gate voltages greater than the

Table 11.2 Parameters extracted for a long-channel n-MOS transistor ($W = 6\,\mu m$, $L = 2.4\,\mu m$) in a 0.13-μm CMOS technology

Parameter	V_{T0}	$2\phi_F$	γ	$\mu_0 C'_{ox}\phi_t^2[W/(2L)]$	θ	n_{ch}
Value	231 mV	0.79 V	0.22 V$^{1/2}$	0.93 μA	0.43 V$^{-1/2}$	1.18

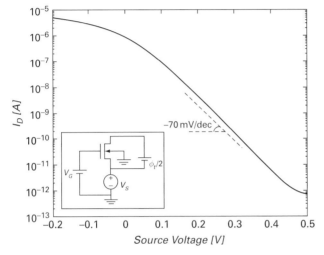

Fig. 11.14 Drain current versus source voltage for a long-channel transistor in a 0.13-μm technology. The value of the gate voltage is $V_G = V_{T0} = 0.231$ V, and $V_{DS} = \phi_t/2 = 13$ mV. The slope of the characteristic gives $n_{ch} \cong 1.18$.

threshold voltage, since for these values of V_G the slope factor extracted using BSIM3 decreases monotonically. Of course, in view of the non-physical slope factor of the BSIM3 model, we cannot expect the ACM model to fit the simulated transistor very well in all operating regions.

The simulated and ACM-modeled curves of the drain current versus gate voltage for $V_{DS} = 13$ mV are shown in Figure 11.15 for a long-channel MOSFET in a 0.13-μm technology. For a variation of six decades in the drain current, the absolute value of the percentage error of the ACM model with respect to BSIM is lower than 50%, as Figure 11.15 shows.

The simulated and modeled transconductance-to-current ratios corresponding to the drain-current-versus-gate-voltage characteristics of Figure 11.15 are shown in Figure 11.16. Once again, the relative error in the transconductance-to-current ratio is very small, except for very high current levels.

11.5 The Early voltage

The Early voltage V_A, which is associated with the output conductance of a transistor through the relation $g_{ds} = I_D/V_A$, is a fundamental design parameter since it affects, for

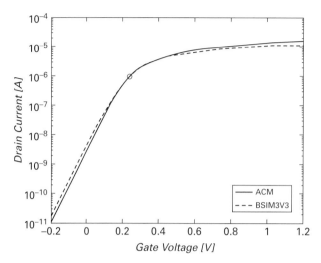

Fig. 11.15   Experimental and modeled currents versus gate voltage of a long-channel ($L = 2.4\,\mu$m) n-MOS transistor in a 0.13-μm technology.

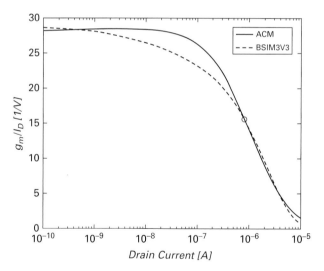

Fig. 11.16   Simulated and modeled transconductance-to-current ratios versus drain current for a long-channel NMOS transistor ($L = 2.4\,\mu$m) in a 0.13-μm CMOS technology.

example, the accuracy of current mirrors and the gain of voltage amplifiers. The output conductance in saturation is difficult to determine due to the need to model the short-channel effects. For hand analysis, the Early voltage is assumed to be proportional to the channel length and independent of both the drain current and the drain voltage. Although in some cases this simplified model of the Early voltage can be rather poor, in particular for very-short-channel and/or pocket-implanted devices, in many cases the designer can use this simplified model to quickly calculate first-order values for specifications such as the amplifier voltage gain and the current-mirror output impedance. The interested reader is referred to [29] and the references therein for a more detailed discussion on the phenomena that contribute to the modulation of the drain current of MOS transistors in saturation.

The main short-channel phenomena responsible for the non-zero output conductance in saturation are the channel-length modulation (CLM) and the drain-induced barrier lowering (DIBL) [14], [29]. Weak avalanche and the substrate-current-induced body effect, which also play a role in the Early voltage, are important only for relatively high drain voltages and high current levels, respectively. In the following, we will summarize the results of reference [29], which describe the dependence of the Early voltage on both CLM and DIBL.

As in Chapter 2, we assume that the dependence of the threshold voltage on the source and drain voltages is

$$V_T = V_{T0} - \sigma(V_{SB} + V_{DB}), \tag{11.5.1}$$

where σ is the DIBL factor.

Assuming that the Early voltage is dominated by DIBL and CLM, we can write that

$$\frac{1}{V_A} = \frac{1}{I_D}\frac{dI_D}{dV_D} = \frac{1}{I_D}\frac{\partial I_D}{\partial V_T}\frac{\partial V_T}{\partial V_D} + \frac{1}{I_D}\frac{\partial I_D}{\partial L}\frac{\partial L}{\partial V_D} = \frac{1}{V_{ADIBL}} + \frac{1}{V_{ACLM}}, \tag{11.5.2}$$

where [29]

$$V_{ADIBL} \approx \frac{n\phi_t}{2\sigma}\left(\sqrt{1 + \frac{I_D}{I_S}} + 1\right). \tag{11.5.3}$$

The derivation of the Early voltage associated with CLM in [29] assumes that the channel of an MOS transistor in saturation is divided into two sections, one closer to the source, where the gradual-channel approximation is valid, and another closer to the drain, in which the two-dimensional nature of the space-charge region must be accounted for [14]. As V_D increases, the depletion region of the drain–channel junction widens, thus shortening the effective channel length. Using this formulation, the current can be calculated using the expression derived under the gradual-channel approximation but considering the effective channel length of the device to be reduced by the length Y_D of the drain section. The derivation of the CLM in weak and moderate inversion can be found in [29]. Here, we will assume that the hypotheses employed therein for the derivation of the model are valid, and we reproduce only the main results of [29] concerning the CLM phenomenon, which are given by the formula

$$Y_D = \sqrt{\left(\frac{F_L}{2a}\right)^2 + \frac{\phi_{bi} + V_{DB} - \phi_{SL}}{a}} - \frac{|F_L|}{2a} \tag{11.5.4}$$

with

$$a = \frac{qN_A}{2\varepsilon_s}, \tag{11.5.5}$$

$$F_L = F_y(y = L - Y_D) \cong -\frac{I_D}{I_S}\frac{\phi_t}{2L} = \frac{-I_D/W}{\mu C'_{ox}n\phi_t}, \tag{11.5.6}$$

$$\phi_{sL} = \phi_s(y = L - Y_D) \cong \phi_{sa} - \frac{I_D/W}{nC'_{ox}v_{sat}} = V_P + 2\phi_F - \frac{I_D/W}{nC'_{ox}v_{sat}}, \qquad (11.5.7)$$

where ϕ_{bi} is the junction built-in potential, while F_L and ϕ_{sL} are the longitudinal electric field and surface potential, respectively, at the interface between the channel and the depletion region at the drain end of the channel. The value of Y_D in (11.5.4) was obtained by application of the Poisson equation to a one-sided n$^+$–p junction with boundary conditions given by (11.5.6) and (11.5.7) at a distance $L - Y_D$ from the source. Recall that ϕ_{sa} is the surface potential when the inversion charge density is equal to zero. N_A is the substrate doping and Y_D is the depletion width of the drain–substrate junction at the semiconductor surface. Note that, when the current level is low, the electric field F_L tends to zero, whereas the surface potential ϕ_{sL} tends to ϕ_{sa}. Thus, for low currents, the value of Y_D in (11.5.4) reduces to that of the conventional n$^+$–p junction. Now, using

$$\frac{1}{V_{ACLM}} = \frac{1}{I_D}\frac{\partial I_D}{\partial L}\frac{\partial L}{\partial V_D} = \frac{1}{I_D}\left(\frac{-I_D}{L}\right)\frac{-\partial Y_D}{\partial V_D} \qquad (11.5.8)$$

and (11.5.4), we find that

$$V_{ACLM} \cong 2aL\left(\sqrt{\left(\frac{F_L}{2a}\right)^2 + \frac{\phi_{bi} + V_{DB} - \phi_{SL}}{a}}\right). \qquad (11.5.9)$$

In this simplified model of the Early voltage, the DIBL component is dependent on the inversion level according to (11.5.3), but is nearly independent of V_{DS} while the CLM component is proportional to the square root of V_D.

In order to extract the Early voltage in terms of current, drain voltage, and channel length for different technologies and both n- and p-channel transistors, a set of test devices was fabricated, as described in [30]. We will show here some of the results obtained for n-channel transistors in a 0.35-μm technology, biased with inversion level $i_f \leq 100$.

Experimental results for the drain and source currents of an n-MOS transistor with $L = 0.4\,\mu$m are shown in Figure 11.17. The reason for measuring both the drain and source currents is to determine the weak avalanche current as the difference between the drain and source currents. Since the avalanche current is caused by high electric fields, the drain current is that affected by weak avalanche due to the high electric field close to the drain region. To reduce the effect of the weak avalanche current on the determination of the DIBL and CLM parameters, the source current, rather than the drain current, was measured.

Figure 11.17 shows that the shape of the curves is the same in weak inversion ($i_f \leq 1$), which means that the relative variation of the current is independent of the value of the inversion level. In other words, the Early voltage is not sensitive to the inversion level in weak inversion. On the other hand, for higher values of i_f, the relative variation of the current becomes progressively smaller, that is, the Early voltage becomes increasingly higher for increasing inversion levels. Also, note that the curve knee (saturation voltage) becomes gradually higher for inversion levels above unity.

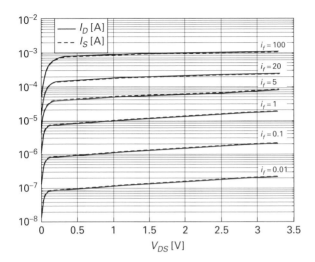

Fig. 11.17 Experimental drain and source currents versus drain-to-source voltage for a minimum-channel-length n-MOS transistor ($L = 0.4\,\mu m$) in a 0.35-μm CMOS technology.

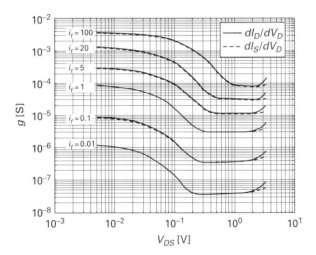

Fig. 11.18 Derivatives of the experimental drain and source currents with respect to the drain voltage versus drain-to-source voltage for a minimum-channel-length n-MOS transistor ($L = 0.4\,\mu m$) in a 0.35-μm CMOS technology.

From the graphs in Figure 11.17, the plots of the transistor output conductance shown in Figure 11.18 were obtained. In weak inversion, the output conductance is proportional to the drain current (or to the inversion level). In moderate inversion, the values of the output conductance increase less than linearly with the current level. Note that, for drain-to-source voltages around 2 V or more, the difference between the derivatives of the source and drain currents increases very rapidly.

Plots of experimental and fitted Early voltages for the minimum-length n-channel device for several inversion levels are shown in Figure 11.19 [29], [30]. As noted

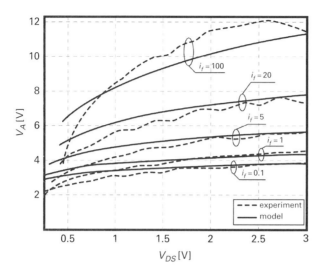

Fig. 11.19   Experimental and modeled Early voltages versus drain-to-source voltage for a minimum-length n-MOS transistor ($L = 0.4\,\mu$m) in a 0.35-μm CMOS technology.

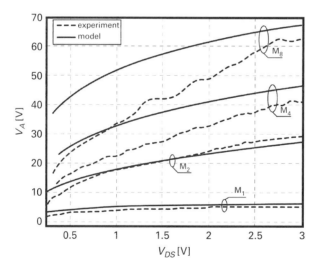

Fig. 11.20   Experimental and modeled Early voltages versus drain-to-source voltage for transistors M_1, M_2, M_4, and M_8, for which the nominal lengths are L_{min}, $2L_{min}$, $4L_{min}$, and $8L_{min}$, respectively, where $L_{min} = 0.4\,\mu$m. The parameters for the model are given in Table 11.3. The inversion level at the source is $i_f = 5$.

previously, just one curve (in this case the curve labeled $i_f = 0.1$) is sufficient to represent the behavior of the device biased in weak inversion. For inversion levels greater than unity, the Early voltage increases slowly with increasing inversion levels and drain-to-source voltages, as predicted in the theoretical model.

Figure 11.20 presents the experimental and fitted Early voltages for transistors with various channel lengths. A first-order approximation of the Early voltage as linearly

Table 11.3 Fitting parameters extracted for the Early voltage of n-MOS transistors in a 0.35-μm CMOS technology

Transistor	σ (mV/V)	a (V/m^2)	$\phi_{bi} - 2\phi_F$ (V)
M$_1$	7	2.5×10^{14}	0.1
M$_2$	0.8	2.5×10^{14}	0.1
M$_4$	0.6	2.5×10^{14}	0.1
M$_8$	0.45	2.5×10^{14}	0.1

related to the channel length is not such a poor approximation, but the figures in this section clearly show that the Early voltage is also significantly affected by the inversion level and the drain-to-source voltage.

Table 11.3 presents the fitting parameters for transistors M$_1$, M$_2$, M$_4$, and M$_8$. The DIBL factor was extracted for each channel length. For the 0.35-μm technology of the fabricated devices, the nominal substrate doping is 2.2×10^{17} cm^{-3}, which gives $a = 1.67 \times 10^{14}$ V/m^2, a value relatively close to that extracted from the Early voltage.

Example 11.4

(a) Determine the intrinsic gain of a transistor in the common-gate configuration in terms of V_A and i_f. (b) Use the theoretical curves in Figure 11.20 to determine the maximum and minimum gains of transistors M$_1$ and M$_8$ for a drain-to-source voltage between 1 V and 2.5 V.

Answer

(a) The intrinsic gain of a transistor in the common-gate configuration is

$$A_{V0} = \frac{g_{ms}}{g_{ds}} = \frac{2I_F / \left[\phi_t\left(\sqrt{1 + i_f} + 1\right)\right]}{I_F / V_A} = \frac{2V_A}{\phi_t\left(\sqrt{1 + i_f} + 1\right)}.$$

(b) For M$_1$, $5\,\text{V} \leq V_A \leq 6.5\,\text{V}$, while, for M$_8$, $53\,\text{V} \leq V_A \leq 65\,\text{V}$. Using $\phi_t = 26\,\text{mV}$ and $i_f = 5$, we find that

$$110 \leq A_{V0}(\text{M}_1) \leq 145,$$

$$1180 \leq A_{V0}(\text{M}_8) \leq 1445.$$

Problems

11.1 Demonstrate that the application of the smoothing function (11.1.4) to (11.1.3) reduces (11.1.3) to either (11.1.1) for weak inversion or (11.1.2) for strong inversion.

11.2 Using the expression for the bulk charge given in (11.1.7) and the potential-balance equation show that, in accumulation, the value of the substrate capacitance is given by

$$C_b' = -\frac{dQ_B'}{d\phi_s} \cong -C_{ox}' \frac{V_G - V_{FB} - \phi_s}{2\phi_t} \cong -C_{ox}' \frac{V_G - V_{FB}}{2\phi_t}.$$

Calculate the resulting gate capacitance in accumulation. Sketch the plot of the gate capacitance versus $(V_G - V_{FB})/\phi_t$ in the accumulation region.

11.3 Verify that expressions (11.1.13) and (11.1.16) lead to the same result for the drain current when the following substitutions are made:

$$n = n_e; \qquad \overline{Q_I'} = (Q_{IS}' + Q_{ID}')/2; \qquad dQ_I' = nC_{ox}' d\phi_s.$$

11.4 Derive the expression below for the slope factor:

$$n \cong 1 + \frac{\gamma}{2\sqrt{V_P + 2\phi_F}}.$$

Also, verify Equation (11.2.11). (Hint: use the results of Section 2.1.5 in Chapter 2.)

11.5 The expression for the inversion capacitance obtained without using the charge-sheet approximation [14] is

$$C_i' = -\frac{Q_I'}{2\phi_t}\left(1 + \frac{Q_B'}{Q_B' + Q_I'}\right).$$

Show that at threshold (pinch-off)

$$\frac{Q_B'}{Q_B' + Q_I'} = \frac{1}{1 + \frac{1}{2}\left(1 + \frac{C_{ox}'}{C_b'}\right)\frac{\phi_t}{\phi_{sa}}}$$

and that classical effects cannot explain the fact that $n_{ch} > 1$ in weak inversion.

11.6 Using the capacitive model in Figure 11.13, derive the differential UCCM given by (11.4.3).

11.7 Determine the effect of the drain and source series resistances on the drain current for low values of the drain-to-source voltage. To simplify the problem, use the MOSFET model in strong inversion (this simplifying assumption is acceptable since the series resistances of drain and source have a negligible effect in weak and moderate inversion). Compare this effect with that of the dependence of the mobility on the transverse field. Verify that, when the inversion level i_f is such that $\sqrt{i_f} = \phi_t/[2I_S(R_S + R_D)]$, the effective transconductance drops to half the value for series resistances equal to zero.

References

[1] Y. Tsividis, "Moderate inversion in MOS devices," *Solid-State Electronics*, vol. **25**, no. 11, pp. 1099–1104, 1982.

[2] J. Watts, C. McAndrew, C. Enz *et al.*, "Advanced compact models for MOSFETs," *Technical Proceedings 2005 Workshop on Compact Modeling*, pp. 3–12.

[3] M. Miura-Mattausch, U. Feldmann, A. Rahm, M. Bollu, and D. Savignac, "Unified complete MOSFET model for analysis of digital and analog circuits," *IEEE Transactions on Computer-Aided Design*, vol. **15**, no. 1, pp. 1–7, Jan. 1996.

[4] R. van Langevelde, A. J. Scholten, and D. B. M. Klaassen, "Physical background of MOS Model 11," Unclassified Report 2003/00239, http://www.semiconductors.philips.com/ Philips_Models/.

[5] G. Gildenblat, H. Wang, T.-L. Chen, X. Gu, and X. Cai, "SP: an advanced surface-potential-based compact MOSFET model," *IEEE Journal of Solid-State Circuits*, vol. **39**, no. 9, pp. 1394–1406, Sep. 2004.

[6] M. A. Maher and C. A. Mead, "A physical charge-controlled model for MOS transistors," in *Advanced Research in VLSI*, ed. P. Losleben, Cambridge, MA: MIT Press, 1987, pp. 211–229.

[7] Y. Byun, K. Lee, and M. Shur, "Unified charge control model and subthreshold current in heterostructure field effect transistors," *IEEE Electron Device Letters*, vol. **11**, no. 1, pp. 50–53, Jan. 1990.

[8] A. I. A. Cunha, M. C. Schneider, and C. -G. Montoro, "An explicit physical model for the long-channel MOS transistor including small-signal parameters," *Solid-State Electronics*, vol. **38**, no. 11, pp. 1945–1952, Nov. 1995.

[9] W. Liu, X. Jin, J. Chen *et al.*, "BSIM3v3.2.2 MOSFET Model User's Manual," http://www-device.eecs.berkeley.edu/~bsim3/arch_ftp.html.

[10] J. R Brews, "A charge sheet model for the MOSFET," *Solid-State Electronics*, vol. **21**, no. 2, pp. 345–355, Feb. 1978.

[11] C. C. McAndrew and J. J. Victory, "Accuracy of approximations in MOSFET charge models," *IEEE Transactions on Electron Devices*, vol. **49**, no. 1, pp. 72–81, Jan. 2002.

[12] M. Miura-Mattausch and H. Jacobs, "Analytical model for circuit simulation with quarter micron metal oxide semiconductor field effect transistors," *Japanese Journal of Applied Physics*, vol. **29**, no. 12, pp. 2279–2282, Dec. 1990.

[13] R. M. D. A. Velghe, D. B. M. Klaassen, and F. M. Klaassen, "MOS Model 9," Unclassified Report 003/94, Philips Electronics N.V. (1994), http://www.semiconductors.philips.com/ Philips_Models/.

[14] C. Galup-Montoro and M. C. Schneider, *MOSFET Modeling For Circuit Analysis and Design*, Singapore: World Scientific, 2007.

[15] G. Gildenblat, X. Li, W. Wu *et al.*, "An advanced surface-potential-based MOSFET model for circuit simulation," *IEEE Transactions on Electron Devices*, vol. **53**, no. 9, pp. 1979–1993, Sep. 2006.

[16] X. Li, G. Gildenblat, G. D. J. Smit *et al.*, PSP 102.3, http://pspmodel.asu.edu/downloads/ psp102p3_summary.pdf.

[17] C. Enz, F. Krummenacher, and E. A. Vittoz, "An analytical MOS transistor model valid in all regions of operation and dedicated to low-voltage and low-current applications," *Journal of Analog Integrated Circuits and Signal Processors*, vol. **8**, pp. 83–114, July 1995.

[18] C. Enz, "A short story of the EKV MOS transistor model," *IEEE Solid-State Circuits Society News*, vol. **13**, no. 3, pp. 24–30, Mar. 2008.

[19] A. I. A. Cunha, M. C. Schneider, and C. Galup-Montoro, "An MOS transistor model for analog circuit design," *IEEE Journal of Solid-State Circuits*, vol. **33**, no. 10, pp. 1510–1519, Oct. 1998.

[20] A. T. Behr, M. C. Schneider, S. Noceti Filho, and C. Galup-Montoro, "Harmonic distortion caused by capacitors implemented with MOSFET gates," *IEEE Journal of Solid-State Circuits*, vol. **27**, no. 10, pp. 1470–1475, Oct. 1992.

[21] C. Galup-Montoro, M. C. Schneider, and I. J. B. Loss, "Series-parallel association of FET's for high gain and high frequency applications," *IEEE Journal of Solid-State Circuits*, vol. **29**, no. 9, pp. 1094–1101, Sep. 1994.

[22] Application notes, in Home-page Dolphin, http://www.dolphin.fr/medal/smash/notes/acm_report.pdf.

[23] A. Arnaud and C. Galup-Montoro, "A compact model for flicker noise in MOS transistors for analog circuit design," *IEEE Transactions on Electron Devices*, vol. **50**, no. 8, pp. 1815–1818, Aug. 2003.

[24] C. Galup-Montoro, M. C. Schneider, H. Klimach, and A. Arnaud, "A compact model of MOSFET mismatch for circuit design," *IEEE Journal of Solid-State Circuits*, vol. **40**, no. 8, pp. 1649–1657, Aug. 2005.

[25] J. He, X. Xi, M. Chan, A. Nikenejad, and C. Hu, "An advanced surface potential-plus MOSFET model," *Technical Proceedings 2003 Nanotechnology Conference*, pp. 262–265.

[26] X. Xi, J. He, M. Dunga *et al.*, "The development of the next generation BSIM for sub-100 nm mixed-signal circuit simulation," *Technical Proceedings 2004 Nanotechnology Conference*, pp. 70–73.

[27] A. I. A. Cunha, M. C. Schneider, C. Galup-Montoro, C. D. C. Caetano, and M. B. Machado, "Unambiguous extraction of threshold voltage based on the transconductance-to-current ratio," *Technical Proceedings 2005 Nanotechnology Conference*, pp. 139–142.

[28] M. Bucher, C. Lallement, and C. C. Enz, "An efficient parameter extraction methodology for the EKV MOST model," *Proceedings of IEEE ICMTS*, 1996, pp. 145–150.

[29] R. L. Radin, G. L. Moreira, C. Galup-Montoro, and M. C. Schneider, "A simple modeling of the Early voltage of MOSFETs in weak and moderate inversion," *Proceedings of IEEE ISCAS* 2008, pp. 1720–1723.

[30] R. L. Radin, "*Modelagem da tensão de Early em transistores MOS nos regimes de inversão fraca e moderada*," M.Sc. Dissertation, Universidade Federal de Santa Catarina, Nov. 2007, http://eel.ufsc.br/~lci/pdf/Dissertacao-Rafael.pdf.

Index

JUL - 7 2010